Agricultural Biodiversity

Agricultural Biodiversity

Edited by **Lester Bane**

SYRAWOOD
PUBLISHING HOUSE
New York

Published by Syrawood Publishing House,
750 Third Avenue, 9th Floor,
New York, NY 10017, USA
www.syrawoodpublishinghouse.com

Agricultural Biodiversity
Edited by Lester Bane

International Standard Book Number: 978-1-68286-225-4 (Hardback)

Printed in China.

Contents

Preface

Biodiversity is an important factor influencing agriculture. To attain sustainable growth and to maintain ecosystem services, the management of agricultural biodiversity is crucial. This book traces the progress of this field and highlights some of the key concepts and applications aimed at enhancing biodiversity. The book provides the reader a detailed insight into the diverse aspects of agricutural biodiversity and farm biodiversity, techniques for influencing the growth of crops, environment, ecosystems, etc. Some sustainable horticultural practices have also been discussed. This book will prove to be a useful tool for the students of agricultural sciences and associated disciplines of study.

Various studies have approached the subject by analyzing it with a single perspective, but the present book provides diverse methodologies and techniques to address this field. This book contains theories and applications needed for understanding the subject from different perspectives. The aim is to keep the readers informed about the progress in the field; therefore, the contributions were carefully examined to compile novel researches by specialists from across the globe.

Indeed, the job of the editor is the most crucial and challenging in compiling all chapters into a single book. In the end, I would extend my sincere thanks to the chapter authors for their profound work. I am also thankful for the support provided by my family and colleagues during the compilation of this book.

Editor

1

The Effects of *Tobacco mosaic virus* Infection on Growth and Physiological Parameters in Some Pepper Varieties (*Capsicum annuum* L.)

Sercan PAZARLAR[1*], Mustafa GÜMÜŞ[1], Gölgen Bahar ÖZTEKİN[2]

[1]*University of Ege, Faculty of Agriculture, Department of Plant Protection, 35100 Bornova, İzmir, Turkey; sercanpazarlar@gmail.com (*corresponding author)*

[2]*University of Ege, Faculty of Agriculture, Department of Horticulture, 35100 Bornova İzmir, Turkey*

Abstract

Changes in some growth (plant leaf number and area, plant biomass, plant height, root length, and plant stem diameter) and physiological (photosynthetic pigments, relative water content (RWC) and proline content) parameters of pepper (*Capsicum annuum* L.) varieties were studied as they were affected by *Tobacco mosaic virus* (TMV) infection. The greenhouse pepper cvs 'Ergenekon F_1', 'Kumsal F_1', and one candidate variety ('497 F_1') were used as material to compare infected versus non-infected control plants. Infected plants showed various degrees of stunting; necrosis on stems, leaves and fruits; mosaic symptoms on leaves; deformations, defoliation of leaves, and reduction in fruit size. Besides these, TMV infection resulted in reduction of vegetative growth parameters, RWC, chlorophyll a and b, plant fresh and dry production in different parts of the plants, but increase of proline content in leaves. The level of response differed depending on varieties.

Keywords: growth, pepper genotypes, physiology, TMV (*Tobacco mosaic virus*), virus

Introduction

Tobacco mosaic disease is one of the major limitations in production of pepper (*C. annuum*), an important vegetable crop in Turkey and all over the world, in both greenhouse and open field. This disease, caused by *Tobacco mosaic virus* (TMV, genus: Tobamovirus), a positive-sense single stranded RNA virus, transmits easily through seed, soil and cultural processes, and can cause important losses of yield and quality properties (Scholthof *et al.,* 1999). Virus-infected pepper plants typically develop symptoms including various degrees of mosaic symptoms, chlorosis on leaves, stems, and fruits, necrosis on several plant parts, and stunting (Baker and Adkins, 2000). All the various macroscopic and microscopic symptoms of diseases must originate in biochemical aberrations induced directly or indirectly by the virus (Hull, 2009).

To date, several studies have focused on interactions between different plants and virus combinations. The level of total chlorophyll (Chl) and carotenoids (Car) in control and *Yellow vein mosaic virus* (YVMV) infected leaves was studied by Palanisamy *et al.* (2009), who reported that total Chl and Car concentrations were significantly reduced in infected leaves by 64% and 62%, respectively. It was recorded that *Zucchini yellow mosaic virus* (ZYMV) infected leaves of pumpkin showed severe symptoms as mosaic, green blisters, size reduction and deformation and the virus infection diminished the Chl a (48%), Chl b (53%) and carotenoid contents (52%). Nevertheless proline contents of ZYMV infected plants were found higher than controls (Radwan *et al.,* 2007). *Tobacco ringspot virus* (TRSV) infection did not significantly affect the number of *Lotus corniculatus* roots produced on the stem cuttings in 14 days; however virus infection significantly reduced the length of the longest root. Also significant yield reduction was observed (Ostazeski, 1970).

As mentioned above there are number of papers that studied the results of different virus-plant combinations, however there is little known about TMV effects on growth and physiological parameters of pepper. The objectives of this study were to clarify the effects of virus infection on host growth and physiological parameters, and to observe the reactions of pepper varieties under TMV disease conditions.

Material and methods

Growing conditions and plant materials

This experiment was carried out in greenhouse conditions at Faculty of Agriculture Ege University (27,09° E, 38,25° N), Turkey in the autumn season of 2011. The greenhouse pepper cvs 'Ergenekon F_1' (bell pepper type, Bircan Tarım, Antalya-Turkey), 'Kumsal F_1' (Banana type, Bircan Tarım, Antalya-Turkey), and one candidate variety ('497 F_1') (Banana type, Bircan Tarım, Antalya Turkey) were used as plant material. Seedlings of peppers (*Capsicum annuum* L.) were transferred in a mixture of soil, farm manure, silt and peat (1:1:1:1) in sterilized plastic pots (200 × 180 millimeter) on October 17, 2011. All cultivars and candidate variety were grown in greenhouse until February 26, 2012 under maintained 20±5 °C, 60-80% relative humidity and standard day light conditions. All pots were irrigated with tap water every three days to maintain soil at the field capacity and all other cultivations were discharged uniformly throughout the whole experimental period.

Pathogen inoculation

The TMV used in this study is a common strain which induces mosaic containing light and dark-green areas in tobacco. The pepper leaves infected by TMV (provided by Plant Protection Department, Ege University, Turkey) were homogenized in 0.05 M phosphate buffer (pH 7.0) and used as inoculum. Plants at 4-6 fully expanded leaf stage were mechanically inoculated on the adaxial surface by using celite powder. The control group was also dusted with celite powder and washed with free inoculum water in order to provide the independence of error.

Measured parameters

TMV damage scale class - Plants were scored on the 35th and 70th days after inoculation (November 30, 2011 and January 03, 2012, respectively) for severity of TMV susceptibility by 1 to 5 scale modified by Leelerg (1967). Accordingly, (1) is scored as undamaged, healthy leaves or slightly involuted normal green leaves, (2) completely involuted green leaves, (3) moderately-excessively damaged dry leaves in addition to involuted green leaves, (4) (50 to 80%) drying damage on most leaves (50 to 80%), (5) drying damages on all leaves.

Growth - Four plants from each replicate were removed 133 days after transplanting in order to determine plant growth parameters. All leaves on the plants were counted and the mean values were given as pieces per plant^{-1}. Plant height was determined by measuring stem length from soil surface to the tip of the stem for all the plants and results were given cm/ plant^{-1}. Stem diameter (mm plant^{-1}) was measured on three parts of stem by digital caliper (Canon, 428830, China) and calculated with arithmetic mean. Following these measurements, harvested plants were separated into plant parts (root, leaves, stems and reproductive parts). The root systems were carefully washed with tap water to separate substrates. The longest root length (cm plant^{-1}) was measured as the distance from soil surface to the end of the longest root. All plant parts were weighted for fresh weight. Then, samples were dried at 65 °C in thermo ventilated oven for 72 hours and their dry weights were recorded. Results were presented as the dry and fresh weight (g plant^{-1}) of root and shoot. Tree plants of each treatment were used for total leaf area according to Oztekin and Tuzel (2011). The area of the mentioned leaves were measured by Sigma Scan Pro 5.0 (Automated Image Analysis, Systat Software Inc., USA) software and detected as pix^2 per plant. These data were converted into cm^2 per plant.

Physiological parameters - The relative water contents (RWC) of leaves were measured according to Smart and Bingham (1974). After sampling, leaf discs were taken and their fresh weight were immediately weighed and then immersed in distilled water for 7 h at room temperature. The leaves were then blotted dry and weighed prior to oven drying at 70°C for 48 h. The leaf relative content was calculated using the following formula;

$$RWC(\%)=[(FW\text{-}DW)/(TW\text{-}DW)]\times 100$$

FW, DW and TW are the fresh, dry and turgid weights of the tissue, respectively.

The procedure for chlorophyll determination was based on the work of Arnon (1949) on the absorption of light by aqueous acetone (80%) extracts of chlorophyll. The determination of chlorophyll a and b were made by measuring a 25 mg leaf, in density of 80% acetone (v/v) chlorophyll extracts with a Varian Carry 100UV-Visible spectrophotometer (Varian Inc, Palo Alto, CA, USA) at 450, 465 and 663 nm and setting up simultaneous equations using the specific absorption coefficients for chlorophyll a and b and carotenoid which were calculated with the following formula and were expressed in terms of mg per L fresh mass:

Chlorophyll a = ((0.0127 × 663nm) - (0.00269 × 645m)) × 1000

Chlorophyll b = ((0.0229 × 645nm) - (0.00468 × 663nm)) × 1000

Cartotenoid = (4.07 × 450nm) × ((0.0435 × *Cl a*) + (0.367 × *Cl b*))

Free proline content was determined according to Bates *et al.* (1973). 0.5 g of leaf samples from each group were homogenized in 3% (w/v) sulphosalycylic acid and the homogenate was filtered through filter paper. After addition of acid ninhydrin and glacial acetic acid, resulting mixture was heated at 100 °C for 1 h in water bath. Reaction was then stopped by using an ice bath. The mixture was extracted with toluene and the absorbance of fraction with toluene aspired from liquid phase was read at 520 nm. Proline concentration was determined using calibration curve and expressed as micromole proline per gram FW.

Experimental design and statistics

Analysis of variance (ANOVA) was carried out to determine statistically significant differences caused by infections in studied parameters. The experimental design was a completely randomized plots design with 4 replicates. Duncan's Multiple Range Test was used to determine significant differences of means.

Results and discussion

While pathogens infect plant's mineral uptake, depending on pathogen and on infected plant tissue, pathogens interfere with different physiological function(s) of the plant which leads to the development of different symptoms (Agrios, 2005). These alterations cause reduction in yield and quality of products. In this study, number of changes was observed in the physiology and morphology of the pepper plants associated with TMV infection. Our results have shown that the growth parameters such as plant height, root length, leaf area, biomass and physiological parameters such as, total photosynthetic pigments, relative water content, were significantly decreased in virus-affected plants. These results were supported by other authors (Ostazeski *et al.*, 1970; Kuhn and Dawson, 1973; Johnson and Main, 1983; Agrios *et al.,* 1985; Funayama *et al.*, 1997; Wintermantel, 2005; Taiwo and Akinjogunla, 2006). In addition, there were differences in the progress of TMV through different pepper varieties.

TMV damages scoring

The first symptoms of infection appeared on the uppermost younger leafs as necrosis and chlorosis along the main veins accompanied by wilting and leaf spots. Generally, infected pepper varieties were stunted, deformed, and they showed chlorotic areas among the leaves. These symptoms are in conformity with the symptoms observed by Rodriguez-Alvarado *et al.* (1994). The varieties responded differently to TMV as judged from their visual appearance. However, they showed similarities in the scale classes 0.2 to 4.0 on 35 days after inoculation and 0.4 to 4.8 on 70 days after inoculation. On November 30, 2011 the highest damage was observed with '497 F_1' plants. A similar situation was also observed on January 03, 2012. Under short and long time stress conditions; ' Kumsal F_1' in scale class-1 were least affected by TMV treatment. Cv. 'Ergenekon' showed moderate tolerance to TMV according to their leaf damages (Tab. 1).

Tab. 1. Leaf damages according to TMV scale

Varieties	30.11.2011**	03.01.2012**
'Kumsal'	0.2 ^{c}	0.4 ^{c}
'Ergenekon'	2.6 ^{b}	3.0 ^{b}
'497'	4.0 ^{a}	4.8 ^{a}

**(p<0.01)

Plant growth

TMV infection decreased leaf number in all varieties (Tab. 2). While the leaf numbers of all varieties were nearly equal before inoculation, leaf number per plant was reduced significantly (p<0.01) by 42.05% for 'Ergenekon', 5.85% for 'Kumsal' and 59.87% for '497' compared to the control. In virus-stressed plants, there were 57.51%, 3.88%, 83.56% decreases ($p \leq 0.01$) in leaf area in 'Ergenekon', 'Kumsal' and '497', respectively, compared to the controls. Under our experimental conditions 'Ergenekon' and '497' varieties were found to be affected more adversely than 'Kumsal' with regards to leaf number and leaf area.

TMV induced reduction in amount of leaf number and leaf area in all tested genotypes contributes to the diminished photosynthesis activity that is closely related with yield and quality of products. Similar results were obtained by Ehinmore *et al.* (2010) and Lehrer *et al.* (2007).

The virus infection caused a decrease (p<0.01) in plant height of 'Ergenekon', 'Kumsal', '497', respectively (Tab. 3). 'Kumsal' gave the highest plant height compared the other varieties. The longest root length was decreased (p<0.01) by TMV infection (Tab. 3). Average decreases in root length of 'Ergenekon' (26.14%) caused by virus were less than that of '497' (35.66%). The average longest root length for KUMSAL had an unexpected increase by 6.2% increase in infected plants compared to control plants. Among the tested plants, '497' gave the thickest stem diameter (reduced 13.55%) followed by 'Ergenekon' (reduced 5.94%) in virus infection (p<0.05). Contrary to

Tab. 2. Effect of TMV infection on leaf number and leaf area of peppers

Varieties	Leaf number (pieces plant^{-1})**			Leaf area (cm^2 plant^{-1})**		
	Control	Infected	Change (%)	Control	Infected	Change (%)
'Ergenekon'	25.17	14.58	-42.05 ^{b}	921.19	391.41	-57.51 ^{b}
'Kumsal'	28.50	26.83	-5.85 ^{a}	829.13	796.93	-3.88 ^{a}
'497'	25.33	10.17	-59.87 ^{b}	1203.30	197.86	-83.56 ^{c}

**(p<0,01), (-) shows decrease

Tab. 3. Effect of TMV infection on plant height, the longest root length and stem diameter of peppers

Varieties	Plant height (cm plant^{-1})**			The longest root length (cm plant^{-1})**			Stem diameter (mm plant^{-1})*		
	Control	Infected	Change (%)	Control	Infected	Change (%)	Control	Infected	Change (%)
'Ergenekon'	11.21	9.13	-18.58 ^{b}	25.50	18.83	-26.14 ^{b}	5.38	5.06	-5.94 ^{a}
'Kumsal'	24.88	22.00	-11.55 ^{a}	17.46	18.54	+6.20 ^{a}	5.14	5.25	+2.12 ^{a}
'497'	19.83	15.30	-23.10 ^{b}	23.83	15.33	-35.66 ^{b}	5.29	4.57	-13.55 ^{b}

*(p<0,05), **(p<0,01, (-) shows decrease, (+) shows increase

Tab. 4. Effect of TMV infection on root and shoot dry and fresh weight of peppers

Varieties	Root						Shoot					
	Dry weight (g plant^{-1})**			Fresh weight (g plant^{-1})**			Fresh weight (g plant^{-1})*			Dry weight (g plant^{-1})*		
	Control	TMV	Change (%)	Control	TMV	Change (%)	Control	TMV	Change (%)	Control	TMV	Change (%)
'Ergenekon'	0.87	0.47	-45.57 ^{b}	7.50	4.25	-43.40 ^{b}	24.35	9.98	-58.99 ^{b}	2.43	1.15	-52.64 ^{b}
'Kumsal'	0.71	0.68	-4.49 ^{a}	5.61	5.56	-10.69 ^{a}	25.31	23.17	-8.42 ^{a}	2.85	2.42	-15.15 ^{a}
'497'	0.65	0.33	-49.50 ^{b}	4.69	2.24	-52.13 ^{b}	31.07	9.66 ^{b}	-68.91 ^{b}	3.39	1.22	-64.10 ^{b}

*(p<0,05), **(p<0,01), (-) shows decrease

expectations stem diameter increased 2.12% compared to control plants in 'Kumsal'.

In virus disease conditions reduction of shoot and root dry weight were statistically significant (p<0.01) (Tab. 4). However, the highest reduction in root dry weight (49.50%) was observed in '497' followed by 'Ergenekon' (45.57%). Root dry weight of infected 'Kumsal' plants was reduced only by 4.49%, which was significantly lower than those of other varieties. There were 43.40%, 52.13%, 10.69 decreases (p<0.01) in root fresh weight in 'Ergenekon', '497', 'Kumsal' respectively in virus-stressed plants as compared to the controls. The results show that TMV caused reduction in shoot fresh and dry weight (p<0.05). Both parameters were affected similarly and '497' showed the most, whereas 'Kumsal' showed the less reduction in shoot fresh and dry weight. The findings of Wani *et al.* (1991), Friess and Maillet (1997), Funayama *et al.* (1997), Pacheco *et al.* (2003), Wintermantel (2005), Kazinczi *et al.* (2006), Kollmann *et al.* (2007) on lsupport these results.

Physiological Parameters

Relative water content in the leaves of plants grown under TMV infection stress decreased significantly in all cultivars and candidate varieties compared to uninfected plants (p<0.05) (Tab. 5). '497' showed greater reduction in the RWC than 'Ergenekon' (3.09%) and 'Kumsal' (2.12%). Similar results found by other authors (Synková *et al.*, 2006; Al-Saleh *et al.*, 2007).

Total proline content results are given in Tab. 6. These results show that leaves from all varieties subjected to TMV infection had increased proline content relative to leaves from control plants. Leaves of 'Kumsal' plants showed proline concentration increases of 0.88% compared to controls, whereas 2.97% and 4.89% increases in proline content were observed in leaves from 'Ergenekon' and '497' respectively.

Tab. 5. Effect of TMV infection on relative water content of peppers

Varieties	Relative Water Content (%)*		
	Control	Infected	Change (%)
'Ergenekon'	90.25	87.46	-3.09 ^{ab}
'Kumsal'	89.53	87.61	-2.15 ^{a}
'497'	89.20	83.65	-6.22 ^{b}

*(p<0,05), NS: Not significant, (-) shows decrease

Proline accumulation is a common metabolic response to both abiotic and biotic stress and when higher plants are exposed to stress; many plants accumulate high amounts of proline in tissues (Mansour, 2000; Mazid *et al.*, 2011). In addition some other amino acids can also be found at higher concentrations (Drossopoulos *et al.*, 1985). For instance, water deprivation (Ain-Lhout *et al.,* 2001), salinity (Jouve *et al.*, 2004), high and low temperatures (Lv *et al.*, 2011), pathogen infection (Fabro *et al.*, 2004) may cause activation in numerous compounds in the cell or the proline production (Verbruggen *et al.*, 1993; Yoshiba *et al.*, 1995). In our experiment, a higher amount of proline in diseased material was observed compared to the respective controls in all genotypes (Tab. 6). The results are supported by Kakani *et al.* (2001), Liu *et al.* (2001), Radvan *et al.* (2007), Chatterjee and Ghosh (2008), Mohamed (2011). When plants are exposed to microbial pathogens, they produce reactive oxygen species (ROS) that induce programmed cell death in the plant cells surrounding the infection site to effectively wall off the pathogen and terminate the disease process (Apel and Hirt, 2004; Adi *et al.*, 2012). The amino acid proline may act as a potent scavenger of ROS and this property of proline might prevent the induction of programmed cell death by ROS (Chen and Dickman, 2005).

The changes in the level of total chlorophyll (*Chl*) and carotenoids (*Car*) in control and TMV infected leaves are shown in Tab. 7. When determined on the basis of unit fresh weight, the *Chl a* and *Chl b* concentrations in infected leaves were significantly reduced in all genotypes (p<0.01).

Chlorophyll a and b contents of infected leaves were more than half of those of uninfected leaves in 'Ergenekon' (54.13%) and '497' (60.08%) genotypes. Kumsal showed less reduction of Chl (9.82%) among the other genotypes. From the data taken, and through analysis, we calculated that TMV infection reduced chlorophyll contents by

Tab. 6. Effect of TMV infection on proline content of peppers

Varieties	Proline (μmol proline g^{-1} FW)**		
	Control	Infected	Change (%)
'Ergenekon'	5.65	5.82	+2.97 ^{b}
'Kumsal'	5.60	5.64	+0.88 ^{c}
'497'	5.54	5.81	+4.89 ^{a}

**(p<0,01), (+) shows increase

Tab. 7. Effect of TMV infection on chlorophyll a, chlorophyll b and carotenoid of peppers

Varieties	Chlorophyll a (mg/L)**			Chlorophyll b (mg/L)*			Carotenoid (mg/L) NS		
	Control	Infected	Change (%)	Control	Infected	Change (%)	Control	Infected	Change (%)
'Ergenekon'	3.85	1.77	-54.13 [b]	0.72	0.59	-18.02 [a]	1.43	0.88	-38.57
'Kumsal'	3.13	2.82	-9.82 [a]	0.57	0.33	-40.86 [a]	1.24	1.04	-16.12
'497'	3.73	1.49	-60.08 [b]	0.77	0.12	-84.26 [b]	1.40	0.86	-38.51

*($p<0,05$), **($p<0,01$), NS: Not significant, (-) shows decrease

about 18.02%, 40.86%, 84.26% in 'Ergenekon', 'Kumsal' and '497' compared to those of control group. The variance analysis of the carotenoid content showed that this particular TMV infection did not affect the carotenoid content. There are various theories about how *Chl a* and *Chl b* contents in infected leaves were decreased. According to Mandahar and Garg (1972) and Roca and Minguez-Mosquera (2003) decreases of chlorophyll was the result of increased chlorophyllase activity. On the other hand Lucas (1998) thought that it could be due to the blocking *Chl* synthesis. Similar results for numerous host plants infected with different viruses were published by Christov *et al.* (2005), Radwan *et al.* (2007), Palanisamy *et al.* (2009) and Kyselakova *et al.* (2011). Carotenoid content was not affected by TMV and that shows differences with the results of carotenoid data published by Palanisamy *et al.* (2009). In addition a remarkable finding of this study is that TMV reduced leaf number and leaf area in all tested genotypes, contributing to the diminished photosynthesis activity (Tab. 2). Similar results obtained by Lehrer *et al.* (2007) and Ehinmore *et al.* (2010).

Conclusions

Based on the results of the present study, leaf number and area, plant height, the longest root length, stem diameter, dry and fresh weights of root and shoot were reduced significantly when peppers were exposed the TMV infection. Because of the fact that plant leaf number and total leaf area are directly related with photosynthetic activities which provide energy for growth and defense against pests and diseases, reduction of these parameters caused decline in overall growth resulting in shorter plants, slender stems, and reduced biomass compared controls. TMV infection also reduced relative water content and photosynthetic pigments except carotenoids. Reduction in relative water content and root length indicates that water transport in the soil-plant-atmosphere and respiratory system of the plants were adversely affected. Results also show that leaves subjected to TMV infection had increased proline content as compared to the leaves of control plants. The variation in the degree of change of this parameter suggests that the extent of damage is not the same in all tested groups. In general, 'Kumsal' showed more vigor and 'Ergenekon' showed moderate tolerance depending on its own characteristics when TMV was inoculated. On the other hand the highest damage was observed with candidate variety '497 F_1'. It is possible to say that, Kumsal is the most tolerant genotype among the other tested groups.

Acknowledgements

The authors wish to thank Prof. Dr. Eftal Düzyaman, Prof. Dr. Semih Erkan and Dr. İsmail Can Paylan for technical supports. And also we thank Bircan Seed company for supplying pepper genotypes.

References

Adi M, Jens P, Brotman Y, Mikhail K, Iris S, Henryk C, Renal G (2012). Stress responses to *Tomato yellow leaf curl virus* (TYLCV) infection of resistant and susceptible tomato plants are different. Metabolomics S1.

Agrios GN, Walker ME, Ferro DN (1985). Effects of Cucumber mosaic virus inoculation at successive weekly intervals on growth and yield of pepper *(Capsicum annuum)* plants. Plant Disease 69:52-55.

Agrios GN (2005). Plant pathology, 5th ed. Elsevier Academic Press, San Diego, 105 p.

Ain-Lhout F, Zunzunegui1 M, Diaz Barradas MC, Tirado R, Clavijo A, Garcia Novo F (2001). Comparison of proline accumulation in two mediterranean shrubs subjected to natural and experimental water deficit. Plant and Soil 230:175-183.

Al-Saleh MA, Melouk HA, Mulder P (2007). Reaction of peanut cultivars to Tomato spotted wilt virus (TSWV) under field conditions and their response to mechanical inoculation by TSWV under greenhouse conditions. Peanut Science 34:44-52.

Apel K, Hirt H (2004). Reactive oxygen species: metabolism, oxidative stress and signal transduction. Plant Biology 55:373-399.

Arnon DI (1949). Copper enzymes in isolated chloroplasts. Polyphenoloxidase in *Beta vulgaris.* Plant Physiology 24(1):1-15.

Baker C, Adkins S (2000). Peppers, tomatoes, and Tobamoviruses. Plant Pathology Section Circulars No:400 (http:// freshfromfl.com /pi/enpp/ pathology/ pathcirc/ pp 400. pdf accessed date 31.10.2012).

Bates LS, Waldren RP, Teare ID (1973). Rapid determination of free proline for water-stress studies. Plant and Soil 39:205-207.

Chatterjee A, Ghosh SK (2008). Alterations in biochemical

components in mesta plants infected with yellow vein mosaic disease. Brazilian Journal of Plant Physiology 20(4).

Chen C, Dickman MB (2005). Proline suppresses apoptosis in the fungal pathogen *Colletotrichum trifolii*. Plant Pathology 102(9):3459-3464.

Christov I, Stefanov D, Velinov T, Goltsev V, Georgieva K, Abracheva P, Genova Y, Christov N (2005). The symptomless leaf infection with Grapevine Leafroll Associated Virus 3 in grown *in vitro* plants as a simple model system for investigation of viral effects on photosynthesis. Journal of Plant Physiology 164:1124-1133.

Drossopoulos JB, Karamanos AJ, Niavis CA (1985). Changes in free amino compounds during the development of two wheat cultivars subjected to different degrees of water stress. Annals of Botany 56:291-305.

Ehinmore I, Kareem KT (2010). Effect of Amaranthus mosaic virus on the growth characters of *Amaranthus hybridus*. Agriculture and Biology Journal of North America 1(2):75-79.

Fabro G, Kovács I, Pavet V, Szabados L, Alvarez ME (2004). Proline accumulation and *AtP5CS2* gene activation are induced by plant-pathogen incompatible interactions in *Arabidopsis*. Molecular Plant-Microbe Interactions 17(4):343-350.

Friess N, Maillet J (1997). Influence of Cucumber mosaic virus infection on the competitive ability and reproduction of chickweed *(Stellariamedia)*. New Phytologist 135:667-674.

Funayama S, Hikosaka K, Yahara T (1997). Effects of virus infection and growth irradiance on fitness components and photosynthetic properties of *Eupatorium makinoi* (Compositae). American Journal of Botany 84(6):823-829.

Hull R (2009). Effects on plant metabolism, 181-185 p. In: Comparative plant virology, 2nd ed. Academic Press, Norwich.

Johnson CS, Main CE (1983). Yield/quality trade-offs of tobacco mosaic virus-resistant cultivars in relation to disease management. Plant Disease 67:886-890.

Jouve L, Hoffmann L, Hausman JF (2004). Polyamine, carbohydrate, and proline content changes during salt stress exposure of aspen (*Populus tremula* L.): involvement of oxidation and osmo regulation metabolism. Plant Biology 6:74-80.

Kakani K, Sgro JY, Rochon D (2001). Identification of specific *Cucumber necrosis virus* coat protein amino acids affecting fungus transmission and zoospore attachment. Journal of Virology 75(12):5576-5583.

Kazinczi G, Lukas D, Takacs A, Horvath J, Gaborjanyi R, Nadasy M, Nadays E (2006). Biological decline of *Solanum nigrum* due to virus infections. Journal of Plant Diseases and Protection 20:325-330.

Kollmann J, Bañuelos MJ, Nielsen SL (2007). Effects of virus infection on growth of the invasive alien *Impatiens glandulifera*. Preslia 79:33-44.

Kuhn CW, Dawson WmO (1973). Multiplication and pathogenesis of Cowpea chlorotic mottle virus and Southern bean mosaic virus in single and double infections in cowpea. Phytopathology 63:1380-1385.

Kyseláková H, Prokopová J, Nauš J, Novák O, Navrátil M, Šafářová D, Špundová M, Ilík P (2011). Photosynthetic alerations of pea leaves infected systemically by Pea enation mosaic virus: A coordinated decrease in efficiencies of CO_2 assimilation and photosystem II photochemistry. Physiological and Molecular Plant Pathology 49:1279-1289.

Leelerg EL (1967). Methodology for disease measurement related to assesment of losses, 11-43 p. In: T.A.O. Symposium on Crop Losses, Rome.

Lehrer AT, Moore PH, Komor E (2007). Impact of Sugarcane yellow leaf virus (ScYLV) on the carbohydrate status of sugarcane: Comparison of virus-free plants with symptomatic and asymptomatic virus-infected plants. Physiological and Molecular Plant Pathology 70:180-188.

Liu H, Lucy AP, Davies JW, Boulton MI (2001). A single amino acid change in the coat protein of *Maize streak virus* abolishes systemic infection, but not interaction with viral DNA or movement protein. Molecular Plant Pathology 2(4):223-228.

Lucas JA (1998). Plant pathology and plant pathogens. 3rd ed. Blackwell Science, 151 p.

Lv WT, Lin B, Zhang M, Hua XJ (2011). Proline accumulation is inhibitory to Arabidopsis seedlings during heat stress. Plant Physiology 156:1921-1933.

Mandahar CL, Garg ID (1972) Effect of Cucumber mosaic virus on chlorophyll content, photosynthesis, respiration and carbohydrates of infected *Luffa aegyptiaca* Mill. Journal of Phytopathology 75(2):181-186.

Mansour MMF (2000). Nitrogen containing compounds and adaptation of plants to salinity stress. Biology Plantarum 43:491-500.

Mazid M, Khan TA, Mohammad F (2011). Role of secondary metabolites in defense mechanisms of plants. Biology and Medicine 3(2):232-249.

Mohamed EF (2011). Changes in protein, amino acids composition and leaf cells of beet plants (*Beta vulgaris* L.) due to Beet mosaic virus (BtMV) infection. Journal of American Science 7(12):845-854.

Ostazeski SA, Henson PR, Garrison CS (1970). Effect of Tobacco ringspot virus on yield persistence, flowering, and seed set of *Lotus corniculatus*. Phytopathology 60:1753-1755.

Öztekin GB, Tüzel Y (2011). Comparative salinity responses among tomato genotypes and rootstocks. Pakistan Journal of Botany 43(6):2665-2672.

Pacheco DA, Rezende JAM, Piedade SMS (2003). Biomass, virus concentration, and symptomatology of cucurbits infected by mild and severe strains of Papaya ringspot virus. Scientia Agricola 60(4):691-698.

Palanisamy P, Michael PI, Krishnaswamy M (2009). Physiological response of Yellow vein mosaic virus infected bhendi (*Abelmoschus esculentus*) leaves. Physiological and Molecular

Plant Pathology 74:129-133.

Radwan DEM, Fayez KA, Mahmoud SY, Hamad A, Lu G (2007). Physiological and metabolic changes of *Cucurbita pepo* leaves in response to Zucchini yellow mosaic virus (ZYMV) infection and salicylic acid treatments. Plant Physiology and Biochemistry 45:480-489.

Roca M, Minguez-Mosquera MI (2003). Involvement of chlorophyllase in chlorophyll metabolism in olive varieties with high and low chlorophyll content. Physiologia Plantarum 117:459-466.

Rodriguez-Alvarado G, Kurath G, Dodds JA (1997). Symptom modification by satellite Tobacco mosaic virus in pepper types and cultivars infected with helper Tobamoviruses. Phytopathology 84(6):617-621.

Scholthof KBG, Shaw JG, Zaitlin M (1999). Tobacco mosaic virus: one hundred years of contributions to virology. American Phytopathological Society Press, St. Paul, MN.

Smart RE, Bingham GE (1974). Rapid estimates of relative water content. Plant Physiology 53:258-260.

Synková H, Semorádová S, Schnablová R, Müller K, Pospíšilová J, Ryšlavá H, Malbeck J, Čeřovská N (2006). Effects of biotic stress caused by Potato virus Y on photosynthesis in ipt transgenic and control *Nicotiana tabacum* L. Plant Science 171:607-616.

Taiwo MA, Akinjogunla OJ (2006). Cowpea viruses: quantitative and qualitative effects of single and mixed viral infections. African Journal of Biotechnology 5(19):1749-1756.

Verbruggen N, Villarroel R, Van Montagu M (1993). Osmoregulation of a pyrroline-5-carboxylate reductase gene in *Arabidopsis thaliana*. Plant Physiology 103:771-781.

Wani SP, Sivaramakrishnan S, Naidu RA, Zambre MA, Lee KK, Pande S (1991). Biochemical changes in sorghum (Sorghum bicolor L. Moench) plants infected with Maize mosaic virus. Indian Journal of Microbiology 31(4):387-395.

Wintermantel WM (2005). Co-infection of Beet mosaic virus with beet yellowing viruses leads to increased symptom expression on sugar beet. Plant Disease 89:325-331.

Yoshiba Y, Kiyosue T, Katagiri T, Ueda H, Mizoguchi T, Yamaguchi-Shinozaki K, Wada K, Harada Y, Shinozaki K (1995). Correlation between the induction of a gene for delta 1-pyrroline-5-carboxylate synthetase and the accumulation of proline in *Arabidopsis thaliana* under osmotic stress. Plant Journal 7:751-760.

2

Effect of Low Frequency Magnetic Field (LFMF) on the Germination of Seeds and Selected Useful Characters of Onion (*Allium cepa* L.)

Roman HOŁUBOWICZ[1], Leszek KUBISZ[2*], Marlena GAUZA[2], Yilin TONG[1], Dorota HOJAN-JEZIERSKA[2]

[1]*Poznań University of Life Sciences, Department of Plant Pathology Seed Science and Technology, Division of Seed Science and Technology, Szamotulska 28, Baranowo 62-081 Przeźmierowo, Poland*

[2]*Poznań University of Medical Sciences, Department of Biophysics, Poland, Fredry 10, 61-710 Poznań, Poland; lkubisz@ump.edu.pl (*corresponding author)*

Abstract

Magnetic fields (MF) have been reported widely as pre-sowing seeds treatment to increase their vigour, seedling growth and the plants' yield. Although, it has been shown that MF could affect seed germination and the seedling root growth, still their further effect on plant's life has been found to be inconsistent, especially under field conditions. Commercial seed lots of onion (*Allium cepa* L.) of the cultivars 'Octavia' and 'Eureka' were soaked for 12 hours in the distilled water at 20 °C. Next, they were exposed to low frequency magnetic field (LFMF) of magnetic induction B=20 mT, for 10, 30 and 60 minutes. The LFMF was generated using a standard medical device for magnetostimulation VIOFOR JPS. Next, the seeds were evaluated for germination and received bulbs were analysed for the selected useful characters including the root length, maximal diameters, dry matter and quercetin content. For the LFMF treatment of 60 minutes, for the cultivars 'Octavia' and 'Eureka', the energy of germination increased from 75.8% and 65% (control) to 88.3% and 87.5%, respectively. For the LFMF treatment of 60 minutes, the germination capacity, for the same cultivars, increased from 85% and 76.3% (control) to 92% and 90%, respectively, Treated for 60 minutes with LFMF onion seeds had longer seedlings than the ones from the control seeds. These values, for both cultivars, increased from 5.3 cm and 4.2 cm (control) to 8.4 cm and 5.7 cm, respectively. LFMF used for 60 minutes on the 'Octavia' onion seeds, increased their emergences in the field and length of the roots on the bulbs. LFMF used had no effect on the dry matter of bulbs grown out from the seeds when exposing them for 10 and 30 minutes. However, when exposing them for 60 minutes, LFMF decreased the dry matter and increased the amount of quercetin.

Keywords: seed germination, seed quality, magnetic field, seed grading

Introduction

Magnetic fields (MF) have been used widely as pre-sowing seed treatments to increase their vigour, seedling growth and yield (Pietruszewski, 1993, Phirke *et al.*, 1996a, Ahmet, 2003).

Experimental data demonstrate various effects of MF on plants both *in vivo* (Kavi, 1977, Lebedev *et al.*, 1977, Garcia and Arza, 2001, Flórez *et al.*, 2007, Ahmad and Azita, 2009) and *in vitro* (Van *et al.*, 2011). There is also no doubt that MF can affect seed germination, root growth and their faster early growth. Still their further effect on plant's life is inconsistent, especially under field conditions (Phirke *et al.*, 1996 a, b). As a result, there is still the necessity to carry out this kind of research.

In 2000, the world's production of onion was about 46.7 million tons of bulbs from 2.7 million hectares. According to the available FAO data, onion is now the third most important vegetable (after tomato and cabbage) produced in the world. For many years, it has also been the most important fresh vegetable exported from Poland. Its seeds are quite expensive as they lose their germination capacity quickly (Black *et al.*, 2006). This causes huge loss to seed companies all over the world. Therefore, for many years, there has been a search for simple and cheap method to improve germination of onion and other agricultural and horticultural crops seeds (George, 1985, Halmer, 2008).

So far, onion seed production has been investigated, but only one experiment on the effect of MF on onion seed quality has been reported (Tkalec *et al.*, 2009). In their experiments, the MF had no effect on germination of onion seeds. Moreover, it negatively affected radicle cells division.

Previous work on onion seeds (Kubisz *et al.*, 2012) has shown that low frequency magnetic field (LFMF) had a positive effect on germination of onion seeds.

The main objectives of the study were to confirm the earlier data about the positive effect of LFMF on the onion germination and to find out if LFMF also affects the onion seedlings' total length and selected plant characters of the bulbs received from the seeds treated with this factor.

Materials and methods

Onion seeds (*Allium cepa* L.) of two cultivars were used in these experiments: 'Octavia', which came from a Polish seed company "Torseed SA", located in Toruń, with the seed lot number PL 804/63/51/002A; and the cultivar 'Eureka', which came from a Polish seed company "CNOS Ożarów Mazowiecki SA - division", "CNOS Poznań", with the seed lot number PL 830/64/13135/408A. Both cultivars are commonly used in professional production in large horticultural farms in Poland. Their seed quality met the actual market quality standards.

First, 1600 full, not damaged and well ripen seeds were selected from each cultivar. Then, they were divided into 4 groups; each treatment had 400 (4 × 100) seeds. Before starting the experiment, the seeds were disinfected by soaking them in the 3% hydrogen peroxide solution for 5 minutes. Then, they were rinsed with sterilized water 3 times and dried out by putting them on the blotting paper.

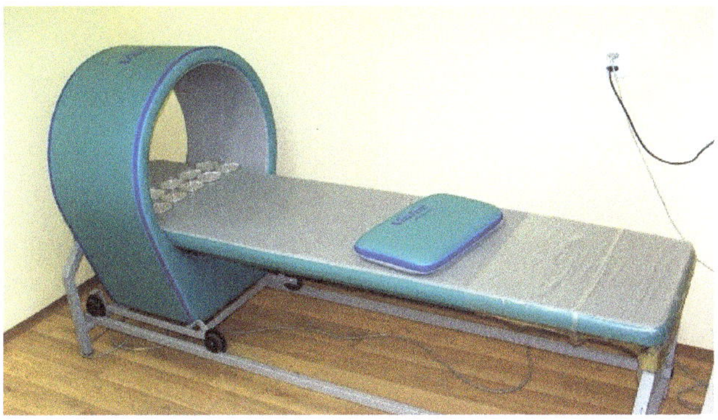

Fig. 1. The Petri dishes with the onion seeds exposed to LFMF generated by the VIOFOR JPS System

After that, they were soaked in the distilled water for 12 hours, and kept in the Elkon CL65 (800W, 220V) thermostat at 20±1 °C. Next, they were drained, surface dried and placed on the blotting paper in Petri dishes. The seeds were then exposed to LFMF (20 mT) using the VIOFOR JPS System for 10, 30 and 60 minutes. The MF was generated using a standard device for magnetic stimulation VIOFOR JPS applied in human medical treatment using application mode M1 (10 minutes) for 1, 3 or 6 times, programme P1 and filed intensity 12 (Fig. 1).

The parameters of the magnetic field used in the experiment were the same as described previously (Kubisz *et al.*, 2012).

After exposing seeds to LFMF, they were tested for germination following the instructions published by the International Seed Testing Association (ISTA) (Anonymous, 2003). The first counting of seedlings took place after 6 days, and the second one after 12 days.

In the second experiment where the total length of the seedlings was measured, the seeds were treated with the magnetic field in this same manner. After exposition to magnetic field, the seeds were placed onto the wet paper towels, which were then rolled up and tied up on both ends. Next, they were placed in upright position into vessels with distilled water and placed into the thermostat at 20±1 °C. After 12 days of each treatment, 240 (4 × 60) germinating seedlings were taken out. Their total length was measured using a ruler.

In a separate experiment, seeds of the cultivar 'Octavia' were treated the same way and were sown in the field at the Marcelin Experiment Station on June 23rd, 2011. For each seed treatment, there were 4 replications of 100 seeds sown in 4 rows on the field plot of 1.5 m × 1 m. The conditions of the seeds sowing in the field were as in routine production of the species by onion growers. After sowing the seeds, the field was then watered triweekly. The seeds were sown in the lessive soil created from clay on strong sands, useful to grow sugar beet or wheat, with pH = 6.5. On July 15th, the counting of the seedlings emergences was done. No plant protection against diseases and pests was used. Plants were collected on October 15th, 2011, 20 (5 for each treatment) and best plants were selected. They were weighed and the root length and the diameter of their bulbs were measured. The bulbs were then taken to the laboratory at the Faculty of Food Science and Nutrition for chemical analyses.

The dry matter was measured directly after harvesting the plants from the field for the bulbs of the cultivar 'Octavia'. The used method has been described by Drzazga (1999), i.e. 0.5 g onion samples were taken and dried at 130±1 °C for 1 hour in automatic analyzer of dry matter OX-35 (ONYX, Fawag . S.A).

The quercetin content was measured using high pressure liquid chromatography (HPLC) method. First, the extraction of the onion samples was made following the method described by Hertog *et al.* (1992) and Nutilla *et al.* (2002) with slight modification when it came to hydrolysis. The extraction process was carried out using 70% solution of acetone employing a single extraction of poly phenols from the examined samples. Acid hydrolysis of 0.5 g of onion tissue was carried out in round bottom flask with 1.2 N HCL for 2 hours at 90±1 °C. After cooling, the solution was filtered and centrifuged. The quercetin content was analyzed using the HPLC separation technique and was carried out using the X-Terra C18 (Waters Corporation, Taunton, MA, USA) column (3.9 L 150 mm, 4 mm). The column was placed in a column oven set at 30±1 °C. The mobile phase consisted of a mixture of methanol - 4.5% formic acid water with flow rate of 1 ml/min, by 75 minutes. Then, the samples were injected using auto-injector with a fixed 10 ml loop. The eluted fractions were monitored by UV-VIS 966 detector (Waters Corporation, Taunton, MA, USA). Standards of the analyzed quercetin (Sigma) were used for the identification procedure and the values were recorded in mg/g of dry matter.

Statistical analysis

The received data was subjected to statistic analysis based on the variance calculated and the significant differences were calculated based on the Duncan's test for $\alpha = 0.05$. Means followed for a given cultivar and character by the same letters are not significantly different according to the Duncan's test for $\alpha = 0.05$. For some data, standard deviation (SD) was calculated.

Results and discussion

Weather conditions in the field during the experiment were measured using the station's equipment. The mean air day temperature during the field experiment in June was higher than the average value for the previous 6 years. However, it was lower for July 2011 which was quite cold and plants grew slower than usually. However, the plant's growth sped up in the late summer and autumn (August and September) because of temperatures higher than usually. Month sums of the rains during the field experiment in June and July were much higher than in the previous years. However, in August, September and October, they were obviously lower than the values at the same time in the years 2005-2010 (Tab. 1).

Tab. 1. The average day air temperature (°C) and month sums of precipitation (mm) in Poznań at the Marcelin Experiment Station in 2011 and the mean values of the same parameter for the years 2005-2010

Month	Average day air temperature		Sums of precipitation	
	2011	2005-2010	2011	2005-2010
June	18.7	14.7	55.0	38.6
July	16.7	20.6	146.4	64.5
August	18.8	18.4	30.0	62.2
September	15.1	14.8	17.4	33.3
October	8.8	8.9	20.6	26.0

The LFMF used, increased germination of the tested onion seeds (Tab. 2). For the energy of germination, for the LFMF treatment of 60 minutes, for the cultivars 'Octavia' and 'Eureka', it increased from 75.8% and 65% (control) to 88.3% and 87.5%, respectively. For the 60 minutes treatments, for the same cultivars, the germination capacity increased from 85% and 76.3% (control) to 92% and 90%, respectively. These phenomena were accompanied by changes in the length of seedlings. Onion seeds treated for 60 minutes with LFMF had longer seedlings. Their length increased from 5.3 cm and 4.2 cm (control) to 8.4 cm and 5.7 cm, respectively.

The LFMF used, increased the field emergences of the seeds of the onion cultivar 'Octavia' (Tab. 2). Especially for the 60 minutes treated seeds, there were differences in comparison with the control seeds. For the 10 and 30 minutes treatments, no differences between the treated seeds and the control ones, in terms of their field emergences, were found. Seeds treated with LFMF for 60 minutes grew out eventually in visually better plants (Fig. 2).

The LFMF used on the onion 'Octavia' seeds had no effect on the dry matter of the 3-month-old bulbs grown out from these seeds when exposed for 10 and 30 minutes (Tab. 3), but when treated for 60 minutes, the dry matter decreased and increased the plant's fresh weight, root length maximal diameters of the bulb and its quercetin content.

The numerous experiments, carried out in the last 80 years, have proven in majority of cases, that MF could increase germination of the seeds (Savostin, 1930; Phirke *et al*., 1996 b). The 60 minutes exposition to the 20 mT of LFMF used in this experiment for onion seeds of the cultivars 'Octavia' and 'Eureka', prior the exposition soaked in the distilled water at 20±1 °C for 12 hours, resulted in the increase of their germination. These results, however,

Tab. 2. The effect of treating the onion 'Octavia' and 'Eureka' seeds with low frequency magnetic field (LFMF) on their germination energy (%) and capacity (%), length of the seedlings (cm) in the laboratory test and field emergences (%)

Cultivars	Treatment	Germination energy	Germination capacity	Length of the seedlings	Field emergences
'Octavia'	Control	75.8 ^{a*}	85.0 ^{a}	5.3 ^{a}	10.3 ^{a}
	10 min.	80.8 ^{ab}	87.5 ^{ab}	6.6 ^{b}	11.3 ^{a}
	30 min.	83.0 ^{ab}	88.5 ^{ab}	7.5 ^{c}	10.8 ^{a}
	60 min.	88.3 ^{b}	92.0 ^{b}	8.4 ^{d}	16.3 ^{b}
'Eureka'	Control	65.0 ^{a}	76.3 ^{a}	4.2 ^{a}	-
	10 min.	72.5 ^{b}	84.8 ^{b}	5.1 ^{b}	-
	30 min.	80.0 ^{c}	85.8 ^{b}	5.5 ^{b}	-
	60 min.	87.5 ^{d}	90.0 ^{b}	5.7 ^{b}	-

* Means followed for a given cultivar and character by the same letters are not significantly different according to the Duncan's test for $\alpha = 0.05$

Tab. 3. The effect of treating the onion 'Octavia' seeds with low frequency magnetic field (LFMF) on the plant's fresh weight (g), the bulb's root length (cm), the bulb's maximal diameter (cm), the bulb's dry matter (%), and the quercetin content (mg/g dry matter) in the bulbs grown out from these seeds

Treatment	Plant's fresh weight	Bulb's root length	Bulb's maximal diameter	Bulb's dry matter content	Bulb's quercetin content
Control	79.8 ^{a*}	13.3 ^{a}	3.86 ^{ab}	16.06 ^{b}	4.58±0.11**
10 minutes	88.7 ^{ab}	14.2 ^{ab}	3.72 ^{a}	14.93 ^{ab}	5.29±0.09
30 minutes	79.8 ^{a}	15.3 ^{b}	4.12 ^{bc}	15.23 ^{b}	4.60±0.11
60 minutes	97.0 ^{b}	17.2 ^{c}	4.21 ^{c}	12.52 ^{a}	6.56±0.12

* See Tab. 2 for the explanation; ** Standard deviation.

Fig. 2. The effect of low-frequency magnetic field (LFMF) treatment on the field emergences of the onion seeds of the cultivar 'Octavia'. Pictures were taken on October 15th, 2011, at the Marcelin Experiment Station. Abbreviations: K - control, M 10 - LFMF for 10 minutes, M 30 - LFMF for 30 minutes, M 60 - LFMF for 60 minutes

depended in high extent on the used cultivar and were not always completely the same, i.e. they varied. The LFMF treated seeds were better germinating and had longer seedlings than the control seeds. This, however, was not always observed due to big variation in the received results (Kubisz *et al.*, 2012). This disadvantage of the MF, as a factor of plant growth, was also reported by Phirke *et al.* (1996 a, b). For the same reason, there was probably no positive effect of MF on onion seeds germination in the works of Tkelec *et al.* (2009).

There is no doubt, that seeds of important horticultural crops, such as tomato or onion, will be a subject of research by seed companies to improve their quality and storability. Even if such research with the MF has been done, no data is available today as to their subsequent useful characters in the field. In this aspect, the results obtained by the authors are new and unique. To the best of the authors' knowledge, no data is available in the literature on this subject.

The LFMF used on the onion 'Octavia' seeds decreased the bulbs dry matter content when treated for 60 minutes and increased the plant's fresh weight, root length, maximal diameter and its quercetin content. The LFMF used on the 'Octavia' onion seeds, for the treatment 10 and 30 minutes, had no effect on the dry matter of the bulbs grown out from the seeds, whereas for the 60 minutes treatment decreased the dry matter. The quercitin content in the bulb has been the crucial useful character for onion breeders because it decides about the taste of the bulb. In turn, dry matter of the bulb plays the same crucial role for the onion bulbs in the processing industry. This result, however, as received with small plots, should be commented on rather carefully and, in our opinion, will need to be verified in a separate, follow up larger plots experiment. The obtained result for the 60-minutes application of the LFMF is in accordance with previous work carried out on the onion seeds by Kubisz *et al.* (2102).

A bulb size is a very important characteristic in onion seed production. It is so because, when planting big bulbs, higher number of seed stalks is received and consequently bigger seed yield can be expected (Morozowska and Hołubowicz, 2009). Root length, in turn, is a crucial factor in breeding onion cultivars with resistance to drought, i.e. longer the roots the better the onion plant's tolerance to water stress (George, 1985). This direction of breeding has been dominating in many seed companies in the world. Higher amount of quercetin found in the bulbs as a result of the LFMF treatment is a positive effect due to the health - promoting influence of this flavonoid on humans.

There is, however, more fundamental question: how does the MF affect the seed germination?

So far, we know that the LFMF could elongate the seedling because MF nanoparticles in a molecular process are an endogenous source of magnetic exposure to living tissues and cells, both in plants and animals.

One of the theories formed by Johnsen and Lohman in 2005 said that all living tissues had special very small parts called "nanoparticles", which are involved in biological reaction to MF exposure. They could be generated from magnetic minerals precipatated in living tissue such as: magnetite (Fe_3O_4), maghernite (Fe_2O_3) or greigite (Fe_3S_4). The crystals of magnetite called magnetosomes were found in the cell membrane. In 2001, Kishvink *et al.* (after Binhi, 2004) proved the relationship between weak magnetic field (200 nT) and responses of biological material.

Mechanisms of the action on LFMF at the initial stages

of germination of wheat seeds have been discovered experimentally. The effect of LFMF at the stage of release of the enzyme called esterase in the course of swelling of wheat seeds greatly speeds up the release of enzymes from the bound state and also the release of seeds from the dormant stage; later, however, the field effect appreciably decreases (Aksenov *et al.*, 1996).

Conclusions

The low frequency magnetic field (LFMF) (20 mT) used for 12 hours soaked in the distilled water at 20 °C onion seeds of the cultivars 'Octavia' and 'Eureka' resulted in increasing their germination. The energy of germination for both cultivars increased from 75.8 % and 65 % (control) to 88.3 % and 87.5 %, respectively. For the germination capacity, for the 60 minutes treatment with the LFMF, for the same cultivars, it increased from 85 % and 76.3 % (control) to 92 % and 90 %, respectively. These phenomena were accompanied by the increase in the length of seedlings, from 5.3 cm and 4.2 cm (control) to 8.4 cm and 5.7 cm, respectively. The used LFMF increased the field emergences of the seeds of the onion cultivar 'Octavia'. Especially for the 60 minutes treated seeds, there were significant differences in comparison with the control plants. For the 10 and 30 minutes treatments, no differences between the treated seeds and the control ones, in terms of their field emergences, were found. The LFMF used for 60 minutes on the onion 'Octavia' seeds increased their emergences in the field and length of the roots on the bulbs. The LFMF used had no effect on the dry matter of bulbs grown out from the seeds when exposed to 10 and 30 minute treatments, but when exposed to a 60 minute treatment, it decreased the dry matter and increased the amount of quercetin content.

References

Ahmad M, Azita S (2009). Effect of seed pretreatment by magnetic fields on seed germination and ontogeny growth of agricultural plants. Electromagnetics Res Symp, March 23rd - 27th 1137-1141.

Ahmet E (2003). Effects of magnetic fields on yield and growth in strawberry 'Camarosa'. Hort Sci Biotech 78:145-147.

Aksenov SI, Bulyechev AA, Gruzina TY, Turovetskii VB (1996). Mechanism of action of a low-frequency magnetic field on the initial stages of germination of wheat seeds. Biophys 41:931-937.

Anonymous (2003). International rules for seed testing. International Seed Testing Association (ISTA), Zürich, Switzerland.

Binhi VN (2004). Stochastic dynamics of magnetic nanoparticles and a mechanism of biological orientation in the geomagnetic field. Physics. Bio-Ph. December 27th 1-8.

Black M, Bewley JD, Halmer P (2006). The Encyclopedia of Seeds Science, Technology and Uses. CAB International Press, 44.

Drzazga B (1999). Techniques of analysis in the food industry. WSiP, Warszawa, 48 (in Polish).

Flórez M, Carbónell MV, Martínez E (2007). Exposure of maize seeds to stationary magnetic fields: effects on germination and early growth. Environ Exp Bot 59:68-75.

Garcia RF, Arza PL (2001). Influence of a stationary magnetic field on water relations in lettuce seeds. Part I: theoretical considerations. Bioelectromag 22:589-595.

George RAT (1985). Vegetable Seed Production. Longman, London-New York.

Halmer P (2008). Seed technology and seed enhancement. Proc. 28th Internat Horticultural Congress - S13 "Seed Enhancement and Seedling Production Technology". (Eds. Changhoo Chun and Jungmyung Lee). Acta Hort 771:17-26.

Hertog MGL, Hollman PCH, Venema DP (1992). Optimisation of a quantitative HPLC determination of potentially anticarcinogenic flavonoiuds in vegetables and fruits. Agric Food Chem 40:1591-1598.

Kavi PS (1977). The effect of magnetic treatment of soybean seed on its moisture absorbing capacity. Sci Cult 43:405-406.

Kubisz L, Hołubowicz R, Li H, Gauza M, Hojan-Jezierska D, Jaroszyk P (2012). Effect of low frequency magnetic field (LFMF) on the germination of seeds of onion (*Allium cepa* L.). Acta Phys Pol 121A:49-53.

Lebedev IS, Litvinenko LG, Shiyan LT (1977). After effect of a permanent magnetic field on photochemical activity of chroloplasts. Sov Plant Physiol 24:394-395.

Morozowska M, Hołubowicz R (2009). Effect of bulbs size on selected morphological characters of seed stalks, seed yield and quality of onion (*Allium cepa* L.) seeds. Folia Hort 21:27-38.

Nutilla AM, Kammiovrta K, Oksman-Caldentey KM (2002). Comparison of methods for the hydrolysis of flavonoids and phenolic acids from onion and spinach for HPLC analysis. Food Chem 76:519-525.

Phirke PS, Kubde AB, Umbakar SP (1996a). The influence of magnetic field on plant growth. Seed Sci & Technol 24:375-392.

Phirke PS, Patil MN, Umbarkar SP, Dudhe YH (1996b). The application of magnetic treatment to seeds: methods and responses. Seed Sci & Technol 24:365-373.

Pietruszewski S (1993). Effect of magnetic seed treatment on yields of wheat. Seed Sci & Technol 21:621-626.

Savostin PV (1930). Magnetic growth relations in plants. Planta 12:327.

Tkalec M, Malarić K, Pavlica M, Pevalek-Kozlina B, Vidaković-Cifrek Z (2009). Effect of radiofrequency electromagnetic field on seed germination and root meristematic cells of *Allium cepa* L. Mutation Res 672:76-81.

Van PT, Teixeira Da Silva JA, Ham LH, Tanaka M (2011). The effects of permanent magnetic fields on *in vitro* growth of *Phalaenopsis* plantlets. J Hort Sci & Biotechnol 86:473-478.

3

Structural Analysis of Reproductive Development in Pistillate Flowers of *Laurus nobilis* L.

Özlem AYTÜRK*, Meral ÜNAL

*Marmara University, Department of Biology, Göztepe Campus, 34730 Istanbul, Turkey; oayturk2004@hotmail.com (*corresponding author)*

Abstract

In *Laurus nobilis* L. (Lauraceae) the development of female flowers (pistillate), between floral meristem differentiation and fruit formation was inspected through histological sections and SEM techniques. The reproductive development of the female flower starts when the apical meristem converts into a floral meristem. Four tepals, four stamens and a carpel are developed from the floral meristem in turn. Filaments emerge however, anther development is arrested, and stamens become nonfunctional staminodes. The stigma is of the dry type. The solid style being short and thick consists of an epidermis, a cortex, a vascular bundle and a core of transmitting tissue composed of elongated cells. In the style a funnel-shaped zone extending from within the stigma to the stylar base is visible. The presence of high amounts of sugars and lipid substances within and around the vascular bundles are identified by histochemical techniques. The ovary contains an anatropous, bitegmic and crassinucellate ovule. Starch grains are present throughout the development of nucellar tissue. The chalazal region of nuclear endosperm forms a short haustorium. Endosperm does not exist in mature seed; the cotyledons are piled with considerably large starch grains. Idioblasts are observed in all stages of development.

Keywords: dioecious, idioblast, *Laurus nobilis* L., pistillate flower development, staminode

Introduction

Although unisexuality is very common in animals, hermaphroditism is the rule in angiosperms (Tanurdzic and Banks, 2004). In flowering plants approximately 10% of species are dioecious, producing flowers of one sex only on each plant (Dellaporta and Calderon-Urrea, 1993). Only stamens or carpels develop to maturity in unisexual flowers.

The start of flower bud development comprises the differentiation of vegetative meristem, its conversion into apical meristem and its transformation over again, into floral meristem in the upcoming stage (Bernier *et al.*, 1993). The flower organs; sepal, petal, stamen and carpel develop from the floral meristem (Atsmon and Galun, 1960; Malepszy and Szczytt, 1991) but further development of stamens or pistils is selective, resulting in unisexual flowers (Haughn and Somerville, 1988; Rastogi and Sawhney, 1988).

The prevention of stamen or carpel development is a significant process that shows dissimilarity among species. The spectrum interval of unisexual flower development evolves from the differentiation of organ primordia until the formation of the sex organs which are completely developed but non-functional (Dellaporta and Calderon-Urrea, 1993). In dioecious Silene species, both the stamen and carpel primordia are present in both sexes, with the developmental arrest of the inappropriate sex occurring at early stages of floral development (Ye *et al.*, 1991). The stage of arrest is later in maize, when the organ primordia are well defined but prior to their full maturation and meiosis (Dellaporta and Calderon-Urrea, 1993). The arrested development of the sex organs is not accidental; it is manipulated by the environmental conditions and genotype (Kinet *et al.*, 1981).

One of the interesting issues in developmental biology is the sex differentiation and development in the flowers of monoecious and dioecious plants which carry unisexual flowers. Even though there is considerable knowledge on sexual differentiation in hermaphrodite plants, little information is present about sex differentiation in dioecious species and the prevention of opposite sex development.

Having gained a significant role in both economic and social life since the early ages, *L. nobilis* L. (sweet bay) is a dioecious species which has an aromatic smell and remains green not only in summer but also in winter. Sweet bay, which remains present mostly in Mediterranean countries, not only originated from Asia but is also a characteristic tree of the Mediterranean maquis.

Embryological features of Lauraceae were clearly reviewed by Kimoto and Tobe (2001). The present study of *L. nobilis* aims to contribute not only to filling the gap in embryological knowledge but also to understanding sexual differentiation and female flower development. We analysed the course of events starting from the initiation of the female flower up to fruit formation based on histological sections and scanning electron microscopy (SEM) techniques. In parallel, we examined the structure of stigma and style of this species that is first presented in this family species.

Materials and methods

Plant Material

Flowers of *L. nobilis* at various stages of development were collected from the Göztepe Campus of Marmara University, from November of 2008 to June of 2009. Firstly, female flowers were morphologically analysed by stereomicroscope (Olympus 970931). The lengths of flower parts were mesaured and the samples were prepared for light and SEM analysis.

Preparation for microscopy

The material was fixed in acetic-alcohol (1:3, v/v) and then placed in a vacuum desiccator to facilitate the penetration of the fixative into the plant tissues. After embedding in paraffin, blocks were sectioned at 8-25 μm by Leica RM2235 rotation microtome and sections were stained with hematoxylin.

Histochemistry

For histochemical analysis, sections were stained with periodic acid-Schiff (PAS) (Feder-O'Brien, 1968) for insoluble polysaccharides, Coomassie brilliant blue (Fisher, 1968) for total proteins and Sudan Black B (Pearse, 1961) for lipids. The preparations were photographed with an Evolution LC color camera and an Olympus BH-2 microscope, and the images were analyzed with Image-Pro Express Version 6.0 scientific image processing and analysis software.

Periodic acid schiff's reagent (pas) treatment

The slides were immersed in a 1% (w/v) periodic acid for 20 min, rinsed in tap water for 10 min and stained in Schiff's reagent for 30 min in the dark. Then they were washed three times for 5 minutes each with 0.5% sodium metabisulfite and in tap water for 5 min. The slides were mounted in Entellan.

Coomassie brilliant blue treatment

The sections were stained in a 0.2% Coomassie brilliant blue (in a mixture of water, methanol and acetic acid (v:v:v, 87:10:3) for 10 min at 60^{0}C and transferred to water for 5 min. The slides were mounted in Entellan.

Sudan black B staining

The slides were immersed in Sudan Black B which was prepared immediately before use for 1 hour at 60^{0}C and rinsed in 70% ethanol for 2 times and one time with water. After drying at room temperature, the sildes were mounted in gelatin-glycerol.

Electron microscopy

For SEM analysis, the plant material was fixed in 2.5% glutaraldehyde in 50 mM cacodylate buffer, pH 7.0 (Platt *et al.*, 1983) and then dehydrated with an increasing ethanol gradient: from 70% up to 100%. Then, the material for drying was kept in various percentages of ethanol-HMDS solution at room temperature (Topçuoğlu *et al.*, 2009). Then, coated with 11 nm of gold by using an automated sputter coater and then examined with a SEM (JEOL JMS-59 10LV).

Results

The early developmental stages of female and male flowers are similar to each other. The visible differences occur after the sex organ primordia emerge.

Flower morphology

The flowering phase in female trees of *L. nobilis* happens between March and May. The potential flowering points are formed in October-November of the previous year. Mostly, flower buds come into existence at the growing points. Flower buds are rounder and larger, while leaf buds are thinner and more tapered (Fig. 1). Female flower buds exist in groups of five, at the tip of shoots (Fig. 1), covered by bud scales. Flowering doesn't start simultaneously at every flowering point. Inside the flower groups flowering develops from the center to outwards, thus the inner flowers mature more quickly than the outers. The length of flower buds, which is 1-2 mm at the early stage, reaches 4-6 mm at maturity.

The mature female flowers (3.9 mm) are smaller than male flowers. There are four tepals which are yellow and approximately 3 mm in length in a mature female flower. Inside the tepals there are four nonfunctional stamens (staminodes) (Fig. 1), whereas a male flower bears 8-10 fertile stamens.

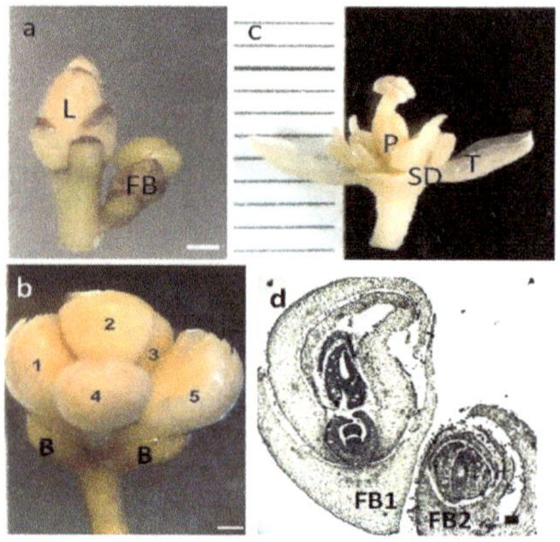

Fig. 1. a) Leaf and flower buds, Scale bar: 1 mm, b) 5 flower buds, Scale bar: 1 mm, c) A recently opened pistillate flower (a-c viewed by stereomicroscope), d) Longitudinal section of flower buds, Scale bar: 100 μm. B: Bracte, L: Leaf, FB: Flower bud, P: Pistil, SD: Staminode, T: Tepal

The nectarium, which is smaller in male flowers, has an approximate length of 1.3 mm and a width of 1.0 mm, and is located on both sides of the filament in the female (Fig. 2).

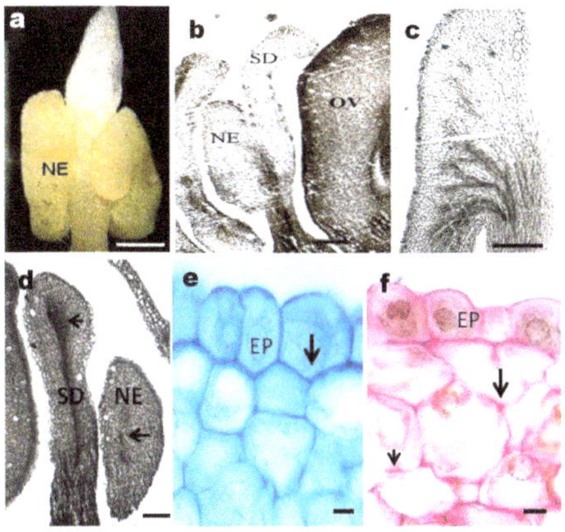

Fig. 2. a) A mature staminode and nectaries viewed by stereomicroscope, Scale bar: 1mm, b) Longitudinal section of staminode and nectarium stained by hematoxyline, Scale bar: 100 µm, c) Extensive networks of vascular bundles in nectarium, Scale bar: 100 µm, d) Longitudinal section of staminode, nectarium and vascular bundles stained by Sudan Black B, Scale bar: 100 µm, e), f) The nectariferous cells surrounded by secretory substances stained by Coomassie Brilliant Blue (e), stained by PAS (f) and starch grains, Scale bar: 10 µm. EP: Epidermis cell, NE: Nectarium, SD. Staminode, OV:ovary

Nectarium

In a pistillate flower, two nectaries are situated close to the base of a staminode (Fig. 2). Vascular bundles in the nectarium of pistillate flowers are present only below the nectariferous tissue (Fig. 2). The epidermis of the nectarium is covered by a cuticle (Fig. 2) and the nectariferous cells differ from those of the neighbouring tissues due to their bigger size and stronger positive reaction to Coomassie Brilliant Blue, PAS and Sudan Black B. They contain a small amount of starch grains at all of the development stages (Fig. 2). In an unopened flower, the whole nectarium is uniformly stained by PAS. In unopened and opened flowers, no changes in staining were distinguished. Subepidermal cells are surrounded by secretory substances giving a positive reaction to Coomassie Brilliant Blue and PAS, accumulating in the intercellular spaces (Fig. 2).

Flower development

While female flower development was examined starting from the apical meristem differentiation to fruit formation, the bud lengths are measured in order to identify the alterations that occur during the development, and being divided into eight stages, the development was analysed by stereomicroscobe, light microscobe and SEM (Tab. 1).

1. Initiation of floral meristem

Initially, the apical meristem differentiates from the vegetative meristem when the female flower starts to develop. Young leaves exist on both sides of the apical meristem, which consists of sequential cell layers (Fig. 3). The width of the floral meristem, which develops from the apical meristem, is more than the width of the apical meristem; however, the depth of the floral meristem is lesser (Fig. 3 and 4). The floral meristem cells in the female flower contain abounding cytoplasm, big nuclei and small and many vacuoles. The nuclei contain numerous nucleoli. The space between cells is considerably small among thin–walled meristem cells which form well-adjusted cell layers. Flower organ primordia, which are tepal, stamen and carpel in order, develop from the floral meristem (Fig. 3).

Tab 1. Morphological indications and respective lengths of floral buds in *Laurus nobilis* L. at various stages of development

Stage	Morphological indications	Bud Length (mm) n:25	Figure Numbers
1	Initiation of floral meristem	< 1.12	3b, 4b
2	Initiation of tepal primordium	1.21	3b, 4b
3	Initiation of stamen primordium	1.35	3c, 4d,5a-c
4	Initiation of carpel primordium	1.42	3d, 4f
5	Pistil development	1.64	6a-c
6	Megasporogenesis and Megagametogenesis	1.78	13 a-d
7	Endosperm and embryo development	2.37	13e
8	Seed and fruit formation	3.78	14a-c

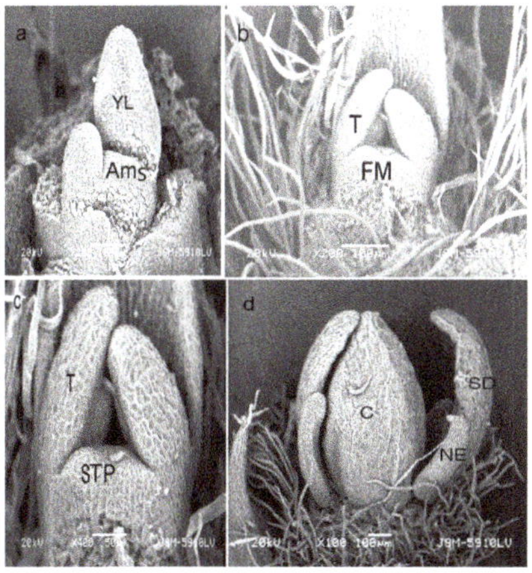

Fig. 3. Scanning electron micrographs of early development stages of the flower (outer flower whorls). a) Apical meristem, b) Floral meristem, c) Stamen iniation, d) Carpel and staminodes. YL: Young leaf, Ams: Apical meristem, FM: Floral meristem, T: Tepal, STP: Stamen primordium, SD: Staminode, NE: Nectarium, C: Carpel

2. Initiation of tepal primordium

Firstly, four light yellow tepals develop from the floral meristem in female and male flowers. Tepal primordia, which are located on the outermost first whorl, initiate from the floral meristem. The floral meristem enlarges and bulges are observed on the upper surface. Four tepal primordia develop consecutively and in whorl form (Fig. 4).

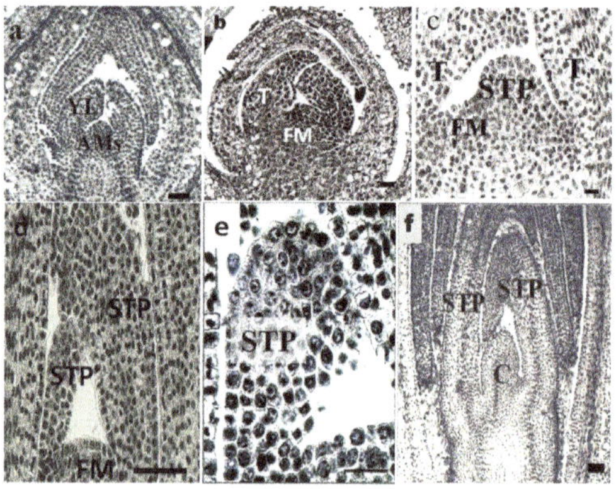

Fig. 4. Some early development stages in longitudinal sections of the flower. All bud scales have been removed in all. a) Apical meristem, Scale bar: 100 μm, b) Floral meristem, Scale bar: 100 μm, c)-f) Initiation of stamens and carpel, Scale bar: 100 μm (d, f), 10 μm (c, e). YL: Young leaf, AMs: Apical meristem, FM: Floral meristem, T: Tepal, STP: Stamen primordium, C: Carpel

3. Initiation of stamen primordium

In the beginning of the stamen primordium in the female flower, a group of cells divide and form a bulge from one part of the floral meristem and this bulge lengthens to give rise to the filament. A group of cells differentiate likewise from the other side of floral meristem and start forming the filament of the other stamen (Fig. 4), while the development of the filament continues (Fig. 4). Filaments keep growing longer in the stamens however the anthers do not differentiate (Fig. 5). It was observed that anther differentiation existed in none of the 200 experimental flowers. Stamen rudiments are visible on mature female flowers and form narrow rods in 2.2 mm in length. Staminodes hold two nectaries and they continue to appear throughout flower development. Staminodes degenerate at the end of the development (Fig. 5).

4. Initiation of carpel primordium

After the emergence of all stamen primordia out of the floral meristem is completed, the cells in the center of the floral meristem divide and differentiate as carpel primordium inside the stamen whorls. The carpel primordium arises as a small protuberance. When the floral meristem converts into the form of carpel, a curved appearance arises (Fig. 3, 4 and 6). The carpel primordium occupies most of the center of the flower bud.

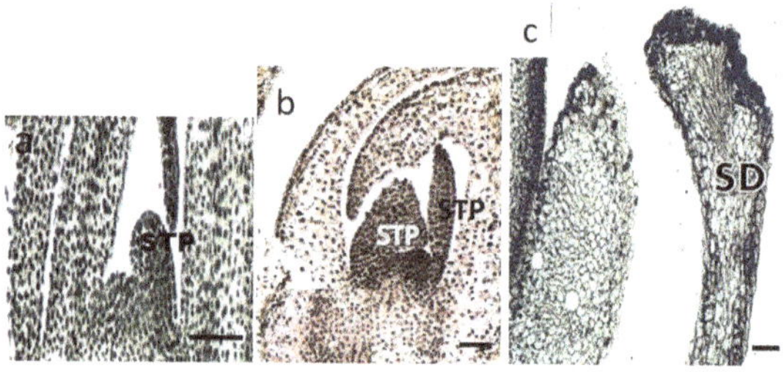

Fig. 5. a), b) Stamen primordium (STP), Scale bar: 100 μm, c) Staminode, Scale bar: 100 μm. STP: Stamen primordium, SD: Staminode, NE: Nectarium

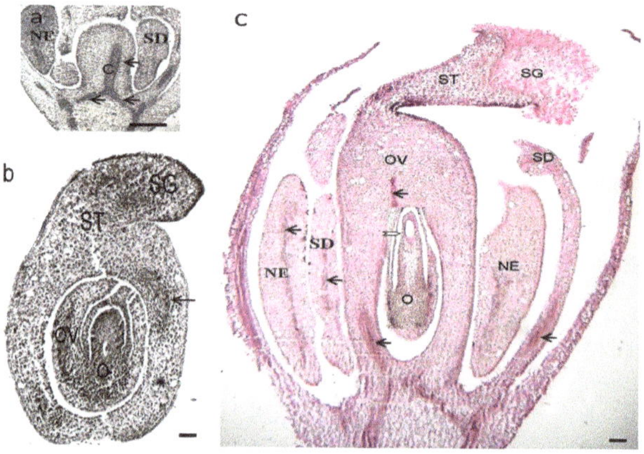

Fig. 6. Pistil development in *L. nobilis*. a), b) Pistil in an unopened flower bud, Scale bar: 100 μm, c) Detail of flower in longitudinal section stained with PAS: A complete pistil containing a stigma, a style and an ovary with one ovule and staminodes with nectarium, Scale bar: 100 μm. ST: Style, SG: Stigma, OV: Ovary, O: Ovule, NE: Nectarium, SD: Staminode, Vascular bundle C: Carpel

5. Pistil development

The carpel primordium starts to grow by active cell divisions and first the ovary then the cylindrical style emerges. The style continues to lengthen, and a wide stigma evolves out of its tip (Fig. 6). In upcoming stages, the ovule comes into existence as a bulge on the upper side of the ovary which differentiates from the base of the carpel (Fig. 6). A big nectarium and staminodes in pistillate flower are shown in Fig. 6. A pistil consists of an ovary with one ovule, a short style and a capitate stigma.

Stigma

In this species, the capitate stigma is papillate and of dry type. The stigma starts developing when the style is approximately 0.3 mm in length, and it surrounds the upper area of the stylar tissue (Fig. 7). In this stage, the megaspore mother cell appears in the ovule. The receptive surface of the stigma is composed of unicellular papillae (Fig. 7). The papilla in young pistils have abounding cytoplasm, small vacuoles and a centrally located nucleus. Its upper surface has a round shape (Fig. 8). While the stigma matures, vacuoles in the papillar cell enlarge, their cytoplasm decline, and the nuclei move to the basal. The papillae enlarge lengthwise, and they gain a cylindrical outlook (Fig. 8). As development proceeds, the boundaries of the papillae become less noticeable and they attain a broken view. The nuclei become invisible in most of them (Fig. 8). The papilla lengths, being on average 38-45 μm when they first appear, reach an average level of 72-84 μm in mature stigma. A noticeable decrease in their diameters is observed from the beginning of the development to the end: the diameters of papillae, which are roughly 20 μm at the young stage, decrease to a level of 15 μm at the mature stage. A thick cuticle layer giving a positive reaction to Sudan Black B surrounds the papillae, and it is covered by a pellicle. This layer reacts positively to Coomassie Brilliant Blue, indicating the presence of proteins. With the maturation, the layer that surrounds the papillae starts thickening. At the young stage, the stigma forms a collapse (funnel shaped zone) which almost reaches the base of the style, inwards through the middle of the style. Long papillae are located, clearly on the surface of this breach. The width of the collapse increases with maturation of the style (Fig. 9).

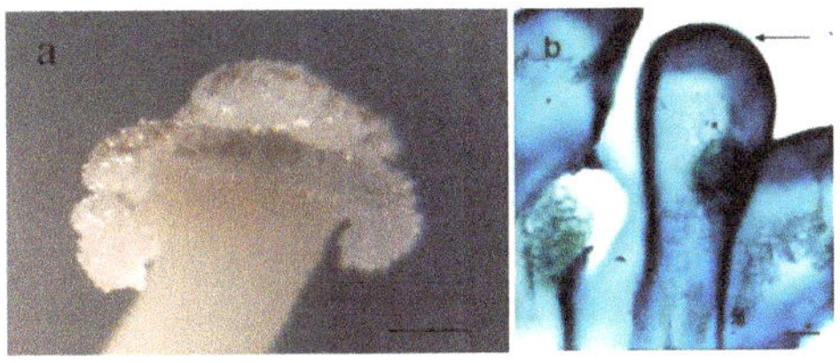

Fig. 7. a) Stereomicrograph of stigma in a mature flower, Scale bar: 250 μm, b) Stigma papilla stained with Coomassie Brilliant Blue, Scale bar:10 μm

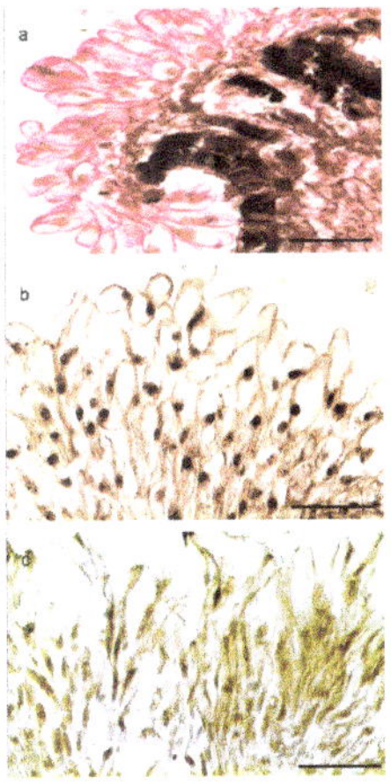

Fig. 8. a)-c) Longitudinal sections of young to old stigma papilla showing gradual changes, Scale bar: 100 μm

A longitudinal section of pistil stained with Coomassie Brilliant Blue and PAS shows that the papillar stigma is basipetally continuous with a closed style and is traversed by a core of transmitting tissue. The inner stigmatic cells are continuous with the stylar transmitting tissue, showing an overall funnel-shaped arrangement which is wider in the stigma and narrows basipetally in the style (Fig. 9).

Style

The style appears short and thick in the beginning of the development in *L. nobilis*. As the development makes progress, the style lengthens and its diameter increases as well (Fig. 9). The diameter of the style, which is roughly 0.4 mm in the early stages of the development, reaches a level of 0.7 mm in the final stages of the maturation and its length increases up to 1.3 mm. It remains bent towards one side of the ovary from the beginning of its development (Fig. 9).

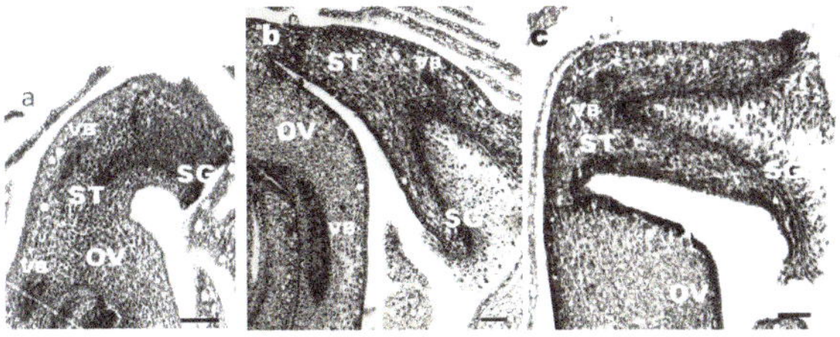

Fig. 9. a)-c) Longitudinal sections of young to old style with stigma showing gradual changes, Scale bar: 100 μm. OV: Ovary, ST: Style, SG: Stigma, VB: Vascular bundle

The style is solid and it is composed of an epidermis, a stylar cortex consisting of larger vacuolated cells organized in layers and a core of transmitting tissue (Fig. 10). One vascular bundle is visible on the side of the transmitting tissue and indicates a positive reaction to PAS (Fig. 6) and Coomassie Brilliant Blue (Fig. 10).

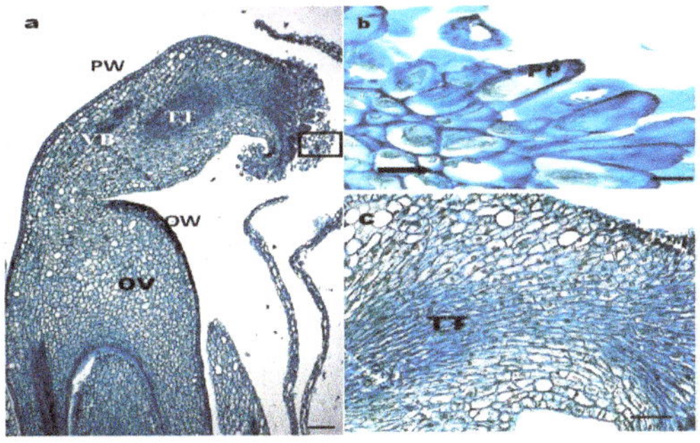

Fig. 10. a) Longitudinal section of a pistil stained with Coomassie Brilliant Blue, Scale bar: 200 μm, b) Details of stigma papillae and inner stigmatic cells. The pellicle and the intercellular spaces show a positive reaction, Scale bar: 10 μm, c) The transmitting tissue is also positive to this stain but shows a strong stained funnel shape, Scale bar: 50 μm. PP: Papillae cells, TT: Transmitting tissue, VB: Vascular bundle, PW: Pistil wall, OW: Ovary wall, OV: Ovary

The transmitting tissue is made-up of elongated, thick-walled cells which are aligned along the longitudinal axis of the pistil. While the cells which form the transmitting tissue are shorter and rounder at the young stage, they lengthen throughout development and become more cylindrical. At the mature stage, vacuolation increases in every cell of the style tissue, and the amount of cytoplasm decreases. The decrement in the stainability of the cells is evident.

Cytochemical tests were applied to distinguish carbohydrates, proteins and lipids within the different tissues of the *L. nobilis* pistil. The cell walls of the pistil stain a pink color with PAS indicating non-soluble polysaccharides (Fig. 6), but staining is more intense in the stylar transmitting tissue and the contiguous non-papillate cells within the stigma (funnel shaped zone) and in the ovary epidermis (Fig. 6). The epidermal cell wall of the style (Fig. 6) and ovary also stain positively with PAS. The pistil vascular bundle shows a stronger reaction with this stain (Fig. 6). The cuticle of the ovary and style are stained with Sudan Black B. Cytochemical analysis indicated that intercellular substances of transmitting tissue show a strong stainability for the total protein (Fig. 10) and insoluble polysaccharides, as revealed by Coomassie Brilliant Blue and PAS reagent, respectively.

Ovary

The ovary is short-stemmed with one chamber with a hanging ovule. The fruit develops on the stem into deep black 2 cm long ovate berries.

Ovule

Ovule development in *L. nobilis* becomes visible subsequent to carpel enlargement. The ovule primordium emerges as a small protuberance of homogenous tissue and it bends in the course of development. The funiculus contains 1-2 vascular bundles, extending to the chalaza.

The ovule is bitegmic. The integuments appear close to the base of nucellus. Although the inner integuments differentiate earlier than the outer integument, the latter overgrows the former. In a fully mature ovule the outer integument is more massive and thicker (5-7 cell layers) as compared to the inner integument (2 cell layers). The micropyle is formed by only the inner integument (endostome). The integuments do not contain vascular bundles.

The ovule is crassinucellate; although, the parietal tissue above the megaspore mother cell is 9-11 cell layers thick, it consists of 16-18 cell layers at maturity. The nucellar tissue can be seen in a mature ovule. The nucellar cells throughout the development contain a large amount of starch grains giving a strong PAS positive reaction.

6. Megasporogenesis and megagametogenesis

Linear megaspore tetrad comes into existence as a result of meiotic division in the megaspore mother cell. The chalazal megaspore is functional. The 2–, 4–, and the 8–nucleated embryo sacs develop from the active megaspore through the sequential mitotic divisions. The mature embryo sac contains one egg cell and 2 synergid cells, 2 polar nuclei and 3 antipodal cells. Synergid cells are rich in cytoplasm, and they degenerate right after fertilization. From the applied PAS staining, it was evident that there was no obvious filliform apparatus.

Three prolonged spindle shaped antipodal cells are located in the chalazal part of the embryo sac. These are smaller compared to the other cells of the embryo sac, and being short-lived; they degenerete and vanish right after fertilization. The central cell contains large vacuoles and the secondary nucleus is round-ellipsoidal shaped. It was identified through the applied PAS reaction that starch grains were observed in the cytoplasm around the secondary nucleus.

7. Endosperm and embryo formation

In *L. nobilis*, the primary endosperm nucleus undergoes subsequent mitotic divisions and forms free nuclear endosperm which persists throughout. In a further stage of development the chalazal end of the endosperm forms a short haustorium reaching up to 0.65 mm in length and it remains free nuclear. In this species, the endosperm is consumed by the developing embryo in the early stage of development and after that the cotyledons support the nutrition of the embryo. The zygote is smaller than the egg cell as a result of the shrinking of the vacuole. Embryogenesis was not observed; however, a mature embryo was only seen in the seed of mature fruit.

8. Seed and fruit formation

The fruit remains as a one-seeded drupa. It matures between October and January, and it is oval-shaped, long-gripped and rigid. First it remains green, and then it turns into red, and ultimately, it becomes glossy black. The embryo is straight and has massive cotyledons in mature fruit. The fruit wall, the outermost epicarp layer of the pericarp, is a dark grey-black colored, rindless, skin-like and straight layer. It consists of thin-walled parenchyma. Mesocarp, the middle layer of the pericarp, is fleshy and thick and it is composed of loosely arranged parenchymatic tissues. The innermost part of the pericarp is the endocarp. The hardened endocarp is derived through the inner area of the ovary wall.

The characteristics of idioblasts

The idioblasts were encountered at every stage of development in the female flower. The idioblasts were evident in every flower part, from the early stage, in which the apical meristem appeared, until the late stages where the fruit tissue came into existence. Although morphological dissimilarity is not evident in idioblasts in various parts of the female flower, it is observed in the idioblasts in young and mature tissues. The idioblasts in young female flower tissues are smaller compared to those at later stages, and cells with big nuclei and vacuoles are located in them. A line of cylindrical shaped cell series are present in every flower tissue, lined up neatly around the idioblasts, in and after the stages where mature embryo sac cells exist.

Vascular bundles

In female flower an ovary, style, staminodes and nectarium contain vascular bundles (Fig. 6). An extensive network of veins feed the nectaria. Furthermore, a distinct vein supplies the staminode even though stamen development is arrested. Vascular bundles in the pistil tissues give the strongest reaction to Coomassie Brilliant Blue, PAS and Sudan Black B, indicating the presence of a high amount of proteins, insoluble polysaccarides and lipids, respectively. Vascular bundles in staminodes and nectarium show strong positive reaction to these stains as well.

Discussion

There are differences between leaf and flower buds which develop out of growing points in *L. nobilis*; leaf buds are pointed and long, and flower buds have small depth and broad expanse. Esau (1977) stated that the vegetation cone in some growing points inside the buds in the beginning of the morphological division period, is broader and swollen, while it is tight and flat in some of them.

At the very beginning of female flower development in *L. nobilis*, the apical meristem develops as a sharp bulge. As in dioecious *Silene latifoloia* (Grant *et al.*, 1994), the apical meristem in *L. nobilis* is composed of uniform cell layers, and young leaves are present on both sides of the apical meristem. The transformation from apical meristem into floral meristem in *L. nobilis* is similar to that in the tomato. The apical meristem encounters many alterations during transformation into tomato: its volume increases, the apical apex comes into existence, the floral meristem appears following the flattening of the apex, and the flower organ primordia differentiate out of the floral meristem (Chandra-Sekhar and Sawhney, 1984; Grant *et al.*, 1994). According to the examinations of Teeri *et al.* (2006), the flattening and broadening of the floral meristem is gained through the increase in the division ratio of the cells in the center, mostly.

Initially, the tepal primordia which exist on the outermost ring are formed out of the floral meristem in *L. nobilis*. Afterwards, the stamen primordia are constituted inside the tepals, and ultimately, the carpel primordium emerges in the center of the female flower. The flower ontogeny of this species is similar to that in Arabidopsis and Antirrhinum. According to this model, the development of flower organ primordia in plants is centripetal and sequential. Firstly, sepal primordia emerge in *Arabidopsis* and *Antirrhinum* flowers. Next, the petals, and immediately following that, the stamen primordia differentiate. Ultimately, the carpel primordium starts developing in the floral meristem center (Smyth *et al.*, 1990; Sommer *et al.*, 1990).

In some species, unisexual flowers might not illustrate any remainder of the atrophic sex. The development in the female flower of *L. nobilis* differs from this situation. For instance, female flowers in *Cannabis sativa* are able to pass directly from the periant development initiation to the start of carpel formation. These flowers do not carry a trace of the stamen primordium (Ram and Nath, 1964). Yet, we observed that filaments which contain 2 nectaries each differentiated in *L. nobilis*; however, anthers did not emerge. Thus; we identified that the staminodes that came into view were effective in the pollination of pistillate flowers.

A similar example to this development of the stamens in *L. nobilis* occurs in *Asparagus officinailis*. The critical phase in the sex determination in dioecious *A. officinalis* arises in the far future stages of flower development. The flower buds in female and male plants can not be distinguished phenotypically until the meiosis initiation (Lazarte and Palser, 1979; Bracale *et al.*, 1991). In this stage, in order for the mature flowers to be unisexual, pollen development is arrested in female flowers, while the embryo sac formation is intercepted in male flowers. One of the breakdowns in stamen maturation in the female flower of *A. officinailis* is the development breakdowns in tapetum cells which emerge in the earlier stage (Lazarte and Palser, 1979). *L. nobilis* sex organ development differs from the development of the *Mercurialis* species, as well. Unisexual flowers in the *Mercurialis* species, which contains monoecious and dioecious species, are missing the sex organ primordia of the opposite sex from the floral meristem stage (Durand and Durand, 1991).

The carpel in *L. nobilis* starts differentiating with a dished appearence in the middle of the flower meristem. A similar development is encountered in the study by Verbeke (1992) on the *Catharanthus roseus*.

Almost all of the characteristics of female gametophyte development in 23 species of the Lauraceae family, which includes *L. nobilis*, were studied. Kimoto and Tobe (2001) demonstrated the embryological characteristics of the Lauraceae family in a reviewed study.

We compared the embryological characteristics of *L. nobilis* to this reviewed study, and we witnessed that a large part of its embryological characteristics were similar to those encountered in most of the Lauraceae members. The similar characteristics are as follows: Firstly, there is an ovary which contains a single anatropous, bitegmic and crassinucellate ovule showing apical placentation (Kimoto and Tobe, 2001). Also, meiosis in the megaspore mother cell results in a linear tetrad of megaspores and the functional megaspore on the chalazal side develops into an 8-nucleate Polygonum type embryo sac.

Antipodal cells are ephemeral. The nucellar tissue surrounding the mature embryo sac remains until the post–fertilization stages. The endosperm is nuclear. No obturator develops at the tip of the outer integument. A micropyle is formed by the inner integument alone. Finally, no endothelium is formed (Kimoto and Tobe, 2001).

We noticed that some characteristics which are seldom encountered in the Lauraceae family were formed in *L. nobilis*. These characteristics are as follows: The parietal cells are 9-12 layered in *L. nobilis*, but in Lauraceae a parietal tissue above the megaspore mother cell is usually 2 to 6 cells thick. In Lauraceae family, usually a 2 to cell-layered nucellar cap is formed by the nucellar epidermis, but this does not exist in *L. nobilis*, as in Cassytha. The occurence of a haustorium is not common in this family; it was reported in *Cryptocarya chinensis* only. We observed a short chalazal endosperm haustorium which is free nuclear in *L. nobilis*, as well.

Idioblasts (fat cells) are encountered at every stage of development in the female flower in *L. nobilis*. Although their taxonomic values have grabbed the attention of botanicians for a long time (Baas and Gregory, 1985), very little information is available about the idioblasts (Platt and Thomson, 1992; Lersten and Curtis, 1993). It is known that their fat cells generate in various plant families such as Araceae, Aristolochiaceae, Calycanthaceae, Lauraceae, Magnoliaceae, Piperaceae and Aururaceae (Fahn, 1979; Baas and Gregory, 1985).

Idioblasts in *L. nobilis* were identified in every flower part, between the young phases where apical meristem tissues developed and the late stages in which the fruit tissue developed. Idioblasts are found intensely in the fruit wall in the Lauraceae family, especially in the mesocarp (Stefan *et al.*, 2009).

There are few studies about the structure and development of the style and stigma in the Lauraceae family members which contain *L. nobilis*. In this species, the capitate stigma (Ghrabi, 2005) is papillate and of dry type (Watson and Dallwitz, 1996).

The pellicle layer surrounding the stigma papil in *L. nobilis* reacts positively with Coomassie Brilliant Blue, which is an indication that it contains protein. Extracellular proteins in the dry stigma are present in the form of an extracuticlar hydrated layer, the pellicle (Mattsson *et al.*, 1974).

In *L. nobilis* the inner stigmatic cells are continuous with the stylar transmitting tissue showing an overall funnel-shaped arrangement which is wider in the stigma and narrows basipetally in the style. In this manner, the transmitting tissue in *L. nobilis* resembles that of the tomato as observed by Kadej *et al.* (1985), who showed it beginning as a broad column in the stigma and gradually narrowing in the style. It has been postulated that the architecture of the stigma and the style play a role in determining the pollen tubes that arrive at the ovule (Heslop-Harrison *et al.*,1985). A development similar to the stylar structure in *L. nobilis* is demonstrated in a study that was carried out in *Olea europaea* L. (Serrano *et al.*, 2008).

Studies conducted on various angiosperm species have shown a correlation between flower abortion and starch content. A strong correlation has often been noted in woody plants between the successful development of fertile sexual organs and the amount of starch available in the flower at various stages of development (Rodrigo *et al.*, 2000; Jean and Lapointe, 2001).

Histochemical analysis of sugars, proteins and lipids within and around vascular bundles in the pistil of *L. nobilis* suggests that they are involved in the transport of these substances, as observed in olives (Serrano *et al.*,2008). In this study (Serrano *et al.*, 2008), the accumulation places of starch in staminate and hermaphrodite flowers are given in comparison, and ultimately it is pointed out that the male flower, in which pistil development is completed, differed from the starch dispersion in hermaphrodite flower, where the fertile pistil is developed.

Starch distribution in the entire pistil in *L. nobilis* was investigated, and although a strong PAS positive reaction was witnessed, it was observed that starch grains did not exist in the style or ovary. At every stage of development, the starch grains dispersed only in the nusellus in an intense manner, being present in a limited number in the nectarium, as well.

The results of this study provide important information about how the sex organ differentiates and develops while the other's development arrests. Moreover, our research contributes to the reproductive biology of this species that can be used in breeding programs.

Acknowledgments

This work was supported by the Research foundation of Marmara University (BAPKO no: FEN–C–YLP–181208–0291).

References

Atsmon D, Galun E (1960). A morphogenetic study of staminate, pistillate and hermaphrodite flowers in *Cucumis sativus* L. Phytomorphology 10:110-115.

Baas P, Gregory MA (1985). Survey of oil cells in the Dicotyledons with comments on their replacement by and joint occurrence with mucilage cells. Israel J of Bot 34:167-186.

Bernier G, Havelange A, Houssa C, Petitjean A, Lejeune P (1993). Physiological Signals That Induce Flowering. Plant Cell 5:1147-1155.

Bracale M, Caporali E, Galli MG, Longo C, Marziani-Longo G, Tassi F (1991). Sex expression and determination in *Asparagus officinalis*. Plant Science 80:67-77.

Chandra SKN, Sawhney VK (1984). A scanning electron microscope study of the development and surface features of floral organs of tomato (*Lycopersicon esculentum*). Canadian J of Bot 62:2403-2413.

Dellaporta SL, Calderon-Urrea A (1993). Sex determination in flowering plants. Plant Cell 5(10):1241-1251.

Durand B, Durand R (1991). Sex determination and reproductive organ differentiation in Mercurialisi. Plant Science 80:49-65.

Esau K (1977). Anatomy Of Seed Plants. Newyork-USA, John Wiley & Sons Press.

Fahn A (1979). Ultrastructure of Nectaries in Relation to Nectar Secretion. American J of Bot 66:977-985.

Feder N, O'Brien TP (1968). Plant microtechnique: some principles and new methods. American J of Bot 55:123-142.

Fisher BB (1968). Protein staining of ribbonedepon sections for light microscopy. Histochemie 16:92-96.

Grant S, Houben A, Vyskot B, Siroky J, Pan WH, Macas J, Saedler H (1994). Genetics of sex determination in flowering plants. Dev Genet 15:214-230.

Ghrabi Z (2005). Morphological description. http://www.uicnmed.org/nabp/database /HTM/PDF/p40.pdf.

Heslop-Harrison JS, Heslop-Harrison J, Heslop-Harrison Y, Reger BJ (1985). The distribution of calcium in the grass pollen tube. Proceedings of the Royal Society B 225:315-327.

Haughn GW, Somerville CR (1988). Genetic control of morphogenesis in Arabidopsis. Dev. Genet 9:73-89.

Jean D, Lapointe L (2001). Limited carbohydrate availability as a potential cause of fruit abortion in *Rubus chamaemorus*. Physiol Plantarum 112:379-387.

Kadej AJ, Wilms HJ, Willemse MTM (1985). Stigma and stigmatoid tissue of *Lycopersicon esculentum*. Acta Bot Neerl 34:95-104.

Kimoto Y, Tobe H (2001). Embryology of Laurales: a review and perspectives. J Plant Res 114:247-261.

Kinet JM, Sachs RM, Bernier G (1981). The physiology of flowering I., The initiation of Flowers, Vol. III. USA, CRC Press Inc.

Lazarte JE, Palser BF (1979). Morphology, vascular anatomy, and embryology of pistillate and staminate flowers of *Asparagus officinalis*. American J Bot 66:753-764.

Lersten NR, Curtis JD (1993). Subepidermal idioblasts in leaflets of *Caesalpinia pulcherrima* and *Parkinsonia aculeata* (Leguminosae; Caesalpinioideae). Bull Torrey Bot Club 120(3):319-326.

Malepszy S, Niemirowicz Szczytt K (1991). Sex determination in cucumber (*Cucumis sativus*) as a model system for molecular biology. Plant Science 80:39-47.

Mattsson O, Knox RB, Heslop-Harrison J, Heslop-Harrison Y (1974). Protein pellicle of stigmatic papillae as a probable recognition site in incompatibility reactions. Nature 213:703-704.

Pearse AGE (1961). Histochemistry-Theoretical and Applied. Boston, J & A Churchill.

Platt Aloia KA, Oross JW, Thomson WW (1983). Ultrastructural study of the development of oil cells in the mesocarp of avocado fruit. Bot Gaz 144(1):49-55.

Platt KA, Thomson WW (1992). Idioblast oil cells of avocado: distribution, isolation, ultrastructure, histochemistry, and biochemistry. International J Plant Sci 153:301-310.

Ram HY, Nath R (1964). The morphology and embryology of *Cannabis sativa* Linn. Phytomorphology 14:414-429.

Rodrigo J, Hormaza JI, Herrero M (2000). Ovary starch reserves and flower development in apricot (*Prunus armeniaca*). Physiol Plantarum 108:35-41.

Rastogi R, Sawhney VK (1988). Suppression of stamen development by CCC and ABA in tomato floral buds cultured in vitro. J Plant Physiol 133:620-624.

Sekhar KN, Sawhney VK (1984). A scanning electron microscopic study of the development and surface features of floral organs of tomato (*Lycopersicon esculentum*). Canadian J Bot 62:2403-2413.

Serrano I, Suárez C, Olmedilla A, Rapoport HF, Rodríguez García MI (2008). Structural organization and cytochemical features of the pistil in Olive (*Olea europaea* L.) cv. Picual at anthesis. Sexual Plant Reproductive 21:99-111.

Smyth DR, Bowman JL, Meyerowitz EM (1990). Early flower development in Arabidopsis. Plant Cell 2:755-767.

Sommer H, Beltran J-P, Huijser P, Pape H, Lönnig W-E, Saedler H, Schwarz Sommer Z (1990). Deficiens, a homeotic gene involved in the control of flower morphogenesis in *Anthirrinum majus*: the protein shows homology to transcription factors. The EMBO J 9:605-613.

Stefan A, Ruth L, Stockey A, Penner B (2009). Anatomy and development of fruits of Lauraceae from the middle eocene princeton chert. American J Bot 96(3):637-651.

Tanurdzic M, Banks J.A 2004. Sex–Determining Mechanisms in Land Plants. Plant Cell 16:S61-S71.

Teeri TH, Uimari A, Kotilainen M, Laitinen R, Help H, Elomaa P, Albert VA (2006). Reproductive meristem fates in Gerbera. Oxford J 57:3445-3455.

Topçuoğlu N, Selvi N, Dokumacı E (2009). Fare dokularının taramalı elektron mikroskop için preparasyonunda kritik nokta kurutma ve hekzametildisilazan ile kurutma yöntemlerinin karşılaştırılması, II– akciğer ve mide bulguları. Anadolu üniversitesi bilim ve teknoloji dergisi 10:127-132.

Verbeke JA (1992). Fusion Events During Floral Morphogenesis. Ann Rev Plant Physiol 43:583-598.

Watson L, Dallwitz MJ (1996). Lauraceae in The families of flowering plants. Website http://biodiversity.uno.edu/delta/.(2010)

Ye ZH, Varner JE (1991). Tissue-Specific Expression of Cell Wall Proteins in Developing Soybean Tissues. Plant Cell 3:23-37.

4

Anatomical and Palynological Studies on *Chaerophyllum astrantiae* and *C. aureum* in Turkey

Gülden YILMAZ[1*], Mehmet TEKIN[2]

[1]*Trakya University, Faculty of Science, Department of Biology, Balkan Campus, 22030, Edirne/Turkey*
[2]*Cumhuriyet University, Faculty of Science, Department of Biology, 58140, Sivas, Turkey;*
*guldenyilmaz2009@yahoo.com (*corresponding author), mtekin2280@gmail.com*

Abstract

In this study, anatomical and palynological properties of *Chaerophyllum astrantiae* Boiss. & Bal. and *C. aureum* L. were determined for the first time. Cross sections of roots, stems, stem leaves and fruits of both species were examined. The leaves are amphistomatic and mesophyll is bifacial. The fruit has two vittae on commissural face and four vittae on lateral and dorsal faces in both species. The palynological studies revealed that, *C. astrantiae* and *C. aureum* have tricolporate pollen type and perforate exine ornamentation under light microscope. Ratio of polar axis to equatorial axis is 2.15 in *C. astrantiae* and 2.05 in *C. aureum*. Thus, pollen shape is perprolate for both of the species.

Keywords: anatomy, *Chaerophyllum*, palynology, Turkey

Introduction

Apiaceae (or Umbelliferae) commonly known as carrot or parsley family is a family of mostly aromatic plants with hollow stems. It is one of the largest plant family in the world, with 446 genus and 3540 species (Simpson, 2006). Apiaceae is the 8th largest family in Turkey with approximately 455 species among which 33% are endemic (Davis *et al.*, 1988; Guner *et al.*, 2000; Ozhatay *et al.*, 2008-2009, 2011). Apiaceae is solidly placed within the Apiales order in the APG III classification system. Traditionally groups within the family have been delimited largely based on fruit morphology (Stevens, 2001). The family is rich in secondary metabolites and embodies numerous genera of high economic and medicinal value yielding flavonoids, essential oils, terpens and coumarins. It is well-known that occurence of oleoresins and essential oils is a characteristic feature of the family. *Chaerophyllum* the largest genus in the subtribe Scandicinae, comprising over 30 species (Downie *et al.*, 2000). It is native to Europe, Asia, North America and northern Africa. Based on a DIVA analysis Chung *et al.* (2005) considered that Eurasia was the ancestral area of the *Scandicinae*. All *Chaerophyllum* species except *C. procumbens* and *C. tainturieri*, are native to Eurasia and North Africa. Spalik and Downie (2001) suggested four sections in *Chaerophyllum* based on ITS internal transcribed spacer and morphological data. The genus is comprised of about 45 species in the world, 35 species in Asia and 16 species in Turkey (Pimenov and Leonov, 2004) and 4 of these species (*C. byzantium* Boiss, *C. aureum* L., *C. bulbosum* L. and *C. temulum* L.) are grown in European part of Turkey (Hedge and Lamond, 1972). In this context, *C. aureum* is widely spread in Turkey, while *C. astrantiae* is locally distributed especially in Rize and Artvin province in Turkey. In recent years, a number of studies have been performed on *Chaerophyllum* L. with particular reference to composition of essential oils (Razavi and Nejad-Ebrahimi, 2010; Chizzola, 2009; Nematollahi *et al.*, 2005; Baser *et al.*, 2000), antimicrobial activity of essential oils (Lakusić *et al.*, 2009), antioxidant compounds (Dall'Acqua and Innocenti, 2004) and flavonoid glycosides (Gonnet, 1985). However, there is not enough study on anatomical and palynological data of this genus. Anatomical structure of root, stem, stem leaves and fruit are very important as biochemical composition of the genus *Chaerophyllum* L. The aim of this study is to observe the anatomical and palynological structures of *C. astrantiae* and *C. aureum*.

Material and methods

The specimens used in this study were collected from natural habitats in Rize and Artvin province (Turkey). Localities of *C. astrantiae*: Artvin province, Borcka, Karcal Mountain, near Adagul upland, N 41° 20' 32.7" E 41° 58' 18.3", 2915 m, 04.08.2010, M. Tekin 1035; Artvin province, Karcal mountain, N 41° 21' 29.1" E 41° 58' 08.7", 2557 m, 02.09.2010, M. Tekin 1050. Localities of *C. aureum*: Rize province, İkizdere to Cimil upland, near stream, N 40° 44' 02.9" E 40° 47' 45.9", 2062 m, 01.08.2010, M. Tekin 1018; Artvin province, Şavşat, Karagöl, Cinoba site, alpine meadow, N 41° 19' 26.6" E 42° 29' 56.4", 1763 m,

02.08.2010, M. Tekin 1019. Some specimens were preperad as herbarium materials and deposited in Cumhuriyet University Herbarium (CUFH).

For anatomical studies, the root, stem, stem leaf and fruit of the some specimens were fixed and conserved in 70% ethyl alcohol. Hand sections were made with a razor blade and sections were stained with Alcian blue (Sigma) for pectic substances, Safranin (Sigma) for lignin in the rate of 3/2 (Yılmaz and Dane, 2013). Sections were left in dye about 5 minutes for staining. The stained sections were mounted in glycerin-gelatine to obtain permanent preparations (Jensen, 1962). Sections were examined using an Olympus light microscope BX51. Images were taken with an Olympus DP 70.

For palynological investigations pollen grains supplied from herbarium specimens and were prepared following (Wodehouse, 1935). For measurements of polar axis, equatorial diameter, exine thickness and lumina size of pollen grains, 100x objective of a Olympus light microscope CX21 were used. Measurements were based on 20 pollen grains. For morphological descriptions of pollen grains, the terminology of Punt *et al.* (1994) was followed. Statistical studies were performed using SPSS 15.0 (SPSS Inc., Chicago, IL, USA).

Results

Morphological properties

C. astrantiae Boiss. & Bal. in Boiss., Fl. Or. 2: 906 (1872) (Hedge and Lamond, 1972) (Figures 1a-9a).

Pilose to ± glabrous perennial with terete stems, branched or not, 30-60 cm. Lower leaves broad ovate-triangular, 10-20 × 7-8 cm, 2-3 pinnate; ultimate segments ovate, deeply pinnatisect with ± linear lanceolate lobes 2-6 × 0.5-1.0 mm. Rays 6-11, ± equal, 1-3 cm, erect. Bracts 0-1. Bracteoles 4-6, conspicuous, white, ciliate, broad elliptic to obovate, 5-9 × 3-4 mm. Pedicels shorter than bracteoles. Fertile flowers 4-10 per umbellule. Petals ± equal. Fruit oblong, 5-8 × c. 1.5 mm. *Fl. 8. Rocky places, 2130-3200 m.*

C. aureum L., Sp. Pl. ed. 2:370 (1762) (Hedge and Lamond, 1972) (Figures 1b-9b).

Almost glabrous to shortly pilose perennial with branched terete or angled, often spotted stems to c. 90 cm. Lower leaves broad-ovate to triangular, 2-3-pinnate, 15-30 × 9-20 cm, ultimate segments ovate, elliptic or broadly lanceolate, acutely lobed, lobes crenate or sharply serrate. Rays 15-25, unequal, 1.5-5 cm, ± erect in fruit. Bracts 0-1. Bracteoles 5-7, lanceolate, ciliate, 3-6 × 1 mm. Pedicels shorter than fruit. Fertile flowers 5-8 (-12) per umbellule. Fruit oblong to ovate-oblong, 7-10 × c. 3 mm, pale brown when mature. *Fl. 6-8. Rocky slopes in Picea-Rhododendron scrub, open grassy meadows, 1000-2300 m.*

Anatomical properties

In root cross sections, roots have secondary structure. Periderm consists of three different tissues: phellem with depressed rectangular cells (3-7 layers in *C. astrantiae*, 5-7 layers in *C. aureum*), phellogen (1 layered in both species) and phelloderm (1 layered in both species), from exterior to interior.

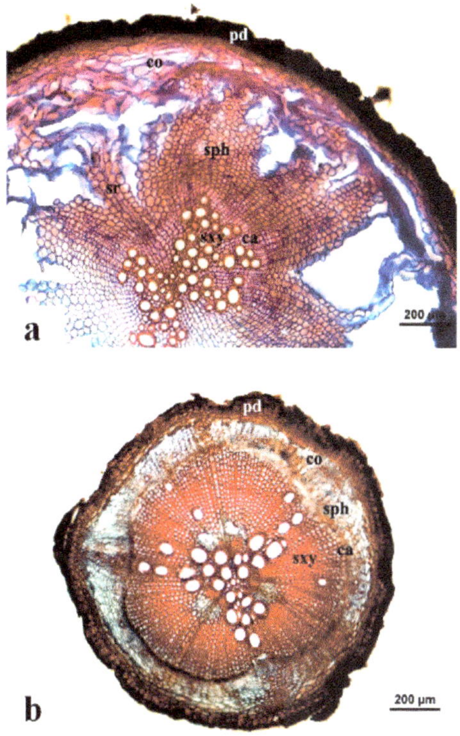

Fig. 1. Cross section of roots: a. *C. astrantiae*; b. *C. aureum* (ca: cambium; co: cortex; pd: peridermis; sph: seconder phloem; sxy: seconder xylem) (Scale 200 µm)

Cortex consists of 3-6 layered parenchymatic tissue in *C. astrantiae* and 4-7 layered parenchymatic tissue in *C. aureum*. Parenchymatic tissues have thickness of 125-288 µm in *C. astrantiae* and 50-115 µm in *C. aureum*. Cambium consists of 2-3 layers in *C. astrantiae* and 3-4 layers in *C. aureum*. In cortex, there are secretory ducts surrounded by 4-8 glandular cells in both of the species (Figs. 1-2).

When cross sections of stems were examined, it is determined that thickness of cuticle were found to be 5-8 µm in *C. astrantiae* and 3-5 µm in *C. aureum*. The epidermis consists of ovoidal cells with 15-24 × 6-10 µm in size in *C. astrantiae* and 12-30 × 10-18 µm in size in *C. aureum*. Under epidermis, parenchymatous cortex is 7-9 layered in *C. astrantiae* while 11-14 layered in *C. aureum*. Cells of cortex are 15-38 × 7-20 µm in size in *C. astrantiae* while 10-55 × 7-45 µm in size in *C. aureum* and are ovoidal in shape. In cortex, it is determined that collenchyma cells have 6-9 layers in *C. astrantiae* and 10-13 layers in *C. aureum*. There are secretory ducts surrounded by 4-8 glandular cells in *C. astrantiae* and 6-13 glandular cells in *C. aureum*.

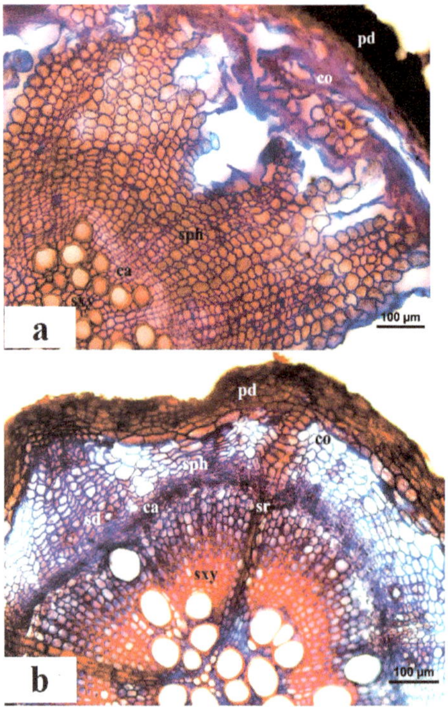

Fig. 2. Cross section of roots in detail: a. *C. astrantiae*; b. *C. aureum* (co: cortex ca: cambium; pd: peridermis; sph: seconder phloem; sr: seconder pith ray; sxy: seconder xylem;) (Scale 100 µm)

Center of the stem (pith region) is made of circular to circular-ovoid parenchymatic cells which have the sizes of 12-93 × 10-82 µm in *C. astrantiae* and 48-125 × 37-105 µm in *C. aureum* (Figs. 3-4).

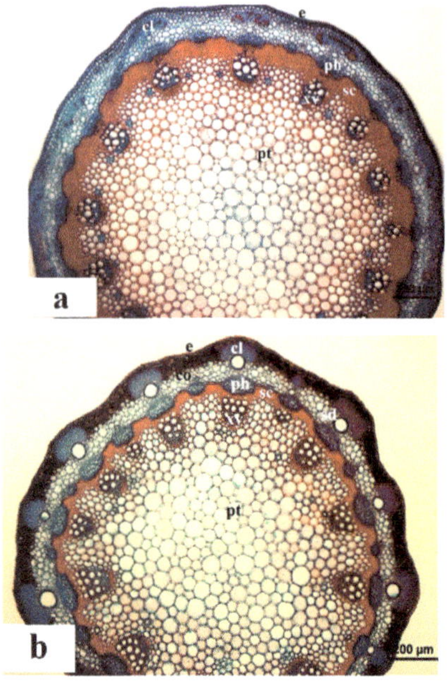

Fig. 3. Cross section of stems: a. *C. astrantiae*; b. *C. aureum* (cl: collenchyma; co: cortex; e: epidermis; ph: phloem; pt: pith; sc: sclerenchyma; sd: secretory duct; xy: xylem; xys: sclerenchyma of xylem) (Scale 200 µm)

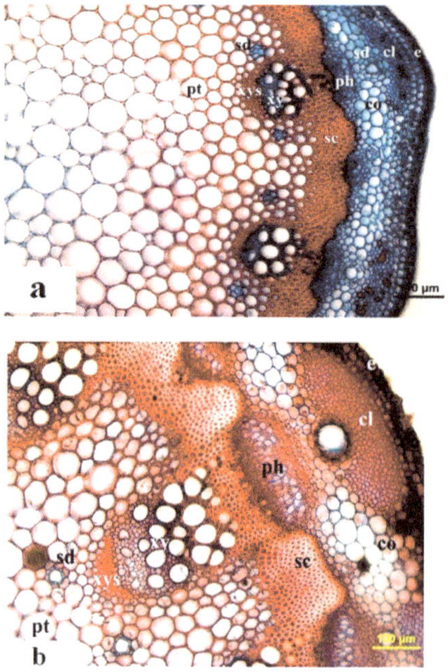

Fig. 4. Cross section of stems in detail: a. *C. astrantiae*; b. *C. aureum* (cl: collenchyma; co: cortex; e: epidermis; ph: phloem; pt: pith; sc: sclerenchyma; sd: secretory duct; xy: xylem xys: sclerenchyma of xylem) (Scale 100 µm)

In stem leaf anatomy, it was observed that the shape of epidermal cells are oblong-ovoidal in both of the species. There are unicellular hairs on leaf surface rarely in *C. aureum*. Upper epidermis cells are 15-25 × 6-12 µm in size in *C. astrantiae* and 12-55 × 7-30 µm in *C. aureum*. The leaves are amphistomatic and mesophyll is bifacial in both species (Fig. 5).

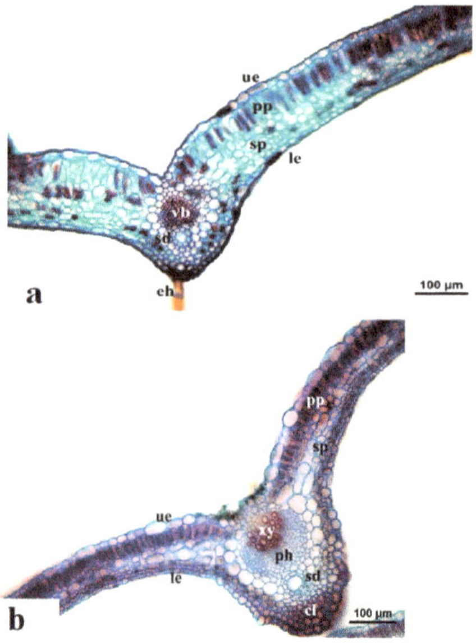

Fig. 5. Cross sections of stem leaves: a. *C. astrantiae*; b. *C. aureum* (cl: collenchyma; eh: eglandular hair; le: lower epidermis; ph: phloem; pp: palisade parenchyma; sd: secretory duct; sp: spongy parenchyma; ue: upper epidermis; vb: vascular bundle; xy: xylem) (Scale 100 µm)

Palisade parenchyma cells generally clyndirical in shape are 1 layered and have the sizes of 35-55 × 7-13 μm in *C. astrantiae* and 17-38 × 7-15 in *C. aureum*. Spongy parenchyma cells are 4-6 layered and 10-38 × 7-23 μm in size in *C. astrantiae* while 3-4 layered and 10-50 × 7-27 μm in *C. aureum*. They are circular, ovoid or rarely irregular shaped. There are secretory ducts consists of 9-13 cells in areas corresponding to the main vascular bundle between lower epidermis and vascular bundle in both species (Figs. 5-6).

Fig. 6. Cross sections of stem leaves in detail: a. *C. astrantiae*;

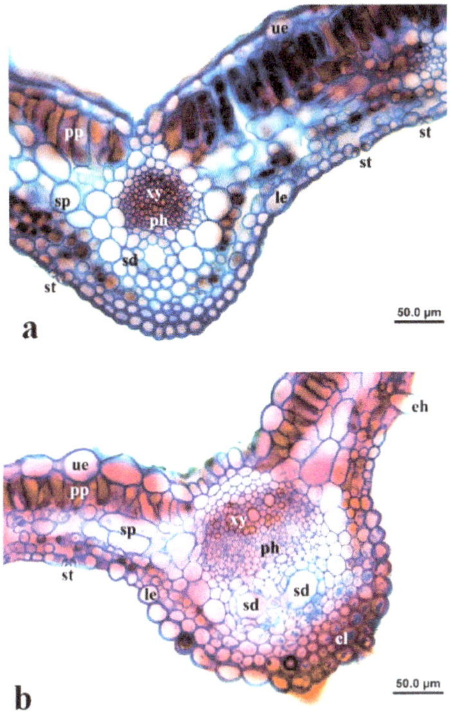

b. *C. aureum* (cl: collenchyma; eh:eglandular hair; le: lower epidermis; ph: phloem; pp: palisade parenchyma; sd: secretory duct; sp: spongy parenchyma; st: stoma; ue: upper epidermis; vb: vascular bundle; xy: xylem) (Scale 50 μm)

Fruits are schizocarpic which made of two mericarp in both of the examined species. (Fig. 7).

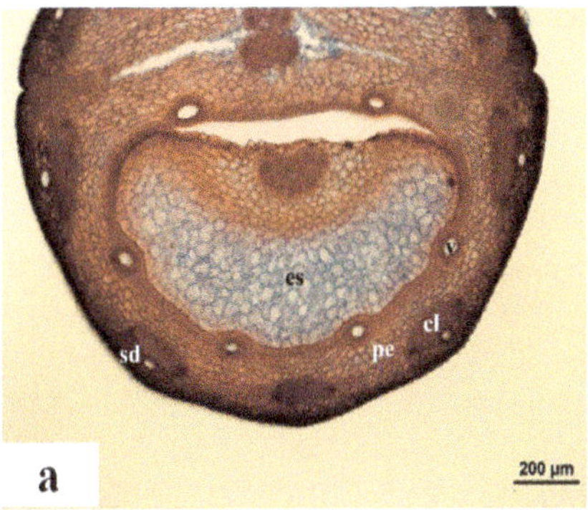

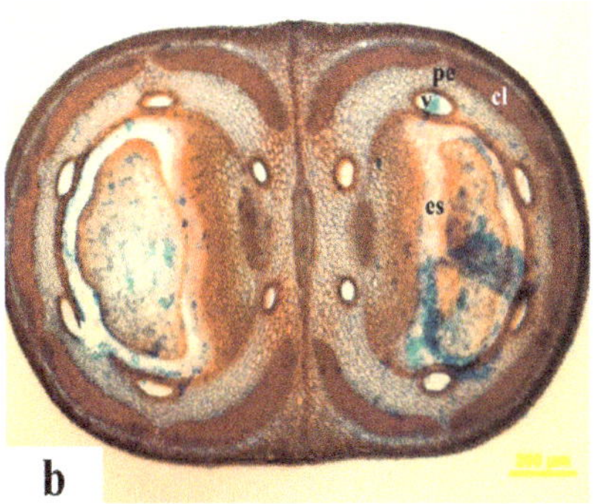

Fig. 7. Cross sections of fruit: a. *C. astrantiae*; b. *C. aureum* (cl: collenchyma; es: endosperm; pe: pericarp; sd: secretory duct; v: vittae) (Scale 200 μm)

Exocarp layer have oblong-ovoidal cells with 15-40 × 7-13 μm in size in *C. astrantiae* and 7-25 × 5-10 μm in *C. aureum*. Above the exocarp cuticula is 3-5 μm in size in both species. Below the exocarp there is mesocarp layer which consists of oblong-ovoidal, pentagonal or hexagonal cells. Mesocarp is 10-14 layered in *C. astrantiae* while 17-23 layered in *C. aureum*.

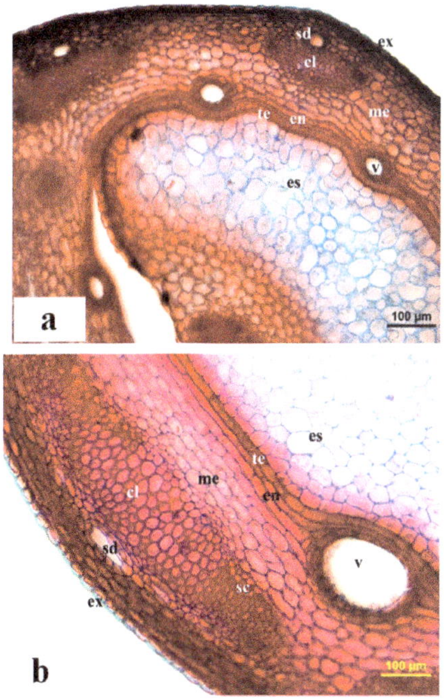

Fig. 8. Cross sections of fruit mericarp in detail: a. *C. astrantiae*; b. *C. aureum* (cl: collenchyma; en: endocarp; es: endosperm; ex: exocarp; me: mesocarp; sc: sclerenchyma; sd: secretory duct; te: testa) (Scale 100 μm)

In the middle of the mesocarp layer there are 7-9 layered collenchyma in *C. astrantiae* and 4-9 layered collenchyma in *C. aureum*. Inner layer of pericarp that endocarp is single layered in both species and endocarp consists of very pressed rectangular cells which have the sizes of 37-90 × 5-10 μm in *C. astrantiae* and 45-74 × 6-9 μm in *C. au-*

reum. Between the mesocarp and endocarp layer 2 vittae (oil duct) are seen on the commissural face while 4 vittae are seen on lateral and dorsal faces in both of the species. Diameter of vittae opening in *C. aureum* is larger than *C. astrantiae*. Testa is single layered in both species and cells are 12-43 × 12-18 µm in size in *C. astrantiae* while 15-50 × 5-13 µm in *C. aureum* (Figs. 7-8).

Palynological properties

In palynological study, *C. astrantiae* and *C. aureum* have tricolporate pollen type, perprolate pollen shape and perforate exine ornamentation. Under the pore there is a costa nearly 1.01 µm in pollen grains of both species. The ratio of P/E is aproximately 2.10 making the pollen grain shape perprolate (Erdtman, 1952) in both species. Exine surface ornamentation is coarsely perforate under light microscope. (Fig. 9, Tab. 1).

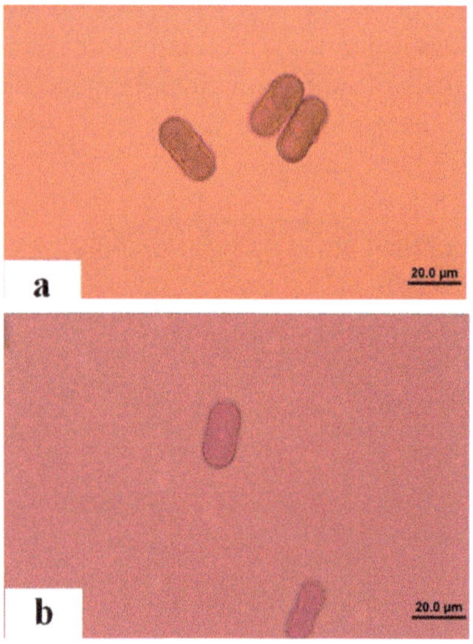

Fig. 9. Morphology of pollen grains under light microscope: a. *C. astrantiae*; b. *C. aureum* (Scale: 20.0 µm)

Discussion

There are many studies about chemical properties of the genus *Chaerophyllum* L., mainly on the composition of essential oils in *C. macrospermum* (Spreng.) Fisch CA Mey, *C. aromaticum*, *C. macropodum* Boiss, *C. critinum* Boiss, *C. aksekiense* (Razavi and Nejad-Ebrahimi, 2010; Chizzola, 2009; Nematollahi *et al*., 2005; Baser *et al*., 2000), antimicrobial activity of essential oils in *C. aureum* (Lakusić *et al*., 2009), antioxidant compounds of *C. hirsutum* (Dall'Acqua and Innocenti, 2004) and flavonoid glycosides in *C. aureum* (Gonnet, 1985). In recent years, there were lots of studies about Apiaceae. Including morphology, anatomy and palynology of some genus in the family. In 2005, Applequist examined the root anatomy of *Ligusticum* species (Apiaceae) which was sold as Osha. The herbal medicine Osha consisted of the roots of *Ligusticum porteri* J. M. Coult. & Rose (Apiaceae). In that study, the anatomy of taproots and rootstocks of several species of *Ligusticum* was described and illustrated (Applequist, 2005). The wood anatomy of *Bupleurum* L. (Apioideae, Apiaceae) in relation to habit, phylogenetic relationships, and infrageneric taxonomy was studied (Stepanova and Oskolski, 2010). Morphological, carpo-anatomical characters, pollen characteristics, mericarp surface features of *Bilacunaria anatolica* sp. nov. (Apiaceae) from southwest Anatolia in Turkey were studied (Duran *et al*., 2012). In recent years, the new monotypic genus *Siculosciadium* (Apiaceae) from Sicily was determined (Brullo *et al*., 2013). Uruşak and Kızılarslan (2013) studied the comparative fruit anatomical properties based on mericarp shape in transverse sections of 27 *Ferulago* species (Apiaceae) grown in Turkey. The studies on anatomical and palynological studies on *Chaerophyllum* taxa both in the world and in Turkey are not much in number. *Chaerophyllum* species including antioxidant compunds and active in terms of microbial activity are pharmacologically important taxa. Palynological and anatomical characteristics of the species rich in essential oils should be defined clearly to make comparative studies between taxa. The present study is therefore important as it provides comparative data to be used in studies within the genus.

Tab. 1. Palynological properties of *C. astrantiae* and *C. aureum* (P: polar axis, E: equatorial axis, Clg: colpus length, Clt: colpus width, Plg: pore length, Plt: pore width)

Pollen features	Mean ± Standard Deviation (Min-Max) (µm)	
	C. astrantiae	*C. aureum*
P	25.87 ± 0.69 (25.37–27.40)	23.95 ± 0.76 (22.33–25.37)
E	12.02 ± 0.37 (11.16–12.18)	11.67 ± 0.52 (11.16–12.18)
P/E ratio	2.15	2.05
Pollen shape	Perprolate	Perprolate
Pollen type	Tricolporate	Tricolporate
Clg	13.09 ± 0.92 (11.16–15.22)	12.07 ± 1.08 (10.15–13.19)
Clt	1.72 ± 0.25 (1.52–2.03)	1.64 ± 0.40 (1.01–2.03)
Plg	3.49 ± 0.32 (3.04–4.06)	3.29 ± 0.26 (3.04–3.55)
Plt	5.47 ± 0.51 (5.07–6.09)	5.24 ± 0.24 (5.07–5.58)
Exine thickness in equatorial axis	1.01 ± 0.00 (1.01–1.01)	1.52 ± 0.00 (1.52–1.52)
Exine thickness in polar axis	1.01 ± 0.00 (1.01–1.01)	1.77 ± 0.26 (1.52–2.03)
Ornamentation type	coarsely perforate	coarsely perforate
Costa	1.01 ± 0.00 (1.01–1.01)	1.01 ± 0.00 (1.01–1.01)

Conclusion

This study is important for the taxonomy of *Chaerophyllum* genus. This is the first report on the anatomy and detailed palynology of *C. aureum* and *C. astrantiae*. In this study, root, stem, stem leaf and fruit anatomy and palynological characters of *C. astrantiae* and *C. aureum* were investigated in detail for the first time.

References

Applequist WL (2005). Root anatomy of *Ligusticum* species (Apiaceae) sold as Osha compared to that of potential contaminants. Journal of Herbs, Spices & Medicinal Plants 11(3):1-11.

Baser KHC, Tabanca N, Özek T, Demirci B, Duran A, Duman H (2000). Composition of the essential oil of *Chaerophyllum aksekiense* A. Duran et Duman, a recently described endemic from Turkey. Flavour and Fragrance Journal 14(5):287-289.

Brullo C, Brullo S, Downie SR, Danderson CA, Giusso del Galdo G (2013). *Siculosciadium*, A new monotypic genus of Apiaceae from Sicily, Annals of the Missouri Botanical Garden 99(1):1-18.

Chizzola R (2009). Composition of the essential oil of *Chaerophyllum aromaticum* (Apiaceae) growing wild in Austria. Nat Prod Commun 4(9):1235-1238.

Chung KF, Peng CI, Downie SR, Spalik K, Schaal BA (2005). Molecular systematics of the Trans-Pacific alpine genus *Oreomyrrhis* (Apiaceae): phylogenetic affinities and biogeographic implications. Am J Bot 92:2054-2071.

Dall'Acqua S, Innocenti G (2004). Antioxidant compounds from *Chaerophyllum hirsutum* extracts. Fitoterapia 75(6):592-595.

Davis PH, Mill RR, Tan K (1988). Flora of Turkey and the East Aegean Islands, vol:10 (Supplement 1). Edinburgh University Press, Edinburgh, 145-154 p.

Downie SR, Katz-Downie DS, Spalik K (2000). A phylogeny of Apiaceae tribe Scandiceae: evidence from nuclear ribosomal DNA internal transcribed spacer sequences. Am J Bot 87:76-95.

Duran A, Öztürk M, Ay H (2012). *Bilacunaria anatolica* sp. nov. (Apiaceae) from southwest Anatolia, Turkey. Nord J Bot 29(6):652-659.

Erdtman G (1952). Pollen Morphology and Plant Taxonomy, Angiosperms. Almquist and Wiksell. 150-170. Stockholm.

Gonnet JF (1985). Individual variation of flavonoid glycosides in *Chaerophyllum aureum*. Biochem Sys Ecol 13(3):313-317.

Guner A, Ozhatay N, Ekim T, Baser KHC (2000). Flora of Turkey and The East Aegean Islands, vol:11 (Supplement 2). Edinburgh University Press, Edinburgh, 136-147 p.

Hedge IC, Lamond JM (1972). *Chaerophylum* L. 312-318 p. In: Davis PH (Eds.). Flora of Turkey and East Aegean Islands, vol. 4. Edinburgh University Press, Edinburgh.

Jensen WA (1962). Botanical histochemistry: principles and practice. WH Freeman and Company, London.

Lakusić B, Slavkovska V, Pavlović M, Milenković M, Stanković JA, Couladis M (2009). Chemical composition and antimicrobial activity of the essential oil from *Chaerophyllum aureum* L. (Apiaceae). Nat Prod Commun 4(1):115-118.

Nematollahi F, Akhgar MR, Larijani K, Rustaiyan A, Masoudi S (2005). Essential oils of *Chaerophyllum macropodum* Boiss. and *Chaerophyllum crinitum* Boiss. from Iran. J of Essent Oil Research 17(1):71-72.

Ozhatay N, Akalın E, Ozhatay E, Unlu S (2008-2009). Rare and endemic taxa of Apiaceae in Turkey and their conservation significance. J Fac Pharm 40:1-10.

Ozhatay N, Kultur S, Gurdal MB (2011). Check-list of additional taxa to the supplement Flora of Turkey V. Turk J Bot 35:589-624.

Pimenov, MG Leonov MV (2004). The Asian Umbelliferae Biodiversity Database (ASIUM) with Particular Reference to South-West Asian Taxa. Turk J Bot 28:139-145.

Punt W, Blackmore S, Nilsson Le Thomas A (1994). Glossary of Pollen and Spores Terminology. Lab. Paleobot. Palynol., Utrecht.

Razavi SM, Nejad - Ebrahimi S (2010). Essential oil composition of *Chaerophyllum macrospermum* (Spreng.) Fisch C.A. Mey seeds. J Essent Oil Bear Pl 13(2):205-210.

Simpson MG (2006). Plant Systematics. Elsevier Academic Press. p. 322.

Spalik K, Downie SR (2001). The utility of morphological characters for inferring phylogeny in Scandiceae subtribe Scandicinae (Apiaceae). Ann Missouri Bot Gard 88:270-301.

Stepanova AV, Oskolski AA (2010). Wood anatomy of *Bupleurum* L. (Apioideae, Apiaceae) in relation to habit, phylogenetic relationships, and infrageneric taxonomy. Plant Div Evol 128(3-4):501-516.

Stevens PF (2001). Angiosperm phylogeny website, version 9, June 2008.

Uruşak EA, Kızılarslan Ç (2013). Fruit anatomy of some *Ferulago* (Apiaceae) species in Turkey. Turk J Bot 37:434-445.

Wodehouse RP (1935). Pollen grains. McGraw Hill, New York.

Yılmaz G, Dane F (2013). Phytotoxic effects of herbicide Attribut and surfactant BioPower on the root, stem and leaf anatomy of *Triticum aestivum* cv. Pehlivan. Turk J Bot 37:886-893.

5

Combined Use of Green Manure and Farmyard Manure Allows Better Nutrition of Organic Lettuce

Sevgi CALISKAN[1]*, Halit YETISIR[2], Sema KARANLIK[3]

[1]*Department of Plant Production and Technologies, Nigde University, 51240, Nigde, Turkey; scaliskan@nigde.edu.tr (*corresponding author)*
[2]*Erciyes University, Faculty of Agriculture, Department of Horticulture, Kayseri, Turkey; yetisir1@erciyes.edu.tr*
[3]*Mustafa Kemal University, Faculty of Agriculture, Department of Soil Science, 31000 Hatay, Turkey; karanlik.sema@gmail.com*

Abstract

Organic crop production has become a major business due to rising consumer demand, price premiums and increasing market opportunities. A field experiment was conducted to compare organic and conventional production systems by evaluating growth, yield and mineral content of lettuce leaves. Six organic productions systems, green manure (GM), farmyard manure (FYM), commercial organic fertilizer (COF) and their combinations were compared with conventional production system (CPS). A non-fertilized control treatment was also included. Plant growth and yield were found higher in the organic production system compare to conventional production system. The lowest plant growth and yield were obtained in the control treatment. Values of vitamin C content of lettuce grown in the organic production systems were higher than those of the conventional production system. The highest lettuce yield (71.8 and 76.5 t ha^{-1}) was obtained with GM + FYM treatment. The results indicated that combined use of green manure and farmyard manure can be successfully used in organic lettuce production.

Keywords: mineral content, plant nutrition, quality, yield, *Lactuca sativa* L.

Introduction

Lettuce (*Lactuca sativa* L. var. longifolia), as a Mediterranean vegetable, is the major salad crop to be cultivated and commercialized internationally (Deshpande and Solunkhe, 1998). It is grown as a leafy vegetable and is usually consumed as raw in salads, hamburgers tacos and a number of other dishes. Leafy vegetables have an important role in human nutrition and are a good source of minerals. It is a leafy cool-season vegetable with a worldwide production area of 1,110,720 ha and production of 24,324,716 t in 2011 (FAO, 2013). It is commonly produced in the Mediterranean-type environments in Turkey (Kamberoglu and Alan, 2011).

The quality and quantity of applied fertilizers are key factors affecting the growth, yield and quality of plants. Generally, excessive mineral fertilizers are applied to vegetables in order to achieve a maximum yield (Stewart *et al.*, 2005). Especially, the excess amount of nitrogen (N) fertilizer increases nitrite accumulation in plant leaves and this could result negative impact on human health (Islam *et al.*, 2012). In recent years, organic farming, which requires utilization of organic fertilizers, has gained popularity to eliminate the negative effects of mineral fertilizers for human health. In organic farming systems, plants are grown without the use of chemical fertilizers, pesticides, fungicides and herbicides. They are usually fertilized solely with organic fertilizers, such as green manures, farmyard manure, crop residues, commercial organic fertilizers or organic wastes. These components maintain soil productivity, supply plant nutrients and help in combination with biological pest management to control insects, weeds and other pests (Polat *et al.*, 2008; Hammad *et al.*, 2011; Talgre *et al.*, 2012). Worthington (2001) showed that organically grown fruits or vegetables contain more mineral and vitamins than conventionally grown fruits or vegetables. The results of previous studies revealed that organically grown lettuce had higher content of vitamin C, iron (Fe), magnesium (Mg), and phosphorus (P) than the conventionally grown ones. Furthermore, Kelly and Bateman (2010) examined the mineral content of organically grown tomato and found more copper (Cu), calcium (Ca), zinc (Zn), Rubidium (Rb), Mg and Fe than in conventional fruit. Citak and Sonmez (2009) investigated the mineral contents of organically and conventionally grown spinach during two successive seasons and reported that the micro nutrient content of spinach was higher with organic fertilizer than compare to mineral fertilizer applications.

The aim of this research was to compare the yield and some of the physico-chemical properties and mineral content of lettuce grown in conventional and organic production systems. Green manure, farmyard manure, commercial organic fertilizer and their combinations were used as plant nutrient sources in the organic production system.

Materials and methods

Location and experimental design

A two-years field experiment were conducted at Mustafa Kemal University, Experimental Farm of Agricultural Faculty in Hatay (36° 39'N, 36° 40'E; 83 m elevation) located in the Eastern Mediterranean region of Turkey. The experimental area had not been cultivated for five years before the research. Five organic production systems, which utilized green manure (GM), farmyard manure (FYM) and commercial organic fertilizer (COF) alone and in combination were compared with a conventional production system (CPS) for lettuce production. A non-fertilized treatment was also included as control. The field experiments were established as a randomized block design with three replications. The each plot had 100 m^2 area (10 x 10 m) and consisted of 25 rows 0.4 m apart. The traditional growing plots were established nearly 200 m away from organic production plots at the same experimental unit.

The FYM was obtained from livestock farm of Mustafa Kemal University and applied at 30 t ha^{-1} a week before transplanting. Some general characteristics of used FYM were given in Tab. 1. The green manure plant, common vetch (*Vicia sativa* L.) was sown as 100 kg ha^{-1}. The common vetch plants were cut above soil level at flowering stage and were incorporated into the soil (at the rate of 27 t ha^{-1} in the first year, and 31 t ha^{-1} in the second year). Poultry manure used as the commercial organic fertilizer (COF) study, was organic, in pellet form (Tab. 1).

Tab. 1. Chemical characteristics of soil, farmyard manure and commercial organic fertilizer used in the experiment

Production systems	Chemical characteristics of production systems						
	E.C. ($mScm^1$)	pH	CaC_3 (%)	O.M .(%)	N (%)	P (mg kg^1)	K (mg kg^1)
Soil	1.0	7.8	21.7	1.73	0.13	305	253
FYM	38.5	8.7	39.0	55.0	2.35	5800	26,000
COF	5.0	6.4	3.0	9.96	2.69	21,500	19,000

The choice of commercial organic fertilizer and application rate was based on a previous study (Vural *et al.*, 2000), in which the fertilizer was approved for organic vegetable growing. For the mineral fertilizer, the total amount of phosphorus (60 kg P ha^{-1} as super phosphate) and potassium (60 kg K ha^{-1} as potassium sulphate) and one-third of nitrogen (80 kg N ha^{-1} as ammonium nitrate) fertilizers were applied before planting. The remaining two thirds of the nitrogen fertilizer was side dressed in two parts during the growing season. The soil used in this study was clayey and some chemical properties of the soil (0-40 cm depth), FYM and COF are given Tab. 1.

The experimental site has a typical Mediterranean climate conditions with hot-dry summers and mild-rainy winter in both years. The mean temperatures were ranged between 7.6 °C (January) and 15.9 °C (April) in the first year, and between 5.6 °C (January) and 19.1 °C (April) in the second year. The total rainfall during experimental periods were 569 mm and 391 mm in the experimental years, respectively.

Cultivation

The lettuce seedlings cv. 'Marco' were pre-grown in plastic multiport trays filled with a mixture of peat and perlite (1:1, v/v) and grown in a greenhouse. Multipot trays were irrigated twice a day with tap water. Plants with three true leaves were transplanted to the soil in an open field at a density of 7 plants m^{-2} on December 10 and on January 2 in the first and second years, respectively. The planting was done by hand and, immediately after transplanting, plants from all treatments were irrigated by a drip irrigation system. Plants were harvested when they reached approximately their highest commercial size in both years. The harvest dates were April 17 and May 5 in the experimental years, respectively.

Analytical methods

Ten plants were chosen randomly from each treatment for physical analysis before the harvest. Head height (cm), stem diameter (mm) and head weight (g) were determined. For mineral analysis, fresh weight (fw) of the harvested ten plants was weighed and then the wrapper leaves were removed. The leaves of lettuce were rinsed with distilled water after washing with tap water and then, blotted dry with paper towels. The leaves were dried in an air-forced oven at 80 °C to a constant weight. The dried samples were ground to pass 20 mesh sieve, and kept in clean glass bottles for analysis. Nutrient (Cu, Fe, Mn and Zn) content of leaves was analysed using atomic absorption spectrophotometer (Kacar and Inal, 2008). Total N content of the lettuce samples were analysed by a modified Kjeldahl method (Nelson and Sommers, 1980). The pH was determined in a 50 ml filtrate obtained from a mixture of 10 g material blended with 100 ml deionized water of fresh leaf tissue and vitamin C was determined according to Pearson (1970). Total soluble solids (TTS) content was measured in the juice of leaves using a digital refractometer (Atago model ATC-1E).

Statistical methods

Data for each year were statistically analysed using a standard analysis of variance technique for randomized block design using the general linear model procedure in the Statistical Analysis System (SAS Institute 1996). The means were compared by Tukey's test ($p \leq 0.05$).

Results

Yield and plant growth

The head height, stem diameter, head weight and yield of lettuce grown under the organic and conventional production systems are presented in Tab. 2. In general, the organic production systems resulted higher values than the conventional production system in both years. Head height and head diameter were measured as indicators of plant growth. Head height was significantly influenced by the treatments in both years (Tab. 2). Organic treatments GM+COF and GM+FYM resulted in significantly higher head height than unfertilized control in the first growing season. The head height values of other organic treatments and conventional mineral fertilizer did not differ significantly than the unfertilized control or GM+COF and GM+FYM treatments. In the second year, the GM+FYM treatment produced significantly higher head height than unfertilized control, conventional mineral fertilizer and sole use of FYM and COF. The organic treatments of GM and GM+COF had also greater head height than conventional control, but they did not significantly differ from other treatments.

The differences between treatments for head diameter were found significant only in the second year (Tab. 2). The head diameter was greater with organic treatments GM+FYM, GM+COF, GM and FYM than the unfertilized control in the second growing season. However, the COF did not differ among these treatments. Conventional mineral fertilizer also significantly increased the head diameter compare to unfertilized control, but it was not significantly different from other organic treatments except GM+FYM for head diameter.

Head weight was significant affected by the treatments, and differences among the treatments displayed more than 90% of the total variation (see test R^2 values) in both years (Tab. 2). The heaviest head weights were obtained from GM + FYM treatment in both years while the lowest head weight values were obtained from the unfertilized control in both years. Head weight of the GM+COF treatment was grouped within the top values, and was significantly greater than unfertilized control and sole use of FYM and COF during the first year. The combined use of GM and COF was also found as productive, and after GM+FYM treatment it produced heavier head height than all other treatments except sole GM. Yields of the lettuce grown under the conventional and organic production systems are presented in Tab. 2. The results showed that lettuce yields were significantly affected by the production systems ($p \leq 0.01$), and differences between treatments displayed 91% and 86% of total variation (see test R^2) in the first and second year, respectively (Tab. 2). The yield values were ranged from 28.9 to 71.8 t ha^{-1} in the first year and from 40.7 to 76.5 t ha^{-1} in the second year. The highest values were obtained from the GM+FYM treatments in both years while the unfertilized control resulted in the poorest yield in corresponding years. Green manure treatment was found as very effective to improve lettuce yield especially when combined with FYM and COF comparing to conventional mineral fertilization or sole use of FYM and COF in both years.

Physico-chemical properties

The results for the main physico-chemical properties of the lettuce grown under the conventional and different organic production systems are presented in Tab. 3. Vitamin C, pH and soluble solids were not significantly affected by the applied treatments in both conventional and organic production systems. Although the differences among treatments were not significant for vitamin C content, all organic treatments had slightly higher Vitamin C content than conventional production system in the first year. However, similar trend was not recorded in the second year, and some organic treatments (Control, COF and GM) resulted with slightly lower vitamin C content than conventional production system. There were no significant differences in pH and soluble solids between organically and conventionally grown lettuce plants (Tab. 3).

Nutrient element contents

The total N contents of the lettuce leaves were not significantly influenced by the treatments in both conventional and organic production system in both years (Tab. 3). In general, the total N content of lettuce leaves in organic production systems was higher than in the conventional production system. Control treatment had the highest total N content (1.58 and 2.60% in 2007 and 2008 respectively) compared with the treatments in the organic production system.

The total Cu, Fe, Mn and Zn contents of the lettuce leaves grown under the conventional and organic production systems are presented in Tab. 4. The results showed that Cu contents of lettuce leaves were significantly affected by the treatments in both years. The GM+COF

Tab. 2. Effects of conventional and organic production systems on head height, head diameter, head weight and yield of lettuce

Treatment	Head height (cm)		Head diameter (mm)		Head weight (g)		Yield (kg/da)	
	Year 1	Year 2	Year 1	Year 2	Year 1	Year 2	Year 1	Year 2
Conventional	27.8 ab	24.8 c	33.8	32.6 bc	820.7 bc	837.0 de	51.3 bc	52.3 bcd
Control	26.1 b	26.5 bc	31.1	28.2 d	462.7 d	652.3 f	28.9 d	40.7 d
FYM	29.7 ab	27.3 bc	40.8	34.7 ab	745.7 c	926.7 cd	46.6 c	57.9 bcd
COF	28.0 ab	26.0 bc	29.6	29.3 cd	717.3 cd	767.0 ef	44.8 cd	47.9 cd
GM	30.5 ab	30.5 ab	32.6	35.5 ab	852.3 bc	973.0 bc	53.3 bc	60.8 abc
GM+FYM	31.8 a	32.3 a	38.5	38.7 a	1149.0 a	1225.7 a	71.8 a	76.5 a
GM+COF	32.2 a	30.3 ab	36.6	36.5 ab	1060.3 ab	1069.0 b	66.3 ab	66.8 ab
Treatment MS	15.3**	23.2**	50.1 ns	44.4**	155069**	110206**	605.3**	429.8**
Test R^2	0.75	0.82	0.53	0.90	0.91	0.96	0.91	0.86
CV	5.5	5.6	14.8	4.6	10.8	5.1	10.7	10.6

FYM - Farmyard manure; COF - Commercial organic fertilizer; O.M. - Organic matter.
ns, (*), (**) - not significant or significant at $P \leq 0.05$ or 0.01, respectively.
Within a column, means followed by the same letter are not significantly different (p=0.05) using Tukey's test.

Tab. 3. Some physicochemical properties and total N content of lettuce grown under conventional and organic production systems

Treatment	Vitamin C (mg 100g^{-1})		pH		Soluble solids (%)		Total N (%)	
	Year 1	Year 2	Year 1	Year 2	Year 1	Year 2	Year 1	Year 2
Conventional	5.60	8.13	5.83	5.97	4.67	4.87	1.97	2.01
Control	6.73	7.33	5.93	5.77	4.00	4.80	2.58	2.60
FYM	7.63	8.30	5.87	5.77	4.33	4.97	2.32	2.37
COF	7.10	7.67	5.83	5.80	4.67	4.87	2.32	2.36
GM	6.30	7.70	5.87	5.83	4.33	4.33	2.38	2.43
GM+FYM	6.70	9.30	5.90	5.77	4.33	4.83	2.35	2.42
GM+COF	6.90	9.43	5.83	5.87	4.00	4.63	2.55	2.42
Treatment MS	1.22 ^{ns}	2.00^{ns}	0.001^{ns}	0.016^{ns}	0.22^{ns}	0.14^{ns}	0.12^{ns}	0.09^{ns}
Test R^2	0.57	0.60	0.58	0.56	0.43	0.62	0.51	0.59
CV	10.2	11.2	0.8	1.4	13.0	5.1	10.3	7.7

FYM - Farmyard manure; COF - Commercial organic fertilizer; O.M. - Organic matter.
ns, (*), (**) - not significant or significant at $p \leq 0.05$ or 0.01, respectively.

Tab. 4. Effects of conventional and organic production systems on micro nutrient (Cu, Fe, Mn and Zn) content of lettuce leaves

Treatment	Cu (mg kg^{-1})		Fe (mg kg^{-1})		Mn (mg kg^{-1})		Zn (mg kg^{-1})	
	Year 1	Year 2	Year 1	Year 2	Year 1	Year 2	Year 1	Year 2
Conventional	12.3 c	16.0 c	108.0	65.0 c	51.7 a	23.3 d	21.0 b	26.7 de
Control	17.0 abc	21.3 a	144.3	145.7 a	42.7 ab	34.7 bc	30.0 ab	46.3 b
FYM	13.7 bc	18.3 b	120.3	106.0 b	37.0 ab	35.7 b	26.7 ab	35.0 c
COF	15.3 abc	22.3 a	119.0	69.3 c	39.3 ab	27.0 cd	29.3 ab	21.3 e
GM	15.0 abc	22.0 a	134.7	119.0 b	38.0 ab	43.7 a	33.3 ab	34.0 cd
GM+FYM	21.3 ab	22.3 a	138.7	149.3 a	39.0 ab	34.7 bc	38.7 ab	55.7 a
GM+COF	23.0 a	22.7 a	135.0	121.3 b	35.3 b	33.3 bc	41.7 a	51.3 ab
Treatment MS	47.1**	19.5**	505.7ns	3359**	89.1*	127.9**	149.3*	493.9**
Test R^2	0.75	0.95	0.53	0.97	0.61	0.90	0.69	0.97
CV	16.9	3.6	13.3	6.1	13.2	8.3	20.9	7.1

FYM - Farmyard manure; COF - Commercial organic fertilizer; O.M. - Organic matter.
ns, (*), (**) - not significant or significant at $p \leq 0.05$ or 0.01, respectively.
Within a column, means followed by the same letter are not significantly different (p=0.05) using Tukey's test.

treatment had significantly higher Cu content than sole FYM treatment and conventional mineral fertilization, but it considered within the same group with other treatments during the first year. The Cu content of lettuce leaves from different treatments divided into three distinct groups in the second year. The lowest Cu content was obtained from conventional production system while all organic treatments except sole FYM application grouped within group of the highest Cu content. Effects of conventional and organic production systems were significant on Fe content of lettuce leaves in the second year, but it was not significant in the first year. Although the differences were not significant among treatments, the lowest value for Fe content were recorded at the leaves of conventionally produced lettuce during the first year. The conventional production system along with the COF treatment had also the lowest Fe content in the second year while the growing lettuce plants in GM+FYM or unfertilized control plots resulted in the highest Fe content on leaves in this year.

The results showed that there were significant differences in Mn content among treatments in the production systems in the first year ($p \leq 0.05$) and the second year ($p \leq 0.01$) Tab. 4. Conventional production system had higher Mn content than GM+COF treatment, but not differed from other treatments in the first year. The highest Mn content was obtained from the GM treatment while the conventional system resulted with the lowest Mn content in the second year.

All treatments had a significant influence on Zn content in both years (Tab. 4). Although the conventional system had the lowest Zn content value in the first year, only the GM+COF treatment had statistically higher Zn content than conventional system. However, the Zn content of lettuce leaves did not differ among organic treatments in the first year. The higher Zn values were recorded in the leaves of lettuce grown in GM+FYM (55.7 mg kg^{-1}) and

GM+COF (51.3 mg kg^{-1}) plots comparing to other treatments in the second year. The considerable lower Zn content was recorded with sole use of COF or conventional production system in the second year.

Discussion

The effects of the conventional and different organic production systems (green manure, farmyard manure, commercial organic fertilizer and their combinations) on the measured parameters differed greatly among the applications. Average total yields were 51.9 and 57.6 t ha^{-1} in the first and second years, respectively. Crop yield was affected by the climatic conditions, length of growing season (Pavlou *et al.*, 2007; Kaplan *et al.*, 2008), the abiotic and biotic stress resistance/tolerance of the variety, production system, crop management (Tüzel *et al.*, 2013) and type of fertilizer used (Masarirambi *et al.*, 2010). In conventional production system, lettuce yields were 51.3 and 52.3 t ha^{-1} in the first and second year, respectively. As shown in Tab. 2, the lettuce yield in the conventional production system was lower than the average total lettuce yield. Previous studies on organic lettuce production reported that inorganic fertilizers resulted in lower yields compared to organic manures in lettuce production (Masarirambi *et al.*, 2010; Maqueda *et al.*, 2011). Similar result was reported by Xu *et al.* (2005), that vegetables grown with organic fertilizers grew better and resulted in a higher total yield than those grown with chemical fertilizers. Since chemical fertilizers do not possess good characteristics of aggregating the soil particles (Masarirambi *et al.*, 2010). In organic production systems, yield reductions for some crops when compared to conventional production system, were also reported (Kaplan *et al.*, 2008). As a result, the plants produced by inorganic fertilizers showed relatively lower yield compared to organic manures.

Organic soil amendments used in organic production systems increase the levels of soil organic matter. Influence of organic matter on soil physical properties dependents on the amount and type of added organic materials (Herencia *et al.*, 2007). Green manure is one of the alternative ways to increase soil organic matter. Green manuring improves soil quality by increasing the organic matter content of soils (Duyar, 2007). In our study, green manure applications and its combinations were found to be more effective in lettuce plant growth and yield than the mineral fertilization in the conventional production system. Adding of green plant material clearly, could be increased the accumulation of soil organic matter during the growth period of lettuce. The leaves and stems of green plant might decompose very quickly. Mandal *et al.* (2003) reported that the organic matter and total soil nitrogen concentrations founded to be higher under green manuring treated plots in 0-15 cm soil layer during the growth of rice and wheat. MacRae and Mehuys (1985) also indicated that the green manuring maintained an elevated soil organic matter level. Similarly, another positive effect of green manure on lettuce could be partly associated with the amounts of plant nutrients in general and N in particular, returned to soil in legumes biomass (Shah *et al.*, 2011). Mandal *et al.* (2003) speculated that possible benefit of green manuring to soil nitrogen may be due to nitrogen released from roots and nodules during their growth. Nitrogen is the key element for higher yield and better quality. It is involved in all of the plant's metabolic processes, its rate of uptake and partition being largely determined by supply and demand during the various stages of plant growth (Delogua *et al.*, 1998). Research showed that the yield effect of green manure in soil depends on the amount of N in biomass, its release rate, the C/N ratio in organic matter (Kumar and Goh, 2002), soil N content and climate. The C/N ratio of organic residues added to soil is very important. Green manure C/N ratios varied from 20 to 37, depending on the species. Green manures with lower C/N ratios decompose rapidly and have a rapid effect on soil physical properties and on biological activity (Wagger, 1989) and therefore could increase crop yields and N nutrition. Salmeron *et al.* (2011) reported that the C/N rate of common vetch is lower than the other green manure (average of 12.8) and it when planted alone it can provide substantial amounts of N to the following crop. Improved soil physical and chemical properties by green manuring can lead to the growth of larger and healthier root systems (Thorup-Kristensen, 2006; Mandal *et al.*, 2003) and nutrients may have been uptaken more efficiently. According Becker *et al.* (1995), even with a short growth period of 45-60 days, green manure legumes can fix 80-100 kg N ha^{-1} of which the major portion (about 80%) is derived from biological N_2 fixation. Sattell *et al.* (1998) also reported that the common vetch can accumulate approximately 6-14 kg N/da^{-1}. Our results indicated that green manuring with common vetch probably resulted in increasing N nutrition of lettuce in following season, and consequently head yield increased. The green manuring also improved the organic matter status of soil, which ultimately increased the crop growth and yield of lettuce in both years.

Effects of different organic and conventional production systems were found to be not significant for vitamin C, pH, soluble solids and total N content of lettuce leaves in both years (Tab. 3). Although the differences among treatments were not significant statistically, the conventional production system resulted in lower vitamin C content comparing to all organic systems in the first year. However, the effects of treatments on vitamin C content were not consistent between years, and some organic treatments had lower values than conventional production system. There are reports of reduced, increased or not changed vitamin C content of different crops grown in organic systems (Worthington, 2001; Masamba and Nguyen, 2008; Citak and Sonmez, 2010). Polat *et al.* (2008) found vitamin C concentration of lettuce between 12.85-18.23 mg 100 ml^{-1}; however, Ismail and Fun (2003) pointed out that organically grown lettuce contains 50 mg $100g^{-1}$ vitamin C. Based on a farming comparison study, Lairon *et al.* (1984) observed no significant difference in vitamin C content of lettuce grown organically and conventionally. Those results are in agreement with the findings of this experiment.

Total N content of lettuce leaves did not differ significantly among the treatments, but it tended to be lower with conventional production system in both years (Tab. 3). It can be expected higher total N content with mineral fertilization due to adding N to soil, but some previous studies with lettuce indicating higher N content in organic systems (Herencia *et al.*, 2007), or no significant differences (Lairon *et al.*, 1984).

Manganese, copper, zinc and iron are considered important plant micronutrient. In this study, all applications in the conventional and different organic production systems resulted in significant effects in mineral composition of lettuce leaves in both years, but no significant differences was found in the first year for Fe content. There was relatively higher copper, iron and zinc contents in plants produced by different organic production systems. Worthington (2001), reported that plants grown in organic environments contained significantly higher Fe than plants grown in conventional agricultural environments. Kaplan *et al.* (2008) reported that the Mn contents of lettuce leaves differed significantly between the different organic fertilization practices and mineral fertilization and decreased with the mineral fertilization in both autumn and spring experiments. Zinc is very important mineral for nourishment. Some earlier studies indicated the higher Zn content of lettuce with organic production systems (Polat *et al.*, 2008; Kelly and Bateman, 2010), or no significant differences between organic and conventional systems (Maqueda *et al.*, 2011). The mean content of Zn in the organically grown lettuce is higher than the conventionally grown in our study. In the current study, Cu contents were significantly higher in the all organic production systems than the conventional system in both years. In earlier studies, there were conflicting reports regarding the effect of organic and mineral fertilization for Cu content of lettuce leaves. Some researchers have reported that mineral fertilization was the most effective application for the Cu content (Hernandez *et al.*, 2010; Kaplan *et al.*, 2008) and this result might be due to synergetic effect between N and Cu (Aydemir and Ince, 1988). However, some other researchers have reported that mean Cu concentrations were lower for the conventionally grown lettuce compared to those grown organically (Polat *et al.*, 2008; Kelly and Bateman, 2010; Maqueda *et al.*, 2011).

Although positive effects of organic amendments on growth and yield traits of lettuce, our study clearly pointed out the synergistic effects of combined use of organic amendments. The combined use of them (GM+FYM or GM+COF) generally resulted in higher values for most of evaluated traits than their sole use (Tab. 2, 3 and 4). Increasing organic matter of soil increases the uptake both macro and micro nutrients from the soil by crops (Herrick and Wander, 1998). The organic amendments used in the study have some advantages over each others. Hence combined use of them probably generated multiple benefits of different contributions from each organic amendments.

Conclusion

The results of current study indicated that lettuce growth and yield were greatly affected by applied agricultural production systems. Lettuce yield of grown in the organic production systems was higher than the conventional production system. The lowest plant growth and yield were recorded in the non-fertilizer applied control treatment. The most significant impact on growth and yield was observed when green manure was applied to soil. Green manure increased lettuce yields when applied with farmyard manure. In addition, lettuce grown in the green manure and farmyard manure applications showed higher mineral content than those grown in conventional production system. Therefore, combined use green manure and farmyard manure applications were best suited for organic lettuce production with high yield and better quality.

Acknowledgements

This study was supported by The State Planning Organization of Turkey (DPT) under the research project number 03 K 120860. The authors wish to thank to the DPT for financial support.

References

Aydemir O, Ince F (1988). Plant nutrition. p. 211-276. Dicle University, Education Faculty, Press, 2. Turkey

Becker M, Ladh JK, Ali M (1995). Green manure technology: potential, usage, and limitations. A case study for lowland rice. Plant Soil 174:181-194.

Citak S, Sönmez S (2009). Mineral contents of organically and conventionally grown spinach (*Spinacea oleracea* L.) during two successive seasons. J Agr Food Chem 57:7892-7898.

Citak S, Sönmez S (2010). Effects of conventional and organic fertilization on spinach (*Spinacea oleracea* L.) growth, yield, vitamin C and nitrate concentration during two successive seasons. Scientia Horticulturae 126:415-420.

Delogua G, Cattivelli L, Pecchioni N, De Falcis D, Maggiore T, Stanca AM (1998). Uptake and agronomic efficiency of nitrogen in winter barley and winter wheat. Eur J Agron 9:11-20.

Deshpande SS, Solunkhe DK (1998). Handbook of vegetable science and technology, p.493-509, Marcel Dekker, New York.

Duyar H (2007). Effects of green manuring on greenhouse vegetable production. PhD, Ege University Graduate School of Natural and Applied Sciences, İzmir, Turkey (in Turkish).

FAO (2013). Food and Agriculture Organization Statistical Database. http://faostat3.fao.org/home (Accessed in August, 2013).

Hammad HM, Khaliq A, Ahmad A, Aslam M, Malik AH, Farhad W (2011). Influence of different organic manures on wheat productivity. International Journal of Agriculture and Biology 13:137-140.

Herencia JF, Ruiz-Porras JC, Melero S, Garcia-Galavis PA, Morillo E, Maqueda C (2007). Comparison between organic and mineral fertilization for soil fertility levels, crop macronutrient concentrations, and yield. Agron J 99:973-983.

Hernandez A, Castillo H, Ojeda D, Arras A, Lopez J, Sanchez E (2010). Effect of vermicompost and compost on lettuce production. Chilean Journal Agricultural Research 70:583-589.

Herrick JE, Wander MM (1998). Relationship between soil organic carbon and soil quality in cropped and rangeland soils: the importance of distribution, composition and soil biological activity. In: Advances in soil science: Soil processes and the carbon cycle (Eds. Lal R, Kimble J, Follet R, Stewart

BA), CRC Press LLC, Boca Raton, FL, pp. 405-425.

Islam MS, Ahmed A, Mahmud S, Tusher TR, Khanom S (2012). Effects of organic fertilizer on the growth and yield of lettuce (*Lactuca sativa* L.) used as vegetables. Int J Agric Sci 2:116-128.

Ismail A, Fun CS (2003). Determination of vitamin C, β – carotene and riboflavin contents in five green vegetables organically and conventionally grown. Malaysian Journal of Nutrition 9:31-39.

Kacar B, Inal A (2008). Analyses of plant, Nobel Press No:1241, Ankara, Turkey.

Kamberoglu MA, Alan B (2011). Occurrence of tomato spotted wilt virüs in lettuce in Cukurova region of Turkey. International Journal of Agriculture and Biology 13:431-434.

Kaplan M, Sonmez S, Polat E, Demir H (2008). Effects of organic and mineral fertilizers on yield and nutritional status of lettuce. Asian J Chem 20:1915-1926.

Kelly SD, Bateman AS (2010). Comparison of mineral concentrations in commercially grown organic and conventional crops – Tomatoes (*Lycopersicon esculentum* L.) and lettuces (*Lactuca sativa* L). Food Chem 119:738-745.

Kumar K, Goh KM (2002). Management practices of antecedent leguminous and non-leguminous crop residues in relation to winter wheat yields, nitrogen uptake, soil nitrogen mineralization and simple nitrogen balance. Eur J Agron 16:295-308.

Lairon D, Termine E, Gautier S, Trouilloud M, Lafont H, Hauton JC (1984). Effects of organic and mineral ferlizations on the contents of vegetables in minerals, vitamin C and nitrates. In: The importance of biological agriculture in a world of diminishing resources. H Vogtmann, E Boehncke & I Fricke (eds). Witzenhausen, Germany.

MacRae RJ, Mehuys GR (1985). The effect of green manuring on the physical properties of temperate-area soils. Advances in Soil Science 3:71-94.

Mandal UK, Singh G, Victor US, Sharma KL (2003). Green manuring: its effect on soil properties and crop growth under rice-wheat cropping system. Eur J Agron 19:225-237.

Masamba KG, Nguyen M (2008). Determination and comparison of vitamin C, calcium and potassium in four selected conventionally and organically grown fruits and vegetables. African Journal of Biotechnology 7:2915-2919.

Masarirambi MT, Hlawe MM, Oseni OT, Sibiya TE (2010). Effects of organic fertilizers on growth, yield quality and sensory evaluation of red lettuce (*Lactuca sativa* L) 'Veneza Roxa'. Agriculture and Biology Journal of North America 1(6):1319-1324.

Maqueda C, Herencia JF, Ruiz JC, Hidalgo MF (2010). Organic and inorganic fertilization effects on DTPA-extractable Fe, Cu, Mn and Zn, and their concentration in the edible portion of crops. J Agr Sci 149:461-472.

Nelson DV, Sommers LE (1980). Total nitrogen analysis of soil and plant tissues. J Assoc Off Anal Chem 63:770-778.

Pavlou GC, EhaliotisCD, Kavvadias VA (2007). Effect of organic and inorganic fertilizers applied during successive crop seasons on growth and nitrate accumulation in lettuce. Scientia Horticulturae 111:319-325.

Pearson D (1970). Analysis. Detemination of L-ascorbic acid. International Federation of Fruit Juice Producers (No:17).

Polat E, Demir H, Onus AN (2008). Comporison of some yield and quality criteria organically and conventionally-grown lettuce. African Journal of Biotechnology 7:1235-1239.

Salmeron M, Isla R, Cavero J (2011). Effect of winter cover crop species and planting methods on maize yield and availability under irrigated Mediterranean conditions. Field Crop Res 123:89-99.

SAS Institute Inc. (1996). SAS/STAT users guide, ver. 6.4th ed. Carry, NC.

Sattell R, Dick R, Luna J, Mc Grath D, Peachey E (1998). Common vetch (*Vicia sativa* L.). In:Oregon cover crops. Oregon State University Extension Service, Corvallis.

Shah Z, Ahmad SR, Rahman HU, Shah MZ (2011). Sustaining rice-wheat system through manrgement of legumes II. Effect of gren manure legumes and fertilizer on wheat yield. Pakistan J Bot 43(4):2093-2097.

Stewart WM, Dibb DW, Jonston AE, Smyth TJ (2005). The contribution of commercial fertilizer nutrients to food production. Agron J 97:1-6.

Talgre L, Lauringson E, Roostalu H, Astover A, Makke A (2012). Green manure as a nutrient source for succeeding crops. Plant Soil and Environment 58: 275-281.

Thorup-Kristensen K (2006). Root growth and nitrogen uptake of carrot, early cabbage, onion and lettuce following a range of green manures. Soil Use Man 22:29-38.

Vural H, Esiyok D, Duman I (2000). Culture vegetables (Vegetables growing). Ege U., Agriculture Faculty, Department of Horticulture, Bornova, Izmir, 400 p.

Wagger MG (1989). Time of dessication effects on plant composition and subsequent nitrogen release from several winter annual cover crops. Agron J 81:236-241.

Worthington V (2001). Nutritional quality of organic versus conventional fruits, vegetables and grains. J Altern Complem Med 7:161-173.

Xu HL, Wang R, Xu RY, Mridha MAU, Goyal S (2005). Yield and quality of leafy vegetables grown with organic fertilizations. Acta Horticulturae 627:25-33.

6

Differential Response of Bean (*Phaseolus vulgaris* L.) Roots and Leaves to Salinity in Soil and Hydroponic Culture

Duygu BAYRAM, Burcu SECKIN DINLER*, Eda TASCI

*Sinop University, Department of Biology, Faculty of Art and Science, 57000, Sinop, Turkey; bseckin@sinop.edu.tr (*corresponding author)*

Abstract

The present study aimed to investigate the response of common bean (*Phaseolus vulgaris* L. Volare) roots and leaves to salinity in different growth mediums (soil and hydroponic culture) through physiologic and biochemical analyses. The relative water content (RWC) and total chlorophyll (CHL) content decreased with 300 mM NaCl treatment in both cultures but did not change with 150 mM treatment in soil culture. Similarly, the malondialdehyde (MDA) content did not change with 150 mM treatment in soil culture, whereas it increased in all other treatments. The highest increase in hydrogen peroxide (H_2O_2) content was observed with 300 mM treatment in hydroponic culture. The highest increase in superoxide dismutase (SOD) activity was observed in plant leaves in the hydroponic culture. Catalase (CAT) activity did not change with 150 mM treatment in soil culture but decreased with 300 mM treatment in both cultures. Ascorbate peroxidase (APX) activity decreased in all treatments, except in the roots in the hydroponic culture. The Na^+ and Cl^- contents were higher in the hydroponic culture than in the soil culture. Salt stress induced more serious oxidative damage in the hydroponic culture compared to the soil culture.

Keywords: antioxidant enzymes, hydroponic, *Phaseolus vulgaris*, salt stress, soil

Introduction

Several environmental stress factors, such as air pollution, climate change, and drought, limit the crop yield and quality around the world. Salinity is also a significant stress factor that limits the plant growth and influences agricultural practices in various parts of the world (Afzal *et al.*, 2006). The ever-increasing salinity problem has already affected 20% of irrigated land resources worldwide (Kader *et al.*, 2008). Salt stress causes osmotic stress and ion toxicity in plants (Hasegawa *et al.*, 2000). Osmotic stress starts after a threshold salt concentration around the roots is reached, leading to inhibition of water uptake and cell expansion, whereas ion toxicity starts with salt accumulation in the leaves, resulting in an increase in chlorosis and necrosis. In addition, salinity affects the photosynthesis rate (Sudhir and Murthy, 2004) because of decreased chlorophyll pigment and closure of stomata (Soussi *et al.*, 1998; Bethke and Drew, 1992). The limitation of photosynthesis causes an over-reduction of photosynthetic electron chains and the production of reactive oxygen species (ROS), which are harmful to plant growth and development. These ROS, including superoxide radicals, hydrogen peroxide, and hydroxyl radicals, are highly reactive and can cause serious damage on membrane lipids, proteins, and nucleic acids. Salinity induces an oxidative stress in plant cells (Imlay, 2003). To protect the cells against oxidative damage, plants develop complex defense mechanisms, such as superoxide dismutase (SOD), catalase (CAT), ascorbate peroxidase (APX), and glutathione reductase (GR) (Noctor and Foyer, 1998). The role of antioxidant enzymes in stress conditions has been well documented for many plant species (Bowler *et al.*, 1992; Shalata *et al.*, 2001; Sairam *et al.*, 2005; Ashraf *et al.*, 2008).

The common bean (*Phaseolus vulgaris* L.) is one of the most important grains for human nutrition. It is also a valuable leguminous crop worldwide because of its ability to fixate atmospheric nitrogen into the root nodules in a symbiotic interaction with soil rhizobia. This plant is classified as relatively more salt-sensitive compared with the other members of the legume family; more than 50% yield losses are experienced at soil salinities of over 2 dS m^{-1} (Gama *et al.*, 2007). Salinity also significantly reduces the growth of these species. Such salinity-induced impacts may result in changes in dry-matter allocation, ion relationships, water status, physiologic processes, biochemical reactions, or a combination of all these parameters (Khadri *et al.*, 2001). The salt tolerance of glycophytes is believed to be related to their ability to avoid accumulating excess monovalent cations in their leaves while taking up Cl^- ions (Lauter *et al.*, 1988). High Cl^- concentrations lead to nutritional imbalance and changes in water relations (Seeman *et al.*, 1985; Gama *et al.*, 2007).

Most of the studies previously conducted on salinity in crops were carried out in hydroponic systems, pots, or fields. Certainly, there are significant differences between the plants grown in each one of these systems. For instance, in soil experiments, the cation exchange mechanism affects soil features by changing the cation and anion activities. However, in hydroponic systems, the plants have to overcome the osmotic stress of the solution. Thus, it is important to elucidate the salt tolerances or responses of plants in different mediums. Although many studies have revealed the effects of salt stress on bean plants at different stages (Bayuelo-Jimenez *et al.*, 2002; Esechie *et al.*, 2002), there is no report in the literature about a comparative analysis of the changes in physiologic (growth and water content) and biochemical (chlorophyll content, antioxidant enzymes, and ion concentration) characteristics of the same bean plants exposed to salinity in different mediums (hydroponic and soil culture). Therefore, the present study aimed to investigate the response of common bean (*Phaseolus vulgaris* L. Volare) roots and leaves to salinity in different growth mediums (soil and hydroponic culture) through physiologic and biochemical analyses, compare the salt tolerance mechanisms of the plants in different mediums to obtain more tolerant cultivars, and underline the degrees of salinity-induced damage in different mediums.

Materials and methods

Plant material and experimental design

Common bean (*Phaseolus vulgaris* L.) seeds were obtained from a commercial seed company (Tasaco) in Antalya, Turkey. The seeds were sterilized in 5% hypochloride solution for 10 minutes, rinsed three times with distilled sterile water, and then sown in plastic trays (10 x 14 cm) filled with soil. After germination, seedlings were taken into a growth chamber at 25 °C with 16/8 h day/night photoperiod and light intensity of 500 μmol $m^{-2}s^{-1}$. The seedlings were watered with Hoagland solution (Hoagland and Arnon, 1950). After 14 days of growth in the chamber, some of the plants were transferred to hydroponic culture. A week later, the seedlings were exposed to saline solutions of 150 mM NaCl and 300 mM NaCl. To reach the desired concentrations, 50 mM was added to Hoagland solution once every 12 h for 48 h in both groups (hydroponic and soil culture). The plants were then harvested and stored at -20 °C until further analyses.

Relative water content

The relative water content (RWC) was calculated in accordance with Smart and Bingham (1974). The leaves were floated on deionized water for 5h in low irradiance; then the turgid tissue was quickly blotted to remove excess water, and the turgid weights (TW) were determined. DW was determined after seedlings were dried in the oven at 70 °C for 72 h.

Chlorophyll content

The chlorophyll contents of the leaves were measured in accordance with the method specified by Lichtenthaler and Wellburn (1983). The pigments from 0.1 g fresh leaves were extracted by using 80% acetone.

H_2O_2 content

The H_2O_2 content was determined according to Velikova *et al.* (2000). Fresh leaves (0.1 g) were homogenized in 5ml of 0.1% trichloroacetic acid (TCA) and centrifuged at 12,000 rpm for 15 minutes. The supernatant (0.5 ml) was then mixed with 0.5 ml of buffer (10 mM potassium phosphate, pH 7) and 1ml of 1M KI. The absorbance reading was taken at 390 nm.

Na^+ and Cl^- ion content

The Na^+ content was determined by flame photometry according to Mathis (1956). The Cl^- concentration was obtained by wet oxidation of dried leaf tissue with nitric and perchloric acids in accordance with the method adapted by Johnson and Ulrich (1959). The digest was diluted with 0.1 N perchloric acid, and Cl^- concentrations were determined by atomic absorption spectrophotometry.

Malondialdehyde content

The level of lipid peroxidation in leaf samples was determined in terms of the malondialdehyde (MDA) content according to the method specified by Madhava Rao and Sresty (2000). The MDA content, an end product of lipid peroxidation, was determined by using the thiobarbituric acid reaction. The MDA concentration was calculated from the absorbance at 532 nm, and measurements were corrected for nonspecific turbidity by subtracting the absorbance at 600 nm. An extinction coefficient of 155 mM^{-1} cm^{-1} was used to determine the MDA concentration.

For protein and enzyme extractions, 0.5 g of fresh leaf samples were homogenized in 1.5 ml of 50 mM sodium phosphate buffer (pH 7.8) containing 1mM ethylenediaminetetraacetic acid (EDTA) Na_2 and 2% (w/v) polyvinylpolypyrrolidone (PVPP). All operations were done at 4 °C. Samples were centrifuged at 14 000×g for 30 min, and supernatants were used to determine the protein content and enzyme activities. All spectrophotometric analyses were done on a UV-visible spectrophotometer.

SOD activity

The superoxide dismutase (EC 1.15.1.1) activity was assayed by its ability to inhibit the photochemical reduction of nitrotetrazolium blue chloride (NBT) at 560 nm (Beauchamp and Fridovich, 1973).

APX activity

The ascorbate peroxidase (EC 1.11.1.11) activity was measured according to Nakano and Asada (1981). The assay depended on the decrease in absorbance at 290nm as ascorbate was oxidized. The reaction mixture contained 50 mM Na-phosphate buffer (pH 7.0), 50 mM ascorbate, 0.1 mM EDTA Na_2, 1.2 mM H_2O_2, and 0.1 ml of enzyme extract in a final assay volume of 1 ml. The concentration of oxidized ascorbate was calculated by using an extinction coefficient of 2.8 mM^{-1} cm^{-1}. One unit of APX was defined as 1mmol ml^{-1} ascorbate oxidized min^{-1}.

CAT activity

The catalase (EC 1.11.1.6) activity was assayed in a reaction mixture (2 ml) containing 50 mM K-phosphate buffer (pH 7.0). Then, 12.2 mM H_2O_2 was added, and the reaction was started by adding 100 μl supernatant. The activity was determined by monitoring the degradation of H_2O_2 at 240 nm over 2 min against a supernatant–free blank. Enzyme-specific activities were expressed as μmol of H_2O_2 oxidized $min^{-1}mg^{-1}$ protein (Bergmeyer, 1970).

Statistical analysis

All data were subjected to analysis of variance, and the mean differences were compared by least significant difference (LSD) testing at $p<0.05$ level. Each data point represented the mean of six replicates

Results and discussion

Salt and water stress are well known to limit the availability of water in tissues and to reduce the relative water content (RWC) in *Phaseolus* plants (Bayuelo-Jimenez *et al.*, 2002; Kabir *et al.*, 2004; Lazcano-Ferrat Lovatt, 1999). In the present study, compared with the control group in soil culture, the RWC did not change significantly in leaves with 150 mM NaCl treatment but decreased by 42.61 % in hydroponic culture. In addition, 300 mM NaCl treatment inhibited the RWC by 31.30 and 58.47% in the soil and hydroponic cultures, respectively (Tab. 1). The decrease in RWC indicated a loss of turgor that resulted in limited water availability for the cell extension process, as also reported by Katerji *et al.* (1997). Besides, high exogenous salt concentrations lead to ion toxicity and osmotic stress damage in plant cells (Cuartero *et al.*, 2006). It could be suggested that the decreased water content was related to osmotic stress in the leaves of *Phaseolus vulgaris* because, in the hydroponic culture, osmotic stress affected the roots by decreasing the water content in the nutrient solution directly and more negatively than in the soil culture at the same NaCl concentration (300 mM NaCl).

Stepien and Klobus (2006) reported strong evidence of salt effects on photosynthetic enzymes, chlorophylls, and carotenoids of plants. This is related to the interference of salt ions with the de novo synthesis of proteins and the structural component, rather than the breakdown, of chlorophyll (Jaleel *et al.*, 2007). In the present study, compared with the control treatment, the total chlorophyll content did not change with 150 mM NaCl treatment in both the soil and hydroponic cultures, whereas it decreased by 50 and 26.30%, respectively, with 300 mM NaCl treatment in the hydroponic and soil cultures (Tab. 1). Similarly, Gadallah (1999) showed that NaCl decreased the chlorophyll content in bean plants. However, Delfine *et al.* (1999) reported no changes in chlorophyll content over 20 days in salt-stressed spinach (*Spinacia oleracea* L.) plants. Such findings agree with our current results. In the present study, the higher chlorophyll reduction rates in the hydroponic culture compared to the soil culture may be related to the higher Na^+ content in leaves with 300 mM NaCl treatment.

Hydrogen peroxide (H_2O_2) is a versatile molecule that may be involved in several cell processes in normal and stress conditions (Quan *et al.*, 2008). In stress conditions, H_2O_2 is produced and accumulated, leading to oxidative stress in plants. Plants possess a battery of antioxidant mechanisms, both enzymatic and nonenzymatic, by which ROS are removed from the cells (Noctor and Foyer, 1998). In the present study, compared with the control treatment, the H_2O_2 content of roots and leaves increased by 5.7 and 5.4%, respectively, with 150 mM NaCl treatment. Moreover, 300 mM NaCl treatment increased the H_2O_2 content of roots and leaves by 39.12 and 48.93%, respectively. In the hydroponic culture, 150 mM NaCl treatment increased the H_2O_2 content by 35.69% in roots and 21.56% in leaves. Treatment with 300 mM NaCl increased the H_2O_2 content by 71.59% in roots and 46.03% in leaves (Tab. 2). These findings are in agreement with the results of Sairam and Srivastava (2002). It can be suggested that the higher H_2O_2 content in roots with 300 mM NaCl treatment in the hydroponic culture may have led to more oxidative damage

Tab. 1. Changes in relative water content and total chlorophyll in leaves of *Phaseolus vulgaris* L. under salt stress for 48 hours in hydroponic and soil culture

Hydroponic	Leaves	
	Relative water content (%)	Total chlorophyll (mg/ g FW)
C	56.88±3.25^{a}	6.5±2.86^{a}
150	32.51±2.67^{b}	6.1±2.14^{a}
300	23.62±3.87^{c}	4.7±2.78^{b}
Soil		
C	84.27±2.98^{a*}	7.8±3.14^{a*}
150	82.13±.3.89^{a}	6.9±2.18^{a*}
300	57.89±21.98^{b*}	3.9±1.67^{b*}

The different letters indicate a significant difference ($p<0.05$) (n=6). C: control; 150 mM NaCl; 300 mM NaCl. *Indicates significant differences between the two mediums at the same concentrations

Tab. 2. Changes in malondialdehyde content and hydrogen peroxide in roots and leaves of *Phaseolus vulgaris* L. under salt stress for 48 hours in hydroponic and soil culture

Hydroponic	Roots		Leaves	
	Malondialdehyde (nmol gr FW^{-1})	Hydrogen peroxide (μM/gr)	Malondialdehyde (nmol gr FW^{-1})	Hydrogen peroxide(μM/gr)
C	24.38±5.46^{a}	70.34±2.76^{a}	23.67±4.20^{a}	82.63±2.13^{a}
150	32.78±4.98^{b}	95.45±1.68^{b}	36.12±3.29^{c}	100.45±1.19^{b}
300	39.17±3.67^{c}	120.7±2.89^{c}	27.24±5.12^{b}	120.67±2.56^{c}
Soil				
C	20.27±5.83^{a*}	72.36±2.56^{a}	21.41±3.29^{a}	80.35±2.78^{a}
150	22.38±4.12^{a}	76.49±1.68^{b*}	23.16±2.87^{a}	84.69±2.67^{b*}
300	30.12±3.87^{b}	100.67±1.56^{c*}	29.35±4.69^{b}	119.67±1.67^{c}

The different letters indicate a significant difference ($p<0.05$) (n=6). C: control; 150 mM NaCl; 300 mM NaCl. *Indicates significant differences between the two mediums at the same concentrations compared to the soil culture. Our results are also in good agreement with the findings of Mittova *et al.* (2004), who reported a decrease in hydrogen peroxide content in the roots of tomatoes.

Plants develop antioxidant enzymes, such as superoxide dismutase, peroxidase, catalase, ascorbate peroxidase, and glutathione reductase, to repair the damage to cellular components caused by reactive oxygen species (ROS), such as superoxide (O_2^-), hydrogen peroxide (H_2O_2), and hydroxyl radicals (OH^-) (Noctor and Foyer, 1998). SOD plays a significant role in the removal of superoxide radicals from the compartments where they are formed (Alscher *et al.*, 2002). In the present study, the SOD activity in the roots and leaves increased by 33.33 and 34.94%, respectively, with 150 mM NaCl treatment in the soil culture. However, with 300 mM NaCl treatment, the SOD activity increased by 36.38% in roots and 36.11% in leaves. In the hydroponic culture, the SOD activity in the roots and leaves increased by 34.12 and 38.23%, respectively, with 150 mM NaCl treatment, whereas with 300 mM NaCl treatment, the activity increased by 71.93% in leaves and 56.45% in roots (Fig. 1a and b). Parallel to the current findings, previous studies also reported increased SOD activities for cotton and tomato plants under salt stress (Meloni *et al.*, 2003; Mittova *et al.*, 2001). It could be suggested that the oxidative damage was greater in the roots than in the leaves in terms of the unchanged SOD activity in both cultures with 300 mM NaCl treatment. However, there was a decrease in SOD activity in the hydroponic culture, which might have led to an induction of MDA content in roots and leaves in the soil culture.

In the present study, the CAT activity in the roots and leaves did not change with 150 mM NaCl treatment in the soil culture. However, it decreased by 39.21 and 27.69%, respectively, in the leaves and roots with 300 mM NaCl treatment. These results are in agreement with the findings on the hydrogen peroxide content of the leaves and roots with 150 mM NaCl treatment. However, 300 mM NaCl treatment clearly increased the hydrogen peroxide content in the roots and leaves by 25.30 and 11.02%, respectively, and decreased CAT activity (Fig. 2a and b). Similar decreases in CAT activity (Cavalcanti *et al.*, 2004) and unchanged CAT activity (Azevedo Neto, 2005) under salt stress have been reported. Similarly to the soil culture, 150 mM NaCl treatment increased the CAT activity by 21.56 and 56%, respectively, in the leaves and roots in the hydroponic culture. Treatment with 300 mM NaCl decreased the CAT activity in the roots and leaves by 23.12

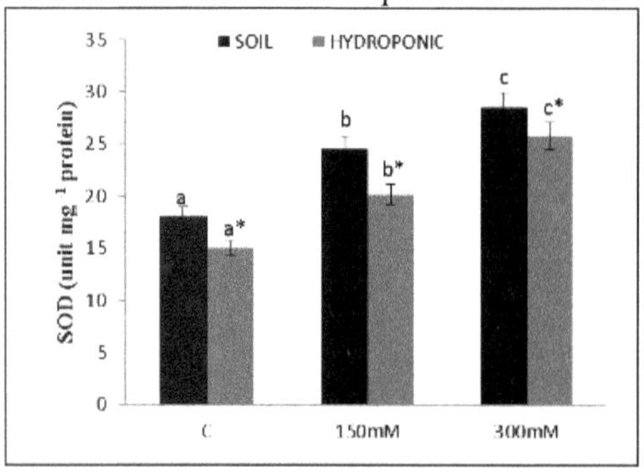

Fig. 1a. Changes in SOD activities in leaves of *Phaseolus vulgaris* L. under salt stress for 48 hours in hydroponic and soil culture. The different letters indicate a significant difference ($p<0.05$). C: control; 150 mM NaCl; 300 mM NaCl. *Indicates significant differences between the two mediums at the same concentrations

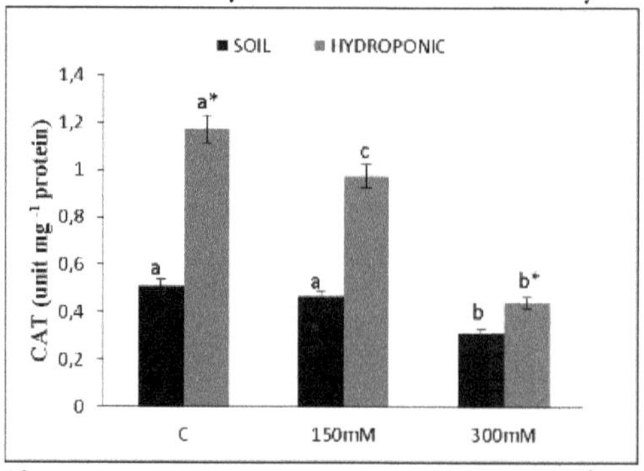

Fig. 2a. Changes in CAT activities in leaves of *Phaseolus vulgaris* L. under salt stress for 48 hours in hydroponic and soil culture. The different letters indicate a significant difference ($p<0.05$). C: control; 150 mM NaCl; 300 mM NaCl. *Indicates significant differences between the two mediums at the same concentrations

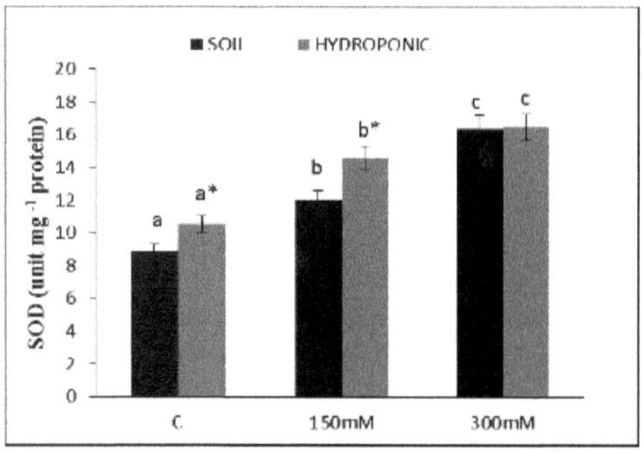

Fig. 1b. Changes in SOD activities in roots of *Phaseolus vulgaris* L. under salt stress for 48 hours in hydroponic and soil culture. The different letters indicate a significant difference ($p < 0.05$) C: control; 150 mM NaCl; 300 mM NaCl. *Indicates significant differences between the two mediums at the same concentrations

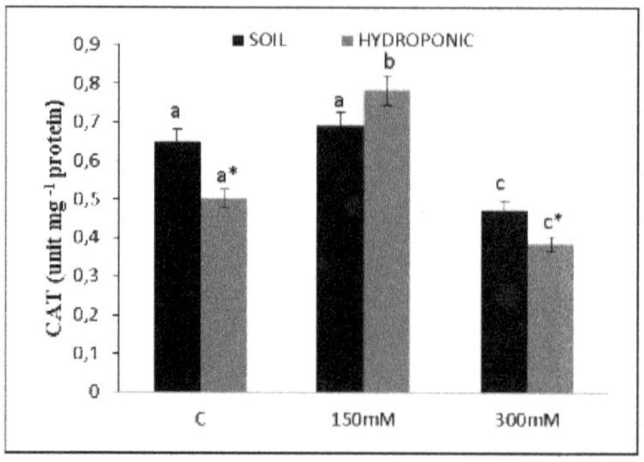

Fig. 2b. Changes in CAT activities in roots of *Phaseolus vulgaris* L. under salt stress for 48 hours in hydroponic and soil culture. The different letters indicate a significant difference ($p<0.05$). C: control; 150 mM NaCl; 300 mM NaCl. *Indicates significant differences between the two mediums at the same concentrations

and 62.39%, respectively. Parallel to the current findings, Vaidyarathan *et al.* (2003) and Shalata *et al.* (2001) also reported increased CAT activities in rice and tomato plants under salt stress.

The predominant peroxidase enzyme is ascorbate peroxidase (APX; EC 1.11.1.11), which catalyzes the oxidation of ascorbate (AsA) by H_2O_2, generating dehydroascorbate radicals (Hideg, 1999). In the present study, compared with the control treatment, the APX activity in the leaves and roots increased by 53.60 and 76.92%, respectively, with 150 mM NaCl treatment in the soil culture. The APX activity in the roots did not change with 300 mM NaCl treatment, but it increased by 45.36% in the leaves. In the hydroponic culture, 150 mM NaCl treatment did not change the APX activity in the leaves but increased that in the roots by 44.11%. In addition, 300 mM NaCl treatment increased the APX activity in the roots and leaves by 36.76 and 33.33%, respectively (Fig. 3a and b). Parallel to the current findings, the APX activity in peas and tomatoes reportedly increased under salt stress (Hernandez *et al.*, 2000; Mittova *et al.*, 2004), whereas it did not change in citrus plants under salinity (Arbona *et al.*, 2003).

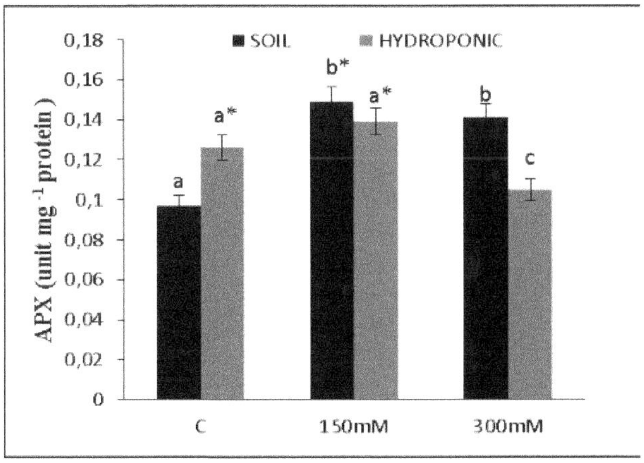

Fig. 3a. Changes in APX activities in leaves of *Phaseolus vulgaris* L. under salt stress for 48 hours in hydroponic and soil culture. The different letters indicate a significant difference ($p<0.05$). C: control; 150 mM NaCl; 300 mM NaCl. *Indicates significant differences between the two mediums at the same concentrations

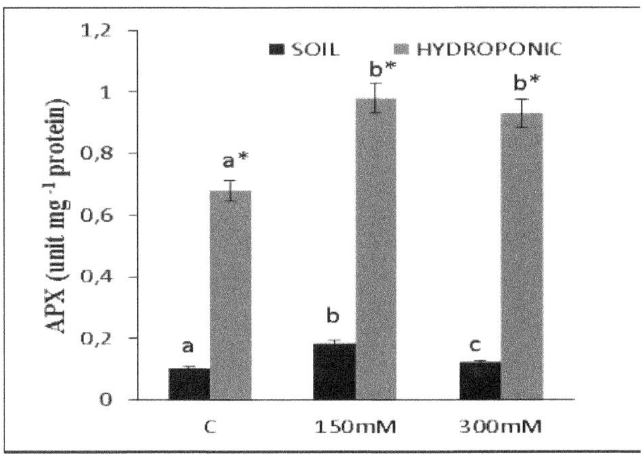

Fig. 3b. Changes in APX activities in roots of *Phaseolus vulgaris* L. under salt stress for 48 hours in hydroponic and soil culture. The different letters indicate a significant difference ($p<0.05$). C: control; 150 mM NaCl; 300 mM NaCl. *Indicates significant differences between the two mediums at the same concentrations

An increase in the level of MDA, produced during the peroxidation of membrane lipids, is often used as an indicator of oxidative damage. Many studies have reported that the lipid peroxidation levels (MDA) in plants under stress conditions are decreased by an effective antioxidant enzyme system. In the current results, compared with the control treatment, the MDA content in the leaves and roots did not change with 150 mM NaCl treatment in the soil culture but increased by 41.42 and 48.59 %, respectively, with 300 mM NaCl treatment. In the hydroponic culture, 150 mM NaCl treatment increased the MDA content by 24.41% in leaves and by 34.35% in roots. In comparison, the MDA content in the leaves and roots increased by 57.32 and 60.66%, respectively, with 300 mM NaCl treatment. The highest increase was observed in the roots and leaves with 300 mM NaCl treatment in the hydroponic culture. The MDA content in mulberries and *Phaseolus vulgaris* has also been reported to increase under saline conditions (Sudhakar and Giridarakumar, 2001; Babu and Devaraj, 2008).

In the soil culture with 150 mM NaCl treatment, the increased APX activity was not efficient in terms of maintaining the hydrogen peroxide content (slight increase) in the leaves and roots because of the unchanged CAT activity. Besides, it could be suggested that the unchanged MDA content was related to the smaller increase in hydrogen peroxide caused by the increased SOD activity. The increase in APX activity with 300 mM NaCl treatment was also inefficient because the decrease in CAT activity might have led to an increase in hydrogen peroxide.

In the hydroponic culture, the increase in hydrogen peroxide with 150 mM NaCl treatment could be related to the decreased CAT activity. Similarly to the soil culture, the increase in APX activity was not sufficient to reduce hydrogen peroxide in the roots. However, at the same concentration, the unchanged CAT and APX activities clearly led to an increase in the hydrogen peroxide content in the leaves. The fact that 300 mM NaCl treatment increased the hydrogen peroxide content in leaves could be due to a decrease in CAT activity; similarly, the induction of APX activity was not enough to protect cells from oxidative damage.

In the present study, compared with the control treatment, the Na^+ content in the leaves increased 4.5- and 13.75-fold with 150 mM and 300 mM NaCl treatment, respectively. However, in the roots, the Na^+ content increased by 98.02% with 150 mM NaCl and 2.62-fold with 300 mM NaCl treatment (Tab. 3). Glycophytic plants adapt to high salt concentrations by lowering their tissue osmotic potentials through an increase in inorganic ions from an external solution (Cacchorro *et al.*, 1995). In *Phaseolus* species, the inorganic ions in leaves, especially Cl^-, Na^+, and K^+, increase as the salt level increases. In the present study, the cultivated samples (*Phaseolus vulgaris* L. Volare) accumulated Na^+ ions in their leaves, as also found

Tab. 3. Changes in Na^+ and Cl^- ion contents in roots and leaves of *Phaseolus vulgaris* L. under salt stress for 48 hours in hydroponic and soil culture

Hydroponic	Roots		Leaves	
	Na^+ (%)	Cl^- (%)	Na^+ (%)	Cl^- (%)
C	0.33 ± 4.21^a	0.67 ± 4.22^a	0.04 ± 5.12^a	0.66 ± 4.99^a
50	4.92 ± 4.12^b	3.95 ± 4.09^b	5.93 ± 5.03^b	7.01 ± 4.32^b
300	11.76 ± 3.97^c	11.46 ± 4.32^c	7.4 ± 5.12^c	8.7 ± 4.81^c
Soil				
C	0.53 ± 3.64^a	$1.01\pm4.35^{a*}$	0.04 ± 5.39^a	$1.76\pm3.56^{a*}$
150	$1.05\pm3.12^{b*}$	0.95 ± 4.23^a	0.18 ± 5.02^a	1.54 ± 3.12^a
300	1.39 ± 3.65^b	0.96 ± 4.12^a	0.55 ± 4.99^b	2.45 ± 3.49^b

The different letters indicate a significant difference ($p<0.05$ (n=6). C: control; 150 mM NaCl; 300 mM NaCl. *Indicates significant differences between the two mediums at the same concentrations

by Bayuelo-Jimenez *et al.* (2002). In the hydroponic culture, compared with the control treatment, the Na^+ content in the leaves increased 14.82-fold with 150 mM NaCl and 18.5-fold with 300 mM NaCl treatment. In the roots, compared with the control treatment, the Na^+ content increased 14.90- and 35.63-fold with 150 mM and 300 mM NaCl treatment, respectively. As a result, there were differences in ion accumulation between the hydroponic culture and the soil culture, as also found by Tavakkoli *et al.* (2010). The higher Na^+ and Cl^- contents in the hydroponic culture, compared to the soil culture, may be related to the roots taking up minerals directly from the external solution more rapidly over 48 h and transmitting them to the leaves in response to salinity. In the soil culture, the soil matrix may have led to a decrease in the Na^+ content by ion exchange mechanisms. It can be suggested that the higher MDA content in the hydroponic culture compared to the soil culture is related to the toxic effects of accumulated Na^+ ions.

In the soil culture, the Cl^- content in the leaves increased by 12.3% with 150 mM NaCl treatment by 39.20% with 300 mM NaCl treatment. However, the Cl^- content in the roots did not change with 150 mM NaCl and 300 mM NaCl treatment. In the hydroponic culture, the Cl^- content in the leaves increased 10.62-fold with 150 mM NaCl and 13.18-fold with 300 mM NaCl treatment. In the roots, the Cl^- content increased 5.89- and 17.10-fold with 150 mM NaCl and 300 mM NaCl treatment, respectively (Tab. 3). The Cl^- ions clearly increased to a greater extent in the hydroponic culture than in the soil culture. Dasgan and Koc (2009) and Tavakkoli *et al.* (2012) also reported that Cl^- ions increased in bean plants under salinity. Because both Na^+ and Cl^- are toxic to plants, and Na^+ in particular causes the deterioration of the physical structure of soil, these ions are considered the most important (Dubey, 1997; Hasegawa *et al.*, 2000). Unchanged Cl^- ions in the roots can be related to the soil matrix, which leads to the cation exchange mechanism. The increase in Cl^- was higher than that in Na^+ in all species, as also found by Bayuelo-Jimenez *et al.* (2002).

Conclusions

Based on the results of this study, it can be suggested that salt stress causes more oxidative damage in hydroponic culture than in soil culture. The changes in antioxidant enzyme responses in different mediums could be explained by a disturbance in the redox balance through the accumulation of toxic ions and osmotic stress, which was more rapid in the hydroponic culture.

Acknowledgements

The authors thank The Scientific and Technological Research Council of Turkey (TUBITAK) 2209 for its financial support.

References

Afzal I, Basra SMA, Farooq M, Nawaz A (2006). Alleviation of salinity stress in spring wheat by hormonal priming with ABA, salicylic acid and ascorbic acid. Int J Agr Biol 8:23-28.

Alscher RG, Erturk N, Heath LS (2002). Role of superoxide dismutases (SODs) in controlling oxidative stress in plants. J Exp Bot 53:1331-1341.

Arbona V, Flors V, Jacas J, Garcia-Agustin P, Gomez-Cadenas A (2003). Enzymatic and non-enzymatic antioxidant responses of carrizo citrange, a salt-sensitive citrus roostoock, to different levels of salinity. Plant Cell Physiol 44:388-394.

Ashraf M, Athar HR, Harris PJC, Kwon TR (2008). Some prospective strategies for improving crop salt tolerance. Adv Agron 97:45-109.

Azevedo Neto AD, Prisco JT, Enéas-Filho J, Medeiros JR, Gomes-Filho E (2005). Hydrogen peroxide pre-treatment induces salt-stress acclimation in maize plants. J Plant Physiol 162:1114-1122.

Babu Nagesh R, Devaraj VR (2008). High temperature and salt stress response in french bean (*Phaseolus vulgaris*). Aust J Crop Sci 2(2):40-48.

Bayuelo-Jimenez JS, Jasso-Plata N, Ochoa I (2012). Growth and physiological responses of *Phaseolus* species to salinity stress. Int J Agr 80:207-222.

Beauchamp C, Fridovich I (1973). Isoenzymes of superoxide dismutase from wheat germ. Biochim Biophys Acta 317:50-64

Bergmeyer HU (1970). Methods of enzymatic analysis. Akademie Verlag, Berlin, Germany, 1: 36-647.

Bethke PC, Drew MC (1992). Stomatal and non-stomatal components to inhibition of photosynthesis in leaves of *Capsicum annum* during progressive exposure to NaCl salinity. Plant Physiol 99:219-226.

Bowler C, Montague MV, Inze D (1992). Superoxide dismutase and stress tolerance. Annu Rev Plant Physiol 43:83-116.

Cachorro P, Martinez R, Ortiz A, Cerda A (1995). Abscisic acid and osmotic relations in *Phaseolus vulgaris* L. under saline conditions. J Plant Growth Regul 14:99-104.

Cavalcanti FR, Oliveira JTA, Martins-Miranda AS, Viegas RA, Silveira JAG (2004). Superoxide dismutase, catalase and peroxidase activities do not confer protection against oxidative damage in salt-stressed cowpea leaves. New Phytol 163:563-571.

Cuartero J, Bolarin MC, Asins MJ, Moreno V (2006). Increasing salt tolerance in the tomato. J Exp Bot 57(5):1045-1058.

Dasgan HY, Koc S (2009). Evaluation of salt tolerance in common bean genotypes by ion regulation and searching for screening parameters. J Food Agric and Environ 7:363-372.

Delfine S, Alvino A, Villani MC, Loreto F (1999). Restrictions to CO_2 conductance and photosynthesis in spinach leaves recovering from salt stress. Plant Physiol 119:101-1106.

Dubey RS (1997). Handbook of photosynthesis, p. 859-975. In: Pessarakli M (Eds). Photosynthesis in plants under stressfull conditions, New York.

Esechie HA, Al-Saidi A, Al-Khanjari S (2002). Effect of sodium chloride salinity on seedling emergence in chickpea. J Agron Crop Sci 188:155-160.

Gadallah MAA (1999). Effects of proline and glycinebetaine on *Vicia faba* response to salt stress. Biol Plantarum 42:249-257.

Gama PBS, Inanana S, Tanaka K, Nakazaw R (2007). Physiological response of common bean (*Phaseolus vulgaris* L.) seedlings to salinity stress. Afr J Biotechnol 6:79-88.

Hasegawa PM, Bressan RA, Zhu JK, Bohnert HJ (2000). Plant cellular and molecular responses to high salinity. Annu Rev Plant 51:463-499.

Hernandez M, Fernandez-Garcia N, Diaz-Vivancos P, Olmos E (2010). A different role for hydrogen peroxide and the antioxidative system under short and long salt stress in *Brassica oleracea* roots. J Exp Bot 61:521-535.

Hideg E (1999). Handbook of plant and crop stress, p. 911-930, In: Pessarakli M (Eds), Free radical production in photosynthesis under stress conditions, New York.

Hoagland DR, Arnon DI (1950). The water culture method for growing plants without soil. Calif AES Bull 347:1-32.

Imlay JA (2003). Pathways of oxidative damage. Annu Rev Microbiol 57:395-418.

Jaleel CA, Manivannan P, Lakshmanan GMA (2007). NaCl as a physiological modulator of proline metabolism and antioxidant potential in *Phyllanthus amarus*. CR Biol 9330:806-813.

Johnson CM, Ulrich A (1959). II. Analytical methods for use in plant analysis. Calif AES Bull 766:30-33.

Kabir ME, Karim MA, Azad MAK (2004). Effect of potassium on salinity tolerance of mungbean (*Vigna radiata* L. wilczek). J Biol Sci 4:103-110.

Kader MA, Seidel T, Golldack D, Lindberg S (2006). Expressions of OsHKT1, OsHKT2, and OsHVA are differentially regulated under NaCl stress in salt-sensitive and salt-tolerant rice (*Oryza sativa* L.) cultivars. J Exp Bot 57:4257-4268.

Katerji N, Van Hoorn JW, Hamdy A, Mastrorilli MM, Karzel E (1997). Osmotic adjustment of sugar beets in response to soil salinity and its influence on stomatal conductance, growth and yield. Agr Water Manage 34:57-69.

Khadri M, Pliego L, Soussi M, Llcuh C, Ocana A (2001). Ammonium assimilation and ureide metabolism in common bean (*Phaseolus vulgaris*) nodules under salt stress. Agronomie 21:635-643.

Lauter DJ, Meiri A, Shuali M (1988). Isoosmotic regulation of cotton and peanut at saline concentrations of K and Na. Plant Physiol 87:911-916.

Lazcano-Ferrat Lovatt CJ (1999). Relationship between relative water content, nitrogen pools, and growth of *Phaseolus vulgaris* L. and *P. acutifolius* a. gray during water deficit. Crop Sci 39:467-475.

Lichtenthaler HK, Wellburn AR (1983). Determinations of total carotenoids and chlorophylls a and b in leaf extracts in different solvents. Biochem Soc Trans 11:591-592.

Madhava RKV, Sresty TVS (2000). Antioxidative parameters in the seedlings of pigeon pea (*Cajanus cajan* L. millspaugh) in response to Zn and Ni stresses. Plant Sci 157:113-128.

Mathis WT (1956). Report on the flame photometric determination of potassium and sodium in plant tissue. J Assoc Off Analyt Chem Int 39:419.

Meloni DA, Oliva MA, Martinez CA, Cambraia J (2003). Photosynthesis and activity of superoxide dismutase, peroxidase and glutathione reductase in cotton under salt stress. Environ Exp Bot 49:69-76.

Mittova V, Tal M, Volokita M, Guy M (2002). Salt stress induces up-regulation of an efficient chloroplast antioxidant system in the salt-tolerant wild tomato species *Lycopersicon pennellii* but not in the cultivated species. Physiol Plant 115:393-400.

Mittova V, Guy M, Tal M, Volokita M (2004). Salinity upregulates the antioxidative system in root mitochondria and peroxisomes of the wild salt-tolerant tomato species *Lycopersicon pennellii*. J Exp Bot 55:1105-1113.

Nakano Y, Asada K (1981). Hydrogen peroxide is scavenged by ascorbate-specific peroxidase in spinach chloroplasts. Plant Cell Physiol 22:867-880.

Noctor G, Foyer CH (1998). Ascorbate and glutathione: keeping active oxygen under control. Annu Rev Plant 49:249-79.

Quan LJ, Zhang B, Shi WW, Li HY (2008). Hydrogen peroxide in plants: a versatile molecule of the reactive oxygen species network. J Integ Plant Biol 50:2-18.

Sairam RK, Roa KV, Srivastava GC (2002). Differential response of wheat genotypes to long term salinity stress in relation to oxidative stress, antioxidant activity and osmolyte concentration. Plant Sci 163:1037-1046.

Sairam RK, Srivastave GC, Agarwal S, Meena RC (2005).

Differences in antioxidant activity in response to salinity stress in tolerant and susceptible wheat genotypes. Biol Plant 49:85-91.

Seeman JR, Critchley C (1985). Effects of salt stress on the growth, ion content, stomatal behavior and photosynthetic capacity on a salt-sensitive species, *Phaseolus vulgaris* L. Planta 164:151-162.

Shalata A, Neuamann PM (2001). Exogenous ascorbic acid (Vitamin C) increases resistance to salt stress and reduces lipid peroxidation. J Exp Bot 52:2207-2211.

Smart RE, Bingham GE (1974). Rapid estimates of relative water content. Plant Physiol 53:258-260.

Soussi M, Ocana A, Lluch C (1998). Effects of salt stress on growth, photosynthesis and nitrogen fixation in chickpea (*Cicer arietinum* L.). J Exp Bot 49:1329-1337.

Stepien P, Klobus G (2006). Water relations and photosynthesis in *Cucumis sativus* L. leaves under salt stress. Biol Plant 50 (40):610-616.

Sudhakar C, Lakshmi A, Giridarakumar S (2001). Changes in the antioxidant enzyme efficacy in two high yielding genotypes of mulberry (*Morus alba* L.) under NaCl salinity. Plant Sci 161:613-619.

Sudhir P, Murthy SDS (2004). Effect of salt stress on basic processes of photosynthesis. Photosynthetica 42(4): 481-486

Tavakkoli E, Rengasamy P, Mcdonald GK (2010). The response of barley to salinity stress differs between hydroponics and soil systems. Funct Plant Biol 37:621-633.

Tavakkoli M, Poustini K, Alizadeh H (2012). Soluble proteins, a biochemical indicator for salinity screening in wheat cultivars (*Triticum aestivum* L.). Elixir Appl Botany 48:9312- 9314.

Vaidyanathan H, Sivakuma P, Chakrabarty R, Thomasm G (2003). Scavenging of reactive oxygen species in NaCl-stressed rice (*Oryza sativa* L.)-differential response in salt-tolerant and sensitive varieties. Plant Sci 165:1411-1418.

Velikova V, Yordanov I, Edreva A (2000). Oxidative stress and some antioxidant system in acid rain treated bean plants: protective role of exogenous polyammines. Plant Sci 151:59-66.

Elimination of *Grapevine fleck virus* by *in vitro* Chemotherapy

Ionela Cătălina GUȚĂ*, Elena-Cocuța BUCIUMEANU, Emilia VIȘOIU

*National Research and Development Institute for Biotechnology in Horticulture, 37 Calea București, Ștefănești-Argeș, România; gutaionelacatalina@yahoo.com (*corresponding author), ebuciumeanu@yahoo.com, visoiu_e@yahoo.com*

Abstract

Grapevine fleck virus produces a ubiquitous disease, latent in European grapevine varieties (*Vitis vinifera* L.) and in most American rootstocks, being found in all viticultural countries, in simple or complex infections with other more dangerous viruses. Different techniques for sanitation showed controversial results regarding fleck elimination in grapevine. *In vitro* culture of 'Tămâioasă românească' 3-2-2 with fleck virus and 'Burgund' 63 Mn with double infection (*Grapevine fleck virus* and *Grapevine virus A*), naturally infected have been submitted to chemotherapy using a combination of ribavirin and oseltamivir in three concentration variants (V1-40 mg/L ribavirin + 40 mg/L oseltamivir; V2-20 mg/L ribavirin + 40 mg/L oseltamivir; V3-20 mg/L ribavirin + 80 mg/L oseltamivir) and three consecutive subcultures. The plants regenerated after each subculture were evaluated by ELISA from the viewpoint of virus elimination and the RT-PCR was used for confirmation of the diagnostic. Due the phytotoxic effect of viricides, the multiplication rate decreased on experimental variants in the next subculture comparatively to the control, but no mortality of explants has been registered. *Grapevine fleck virus* has been 100% eliminated both from simple and mixed infections on all variants by *in vitro* chemotherapy, under the simultaneous action of two viricides. Unsatisfactory results have been achieved with *Grapevine virus A* elimination.

Keywords: ELISA, oseltamivir, ribavirin, RT-PCR, sanitation, *V. vinifera* L.

Introduction

Grapevine fleck virus (GFkV) produces a ubiquitous disease, latent in European grapevine varieties (*Vitis vinifera* L.) and in most American rootstocks, being found in all grapevine producing countries of the world (Martelli and Boudon-Padieu, 2006). The prevalence of the virus is very high in both simple and mixed infections with other more dangerous viruses (Cretazzo *et al.*, 2010; Komínek, 2008). Some comparative investigations of grapevine showed the negative influence of GFkV presence on *in vivo* and *in vitro* propagation, vigour, quality of the wood, agronomic and oenological parameters (Credi and Babini, 1996; Mannini *et al.*, 2012; Vișoiu *et al.*, 2000).

It is known that GFkV can be eliminated by heat therapy, meristem tip or fragmented shoot apex culture (Bovey and Martelli, 1992). However, while some researchers have reported its successful elimination by thermotherapy (Boulila *et al.*, 1990), others have found the heat treatment as unsatisfactory (Savino *et al.*, 1985; Panattoni and Triolo, 2010). GFkV-free plants (91%) have been obtained by meristem culture without thermotherapy and, the elimination of the virus by shoot tip culture also increased proportionally with the heat treatment period (Buciumeanu and Vișoiu, 2000). The achieving of healthy grapevine through somatic embryogenesis provided promising results, despite the high proportion of abnormal embryos and their low capacity of conversion into plants (Popescu *et al.*, 2003).

Ribavirin and oseltamivir are chemical drugs used in virus elimination in various horticultural species. Ribavirin has been successfully used in *Grapevine virus A* (GVA) (Panattoni *et al.*, 2007), *Grapevine fanleaf virus* (GFLV) (Weiland *et al.*, 2004), *Grapevine rupestris stem pitting-associated virus* (GRSPaV) (Skiada *et al.*, 2013) elimination and oseltamivir in *Grapevine leafroll-associated virus 3* (GLRaV-3) effectively removing (Panattoni *et al.*, 2006).

The use of viricides in media led to 37.5% GFkV-elimination with 10 mg/L ribavirin and 25% virus removing in the presence of 40 mg/L oseltamivir, as the highest sanitation rate achieved after one subculture in each case (Guță, 2010).

The aim of the present work was to eliminate GFkV in simple and mixed infections in two *Vitis vinifera* L. genotypes, by chemotherapy with ribavirin and oseltamivir used concomitantly, in controlled conditions of *in vitro* culture.

Materials and methods

Biological material. In vitro stabilized grapevine (*Vitis vinifera* L.) belonging to 'Tămâioasă românească' 3-2-2 proposed for approval and 'Burgund' 63 Mn approved clones of economic interest in Romanian viticulture have been used in the experiment. These genotypes have been identified as naturally infected with GFkV, and GFkV in addition with GVA, respectively.

The explants collected from infected plants have been cultured on solidified M&S (Murashige and Skoog, 1962) basic medium supplemented with 1mg/L benzylaminopurine (BAP) and 0.5 mg/L β-indolylacetic acid (β-IAA). The multiplication and regeneration processes carried out in the growth chamber, at 22 ± 1 °C, 16 hs photoperiod, 3000-3500 lx. Infected cultures were the source of inocula for grapevine chemical drug treatment.

Treatment with viricides. Ribavirin (1-β-D-ribofuranosyl-1, 2,4 triazone-3-carboxamide) and oseltamivir ([(3E,4R,5S)-4-acetamido-5-amino-3-(1-ethylpropoxy)-1-cyclohexane-1-carboxylic acid)]) as viricides belonging to inosin monophophate dehydrogenase (IMPDH) and neuraminidase inhibitors (NI) group, respectively, have been used. Ribavirin as pure substance and oseltamivir as capsules containing oseltamivir phosphate have been available. The viricide mixtures have been experienced in three variants (V1, V2, V3) of concentration during three consecutive subcultures (S1, S2, S3), as follows: V1= 40mg/L ribavirin + 40 mg/L oseltamivir; V2= 20mg/L ribavirin + 40 mg/L oseltamivir; V3 = 20mg/L ribavirin + 80 mg/L oseltamivir. The chemical drugs have been added to the medium before autoclaving.

The microshoot originated from stabilized cultures have been cut in single node fragments and subcultured on media containing viricides. Each inoculum constituted one repetition and has been individually evaluated.

The phytotoxic effect of viricides expressed by multiplication rate – number of formations specific for multiplications (adventive buds + primordia of shoots + shoots)/ initiated explants has been assessed after each subculture.

The treatment with viricides was followed by a subcultivation on proliferating medium without chemical drugs, than on rooting medium, rooted shoots being acclimatized.

Virus detection. In order to establish the efficiency of the method, expressed by virus elimination rate (no. of virus-free plants/no. of regenerated plants/initiated inoculum x 100), three months old acclimated plants, resulted from each subculture and variant, have been analyzed individually by ELISA and several ELISA negative, by RT-PCR.

ELISA was performed according to the method described by Clark and Adams (1977), using detection commercial kits.

RNA extraction and reverstranscription steps have been done using commercial kits protocols. Two pairs of primers have been used: GFkV585F (CTC AGC CTC CAC CTT GCC CCG T), GFkV 1117R (CAA TTT GGC TGG GCG AGA AGT ACA) and GFkV 6351F (CTC TCC GCC TCG TCT GAT GA), GFkV 7064R (TCG GTT CAT GAC GAG GGA GT) (Naidu et al., 2010). The initial denaturing was performed at 95°C for 10 min, followed by 40 cycles of 95 °C for 30 s, annealing (at 55 °C for the first primer set and 50 °C for the second) for 30 s, extension at 72°C for 60 s and final polymerization at 72°C for 7 min. The size of amplified DNA has been determined using the 100 bp molecular marker.

Results and discussion

The phytotoxic effect of viricides. Ribavirin has a phytotoxic effect expressed by a multiplication rate and mortality of inocula that is more pronounced comparatively to oseltamivir (Guţă *et al.*, 2009). The multiplication rate decreased after the first and the second subculture with 4.30 and 3.03 respectively, in the case of GFkV-infected Caner variety growth on media containing 40 mg/L ribavirin, so that the inocula did not survive after three subcultures. On the other hand, the oseltamivir induced an intense proliferation of adventive buds after the first subculture, formations that did not totally evolve in shoot elongation and differentiation, registering no mortality of inocula on any concentration of viricide (Guţă, 2010). The simultaneous use of ribavirin and oseltamivir for GFkV-elimination in 'Tămâioasă românească' 3-2-2 genotype allowed the triggering of caulogenesis processes on all variants and no mortality of explants after three consecutive subcultures has been registered. The visual observations revealed the shoot elongation and differentiation, as well as, their chlorosis depending on the viricides concentration, especially of ribavirin, whose phytotoxic effect is more evident as compared with that of oseltamivir.

The multiplication rate decreased both on experimental variants and subcultures comparatively to the control. After the first subculture, the highest concentration of ribavirin in V1 determined a higher degrease of the multiplication rate (2.2) in comparison with the control (3.0), the reducing by half of the viricide concentration, completed by the presence of oseltamivir causing an intense proliferation of adventive buds. Even a concentration of 40 mg/L ribavirin combined with 40 or 80 mg/L oseltamivir determined a decrease of the multiplication rate, but it did not induce the explants mortality, not even after three subcultures of treatment (Guţă, 2010). This behaviour is due, first of all to

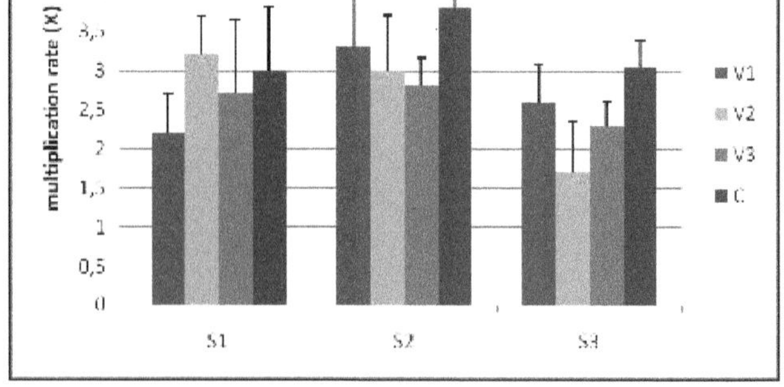

Fig. 1. The influence of viricides on grapevine multiplication rate to GFkV- infected 'Tămâioasă românească' 3-2-2 cv. The values are averages, the bars indicate the standard deviations

the different phytotoxic effect of viricides and, secondly, to the different tolerance to viricides and regenerative potential of the studied grapevine genotypes. The multiplication rate values demonstrated that the correlations regarding the influence of viricide concentrations and exposure time can hardly be made on the evolution of the cultures (Fig. 1).

The treatment with viricides applied to GFkV + GVA

naturally infected 'Burgund' 63 Mn genotype determined a striking increase of the multiplication rate after the first subculture, in all variants, as compared to the control.

The multiplication rate decreased by half in the second subculture to all variants, without differences brought about by the viricides concentration. This behaviour may be explained by the presence of viricides in media which induce a stimulation of adventive buds formation that, in the following subculture did not evolve in shoots differentiation and elongation.

Tab. 1. Grapevine plants 'Tămâioasă românească' 3-2-2 regenerated by *in vitro* chemotherapy with ribavirin and oseltamivir

Sub-cultures	Variants		
	V1	V2	V3
	Regenerated plants no. (average ±S.D, n=5)		
S1	25.6±0.8944	26.6±1.1401	25.8±0.8366
S2	30.0±1.5811	29.0±1.5811	25.4±0.5477
S3	28.4±1.1401	25.8±1.3038	26.6±1.1401

Virus elimination. Following the *in vitro* chemotherapy, there have been regenerated, on average, between 25 and 32 grapevines on each variant of 'Tămâioasă românească' 3-2-2 genotype (Tab.1); the plants have been individually analyzed by ELISA and all of them identified as GFkV-free.

The elimination of GFkV by *in vitro* chemotherapy started with the separate use of the ribavirin and oseltamivir viricides at different concentrations and treatment period. The highest elimination rate has been obtained at 10 mg/L ribavirin (37.5 %) or 40 mg/L oseltamivir (25 %) (Guţă, 2010). In the present work, the simultaneous use of ribavirin and oseltamivir led to the total elimination of GFkV in all experimental variants.

The management of chemotherapy efficiency with the aim of inactivation of viral multiplication led to the conclusion that a concentration of 20 mg/L ribavirin combined with 40 mg/L oseltamivir (V2) and one subculture of treatment is sufficient for a maximum effect of virus elimination.

In the viral complex infecting 'Burgund' 63 Mn genotype, GFkV has been 100% eliminated, while GVA registered the highest virus elimination rate of 9% after the first subculture of V1 (Tab. 2).

GFkV has been completely eliminated both from simple and mixed infections with GVA by the simultaneous use of ribavirin and oseltamivir viricides added in the medium before autoclaving, while some authors were successful in using the same substances ultrafiltered, added in the medim after autoclaving (Panattoni *et al.*, 2011). This has been based on the results of ribavirin FTIR (Fourier Transform Infrared Spectroscopy) spectral analysis at different temperatures, which demonstrated that at an autoclaving temperature of 120°C, the viricide is in a stable form (Topală, 2013).

Electrophoresis analysis of RT-PCR product showed a single amplified fragment of about 500 bp and 700 bp, in the infected plant used as control and no amplification product has been revealed in healthy regenerated vines.

In addition, *in vitro* chemotherapy as virus-elimination method in grapevine can be successfully applied, because the RAPD (Random Amplified Polymorphic DNA) results emphasized elsewhere that the viricide used in medium did not induce the genetic variability of virus-free regenerated plants (Guţă, 2010).

Conclusions

The concomitant use of ribavirin and oseltamivir led to the effective elimination of GFkV in 'Tămâioasă românească' 3-2-2 grapevine genotype. In the complex infection together with GVA, GFkV behaved similarly, being 100% eliminated in 'Burgund' 63Mn clone. The viricides mixture allowed the evolution of the cultures and the plants regeneration on all variants. Three month old acclimated plants have been analyzed by ELISA and the GFkV- free status has been confirmed by RT-PCR.

Acknowledgements

This work was supported by the Romanian National Authority for Scientific Research, CNDI-UEFISCDI, project number 104/2012 (PN-II-PT-PCCA-2011-3.1).

References

Boulila M, Boscia D, Di Terlizzi B, Castellano MA, Minafra A, Savino V, Martelli GP (1990). Some properties of a phloem limited non mechanically-transmissible grapevine virus. J Phytopathol 129:151-158.

Bovey R, Martelli GP (1992). Directory of major virus and virus-like diseases of grapevine. Mediterranean Fruit Crop Improvement Council/ICVG, 111 p.

Buciumeanu E, Vişoiu E (2000). Elimination of grapevine viruses in *Vitis vinifera* L. cultivars. Proc. 13th ICVG, 165-166.

Clark MF, Adams AN (1977). Characteristics of the microplate method of the enzyme-linked immunosorbent assay for the detection of plant viruses. J Gen Virol 34:475-483.

Tab. 2. Efficiency of *in vitro* chemotherapy with ribavirin and oseltamivir for GFkV+GVA elimination in grapevine plants 'Burgund' 63 Mn. Number of plants represents average±S.D. (n=5) of regenerated plants/inoculum initiated, ELISA analyzed

Subcultures	Variants								
	V1			V2			V3		
	Regenerated plants (no.)	GFkV free (%)	GVA Free (%)	Regenerated plants (no.)	GFkV Free (%)	GVA free (%)	Regenerated plants (no.)	GFkV Free (%)	GVA Free (%)
S1	19.2±1.3038	100	9	20.0±1.5811	100	3	19.8±1.4832	100	1
S2	9.4±1.5165	100	5	9.6±1.9493	100	1	10.0±2.2360	100	0
S3	9.0±1.5811	100	6	10.0±2.2360	100	2	9.8±1.9235	100	1

Credi R, Babini AR (1996). Effect of virus and virus-like infections on the growth of grapevine rootstocks. Adv Hort Sci 10:95-98.

Cretazzo E, Tomás M, Padilla C, Rosselló J, Medrano H, Padilla V, Cifre J (2010). Incidence of virus infection in old vineyards of local grapevine varieties from Majorca: implications for clonal selection strategies. Span J Agric Res 8(2):409-418.

Guţă IC (2010). Alternative method for obtaining virus-free grapevine propagating material. Univ. of Agronomic Science and Veterinary Medicine Bucharest, PhD Thesis, 208 p.

Guţă IC, Buciumeanu EC, Gheorghe RN, Teodorescu Al (2009). The assessement of viricides phytotoxicity used in grapevine *in vitro* chemotherapy. Bulletin UASVM 66(1-2):661.

Komínek P (2008). Distribution of grapevine viruses in vineyards of the Czech Republic. J Plant Pathol 90(2):357-358

Mannini F, Mollo A, Santini D, Gambino G, Tragni R (2012). Field Performances and Wine Quality Modification in a Clone of 'Nebbiolo' (*Vitis vinifera* L.) after Grapevine Fleck Virus Elimination. Proc. 17th ICVG:156-157.

Martelli Gp, Boudon-Padieu e (Eds.) (2006). Directory of infectious diseases of grapevines. In: Options méditeranéenes. Serie B: Studies and Research 55, 280 p.

Murashige T, Skoog F (1962). A revised medium for rapid grow and bioassays with Tabacco tissue culture. Physiol Plant 15:473-497.

Naidu RA, Mekuria TA (2010). Remove from marked Records First report of Grapevine fleck virus Washington vineyards. Plant Disease 94(6):784.

Panattoni A, D'Anna F, Cristani C, Triolo E (2007). Grapevine virus A eradication in *Vitis vinifera* explants by antiviral drugs and thermotherapy. J Virol Methods 146:129-135.

Panattoni A, D'Anna F, Triolo E (2006). Improvement in grapevine chemotherapy. Proc. 15th ICVG:139-141.

Panattoni A, Triolo E (2010). Susceptibility of grapevine viruses to thermotherapy on *in vitro* collection of Kober 5BB. Scientia Horticulturae 125:63-67.

Panattoni A, Luvisi A, Triolo E (2011). Selective chemotherapy on Grapevine leafroll-associated virus -1 and -3. Phytoparasitica 39:503-508.

Popescu CF, Buciumeanu E, Visoiu E (2003). Somatic embryogenesis, a reliable method for grapevine fleck virus free grapevine regeneration. Proc. 14th ICVG, 243.

Savino V, Boscia D, Martelli GP (1985). Incidence of some graft-transmissible virus-like diseases of grapevine in visually selected and heat-treated stokes from Southern Italy. Phytopathol Mediterr 24:204-207.

Skiada FG, Maliogka VI, Katis NI, Eleftheriou EP (2013). Elimination of Grapevine stem pitting–associated virus (GRSPaV) from two *Vitis vinifera* cultivars by *in vitro* chemotherapy. Eur J Plant Pathol 135:407-414.

Topală CM (2013). Temperature Effects On The FTIR Spectra Of Ribavirin. Rev Chim 64(2):132-135.

Vişoiu E, Buciumeanu E, Teodorescu Al (2000). A comparative investigation on the the *in vitro* behaviour of the virus-free and virus-infected vine biological material. Lucrări ştiinţifice UŞAMVB, B(XLIII):227-229.

Weiland CM, Cantos M, Troncosso A, Perez-Camacho F (2004). Regeneration of virus-free plants by *in vitro* chemotherapy of GFLV (Grapevine Fanleaf Virus) infected explants of *Vitis vinifera* L. cv. Zalema. Acta Hort 652:463-466.

8

Different Habitats Show Similar Genetic Structure of *Bunias orientalis* L. (Brassicaceae) in Lithuania

Jolanta PATAMSYTĖ[1], Tatjana ČĖSNIENĖ[1], Donatas NAUGŽEMYS[2], Violeta KLEIZAITĖ[1], Virginija TUNAITIENĖ[1], Virginija VAITKŪNIENĖ[1], Vytautas RANČELIS[1], Rita MIKALIŪNAITĖ[3], Donatas ŽVINGILA[1*]

[1]*Vilnius University, Department of Botany and Genetics, 21 M.K. Čiurlionio Str., LT-03101 Vilnius, Lithuania; donatas.zvingila@gf.vu.lt (*corresponding author)*

[2]*Vilnius University, Botanical Garden, 43 Kairėnų Str., LT-10239 Vilnius, Lithuania*

[3]*Šiauliai University, Department of Environmental Research, 19 P. Višinskio Str., LT-77156, Šiauliai, Lithuania*

Abstract

We studied genetic diversity within and among populations of warty cabbage (*Bunias orientalis* L.), which is an alien species in Lithuania and other Baltic countries. In Lithuania, this weed colonises two main types of habitats: railway/roadsides and meadows on riversides. The aim of this study was to assess the genetic structure of invasive populations of *B. orientalis* in Lithuania and consider the impact of diverse habitats on the partitioning of genetic diversity using inter-simple sequence repeat (ISSR) markers. An analysis of molecular variance (AMOVA) carried out on the basis of ISSR showed that there is high genetic differentiation (46%) among populations of *B. orientalis*, which is probably caused by the founder effect and limited gene flow. However, we observed no impact of habitat on the genetic difference among populations. Similar levels of ISSR polymorphic loci were observed in riverside (P = 31.67%) and railway/roadsides (P = 30.51%) populations. UPGMA cluster analysis and principal coordinate analysis (PCoA) also did not show grouping of studied populations according to habitat type. High genetic differentiation among populations, as indicated by ISSR markers, confirm multiple independent introductions of this species in Lithuania.

Keywords: alien species, biological invasions, invasion history, molecular markers

Introduction

Biological invasion is a multistage process influenced by many factors (Sakai *et al.*, 2001). Intra-specific genetic variation may contribute significantly to the invasiveness of an alien species (Wang *et al.*, 2012; Ward *et al.*, 2008). The founding events usually establish only a fraction of the genetic variance that occurs in the source population(s) (Dlugosch and Parker, 2008; Nei *et al.*, 1975). In spite of frequent loss of genetic diversity, populations of some invaders may evolve rapidly (Andrew *et al.*, 2012; Dlugosch and Parker, 2008; Ellstrand and Schirenbeck 2000) and adapt to local conditions (Riis *et al.*, 2010).

Our study focuses on *B. orientalis*, which is commonly referred to as warty cabbage, hill mustard or Turkish rocket. It is native to the south-west part of Russia, the Caucasus, and western Siberia, where it grows in grasslands, at the sunny edges of forests, in dry meadows and near rivers (Birnbaum, 2006; Laiviņš *et al.*, 2006). *B. orientalis* is invasive agricultural weed in Central Europe and expands to other parts of the continent (Dietz *et al.*, 1999; Mirek and Piękoš-Mirkowa, 1996). The species is also spreading in North America and Asia. In Baltic countries, especially Latvia and Estonia, *B. orientalis* is common (Birnbaum, 2006; Laiviņš *et al.*, 2006). The first record of this species in Lithuania is from 1898 (Gudžinskas, 1997). At present, *B. orientalis* is not evenly spread through the whole country. It usually colonises two types of habitats: railways/roadsides and meadows on riversides. These two *B. orientalis* locations could be influential in determining the means by which the species spreads, which may result in local adaptation of genotypes from populations localised in these habitats. The first type of habitat, railways/roadsides, is strongly influenced by human activity and is more or less artificial, while a fraction of the populations in meadows on riversides in Lithuania are localised in more natural or semi-natural habitats. Ecological differences between habitats may result in genetic structuring of populations through the action of selection or through limited gene flow (Ward, 2006). On the other hand, roads and river valleys are well known as corridors for the spreading of invasive plant species (Christen and Matlack, 2006; Galil *et al.*, 2007; Hejda and Pyšek, 2006). To assess the level of genetic differentiation among populations and the possible impact of habitat on genetic differentiation of *B. orientalis* populations, we used DNA markers as valuable tools in invasive plant population studies (Jasieniuk and Maxwell, 2001). These techniques have also been applied in studies of genetic differentiation among populations existing in different ecological conditions (Owuor *et al.*, 1997; Patamsytė *et al.*; 2010; Reisch *et al.*, 2003, 2005; Trtikova *et al.*, 2011). In our previous study, we identified molecular markers suitable for genomic DNA analysis of *B. orientalis* using the RAPD and ISSR techniques (Patamsytė *et al.*, 2011). RAPD and ISSR assays do not require sequence information for primer synthesis; these techniques are quick and usually identify many polymorphic loci. Moreover, ISSRs exhibit the specificity of microsatellite markers and amplification at higher-stringency conditions, which enables high reproducibility of this assay (Ge *et al.*, 2003; Słomka *et al.*, 2011). Microsatellite markers usually considered to be selectively neutral, are known to be linked to coding regions and possibly mark gene rich regions (Kojima *et al.*, 1998; Reddy *et al.*, 2002). In this study, we used ISSR to answer the following questions: Is there genetic differentiation among populations of *B. orientalis*? How does the genetic diversity correlate with habitat type of study populations? Is there differentiation among populations from different habitats that differ in ecological conditions? Can we consider the spreading of *B. orientalis* in Lithuania to be the result of multiple introductions?

Material and methods

Plant material

Bunias orientalis (L.) is a diploid (2n=14), mainly outcrossing perennial grass (Dietz *et al.*, 1999). To study the genetic diversity within and among populations of *B. orientalis*, nineteen populations from the central and eastern regions of Lithuania (where this species is mainly spread) were chosen (Tab. 1). Plants from populations that were predominantly localised in two different habitats – riverside meadows and railway/roadsides – were collected. These two types of habitat vary according temperature, water fluctuations and environmental pollution. Characteristic properties of studied riverside populations are high water and low temperature fluctuations during growing season and low environmental pollution. In contrast high fluctuations of temperature, environmental pollution, and low fluctuations of water are typical for railway/roadside habitats. Young, disease-free, undamaged leaves were collected from at least 8 different plants per population. A total 309 of plants were sampled and analysed in this study.

DNA isolation and ISSR-PCR analysis

The samples (approximately 100 mg fresh leaves) were processed on-site using Bashing Bead™ and Xpedition™ Sample Processor Technologies (Zymo Research). The samples were homogenised in Xpedition™Lysis/Stabilization solution, and genomic DNA was isolated in a laboratory using a Xpedition$_{TM}$ Plant/Seed DNA Mini prep kit (Zymo Research) following the manufacturer instruc-

Tab. 1. Populations of *Bunias orientalis* analysed in this study and they habitats

No.	Population Name	Code	Site coordinates	Sample size	Habitat
Riverside					
1	Vilnius A	VILA	54°40'51", 25°20'45"	20	Meadow near river
2	Seredžius A	SERA	55°04'47", 23°25'31"	20	Meadow near river
3	Darsūniškis	DAR	54°44'14", 24°07'06"	14	Meadow near river
4	Seredžius B	SERB	55°04'19", 23°21'21"	15	Meadow near river
5	Pamerkiai	PAM	54°24'59", 24°54'45"	20	Riverside
Railway/roadside					
6	Viduklė	VID	55°24'56", 22°50'52"	14	Meadow near railway line
7	Vilnius B	VILB	54°38'09", 25°12'40"	14	Railway bed
8	Vokė	VOK	54°38'01", 25°07'45"	12	Meadow near railway line
9	Kaišiadorys	KAI	54°51'46", 24°28'01"	11	Railway bed
10	Lazdėnai	LAZ	54°44'45", 24°56'45"	18	Meadow near railway line
11	Kariotiškės	KAR	54°40'35", 25°00'22"	9	Railway bed
12	Marijampolė	MAR	54°33'31", 23°21'59"	8	Railway bed
13	Obeliai	OBE	55°57'11", 25°49'15"	20	Meadow near road
14	Molėtai	MOL	55°13'17", 25°26'10"	20	Meadow near highway
15	N. Vilnia	NVIL	54°41'39", 25°26'53"	20	Railway bed and meadow near railway
16	Žąsliai	ZAS	54°50'16", 24°34'59"	22	Railway bed and meadow near railway
17	Lentvaris	LEN	54°38'40", 25°03'19"	20	Bush and meadow near railway
18	Kaunas	KAU	54°52'51", 23°56'04"	20	Railway bed
19	Kurniškės	KUR	54°51'45", 24°30'47"	12	Railway bed

tions. Five ISSR primers, selected in previous study and generating clear and reproducible ISSR banding patterns, were chosen for the analysis of *B. orientalis* (Patamsytė *et al.*, 2011). ISSR-PCR was performed in a 10 µl reaction volume containing approximately 10 ng of genomic DNA, 1 unit recombinant Taq DNA polymerase (Thermo Fisher Scientific Baltics), 1µl 10·PCR buffer with KCl (100 mM Tris-HCl, 500 mM KCl, 0.8% Nonidet P40), 300 µM $MgCl_2$, 200 µM dNTPs and 0.4 µM of the primer. The PCR program consisted of an initial 7 min denaturation at 94 °C, followed by 32 cycles of 30 s at 94 °C, 45 s at a primer-specific temperature, and 120 s at 72 °C. The last reaction extension step was performed at 72 °C for 7 min. ISSR-PCR products were separated on 1.5% agarose gels stained with ethidium bromide. The GeneRuler™ DNA Ladder Mix, 100-10000 bp (Thermo Fisher Scientific Baltics), was used for DNA band sizing. All of the samples were analysed at least twice in different ISSR-PCR experiments. Samples from a few different populations were usually analysed within the same gel, and the monomorphic bands, though not included in population variability data analysis, were used as size and reproducibility criteria to ensure correct loci scoring among populations. All of the amplifications also contained negative control (without DNA) samples. We also applied the data evaluation recommendations of Bonin *et al.* (2004) to calculate the error rate in the genotyping analysis and followed the suggestions of these authors to minimise this error. The error rate was 6.4% (169 differences in 2640 comparisons).

Statistical analysis

ISSR profiles were documented using the BioDocAnalyse gel documentation system (Biometra, Germany). An example of typical ISSR profiles detected with different primers is shown in Fig. 1. Each clear and reproducible DNA fragment was scored across all samples as either present (1) or absent (0). Based on these data, a binary matrix was generated. The parameters of molecular diversity within populations (the percentage of polymorphic loci (P), Nei's gene diversity ($h=1-\Sigma p^2_i$), Shannon's information index ($I=-\Sigma p_i \log_2 p_i$), coefficient ofgenetic differentiation ($G_{ST} = (H_T-H_S)/H_T$) and gene flow ($Nm=0.5(1-G_{ST})/G_{ST}$) were estimated using the computer program POPGENE version 1.32 (Yeh and Boyle, 1999). This program was also used for cluster analysis, which was carried out using Nei's genetic distance with an unweighted pair group method and arithmetic mean (UPGMA).

To estimate pairwise genetic distances (Φ_{ST}) between populations and their levels of significance, analysis of molecular variance (AMOVA) was carried out. The binary matrix based on ISSR profiles was used in AMOVA with the computer program GenAIEx version 6.5 (Peakall and Smouse, 2012). For each analysis, 999 permutations were carried out to obtain significance levels. To assess possible habitat influence on genetic differentiation among populations, an additional AMOVA was performed, in which the grouping of populations was based on their reliance on particular habitats. For correlation analyses, the STATIS-

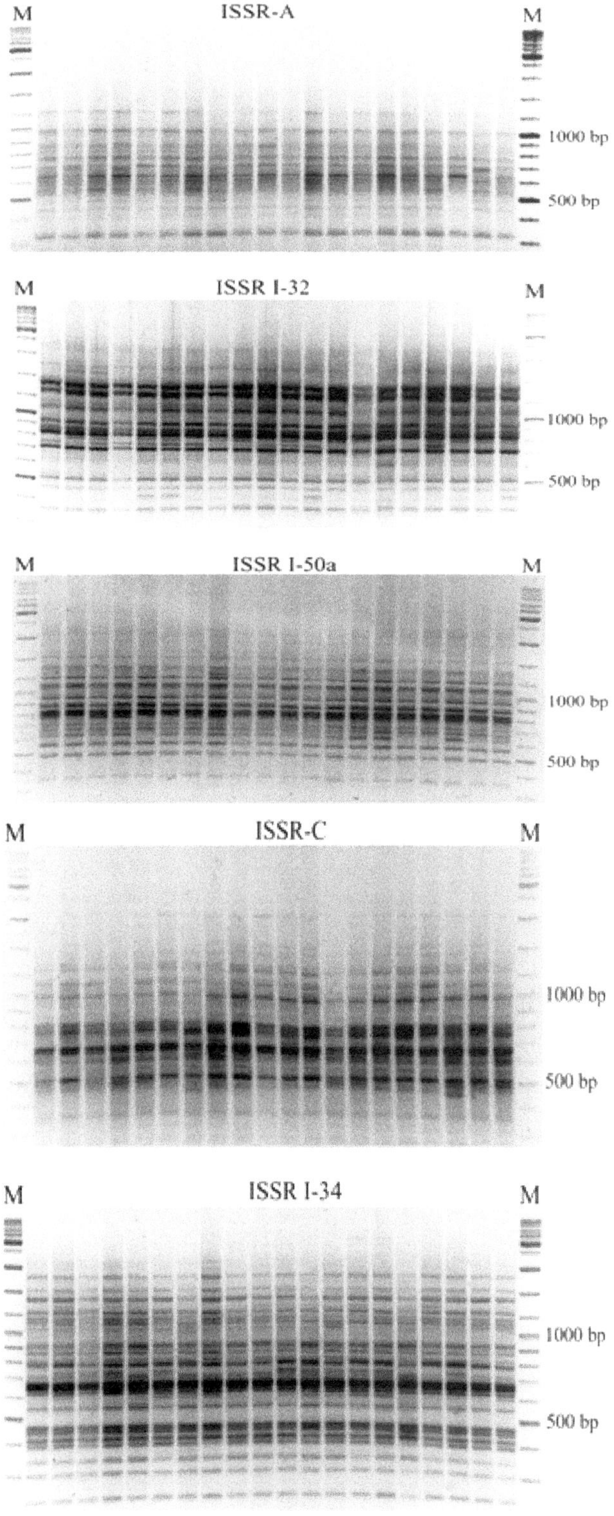

Fig. 1. ISSR profiles obtained from individual plants of *B. orientalis* using five oligonucleotide primers (ISSR-A, ISSR-C, ISSR I-32, ISSR I-34, ISSR I-50a). M-molecular size marker (The GeneRuler™ DNA Ladder Mix, 100-10000 bp)

TICA version 8.0 program (StatSoft. Inc., 2008; www.statsoft.com) was used.

Results

ISSR polymorphism at species level

The genetic variability study of *B. orientalis* was based on an analysis of 309 individuals from 19 populations (approx. 16 individuals/population) located in two main habitat types. Eighty-eight reliable ISSR loci were identified in this study using five preselected primers (Tab. 2). The average percentage of polymorphic loci at species level was 55%. In spite of moderate genetic diversity, 299 genotypes were identified among all studied plants. Ten genotypes that had two identical clones were identified in six populations (MOL, ZAS, PAM, VID, LEN and KUR). No clones were shared between populations. The size of scored DNA fragments ranged from 390 to 1800 bp. The used primers generated between 40% and 64.71% polymorphic loci. Primers based on dinucleotide repeats (ISSR-A and ISSR-C) revealed the highest polymorphism. This parameter was highest (64.71%) with the ISSR-C primer based on (AG) repeat.

Molecular diversity within populations

The number of polymorphic loci displayed per population ranged from 5 (KUR) to 21 (LAZ), and the percentage of polymorphic bands per population ranged from 10.42% to 43.75% respectively (Tab. 3). The mean level of polymorphic loci in analysed populations of *B. orientalis* was 30.81%. The Nei's gene diversity values and Shannon's information index values were also lowest in the KUR population ($h = 0.039$; $I = 0.058$) and highest in the LAZ population ($h = 0.164$; $I = 0.243$). In the small KUR population, only nine different genotypes were identified among twelve studied individuals. In this population, three distinct ISSR patterns were each shared by a different pair of plants sampled near each other. In the LAZ population, all 18 individuals were genetically different. The MOL population also showed low genetic diversity ($P = 20.83\%$; $h = 0.081$; $I = 0.119$). The ZAS population contained more genotypes than any other population examined in this study: 21 genotypes were identified among 22 studied individuals. Some of the 48 polymorphic loci were fixed in certain populations. For example, the ISSR-A_{970} locus was fixed in 15 populations, and the ISSR-C_{850} locus was fixed in all except one MAR population (data not shown).

Genetic differentiation of populations

The mean G_{ST} for all loci was 0.507, which indicates that 50.7% of the genetic diversity occurs among populations and 49.3% occurs within populations (data not shown). The gene flow estimated from G_{ST} was 0.487. Similar results were obtained with AMOVA. AMOVA revealed that 46% of the total genetic variance occurred among populations and that 54% occurred among individuals within populations (Tab. 4). The variation in ISSR profiles among and within populations was highly significant ($P < 0.001$). The pairwise genetic differentiation of populations did not reveal reliance on geographic distance ($r = -0.08$; $P = 0.05$). The lowest $\Phi_{ST} = 0.10$ was established between the SERA and ZAS populations (Tab. 5). The largest Φ_{ST} value was 0.72, between VOK and KUR populations. Upon arranging the *B. orientalis* populations in two groups according to their origin (meadows on riversides and railway/roadsides) (Tab. 1), no molecular differentiation among populations from different habitats was observed. This same result was obtained when these two groups were different in size (5 riverside populations versus 12 railway/roadsides populations) and when they were equal in size (5 riverside and 5 arbitrarily chosen railway/roadsides populations). Similar levels of ISSR polymorphism were observed in riverside ($P = 31.67\%$) and railway/roadsides populations ($P = 30.51\%$) (Tab. 3).

The relationships among individuals and populations were also analysed using UPGMA cluster analysis and principal coordinate analysis (PCoA). Neither analysis revealed any grouping of populations according to geographic location or habitat type. The UPGMA dendrogram produced using Nei's genetic distance is shown in Fig. 2. The dendrogram consists of three clusters and a group of three populations (KUR, SERB and MAR) located separately from these clusters. All clusters are heterogeneous and

Tab. 2. Oligonucleotide primers used and ISSR loci revealed in *Bunias orientalis*

Primer	Primer sequence (5'-3')	Annealing temperature (°C)	Approx. size of DNA fragments (bp)	No. of loci	No. of polymorphic loci	P (%)
ISSR-A	$CTC(GT)_8$	51	570-1200	11	7	63.63
ISSR-C	$(AG)_8TG$	51	490-1250	17	11	64.71
ISSR I-50a	$CCA(GCT)_4$	46	480-1500	20	8	40.00
ISSR I-34	$(AGC)_4GG$	46	390-1800	25	15	60.00
ISSR I-32	$(AGC)_4C$	39	470-1450	15	7	46.67
			Total	88	48	
			Mean ± SE	17.60 ± 2.36	9.60 ± 1.54	55.00 ± 4.94

P (%): percentage of polymorphic loci; SE: standard error

consist of populations from different habitat types. For example, the first cluster includes two riverside populations (VILA and DAR) and four railway/roadside populations (VID, KAI, NVIL and VOK). The second cluster includes two riverside populations (SERA and PAM), three railway populations (ZAS, VILB and LAZ) and two populations sampled in meadows near highways (MOL and OBE). The third small cluster includes exclusively railway/roadsides populations (KAR, LEN and KAU). The remaining unclustered populations also originate from different habitat types. Populations sampled in KUR and MAR are located near the railway, while the SERB population is located near a river. The PCoA also shows a similar heterogeneous pattern of population grouping (Fig. 3). The population grouping in the PCoA plot was similar to that in the UPGMA dendrogram. PCoA revealed no grouping by habitat type. For example, two geographically related riverside populations, SERA and SERB, are located separately in both the UPGMA dendrogram and PCoA plot. Likewise, the MAR, SERB and KUR populations are most differentiated from the main group of populations according to both analyses. In the PCoA of individual genotypes, approximately 44.81% of the total variation is described by the first two axes. This assay revealed that genotypes of the same population are grouped in rather compact groups; however, there is considerable population overlap (data not shown).

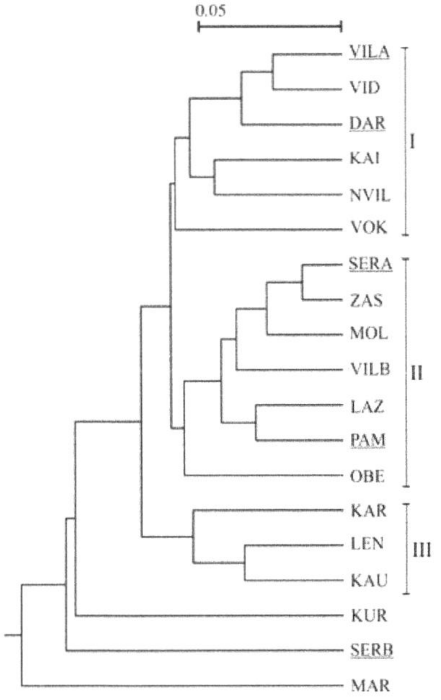

Fig. 2. UPGMA dendrogram constructed using genetic distance matrix based on data from 48 polymorphic ISSR markers identified on 309 *B. orientalis* individuals. Riverside populations codes are underlined. Population codes are given as in Tab. 1

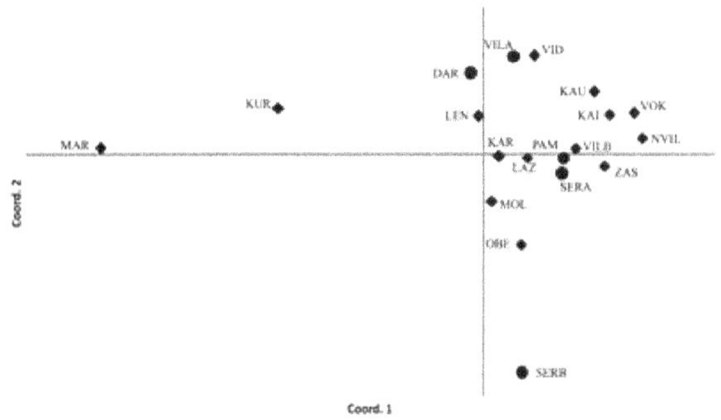

Fig. 3. Principal coordinate analysis (PCoA) plot based on ISSR markers showing similarity and genetic variation among nineteen populations of *B. orientalis*. Diamonds indicate railway/roadside populations, circles – riverside populations. Population codes are given as in Tab. 1

Tab. 3. Estimates of genetic diversity parameters of nineteen populations of *B. orientalis*. Population codes are given as in Tab. 1

Populations	P (%)	Number of polymorphic loci	h	I
		Riverside		
VILA	31.25	15	0.124	0.182
SERA	41.67	20	0.167	0.245
DAR	29.17	14	0.104	0.155
SERB	33.33	16	0.109	0.165
PAM	22.92	11	0.096	0.139
Mean ± SE	31.67 ± 3.05	15.20 ± 1.46	0.120 ±0.013	0.177 ± 0.018
		Railway/roadside		
VID	31.25	15	0.121	0.178
VILB	27.08	13	0.110	0.161
VOK	27.08	13	0.109	0.159
KAI	29.17	14	0.111	0.163
LAZ	43.75	21	0.164	0.243
KAR	25.00	12	0.105	0.151
MAR	27.08	13	0.115	0.165
OBE	33.33	16	0.146	0.209
MOL	20.83	10	0.081	0.119
NVIL	39.58	19	0.141	0.211

Populations	P (%)	Number of polymorphic loci	h	I
ZAS	29.17	14	0.112	0.163
LEN	41.67	20	0.137	0.206
KAU	41.67	20	0.162	0.239
KUR	10.42	5	0.039	0.058
Mean ± SE	30.51 ± 2.44	14.64 ± 1.17	0.118 ± 0.009	0.173 ± 0.013
Mean of all populations	30.81 ± 1.93	14,79 ± 0.93	0.119 ± 0.007	0.174 ± 0.011

P (%): percentage of polymorphic loci; h: Nei's gene diversity; I: Shannon's information index; SE: standard error

Tab. 4. Analysis of molecular variance (AMOVA) for 309 individuals of *B. orientalis* sampled from nineteen populations using 48 markers

Source of variation	df	SS	MS	Est. Var.	P-value	%
Among regions	1	44.934	44.934	0.000		0
Among populations	17	692.073	40.710	2.357	<0.001	46
Within populations	290	798.016	2.752	2.752	<0.001	54
Total	308	1535.023		5.109		100

df: degrees of freedom; SS: sum of squares; MS: mean squares; Est. Var.: variance component estimates; P-value: significance ($P < 0,001$) based on permutation (999 replicates); %: percentage of total variance

Tab. 5. Pairwise genetic distance (Φ_{ST}) and geographic distance among populations of *B. orientalis* included in this study (Φ_{ST} values are shown below the diagonal and geographic distance in kilometres are shown above diagonal)

	VILA	SERA	DAR	SERB	PAM	VID	VILB	VOK	KAI	LAZ	KAR	MAR	OBE	MOL	NVIL	ZAS	LEN	KAU	KUR	
VILA		131.6	79.5	135.0	40.8	179.4	10.2	15.0	60.3	26.8	22.6	128.7	144.1	59.6	6.7	52.0	19.7	93.6	57.3	VILA
SERA	0.39		59.4	4.6	121.6	51.7	125.7	120.9	71.3	104.9	111.5	59.5	179.4	129.1	137.0	79.6	115.2	39.8	74.3	SERA
DAR	0.24	0.35		61.5	62.5	110.9	71.7	66.4	26.9	53.6	57.3	52.5	173.0	99.9	85.8	32.0	60.8	20.4	29.7	DAR
SERB	0.61	0.37	0.59		124.0	50.3	129.0	124.1	74.8	108.3	114.7	57.5	184.0	133.6	140.5	83.0	118.4	42.5	77.8	SERB
PAM	0.41	0.35	0.49	0.55		172.8	30.9	27.9	57.9	37.1	28.3	101.3	180.0	95.0	46.7	51.1	27.4	81.9	56.3	PAM
VID	0.20	0.42	0.39	0.60	0.436		174.4	169.8	119.7	153.0	160.5	102.0	195.4	165.1	184.0	128.0	164.3	91.0	122.4	VID
VILB	0.41	0.26	0.44	0.53	0.47	0.42		5.4	54.7	21.5	14.4	119.7	151.4	66.4	16.8	46.4	10.9	87.0	51.9	VILB
VOK	0.41	0.34	0.40	0.57	0.58	0.42	0.40		50.1	17.5	9.4	114.3	152.8	67.7	21.7	41.8	5.7	82.0	47.4	VOK
KAI	0.38	0.35	0.35	0.54	0.51	0.41	0.32	0.36		33.6	40.9	79.0	147.7	73.0	65.8	8.4	44.6	34.2	3.2	KAI
LAZ	0.40	0.26	0.33	0.47	0.31	0.42	0.35	0.37	0.40		9.8	104.4	144.4	60.4	32.6	25.3	12.9	66.9	30.7	LAZ
KAR	0.48	0.28	0.46	0.55	0.54	0.51	0.34	0.46	0.44	0.37		106.1	152.0	67.2	29.2	32.6	3.7	72.6	38.2	KAR
MAR	0.56	0.56	0.54	0.65	0.66	0.55	0.58	0.64	0.61	0.52	0.57		220.1	151.7	135.3	84.4	109.1	52.0	81.9	MAR
OBE	0.44	0.31	0.45	0.44	0.50	0.44	0.29	0.48	0.38	0.37	0.41	0.44		85.1	141.1	146.7	152.6	168.1	145.9	OBE
MOL	0.52	0.21	0.46	0.40	0.50	0.55	0.42	0.51	0.50	0.39	0.40	0.59	0.37		57.5	69.0	67.7	102.6	70.3	MOL
NVIL	0.31	0.37	0.40	0.50	0.42	0.27	0.41	0.41	0.27	0.44	0.49	0.57	0.34	0.48		57.6	26.3	99.5	62.7	NVIL
ZAS	0.47	0.10	0.41	0.43	0.35	0.47	0.29	0.39	0.36	0.32	0.40	0.61	0.39	0.29	0.37		36.3	42.0	5.7	ZAS
LEN	0.59	0.44	0.51	0.61	0.58	0.59	0.53	0.53	0.49	0.47	0.41	0.61	0.55	0.49	0.53	0.43		76.3	41.9	LEN
KAU	0.44	0.38	0.37	0.57	0.56	0.45	0.43	0.37	0.36	0.42	0.31	0.58	0.47	0.46	0.43	0.42	0.31		37.4	KAU
KUR	0.67	0.57	0.60	0.70	0.69	0.69	0.71	0.72	0.71	0.57	0.67	0.70	0.66	0.65	0.63	0.58	0.60	0.60		KUR

Discussion

In the study of *B. orientalis*, we observed rather different levels of genetic variation in the populations ranging from 10.42% (KUR population) to 43.75% (LAZ population). The mean intra-populational ISSR polymorphism was rather low (30.81%), which implies a possible reduction of genetic diversity during the process of invasion (Dlugosch and Parker, 2008). It is worthwhile to mention that there is no published analysis of the genetic structure of *B. orientalis* in its native range. However, previous plant population studies indicate that genetic variability strongly depends on the species life history traits (Nybom and Bartish, 2000). *B. orientalis*, as an allogamous and perennial plant species, should maintain a high level of genetic variability in its native range. Additionally, high polymorphism of RAPD loci in a single invasive population of *B. orientalis* was reported by Dietz *et al.* (1999). In our study, we determined that genetic variation was almost equally partitioned between and within populations using ISSR markers. The average G_{ST} estimate was 0.507, ranging from 0.172 to 0.992 for different loci (data not shown).

AMOVA also detected significant population differentiation. The obtained distribution of genetic variation shows deviation from the variance pattern established previously for outcrossing plant species. The estimate of population differentiation obtained in our study (46%) is higher than the average Φ_{ST} values reported by Nybom (2004) for outcrossers (0.27) and short-lived perennials (0.41). Increased genetic differentiation in *B. orientalis* populations can be explained by the impact of random genetic processes during invasion. Though *B. orientalis* is an outbreeding species, the plants are self-fertile (Birnbaum, 2006). The founder effect and lack of gene flow in the new range may increase the probability of inbreeding because a low density of plants at a given site restricts mating opportunities. In this situation, some alleles associated with incompatibility may also be lost (Byers, 1995). In our study, the mean gene flow (N_m) estimated from G_{ST} was 0.487. This N_m estimate indicates limited gene flow among populations (Li and Jin, 2006). The absence of correlation between geographic distance and pairwise Φ_{ST} of populations suggests multiple introductions of the species (Ge *et al.*, 2003).

B. orientalis occupies two main types of habitats in Lithuania: riversides and railways/roadsides. The colonisation of these habitats may be associated with different ways of spreading of this alien species and/or the selection mechanism. The spreading of *B. orientalis* on riversides in Lithuania may be associated with the rather wide network of flow mills that were exploited until the middle of the 20th century and that employed water engines. The transportation and processing of grain and forage contaminated with warty cabbage seeds also may have promoted the establishment of founding populations of this species near railways and roadsides. The role of the transport network as one of the most important factors in spreading has been emphasised for *Impatiens glandulifera*, another alien species in Lithuania (Zybartaite *et al.*, 2011). The two types of *B. orientalis* habitat are rather different regarding water availability, light and the impact of environmental pollution. In some previous studies, it was found that the ecological properties of habitats influenced population genetic structure (Owuor *et al.*, 1997; Patamsytė *et al.*, 2010; Reisch *et al.*, 2003). On the other hand, other publications do not show evidence of adaptive genetic differentiation by means of molecular markers (Holderegger *et al.*, 2006;Volis *et al.*, 2003). The finding that there was no evidence of differentiation (AMOVA, UPGMA and PCoA results) in populations of *B. orientalis* colonising different habitats could also mean that both habitats are equally prone to plant invasions, and show the same "corridor" structure.

The specific vegetation conditions of railway/roadside populations, which experience increased environmental pollution, may result in the appearance of new DNA banding patterns due to a genotoxic effect of environmental pollutants (Česnienė *et al.*, 2010; Fedorova *et al.*, 2007; Słomka *et al.*, 2011). However, we did not find higher ISSR variation in plants growing near railways or roads in comparison with plants growing on riversides.

Conclusions

Our study confirms multiple introductions of *B. orientalis* in Lithuania. An analysis carried out using ISSR markers revealed that there is high genetic differentiation among *B. orientalis* populations, which is probably caused by the founder effect and limited gene flow. However,we observed no impact of different ecological conditions on the distribution of ISSR markers and genetic differentiation among populations in two habitat types. Additional studies of genetic diversity in native populations of *B. orientalis* are necessary to elucidate the mechanisms of invasion for this species.

Acknowledgments

This research was founded by grant (LEK-07/2012) from the Research Council of Lithuania.

References

Andrew RL, Ostevik KL, Ebert DP, Rieseberg LH (2012). Adaptation with gene flow across the landscape in a dune sunflower. Mol Ecol 21:2078-2091.

Birnbaum C (2006). NOBANIS – Invasive Alien Species Fact Sheet – *Bunias orientalis*. – From: Online Database of the North European and Baltic Network on Invasive Alien Species – NOBANIS www.nobanis.org, Date of access 1/10/2012.

Bonin A, Bellemain E, Bronken Eidesen P, Pompanon F, Brochmann C, Taberlet P (2004). How to track and assess genotyping errors in population genetics studies. Mol Ecol 13:3261-3273.

Byers DL (1995). Pollen quantity and quality as explanations for low seed set in small populations exemplified by *Eupatorium* (Asteraceae). Am J Bot 82:1000-1006.

Christen D, Matlack G (2006). The role of roadsides in plant invasions: a demographic approach. Conserv Biol 20:385-391.

Česnienė T, Kleizaitė V, Ursache R, Žvingila D, Radzevičius A, Patamsytė J, Rančelis V (2010). Soil-surface genotoxicity of military and urban territories in Lithuania, as revealed by *Tradescantia bioassays*. Mutat Res Genet Toxicol Environ Mutagen 697:10-18.

Dietz H, Fischer M, Schmid B (1999). Demographic and genetic invasion history of a 9-year-old roadside population of *Bunias orientalis* L. (Brassicaceae). Oecologia 120:225-234.

Dlugosch KM, Parker IM (2008). Founding events in species invasions: genetic variation, adaptive evolution, and the role of multiple introductions. Mol Ecol 17:431-449.

Ellstrand NC, Schierenbeck KA (2000). Hybridization as a stimulus for the evolution of invasiveness in plants. Proc Natl Acad Sci USA 97:7043-7050.

Galil B, Nehring S, Panov VE (2007). Waterways as invasion highways – impact of climate change and globalization, p.

59-74. In: Netwig W (Ed.). Biological invasions, Ecological Studies. Berlin, Heidelberg.

Ge XJ, Yu Y, Zhao NX, Chen HS, Qi WQ (2003). Genetic variation in the endangered Inner *Mongolia* endemic shrub *Teteaena mongolica* Maxim. (Zygophyllaceae). Biol Conserv 111:427-434.

Gudžinskas Z (1997). Conspectus of alien plant species of Lithuania. 3. Brassicaceae. Botanica Lithuanica 3:215-249.

Fedorova AI, Kalaev VN, Prosvirina YuG, Goryainova SA (2007). Mutagenic activity of heavy metals in soils of wayside slopes. Eurasian Soil Sci 40:998-1005.

Hejda M, Pyšek P (2006). What is the impact of *Impatiens glandulifera* on species diversity of invaded riparian vegetation? Biol Conserv 132:143-152.

Holderegger R, Kamm U, Gugerli F (2006). Adaptive vs. neutral genetic diversity: implications for landscape genetics. Landscape Ecol 21:797-807.

Jasieniuk M, Maxwell BD (2001). Plant diversity: new insights from molecular biology and genomics technologies. Weed Sci 49:257-265.

Kojima T, Nagaoka T, Noda K, Ogihara Y (1998). Genetic linkage map of ISSR and RAPD markers in Einkorn wheat in relation to that of RFLP markers. Theor Appl Genet 96:37-45.

Laiviņš M, Priede A, Krampis I (2006). Distribution of *Bunias orientalis* in Latvia. Botanica Lithuanica 12:69-77.

Li JM, Jin ZX (2006). High genetic differentiation revealed by RAPD analysis of narrowly endemic *Sinocalycanthus chinensis* Cheng et S.Y. Chang, an endangered species of China. Biochem Syst Ecol 34:725-735.

Mirek Z, Piękoś-Mirkowa H (1996). Distribution and habitats of *Bunias orientalis* in the northern sub-Tatra region. Oecologia Montana 5:79-82.

Nei M, Maruyama T, Chakraborty R (1975). The bottleneck effect and genetic variability in populations. Evolution 29:1-10.

Nybom H (2004). Comparison of different nuclear DNA markers for estimating intraspecific genetic diversity in plants. Mol Ecol 13:1143-1155.

Nybom H, Bartish IV (2000). Effects of life history traits and sampling strategies on genetic diversity estimates obtained with RAPD markers in plants. Perspect Plant Ecol Evol Syst 3:93-114.

Owuor ED, Fahima T, Beiles A, Korol A, Nevo E (1997). Population genetic response to microsite ecological stress in wild barley, *Hordeum spontaneum*. Mol Ecol 6:1177-1187.

Patamsytė J, Čėsnienė T, Naugžemys D, Kleizaitė V, Vaitkūnienė V, Rančelis V, Žvingila D (2011). Genetic diversity of warty cabbage (*Bunias orientalis* L.) revealed by RAPD and ISSR markers. Žemdirbystė Agriculture 98:293-300.

Patamsytė J, Kleizaitė V, Čėsnienė T, Rančelis V, Žvingila D (2010). The genetic structure of red raspberry (*Rubus idaeus* L.) populations in Lithuania. Cent Eur J Biol 5:496-506.

Peakall R, Smouse PE (2012). GenAlEx 6.5: genetic analysis in Excel. Population genetic software for teaching and research. Bioinformatics 28:2537-2539.

Reddy MP, Sarla N, Siddiq EA (2002). Inter simple sequence repeat (ISSR) polymorphism and its application in plant breeding. Euphytica 128:9-17.

Reisch C, Anke A, Röhl M (2005). Molecular variation within and between ten populations of *Primula farinosa* (Primulaceae) along an altitudinal gradient in the northern Alps. Basic Appl Ecol 6:35-45.

Reisch C, Poschlod P, Wingender R (2003). Genetic differentiation among populations of *Sesleria albicans* Kit. ex Schultes (Poaceae) from ecologically different habitats in central Europe. Heredity 91:519-527.

Riis T, Lambertini C, Olesen B, Clayton JS, Brix H, Sorrell BK (2010). Invasion strategies in clonal aquatic plants: are phenotypic differences caused by phenotypic plasticity or local adaptation? Ann Bot 106:813-822.

Sakai AK, Allendorf FW, Holt JS, Lodge DM, Molofsky J, With KA, Baughman S, Cabin RJ, Cohen JE, Ellstrand NC, McCauley DE, O'Neill P, Parker IM, Thompson JN, Weller SG (2001). The population biology of invasive species. Ann Rev Ecol Syst 32:305-332.

Słomka A, Sutkowska A, Szczepaniak M, Malec P, Mitka J, Kuta E (2011). Increased genetic diversity of *Viola tricolor* L. (Violaceae) in metal-polluted environments. Chemosphere 83:435-442.

StatSoft Inc. 2008.STATISTICA (data analysis software system). Version 8.0. www.statsoft.com.

Trtikova M, Güsewell S, Baltisberger M, Edwards PJ (2011). Distribution, growth performance and genetic variation of *Erigeron annuus* in the Swiss Alps. Biol Invasions 13:413-422.

Volis S, Shulgina I, Ward D, Mendlinger S (2003). Regional subdivision in wild barley allozyme variation: Adaptive or neutral? J Hered 94:341-351.

Wang XY, Shen DW, Jiao J, Xu NN, Yu S, Zhou XF, Shi MM (2012). Genotypic diversity enhances invasive ability of *Spartina alterniflora*. Mol Ecol 21:2542-2551.

Ward SM (2006). Genetic analysis of invasive plant populations at different spatial scales. Biol Invasions 8:541-552.

Ward SM, Reid SD, Harrington RJ, Sutton J, Beck KG (2008). Genetic variation in invasive populations of yellow toadflax (*Linaria vulgaris*) in the western United States. Weed Sci 56:394-399.

Yeh FC, Boyle TJB (1999). Popgene version 1.31. Microsoft Window-based Freeware for population Genetic Analysis, Department of Renewable Resources, University of Alberta, Edmonton, Alberta, Canada.

Zybartaite L, Zukauskiene J, Jodinskiene M, Janssens SB, Paulauskas A, Kupcinskiene E (2011). RAPD analysis of genetic diversity among Lithuanian populations of *Impatiens glandulifera*. Žemdirbystė Agriculture 98:391-398.

Evaluation of Drought Tolerance in Safflower Genotypes Based on Drought Tolerance Indices

Marouf KHALILI[1], Alireza POUR-ABOUGHADAREH[2*], Mohammad Reza NAGHAVI[3], Esmail MOHAMMAD-AMINI[4]

[1]*University of Payame Noor, Department of Agriculture, PO BOX 19395-3697, Tehran, Iran; ma_khalili@yahoo.com*
[2]*Imam Khomeini International University, Faculty of Engineering and Technology, Department of Plant Breeding and Production, Qazvin, Iran; s926198001@edu.ikiu.ac.ir (*corresponding author)*
[3]*University of Payame Noor, Department of Agriculture, PO BOX 19395-3697, Tehran, Iran; mr_naghavi@ymail.com*
[4]*University of Tabriz, Faculty of Agriculture, Department of Biotechnology and Plant Breeding, Tabriz, Iran; e.amini21@yahoo.com*

Abstract

Among the different environmental stresses, drought is the constraint that induces a highly negative effect on crop production. In order to identify drought tolerant genotypes in safflower, an experiment was conducted in a randomized complete block design with three replications under rainfed and irrigated conditions during 2011-2012 cropping season at Agricultural and Natural Resources Research Center of West Azerbaijan, Iran. Seven drought resistance indices including Stress Tolerance Index (STI), Tolerance Index (TOL), Stress Susceptibility index (SSI), Mean Productivity (MP), Geometric Mean Productivity (GMP), Yield Index (YI) and Yield Stability Index (YSI) were calculated for each genotype based on grain yield under stress and normal conditions. Analysis of variance showed significant difference among genotypes in the terms of grain yield under rainfed and irrigated conditions. Result of correlation analysis between grain yield in both conditions and calculated drought resistance indices revealed that STI, MP, GMP and YI were the best indices for identifying high yielding genotypes in both conditions (drought tolerant genotypes). Gabriel Biplot Multivariate chart showed that 'Syrian', 'PI-537530' and 'PI-537636-S' genotypes relatively identified as drought tolerant and 'Hartman', 'Gila', 'PI-537636', 'PI-537598', 'LRV-51-51' and 'Yenice' genotypes identified as susceptible genotypes. Cluster analysis based on grain yield and drought tolerance indices classified the genotypes into four clusters, therefore this result was consistence with results of biplot analysis.

Keywords: Biplot, drought stress, grain yield, safflower, tolerance indices

Introduction

Safflower (*Carthamus tinctorius* L.) is a member of the family Compositae or Asteraceae, cultivated mainly for its seed, which is used as edible oil and as birdseed (Dordas and Sioulas, 2008; Istanbulluoglu, 2009). This plant is native to the Middle East and it's resistant to saline conditions and to drought stresses (Bassil and Kaffka, 2002). All the alive and non-alive stresses are the most factors to reducing production nonetheless; drought stress is the most important factor limiting crops production in agricultural systems in arid and semi-arid regions (Mollasadeghi *et al.,* 2011). In general, of 18.5 million hectares of agricultural lands, 6.2 million hectares (33.5%) is devoted to dry cultivation, and about 1.2 million/ha of lands under dry cultivation, more than 400 (mm) rainfall will receive (Mohammadi *et al.,* 2006). Thus, in these areas loss of yield is the main concern of plant breeders. Breeding for drought resistance is difficult by the lack of fast, reproducible screening techniques and the inability to routinely create defined and repeatable water stress conditions when a large amount of genotypes are to be evaluated efficiently (Ramirez and Kelly, 1998). Several indices have been utilized to evaluate genotypes for drought tolerance based on grain yield such as geometric mean productivity (Fernandez, 1992), mean productivity (Rosielle and Hamblin, 1981), stress susceptibility index (Fischer and Maurer, 1978), stress tolerance index (Fernandez, 1992), yield index (Gavuzzi *et al.,* 1997), yield stability index (Bouslama and Schapaugh, 1984) and tolerance (Rosielle and Hamblin, 1981). On the other hand, according to Fernandez (1992) theory genotypes can be categorized into four groups based on their performance in stress and non-stress environments: genotypes which express uniform superiority in both stress and non-stress environments (Group A); genotypes which perform favorably only in non-stress environments (Group B); genotypes which yield relatively well only in stress environments (Group C) and genotypes which perform poorly in both stress and non-stress environments (Group D). The optimal selection criteria should distinguish group A from the other three groups. Therefore, a suitable index (indices) must be able to discriminate the genotype in group A from the other groups. The objectives of the present investigation were to

(i) identify drought tolerant safflower genotypes under rainfed conditions in the northwest of Iran, (ii) determine the efficiency of tolerance indices to classify genotypes into sensitive and tolerant groups.

Materials and methods

Experimental design and plant material

Fifteen genotypes of safflower (Tab. 1) were cultivated in a randomized complete block design with three replications under two different environments (irrigated and rainfed) in 2011-2012, at Agricultural and Natural Resources Research Center of West Azerbaijan, Iran (latitude 36.58°N, longitude 46.09°E, Altitude 1310 m above sea level). Each plot consisted of 6 rows, 5 m long and spaced 50 cm apart. Fertilizers were applied prior to sowing at a rate of 80 kgha^{-1} Ammonium Phosphate ($NH_4H_2PO_4$) and 65 kgha^{-1} Urea ($CO(NH_2)_2$). Non-stressed plots were irrigated after sowing and subsequent irrigations were carried out after 75 mm evaporation from class A pan. Weeds were controlled by hand during crop growth and development. At maturity stage after separation of border effects from each plot, grain yield were measured and seven drought tolerance indices such as SSI (Fischer and Maurer, 1978), GMP (Fernandez, 1992), MP (Rosielle and Hamblin, 1981), TOL (Rosielle and Hamblin, 1981), STI (Fernandez, 1992), YI (Gavuzzi *et al.*, 1997), YSI (Bouslama and Schapaugh, 1984) were calculated using the following formula:

(1) $SSI = (1-(Ys/Yp))/(1-(\overline{Ys})/-(\overline{Yp}))$
(2) $GMP = \sqrt{Ys \times Yp}$
(3) $MP = (Ys + Yp)/2$
(4) $TOL = Yp - Ys$
(5) $STI = (Ys \times Yp)/(\overline{Yp})^2$
(6) $YI = (Ys)/(\overline{Ys})$
(7) $YSI = Ys/Yp$

Tab. 1. The genotypes used in the experiment

No.	Genotype	No.	Genotype	No.	Genotype
1	'PI-1989290'	6	'Dincer'	11	'LRV-51-51'
2	'CW-4440'	7	'Hartman'	12	'KINO-76'
3	'Syrian'	8	'Gila'	13	'Yenice'
4	'PI-537530'	9	'PI-537636'	14	'PI-537636-S'
5	'CH-5'	10	'PI-537598'	15	'IL-111'

In the above formulas Ys, Yp, $\overline{Ys}$ and $\overline{Yp}$ represent yield under stress, yield non-stress for each cultivar, yield mean in stress and non-stress conditions for all cultivars, respectively. The data were analyzed using SPSS ver.16 software for analysis of variance and cluster analysis of genotypes based on Square Euclidean distance and Ward method. The principle component analysis (PCA) and biplot display were also used to identify tolerant and high yielding genotypes using MINITAB ver.16 software.

Results and discussion

Results of analysis of variance for grain yield over both conditions, showed that difference significant between genotypes (Tab. 2). In the both conditions highest values for grain yield belonged to genotypes No. 3, 4 and 15. On the other hand, genotypes No. 7, 9 and 11 had lowest grain yield in stress condition, however the lowest grain yield under non-stress condition belonged to genotypes No. 8, 11 and 14 (Tab. 3). The drought tolerance indices for genotype based on grain yield in non-stress and stress conditions are presented in Tab. 3. According to TOL genotypes No. 8, 13 and 14 exhibited the lowest value. Also, based on SSI the lowest value belonged to genotype No. 3, 4 and 14, thus these genotypes were found as drought tolerance. Based on STI and GMP values, genotypes No. 6, 9, 12, 14 and 19 identified as drought tolerant genotypes. In the case of the indices MP and GMP genotypes No.8 followed by No. 9, 10 and 11 were found to be drought tolerant genotypes. Also, STI, YI and YSI indices selected genotypes No. 3, 4 and 15 as the most relatively tolerant genotypes. Albeit selection based on a combination of indices may provide a more useful criterion for improving drought tolerance, however correlation analysis between grain yield and drought tolerance indices can be a good criterion for screening the best genotypes and indices. Thus, a suitable index must significantly correlated with grain yield under both the conditions (Mitra, 2001). Results of correlation analysis between grain yields in both conditions with drought tolerance indices showed that STI, GMP, MP and YI had positive and significant correlations with Yp and Ys. Therefore these indices were able to discriminate group A genotypes from other genotypes. A significant negatively correlation was observed between SSI and grain yield under stress condition. Hence, as for the positive correlation between SSI and yield under non-stress (Yp) condition and a negative correlation between SSI and yield under stress condition suggested that selection based on SSI will result in increased yield under non-stress conditions (Sio-Se Mardeh *et al.*, 2006). The YSI was positively correlated with Ys and had negative correlation with Yp, so highest values of YSI belonged to genotypes that exhibited least yield in irrigated conditions, but exhibited high yield under rain-fed conditions. This result were consistent with Sio-Se Mardeh *et al.* (2006) and Zebarjadi *et al.* (2012). Naghavi *et al.* (2013) showed that correlation between MP, GMP, Ys and Yp was positive. In the study conducted by Ilker *et al.* (2011) STI, MP and GMP were positively significant correlated with grain yield over both conditions, whereas of the three primary grain yield components, with either positive or negative correlation between grain yield and plant height. Khalili *et al.* (2012) reported that GMP, MP, and STI were significantly and positively correlated with grain yield in non-stress and stress conditions.

To better understand the relationships, similarities and dissimilarities among the indicators of drought tolerance, biplot displayed based on principal component analysis (PCA). This analysis revealed that the two first PCA explained 99.7% of total variation. The first PCA explained 62.51% of the variation with Yp, Ys, MP, GMP, YI and STI. Hence, this component can be named as the yield stability and tolerance to drought. The genotypes with higher values of PC1 are expected to be drought tolerant and high yielding genotypes. The second PCA explained 37.18% of the total variability and had high positive correlation with Yp, TOL and SSI. Therefore, the second component can be named as a stress susceptibility and stability yield. Also, the genotypes No. 3, 4 and 14 had higher values for PC1 and low values for PC2 thus, these genotypes identified as drought tolerance. Genotypes No. 1, 2, 5, 6 and 12 for the reason that placed among of PC1 and PC2 identified as semi tolerance to drought stress. In spite of the fact that genotypes No. 15 had high grain yield over both conditions but this genotype were sensitive to drought stress in respect of their high values for both PC1 and PC2. Genotypes No. 7, 8, 9, 10, 11 and 13 were susceptible to drought stress and had low grain yield performance, because this genotypes had lower amounts of both PC1 and PC2 in comparison to other genotypes. Finally, our results were consistent with those reported by Golabadi *et al.* (2006) Kaya *et al.* (2002), Talebi *et al.* (2011) and Zebarjadi *et al.* (2012).

The result of cluster analysis for studied genotypes has been presented in Fig. 2. The 15 safflower genotypes based on grain yield and STI, MP, GMP and YI were classified in four clusters. Genotypes No. 7, 8, 9, 10, 11 and 13 classified in the first cluster. In this cluster, all genotypes had high SSI, thus they were susceptible to drought and only suitable for irrigated conditions. The second cluster comprising genotypes No. 1, 2, 5, 6 and 12. These genotypes had mean indicators values therefore, they identified as semi-tolerance or semi-sensitive genotypes. The third cluster consists of genotypes No. 3, 4 and 14. These genotypes had high STI, MP, GMP, YI and YSI values, thus they considered to be the most desirable genotypes for both growth conditions. Finally, genotype No. 15 belonged to fifth cluster and this genotype have high grain yield in both conditions. In general, these results were agreement with results obtained from biplot analysis. Cluster analysis has been generally utilized for description of variation between genotypes and grouping based on drought tolerance indices (Golabadi *et al.*, 2006; Malek Shahi *et al.*, 2009; Khalili *et al.*, 2012; Naghavi *et al.*, 2013).

Tab. 2. Analysis of variance for grain yield (kg h^{-1}) in stress and non-stress conditions

Source of variation	d.f.	Mean of squares	
		Irrigated	Rainfed
Replication	2	133763.89	268763.89**
Genotype	14	93240.08*	54805.56*
Error	28	48689.48	25609.13
CV%		11.34	14.22

* and **; significant at the 5% and 1% levels probability, respectively.

Tab. 3. Drought tolerance indices and comparison of grain yield based on Duncan method in non-stress and stress conditions

No.	Yp (kg h^{-1})	Ys (kg h^{-1})	SSI	STI	TOL	MP	GMP	YSI	YI
1	1991.67 ab	1150abcd	1.00	0.60	841.67	1570.83	1513.41	0.57	1.02
2	1983.33 ab	1075 bcd	1.08	0.56	908.33	1529.16	1460.16	0.54	0.95
3	2050.00 ab	1300 abc	0.86	0.70	750.00	1675.00	1632.48	0.63	1.15
4	2116.67 ab	1283.33 abcd	0.93	0.71	833.34	1700.00	1648.14	0.60	1.14
5	2041.67 b	1050 bcd	1.15	0.56	991.67	1545.83	1464.15	0.51	0.93
6	2050.00 b	1108.33 abcd	1.09	0.60	941.67	1579.16	1507.34	0.54	0.98
7	1866.67 b	1025bcd	1.07	0.50	841.67	1445.83	1383.23	0.55	0.91
8	1783.33 b	1058.33 cd	0.96	0.49	725.00	1420.83	1373.81	0.59	0.94
9	1850.00 b	991.67 cd	1.10	0.48	858.33	1420.83	1354.46	0.53	0.88
10	1808.33 b	1033.33 cd	1.01	0.49	775.00	1420.83	1366.96	0.57	0.92
11	1733.33 b	983.33 d	1.02	0.45	750.00	1358.33	1305.54	0.56	0.87
12	1966.67 b	1025.1 dc	1.13	0.53	941.57	1495.88	1419.87	0.52	0.91
13	1791.67 b	1066.67 bcd	0.96	0.50	725.00	1429.17	1382.43	0.59	0.94
14	1758.33 b	1358.33ab	0.54	0.63	400.00	1558.33	1545.44	0.77	1.20
15	2400.00 a	1375 a	1.01	0.87	1025.0	1887.50	1816.59	0.57	1.22

*For each trait, means with the same letters have no significant difference at 5% level of probability. Yp: Yield in non-stress condition; Ys: Yield in stress condition; SSI: Stress Susceptibility Index; STI: Stress Tolerance Index; TOL: Tolerance Index; MP: Mean Productivity; GMP: Geometric Mean Productivity; YSI: Yield Stability Index and YI: Yield Index. For Genotypes name (No.) see Tab. 1.

Tab. 4. Correlation coefficients between drought tolerance indices and grain yield in irrigated and rainfed conditions

Trait	Yp	Ys	TOL	MP	GMP	SSI	STI	YSI	YI
Yp	1								
Ys	0.56*	1							
TOL	0.66**	-0.24	1						
MP	0.91**	0.84**	0.31	1					
GMP	0.85**	0.91**	0.19	0.99**	1				
SSI	0.23	-0.67**	0.87**	-0.17	-0.29	1			
STI	0.86**	0.90**	0.19	0.99**	0.99**	-0.28	1		
YSI	-0.23	0.67**	-0.87**	0.18	0.298	-1**	0.28	1	
YI	0.55*	1**	-0.24	0.85**	0.90**	-0.67**	0.90**	0.67**	1

* and **; significant at the 5% and 1% levels probability, respectively.
Ys: Yield in stress condition; Yp: Yield in non-stress condition; SSI: Stress Susceptibility Index; STI: Stress Tolerance Index; TOL: Tolerance Index; MP: Mean Productivity; GMP: Geometric Mean Productivity; YSI: Yield Stability Index and YI: Yield Index.

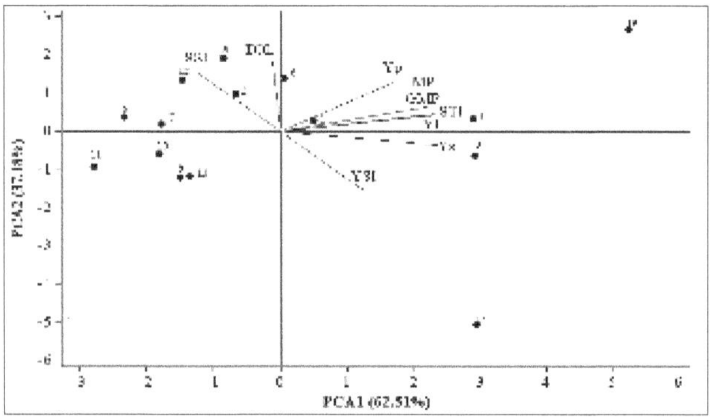

Fig. 1. Biplot for drought tolerance indices in 15 spring safflower genotypes based on first two components. Numbers inside the figure are genotypes number. Ys: Yield in stress condition; Yp: Yield in non-stress condition; SSI: Stress Susceptibility Index; STI: Stress Tolerance Index; TOL: Tolerance Index; MP: Mean Productivity; GMP: Geometric Mean Productivity; YSI: Yield Stability Index and YI: Yield Index.

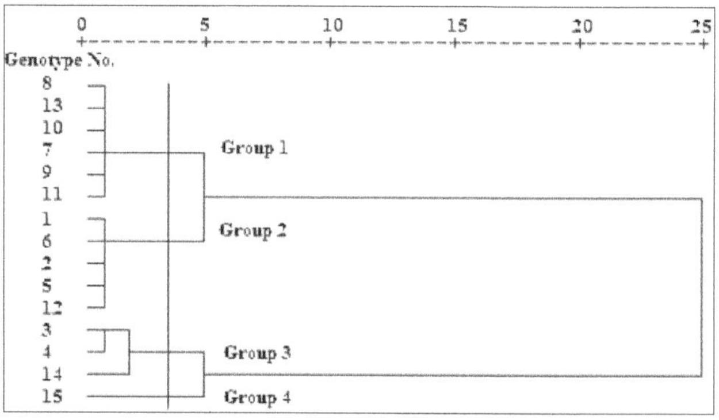

Fig. 2. Dendrogram of cluster analysis of spring safflower genotypes based on yield in non-stress condition (Yp), yield in stress condition (Ys), Mean Productivity (MP), Geometric Mean Productivity (GMP), Stress Tolerance Index (STI) and Yield Stability Index (YI) indices using Ward method. Numbers inside the figure are genotypes number.

Conclusions

Results of present study indicated that STI, MP, GMP and YI are the suitable indices for screening genotypes that produce higher yields in both stress and non-stress conditions (drought tolerant genotypes). Screening drought tolerant genotypes using biplot and cluster analysis discriminated 'Syrian', 'PI-537530' and 'PI-537636-S' genotypes as the most drought tolerant. Therefore they are recommended to be used as parents for improvement of drought tolerance in other cultivars.

References

Bassil ES, Kaffka SR (2002). Response of safflower (*Carthamus tinctorius* L.) to saline soils and irrigation. I. Consumptive water use. Agricultural Water Management 54:67-80.

Bouslama M, Schapaugh WT (1984). Stress tolerance in soybeans. Part 1: evaluation of three screening techniques for heat and drought tolerance. Crop Science 24:933-937.

Clarke JM, De-Pauw RM, Townley-Smith TF (1992). Evaluation of methods for quantification of drought tolerance in wheat. Crop Science 32:423-428.

Dordas CA, Sioulas C (2008). Safflower yield, chlorophyll content, photosynthesis, and water use efficiency response to nitrogen fertilization under rainfed conditions. Industrial Crops and Products 27:75-85.

Fernandez GCJ (1992). Effective selection criteria for assessing stress tolerance. Proceedings of the International Symposium on Adaptation of Vegetables and Other Food Crops in Temperature and Water Stress Tolerance. Asian Vegetable Research and Development Centre, Taiwan, 257-270 p.

Fischer RA, Maurer R (1978). Drought resistance in spring wheat cultivars. I. Grain yield responses. Australian Journal of Agricultural Research 29:897-912.

Gavuzzi P, Rizza F, Palumbo M, Campanile RG, Ricciardi GL, Borghi B (1997). Evaluation of field and laboratory predictors of drought and heat tolerance in winter cereals. Canadian Journal of Plant Science 77:523-531.

Golabadi M, Arzani A, Maibody M (2006). Assessment of drought tolerance in segregating populations in durum wheat. African Journal of Agricultural Research 1(5): 162-171.

Ilker E, Tatar O, Aykut Tonk F, Tosun M (2011). Determination of Tolerance Level of Some Wheat Genotypes to Post-Anthesis Drought. Turkish Journal of Field Crops 16(1):59-63.

Istanbulluoglu A (2009). Effects of irrigation regimes on yield and water productivity of safflower (*Carthamus tinctorius* L.) under Mediterranean climatic condition. Agricultural Water Management 96:1792-1798.

Kaya Y, Palta C, Taner S (2002). Additive main effects and multiplicative interactions analysis of yield performances in bread wheat genotypes across environments. Turkish Journal of Agriculture and Forestry 26:275-279.

Khalili M, Naghavi MR, Pour-Aboughadareh AR, Talebzadeh SJ (2012). Evaluating of drought stress tolerance based on selection indices in spring canola cultivars (*Brassica napus* L.). Journal of Agricultural Science 4(11):78-85.

Malek-Shahi F, Dehghani H, Alizadeh B (2009). Study of drought tolerance indices in some cultivars of winter rapeseed (*Brassica napus* L.). Journal of Science and Technology of Agriculture and Natural Resources, Water and Soil Science 48:78-89.

Mitra J (2001). Genetics and genetic improvement of drought resistance in crop plants. Current Science 80:758-763.

Mohammadi R, Haghparast R, Aghaei-Sarbarzeh M, Abdollahi A (2006). Evaluation of drought tolerance rate of advanced genotypes of Durum wheat on the basis of physiologic standards and other related indices. Iranian Agriculture Sciences 3(1):561-567.

Mollasadeghi V, Valizadeh M, Shahryariand R, Imani AA (2011). Evaluation of end drought tolerance of 12 wheat genotypes by stress in dices. Middle-East Journal of Scientific Research 7(2):241-247.

Naghavi MR, Pour-Aboughadareh AR, Khalili M (2013). Evaluation of drought tolerance indices for screening some of corn (*Zea mays* L.) Cultivars under environmental conditions. Notulae Scientia Biologicae 5(3):388-393.

Ramirez-Vallejo P, Kelly JD (1998). Traits related to drought resistance in common bean. Euphytica 99: 127-136.

Rosielle AA, Hamblin J (1981). Theoretical aspects of selection for yield in stress and non-stress environments. Crop Science 21:943-946.

Sio-Se Mardeh A, Ahmadi A, Poustini K, Mohammadi V (2006). Evaluation of drought resistance indices under various environmental conditions. Field Crops Research 98:222-229.

Talebi R, Baghebani N, Karami E, Ensafi MH (2011). Defining Selection Indices for Drought Tolerance in Chickpea under Terminal Drought Stresses. Journal of Applied Biological Sciences 5(3):33-38.

Zebarjadi A, Mirany T, Kahrizi D, Ghobadi M, Nikoseresht R (2012). Assessment of drought tolerance in some bread wheat genotypes using drought resistance indices. Biharean Biologist 6 (2):94-98.

10

Effect of BA and GA_3 on the Morphological Features of Stomata in the Leaf Epidermis of the *Zantedeschia albomaculata* cv. 'Albomaculata'

Beata JANOWSKA[1*], Natalia MANSFELD[1], Roman ANDRZEJAK[2]

[1]*Department of Ornamental Plants, Faculty of Horticulture and Landscape Architecture, Poznań University of Life Sciences, Dąbrowskiego 159, Poznań, Poland; beataj@up.poznan.pl (*corresponding author)*

[2]*Department of Phytopathology, Seed Science and Technology, Faculty of Horticulture and Landscape Architecture, Poznań University of Life Sciences, Dabrowskiego 159, Poznań, Poland*

Abstract

The study focused on the effect of benzyladenine at concentrations of 100, 350 and 600 mg·dm^{-3} combined with gibberellic acid at the same concentrations on the morphological features (length and width) and number of stomata in the lower and upper leaf epidermis in *Zantedeschia albomaculata* cv. 'Albomaculata'. To determine changes in the structure of stomata, rhizomes were soaked before planting in mixtures of the growth regulators at the above concentrations for 30 minutes. After the application of BA and GA_3 at concentrations of 100+100 mg·dm^{-3}and 350+350 mg·dm^{-3} the stomata in the upper leaf epidermis were larger than in the control treatment and fewer in number. In the lower epidermis a mixture of BA and GA_3 at concentrations of 100+100, 350+350 and 600+600 mg·dm^{-3} caused the formation of bigger stomata, and their number declined when the concentration employed was 350+350 mg·dm^{-3}.

Keywords: *Zantedeschia albomaculata*; BA; GA_3; stomata; leaves

Introduction

Species of the genus *Zantedeschia* belong to the order Arales and the family Araceae (Erhardt *et al.*, 2000) embracing 2,823 species (Govaerts *et al.*, 2002). They have a characteristic inflorescence - a spadix, composed of a thick, fleshy main axis on which flowers are borne. They are usually dioecious, with the perianth reduced or none, two whorls of stamens (2+2 or 3+3), and an upper pistil, usually built of three carpels. The spadix sits in a leaf sheath corresponding to the inflorescence spathes of palms. With advancing specialisation, the sheath has taken the form of a protective casing of the spadix, often brightly coloured, and with time in many taxa has become part of a characteristic inflorescence which traps insects until the flowers have been pollinated. In calla lilies the spathes are 12 to 25 cm long and white, yellow and pink to dark purple in colour; they surround spadices coloured yellow, golden-yellow and creamy-yellow. In morphological terms, all species are similar. Petiole bases closely adhere to one another, forming 'an apparent stem'. Successive leaves and peduncles grow from their midst. Within the genus, leaf blades are lanceolate in shape, narrowly to broadly sagittate or cordate. Medium- to dark-green in colour, they are covered or not in transparent, white spots. The stomatal apparatus types of Araceae vary from anomocytic, actinocytic to paracytic, cyclocytic and transitionary type between paracytic and cyclocytic; the numbers of subsidiary cells from zero to twelve. Epidermal cells are nearly isodiametric in outline with straight, arched and undulate anticlinal walls. Striate ornamentation occurs on periclinal walls of epidermal cells in some species. Although the stomatal apparatus types in Araceae are of little taxonomical significance at infra-family level of Araceae, the combined characters of stomatal apparatuses, the shape of anticlinal wall and ornamentation of cuticules in guard cells may be useful for species identification (Wang and Zhao, 2002).

The calla lily is a native of the Congo River delta and seasonally inundated or swampy areas in South Africa. Funnell (1993) mentions six species and two subspecies of the genus *Zantedeschia*, Kuehny (2000) describes seven species and two subspecies. However, only four have been found useful in floricultural production: *Zantedeschia aehtiopica* (L.) Spreng., *Z. elliottiana* Engl. W. Wats., *Z. rehmanii* Engl., and *Z. albomaculata* (Kuehny, 2000).

Initially, in cultivation the dominant species was the common florist's calla (*Z. aethiopica* L. Spreng.). Today, however, its significance has declined, while cultivars with colourful spathes have grown in importance. Flowering in the calla lily depends on the cultivar, the size of rhizomes, and the duration of their storage (Corr and Widmer, 1991, Funnell *et al.*, 1992). The research conducted in the world focuses on how to improve flowering with the use of growth regulators. For the calla lily, those shown to be effective are gibberellins, in particular gibberellic acid. The application of mixtures of growth regulators in ornamental plants is only rarely employed as a method to improve their flowering. The study of the mixture of benzyladenine and gibberellic acid in the calla lily is based on the assumption that the effects of the growth regulators could add up or reinforce each other. The aim of the research was to determine changes occurring in the structure and number of stomata in the leaf epidermis of the cultivar 'Albomaculata' under the influence of benzyladenine and gibberellic acid.

Material and methods

The research was conducted in the years 2009-2010. The cultivar employed was 'Albomaculata' deriving from the spotted calla (*Zantedeschia albomaculata* Hook. Baill.).

The study focused on the effect of benzyladenine at concentrations of 100, 350 and 600 mg·dm^{-3} combined with gibberellic acid at the same concentrations on the morphological features (length and width) and number of stomata in the lower and upper leaf epidermis. To determine changes in the structure of stomata, rhizomes were soaked before planting in mixtures of the growth regulators at the above concentrations for 30 minutes. They were then planted into 20-cm pots in a medium consisting of peat with a pH of 6.2, enriched with a slow-release fertiliser Osmocote Plus (3-4M) in the amount of 3 g·dm^{-3} and mixed with fresh, shredded pine bark at a rate of 3:1 (v:v). Grown in a plastic tunnel, the plants were fed starting with the fifth week of cultivation. Every 10-14 days, solutions of mixed fertilisers, Peters Professional and brown Superba, were applied at a concentration of 0.2%. At the start of vegetation, when the leaves were fully developed, lime saltpetre was foliar applied once at a concentration of 0.2%.

In both years, ten leaves were selected from each rhizome treated with solutions of the growth regulator mixture at the various concentrations. Tested characteristics determined on 10 samples of each treatment. Leaf segment of an area of 0.5 cm square from each specimen was cut. The upper and lower surface were separated with dissecting needle and rinsed with clean water. The number of stomata per 1 mm^2 of the upper and lower leaf epidermis was determined as well as their length and width. Those observations and measurements were carried out in a light microscope under 400x magnification. The measurements were conduced using the micrometric method with the help of an eyepiece scale. The density of stomata was calculated using a calibrated eyepiece graticule under 100x magnification (LM). The results were processed with the help of analysis of variance. The means were grouped using Duncan's test at the α=0.05 significance level.

Results

The application of BA+GA$_3$ at concentrations of 100+100 mg·dm^{-3} and 350+350 mg·dm^{-3} changed both the length and the width of stomata in the upper leaf epidermis of the cultivar 'Albomaculata'. They were significantly bigger than in the control treatment (Tab. 1, Fig. 1 A-D). It was found that the bigger the stomata grew, the smaller their number was. At a concentration of 600+600 mg·dm^{-3}, BA+GA$_3$ did not affect the size of the stomata, but they influenced their density: there were more of them per 1 mm^2 than in the control treatment (Tab. 1, Fig. 2 E-H).

BA+GA$_3$ in the concentration variants examined had a significant effect on the length of stomata in the lower leaf epidermis of the cultivar 'Albomaculata'. The longest stomata were observed to appear in the treatments with BA+GA$_3$ at concentrations of 100+100 mg·dm^{-3} and 350+350 mg·dm^{-3}. When the growth regulators were used at a concentration of 600+600 mg·dm^{-3}, the stomata were a bit shorter, but still longer than in the control. Their width also changed. BA+GA$_3$ at those concentrations had a similar effect on the formation of wider stomata (Tab. 1, Fig. 3 I-K). As to the number of stomata in the lower leaf epidermis, the results show that the growth regulators had no significant effect here, with the exception of the variant of BA+GA$_3$ at a concentration of 350+350 mg·dm^{-3}, which caused a decline in their number (Tab. 1, Fig. 4 M-P).

Discussion

Growth regulators from cytokinin and gibberellin groups have significant effect on flowering of *Zantedeschia* with colourful spathes increasing yield of cut flowers (Kozłowska *et al.*, 2007; Janowska, 2013; Janowska and Stanecki, 2013; Ranwala and Miller, 1998; Rodriquez-Perez *et al.*, 2009). Regulators have also effect on post-harvest longevity

Tab. 1. Characteristic of stomata in upper and lower epidermis in leaves of *Zantedeschia* 'Albomaculata' after soaking in mixture of benzyladenine and gibberellic acid

Concentration of BA+GA3 (mg·dm^{-3})	Length (µm)		Width (µm)		Number of stomata on 1 mm^2	
	2009	2010	2009	2010	2009	2010
			Upper epidermis			
0	42.3 a	41.9 a	27.5 b	27.8 b	277.5 b	270.0 b
100+100	44.1 c	43.7 b	28.5 c	28.7 c	222.0 a	218.5 a
350+350	45.7 d	44.3 b	28.9 c	28.3 c	213.0 a	220.0 a
600+600	43.4 b	42.0 a	27.8 b	27.4 b	290.5 c	287.0 c
			Lower epidermis			
0	42.1 a	41.9 a	26.7 a	27.0 a	305.0 b	300.5 b
100+100	44.6 c	45.0 c	29.0 b	27.9 b	310.5 b	299.0 b
350+350	45.2 c	49.9 c	29.3 b	28.5 b	210.0 a	215.0a
600+600	43.2 b	42.9 b	28.7 b	28.1 b	314.0 b	303.0 b

Means followed by the same letter do not differ significantly at α = 0.05.

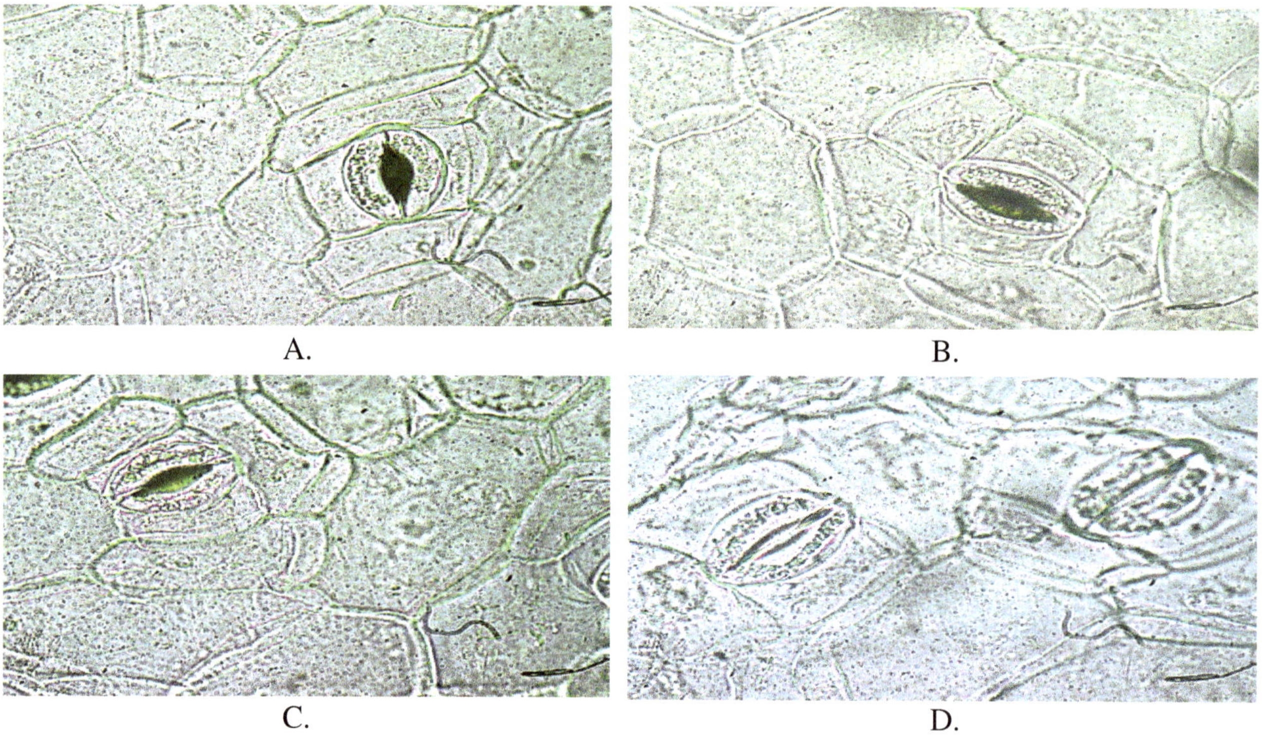

Fig. 1. Stomata in upper epidermis in leaves of *Zantedeschia*: A. Control plants
B. Plants treated with mixture of BA+GA_3 at concentration of 100+100 mg·dm^{-3}
C. Plants treated with mixture of BA+GA_3 at concentration of 350+350 mg·dm^{-3}
D. Plants treated with mixture of BA+GA_3 at concentration of 600+600 mg·dm^{-3}

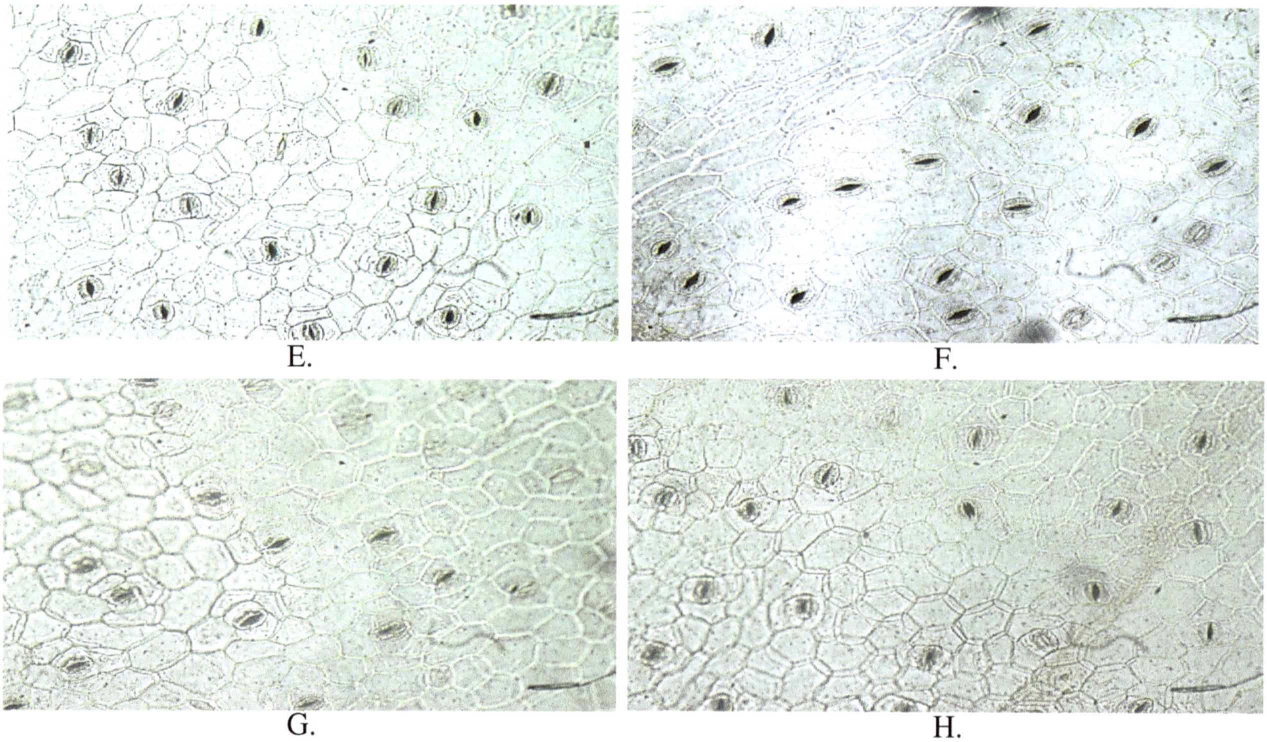

Fig. 2. Number of stomata in upper epidermis in leaves of *Zantedeschia*: E. Control plants
F. Plants treated with mixture of BA+GA_3 at concentration of 100+100 mg·dm^{-3}
G. Plants treated with mixture of BA+GA_3 at concentration of 350+350 mg·dm^{-3}
H. Plants treated with mixture of BA+GA_3 at concentration of 600+600 mg·dm^{-3}

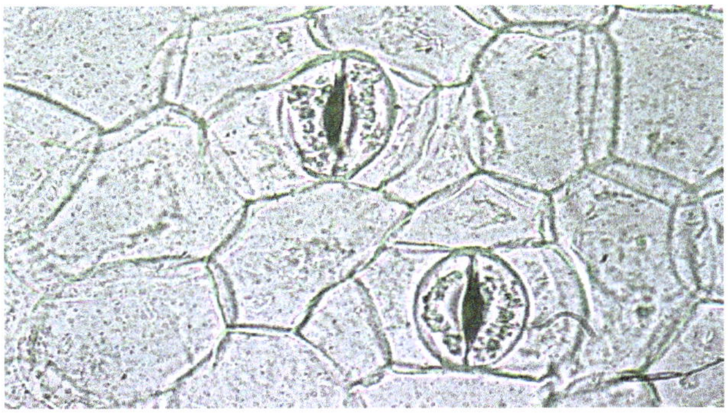

I.

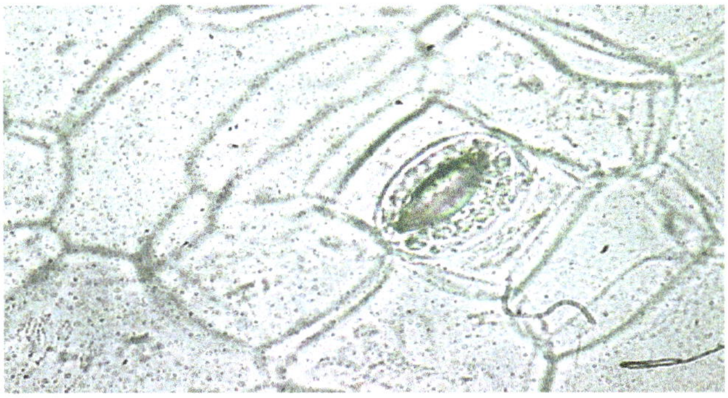

J.

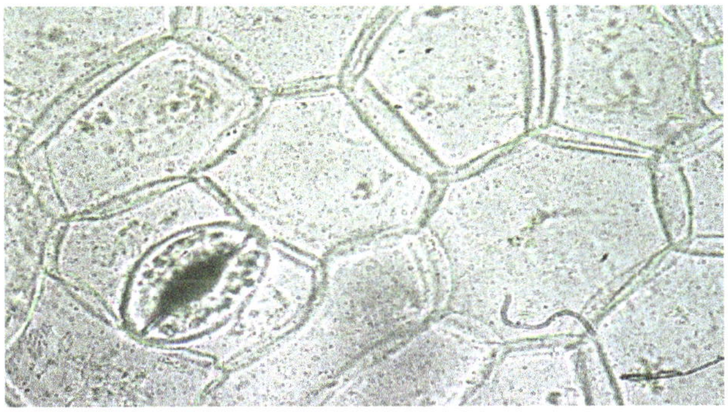

K.

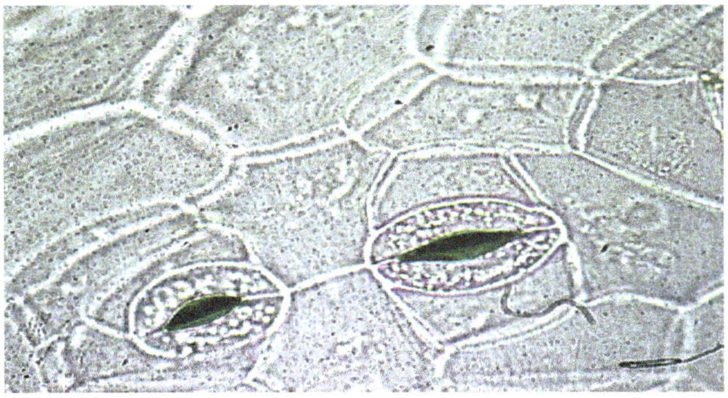

L.

Fig. 3. Stomata in lower epidermis in leaves of *Zantedeschia*: I. Control plants
J. Plants treated with mixture of $BA+GA_3$ at concentration of 100+100 $mg \cdot dm^{-3}$
K. Plants treated with mixture of $BA+GA_3$ at concentration of 350+350 $mg \cdot dm^{-3}$
L. Plants treated with mixture of $BA+GA_3$ at concentration of 600+600 $mg \cdot dm^{-3}$

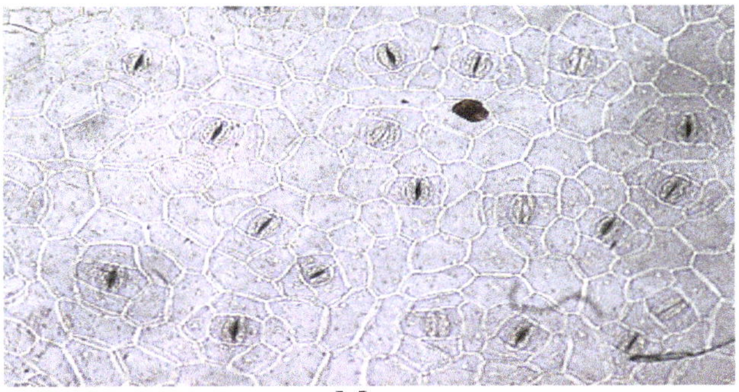

M.

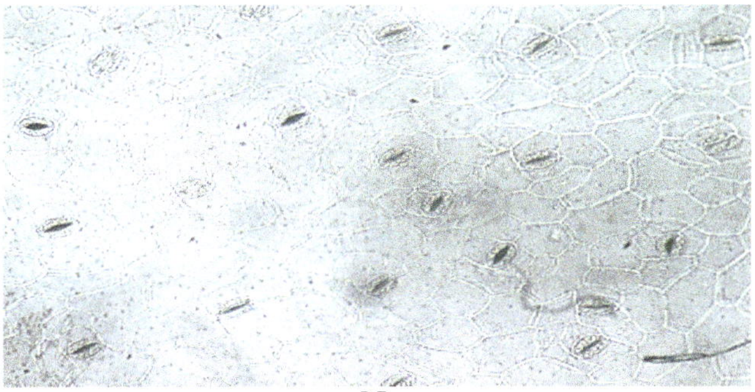

N.

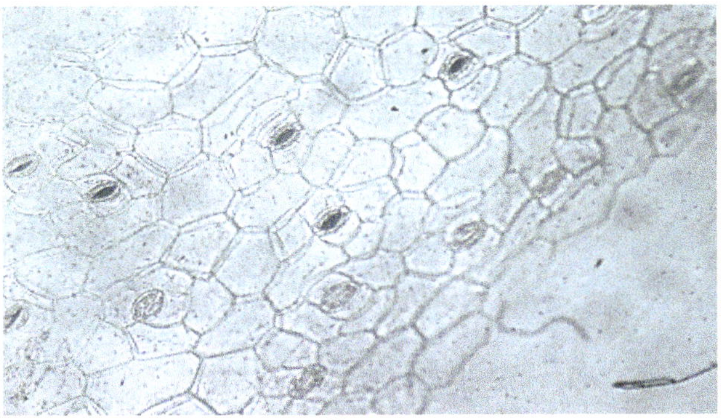

O.

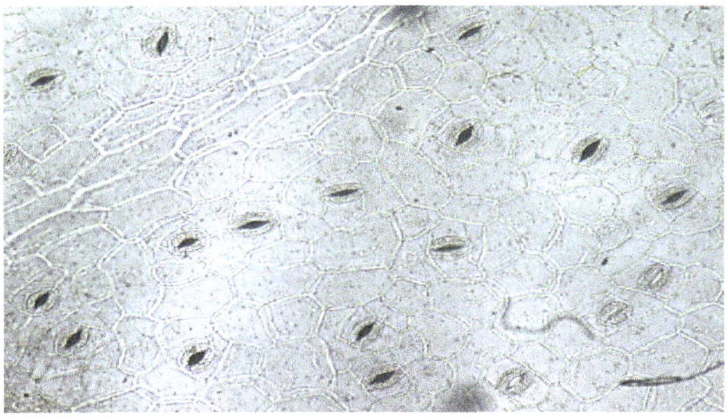

P.

Fig. 4. Number of stoma in lower epidermis in leaves of *Zantedeschia*: M. Control plants
N. Plants treated with mixture of BA+GA3 at concentration of 100+100 $mg \cdot dm^{-3}$
O. Plants treated with mixture of BA+GA3 at concentration of 350+350 $mg \cdot dm^{-3}$
P. Plants treated with mixture of BA+GA3 at concentration of 600+600 $mg \cdot dm^{-3}$

of flowers and leaves of *Zantedeschia* cultivars (Janowska and Stanecka, 2011; Wachowicz *et al.*, 2006).

This experiment determined the effect of gibberellic acid and benzyladenine on size and number of stomata in epidermis of 'Albomaculata' leaves.

As the results show, in the cultivar 'Albomaculata' after the application of a mixture of BA+GA_3 at concentrations of 100+100 and 350+350 mg·dm^{-3}, stomata in the upper leaf epidermis were bigger than in the control treatment, while their number declined. In the lower epidermis a mixture of benzyladenine and gibberellic acid at those concentrations also led to the formation of bigger stomata, but their number dropped only when the concentration employed was 350+350 mg·dm^{-3}. Gupta (1992) found less stomata in epidermis of *Vicia faba* grown from seeds soaked in solution of gibberellic acid but the stomata were bigger than stomata in control plants. The effect of gibberellic acid on the morphology of stomata was studied by Pogroszewska (2002). The author did not show GA_3 to have an effect on the length of stomata in the organs examined or on their width in the lower epidermis of the topmost leaf. In turn, in the lower epidermis of the spathe GA_3 made stomatal pores significantly narrower than in the control. The application of the acid did not bring about any significant change in the density of stomata in the lower leaf epidermis, but it did affect their number per 1 mm^2 in the lower bract epidermis. Gibberellic acid produced a decline in the number of stomata in the spathe epidermis. The author also assessed the aperture in stomata of the lower spathe epidermis. She observed them to be opened more widely in plants treated with GA_3 at a concentration of 1000 mg·dm^{-3} than in the control. Besides, in the bract of those plants there developed morphologically transformed stomata with four or three guard cells. She also noted asymmetrically built stomata and a narrowing of guard cells in the upper bract epidermis. In the present research no anomalies were observed in the structure of stomata. Carey (2008) suggests that the stomatal aperture is stimulated by cytokinins. These growth regulators induce stomata opening, the effects are species specific and correlated with cytokinin centration and method of application (Pospíšilová, 2003). Wachowicz *et al.* (2006) reported that GA_3 in *Zantedeschia* and BA in *Hosta* enhanced stomata opening by 21% and 44%, respectively. In addition, different cytokinins could effect stoma and cause abnormal stomata development in tissue culture studies (Namli and Ayaz, 2007).

Conclusion

After the application of BA and GA_3 at concentrations of 100+100 ppm and 350+350 mg·dm^{-3} the stomata in the upper leaf epidermis were larger than in the control treatment and fewer in number. In the lower epidermis a mixture of BA and GA_3 at concentrations of 100+100, 350+350 and 600+600 mg·dm^{-3} caused the formation of bigger stomata, and their number declined when the concentration employed was 350+350 mg·dm^{-3}.

References

Carey DJ (2008). The effects of benzyladenine on ornamental crops. Horticultural Science. Raleigh, North Carolina, 405 p.

Corr BE, Widmer RE (1991): Paclobutrazol, gibberellic acid and rhizome size affect growth and flowering of *Zantedeschia*. HortScience 26(2):133-135.

Funnell KA (1993). *Zantedeschia*, p. 683-704. In De Hertogh A, Le Nard. M (Eds.). The physiology of flower bulbs. Amsterdam-London-New York-Tokyo.

Funnell KA, MacKay BR, Lawoko CRO (1992). Comparative effects of Promalin and GA_3 on flowering and development of *Zantedeschia* 'Galaxy'. Acta Hort 292:173-179.

Govaerts R, Frodin DG, Bogner J, Boyce P *et al.* (2002). World checklist and bibliography of *Araceae* (and *Acoracaceae*). Royal Botanic Garden 560 p.

Gupta M (1992). Effect of growth regulators on foliar stomata of *Vicia faba* L. Adv Plant Sci 5(2):483-495.

Janowska B (2013). Effect of growth regulators on flower and leaf yield of the calla lily (*Zantedeschia* Spreng.). Hort Sci (Prague) 40(2):78-82.

Janowska B, Stanecka A (2011). Effect of growth regulators on the postharvest longevity of cut flowers and leaves of the calla lily (*Zantedeschia* Spreng.). Acta Agrobot 64(4):91-98.

Janowska B, Stanecki M (2013). Effect of rhizome soaking in a mixture of BA and GA_3 on the earliness of flowering and quality of the yield of flowers and leaves in the calla lily (*Zantedeschia* Spreng.). Acta Sci Pol, Hort Cult 12(2):3-12.

Kozłowska M, Rybus-Zając M, Stachowiak J, Janowska B (2007). Changes in carbohydrate contents of *Zantedeschia* leaves under gibberellin-stimulated flowering. Acta Physiol Plant 29:27-32.

Kuehny JS (2000). Calla history and culture. HortTechnology 10(2):267-274.

Namli S, Ayaz E (2007). Influence of different cytokinins used in *in vitro* culture on the stoma morphology of pistachio (*Pistacia vera* L. cv. Siirt). Afr J Biotechnol 6(5):561-563.

Pogroszewska E (2002). Studia nad wzrostem i kwitnieniem skrzydłokwiatu [A study on the growth and flowering of *Spathiphyllum* Schott]. Rozp Nauk AR w Lublinie 263:116 p.

Pospíšilová J (2003). Interaction of cytokinins and abscisic acid during regulation of stomata opening in bean leaves. Photosynthetica 41(1):49-56.

Ranwala AP, Miller WB (1998). $Gibberellin_{4+7}$, benzyladenine and supplemental light improwe postharvest leaf and flower quality of cold-stored 'Star Gazer' hybrid lilies. J Amer Soc Hort Scie 3(4):563-568.

Rodriguez-Pérez JA, de Leon-Hernández AM., Vera-Batista MC, Rodriguez-Hernández J. Alberto-Rodriguez P (2009). Effect of pretreatment with gibberellic acid (GA_3) and Promalin (GA_{4+7} + BA) on germination of *Protea aristata and P. repens*. Acta Hort 813:441-444.

Wachowicz M, Skutnik E, Rabiza-Świder J (2006). The effect of growth regulators on stomatal aperture in senescing cut leaves of *Zantedeschia aethiopica* Spr. and *Hosta* Tratt. 'Undulata Erromena'. Folia Hort 10(2):65-75.

Wang W, Zhao N-H (2002). Epidermal characters of leaves in *Araceae*. Plant Sci J 20(5):343-349.

11

Genetic Variability and Association Analysis of Some Quantitative Characters in Sweet Corn

Ozlem ALAN[1*], Gulcan KINACI[2], Engin KINACI[2], Imren KUTLU[3] Zekiye BUDAK BASCİFTCİ[2], Kenan SONMEZ[4], Yasemin EVRENOSOGLU[4]

[1]*Ege University, Odemis Vocational Training School, Izmir, Turkey; ozlem.alan@ege.edu.tr (*corresponding author)*

[2]*Eskisehir Osmangazi University, Faculty of Agriculture, Department of Field Crops, Eskisehir, Turkey; gkinaci@ogu.edu.tr; ekinaci@ogu.edu.tr; zbudak@ogu.edu.tr*

[3]*Eskisehir Osmangazi University, Faculty of Agriculture, Eskisehir, Turkey; ikutlu@ogu.edu.tr*

[4]*Eskisehir Osmangazi University, Faculty of Agriculture, Department of Horticulture, Eskisehir, Turkey; ksonmez@ogu.edu.tr; yevrenosoglu@ogu.edu.tr*

Abstract

The objective of this study was to determine genetic variability, heritability, genetic advance, genotypic and phenotypic correlations of yield, yield components and kernel quality traits in seven sweet corn varieties. The present research was conducted during 2009 and 2010 growing season in Eskisehir, midwestern Turkey. The trials were set up in randomised complete block design with four replications. Analysis of variance observed highly significant differences for all the examined traits in both years. Sugar content, soluble solid concentration and number of leaves per plant revealed the highest genotypic and phenotypic coefficient of variation values. The high heritability estimates coupled with high genetic advance for sugar content, soluble solid concentration and starch content. Positive correlations were revealed between yield (husked, dehusked and fresh kernel) and yield components except plant height and 1000 seed weight. Negative correlations were found between kernel quality and yield and yield related traits. It can be concluded that, husked ear weight and dehusked ear weight could be used as the main criteria for yield improvement. It should be unfeasible to develop sweet corn varieties with satisfactory yield potential and improved kernel quality for the different sweet corn markets.

Keywords: correlation coefficient, genetic advance, heritability, quality, sweet corn, yield

Introduction

Sweet corn (*Zea mays* L. *var. saccharata*) is a cultivated plant grown for fresh human consumption and raw or processed material of food industry throughout the world. Sweet corn differs from other corns (field, pop corn etc.). The primary difference is gene expression that determines endosperm carbohydrate content as well as many other genes that affect maize growth (Erdal *et al.*, 2011; Najeeb *et al.*, 2011; Znidarcic, 2012). It is produced for three distinct markets; fresh, canning and freezing. Production within these markets is largely independent of each other (Kleinhenz, 2008). Both total production and value of processed sweet corn has increased 60% over the last 25 years (Williams II, 2006). In Turkey, no information is available regarding the acreage of sweet corn grown. Despite the above-mentioned knowledge, interest in sweet corn has grown in recent years. Frozen sweet corn kernels are in the first rank among the fresh vegetables for import and frozen sweet corn export value was 36 tons in 2010 in Turkey (Aydın, 2011).

Sweet corn crops must meet strict market requirements for quality and appearance. While marketable yield, plant and ear height, and other characteristics are important to growers, the appearance and dimensions of ears and the sensory properties of kernels are important to consumers and these traits may be influenced by genotype, environment and in-field management (Rangarajan *et al.*, 2002; Kleinhenz, 2003).

The sweet corn breeders have often focused on improving quality and ear appearance, rather than on enhancing yield. Eating quality is a very complicated trait because of the effects of the individual gene influencing this trait, which is difficult to isolate and quantify. The polygenic nature of this trait has directed the breeding efforts towards the development of new sweet corn hybrids with improved eating quality and favourable ear and kernel traits (Tracy, 2003).

Knowledge of the relative magnitude of genetic parameters of yield, yield components and quality characters are important for an efficient breeding program. A character which have higher range of genetic variability, high heritability and high genetic advance would be an effective tool to improve yield. Therefore, for plant breeders it is essential to examine correlations to see the relationships between characters in order to decide for suitable selection criteria for a breeding program. Also, many sweet corn varieties are developed under environments which differ from those in target production areas. Therefore, assessing the relationships between characters in different locations and/or time is important to breeders. Yet, such studies are needed to develop better varieties with an emphasis on kernel quality and agronomic requirements to each specific market is required. The objective of this research was to determine genetic parameters, genotypic and phenotypic correlations of yield, yield components and kernel quality traits in sweet corn varieties.

Materials and methods

The field experiments were carried out in the experimental fields of Faculty of Agriculture, Eskişehir Osmangazi University, Eskişehir (39°46'N; 30°33'E; 801 m above sea level) during 2009 and 2010 growing season.

Climatic conditions during the experiment were given in Fig. 1(a) and 1(b). The soil was sandy clay (44% sand; 20% clay) with 1% organic matter, 0.05% total salt, 5% lime and pH 7.6. Seven sweet corn varieties (*Lumina, Merit, Sunshine, Jubile, Challenger Yellow Baby* and 2201) were used as materials.

Each cultivar was sown in randomized complete block with four replications and each experimental plot was 29.4 m^2. Seeds were sown on April 29 in 2009 and on May 19 in 2010, in a spacing of 70 cm x 25 cm. Plants were fertilized with equivalent to 280 kg N, 110 kg P_2O_5 and 110 kg K_2O per hectare (Turgut, 2000) during growing season. The drip irrigation was applied as needed and weeds were controlled mechanically by hand. No fungicide and insecticide were applied during cultivation.

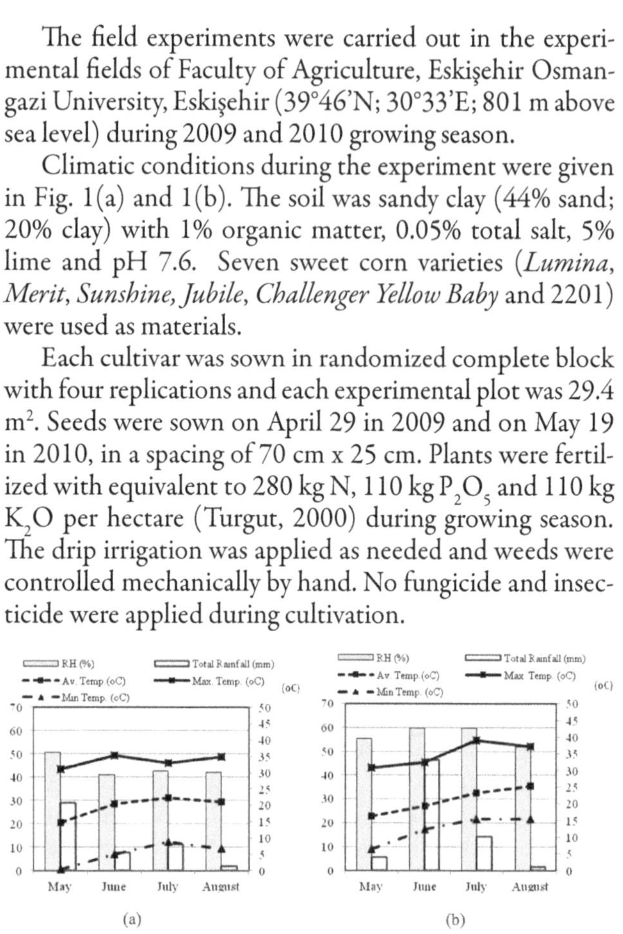

Fig. 1. Climatic conditions during (a) 2009 and (b) 2010

Plant height and number of leaves per plant were recorded on the whole plot basis. When the sweet corn reached harvest maturity (juice consistecy of kernels) twenty ears from in the centre of each replication were harvested randomly by hand in the morning.

These ears were taken to the processing lab and divided into the following categories for measurements: ten for ear traits, ten for kernel quality traits. Husked ear weight, dehusked ear weight, ear length, ear diameter, number of rows per ear, number of kernels per row were recorded for ear traits. Husked ear yield was calculated from husked ear weight, dehusked ear yield was calculated from dehusked ear weight. Approximately 2 months after harvesting as International Seed Testing Association (ISTA, 1999) suggests, 1000 seed weight was calculated by counting 8 replicated 100 seeds from each plot and was weighted in g and the mean was multiplied with 10.

For kernel quality traits, within one hour after harvest, kernels were cut from the ear and the following measurements are recorded; fresh kernel weight, dry matter content, soluble solid concentration, protein content, sugar content and starch content. Fresh kernel yield was calculated from fresh kernel weight. Kernels dried in an oven at 65 °C weight loss between measurements was < 0.05 g. The percentage difference between fresh and dry weights was used as the dry matter content of the kernel. Extract was prepared for soluble solid concentration using modified methods of Hale *et al.* (2005) and determinated with the use of a digital refractometer. Kernel protein content was determinated by kjeldahl method (Kirk and Sawyer, 1991). Kernel sugar content was determinated by the Lane-Eynon Method (Kirk and Sawyer, 1991). Kernel starch content was determinated by polarimetric method (Earle and Milner, 1944).

To estimate the extent of magnitude of variation among examined traits, all data were subjected to analysis of variance. Mean, standard error, range were analyzed according to Singh and Chaudhary (1985). Components of variance $\sigma^2 g$= genotypic variance and $\sigma^2 p$= phenotypic variance were estimated using the following formula (Wricke and Weber, 1986).

The genotypic and phenotypic coefficient variations were calculated as the square root of genotypic and phenotypic variances divided by the mean of traits and multiplying by 100.

Genotypic coefficient of variation:

$$(GCV) = \sqrt{\sigma^2 g}/X \times 100;$$

Phenotypic coefficient of variation:

$$(PCV) = \sqrt{\sigma^2 p}/X \times 100$$

where, $\sigma^2 p$, $\sigma^2 g$ and X are the phenotypic and genotypic standard deviation and mean value of the traits, respectively. Heritability in broad sense (h^2) was estimated as the ratio of genotypic variance to the phenotypic variance described by Hanson *et al.* (1956) as: Heritability (h^2)= $(\sigma^2 g)/(\sigma^2 p)$

Expected genetic advance (GA) and GA as percent of the mean calculated according to Shukla *et al.* (2006).

$GA = i\,\sigma^2 p\, h^2$ GA(%) =GA/X*100

Where, i= efficacy of selection which is 2.06 at 5% selection intensity; $\sigma^2 p$= phenotypic standard deviation; h^2= heritability in broad sense. Genotypic (r_g) and phenotypic (r_F) correlation coefficient between x and y traits were calculated based on the procedure described by Kempthorne (1973).

$$rg(x,y) = \frac{COVg(x,y)}{\sqrt{\sigma 2g(x) . \sigma 2g(y)}}$$

where, $COV_{xy}(g)$ and, $COV_{xy}(p)$ are genotypic and phenotypic covariance between x and y characters, σ^2g= genotypic variance, σ^2p= phenotypic variance.

Results and discussions

Genetic variability: The range, mean, standard error, critical differences, coefficient of variation, genotypic and phenotypic variance, genotypic coefficient of variability (GCV) and phenotypic coefficient of variability (PCV), broad sense heritability (h^2) and genetic advence in a percentage mean (GA) were given in Tab. 1 and 2.

Highly significant differences were observed for all the examined traits in both years. In 2009, husked ear yield ranged from 19.4-29.3 ton per hectare, dehusked ear yield 19.4-24.6 ton per hectare, fresh kernel yield 14.2-20.9 ton per hectare, plant height 176-232 cm, number of leaves per plant 5.4-11.7, husked ear weight 418-525 g, dehusked ear weight 340-430 g, ear length 20-25 cm, ear diameter 45-55 mm, number of rows per ear 16-22, number of kernels per row 38-47, 1000 seed weight 120-178 g, dry matter content 32.4-40.0%, soluble solid concentration 14.4-28.5%, protein content 9.6-13.5%, sugar content 3.1-10.0% and starch content 13.0-20.0%. In the second year of the experiment, husked ear yield ranged from 21.5-30.8 ton per hectare, dehusked ear yield 18.4-23.1 ton per hectare, fresh kernel yield 13.3-17.8 ton per hectare, plant height 182-237 cm, number of leaves per plant 8.8-12.2, husked ear weight 393-539 g, dehusked ear weight 321-403 g, ear length 21-25 cm, ear diameter 49-56 mm, number of rows per ear 15-21, number of kernels per row 38-46, 1000 seed weight 127.3-189-8 g, dry matter content 29.5-44.2%, soluble solid concentration 11.6-31.0%, protein content 8.9-14.7%, sugar content 3.2-8.6% and starch content 11.6-19.1% (Tab. 2). The presented results revealed that the genotypes were differed significantly for investigated yield, yield related traits and kernel quality traits indicating a considerable range of genetic variability.

The values of PCV were higher than the corresponding GCV values for all characters, the differences between them were low for the most of the studied traits (Tab. 1 and 2). This indicates that almost all of the characters are more influenced by the environment. Similar results have been reported by Asghar and Mehdi (1999), Saleh *et al.* (2002) in sweet corn and Hefny (2011) in corn. The highest GCV and PCV values were recorded for sugar content, soluble solid concentration and number of leaves per plant in 2009 (Tab. 1); sugar content and soluble solid concentration in 2010. In both years, the lowest GCV and PCV values were obtained from ear length, ear diameter, dry matter content and number of kernels per row. (Tab. 1 and 2). Hefny (2011) reported that high GCV estimates are an indicative of less amenability of these traits to environmental fluctuations and greater emphasis should be given to these characters, while breeding cultivars from the present material.

Estimates of heritability in broad sense changed between 41.4% for ear length to 99% for dehusked ear yield and dehusked ear weight in 2009 (Tab. 1); and between 32.5% for dry matter content to 97.9% for sugar content in 2010 (Tab. 2). GA was highest for sugar content (2.76%), which was followed by soluble solid concentration (1.33%), number of leaves per plant (1.10%), protein content (0.76%), starch content (0.73%), husked ear yield (0.65%), dehusked ear yield (0.60), while it was lowest for ear length (0.16%) in 2009 (Tab. 1).

In 2010, GA was highest for sugar content (2.70%), which was followed by soluble solid concentration (1.89%), starch content (0.92%), number of rows per ear (0.63%) etc., while it was lowest for dry matter content (0.19%) (Tab. 2). Johnson *et al.* (1955) suggest that estimates of heritability and genetic advance should always be considered simultaneously because high heritability will not always associate with high GA. The sugar content and soluble solid concentration observed high heritability (>80%) with high GA; protein content, starch content, husked ear yield, dehusked ear yield, husked ear weight, dehusked ear weight accompanied with high heritability (>80%) and moderate GA in 2009 (Tab. 1). Similar results were obtained in 2010; sugar content, soluble solid concentration and starch content observed high heritability (>80%) with high GA; husked ear weight, dehusked ear weight, dehusked ear yield, fresh kernel yield and number of rows per ear accompanied with high heritability (>80%) and moderate GA (Tab. 2). Low heritability was associated with low GA for dry matter content and 1000 seed weight, in both years (Tab. 1 and 2).

In this study, the high heritability estimates coupled with high GA for sugar content, soluble solid concentration and starch content indicating that these characters can be considered as favorable attributes for improvement through selection and this may due to additive gene action (Panse 1957). Rosenbrook and Andrew (1971) also reported the additive gen action could be significant for sucrose and phytoglycogen accumulation in kernel. Traits with high heritability and moderate genetic advance (husked ear yield, dehusked ear yield, husked ear weight, dehusked ear weight, number of rows per ear) are considered not suitable for genetic improvement through selection under such conditions. Traits with low heritability and GA (dry matter content and 1000 seed weight) are limited for genetic improvement through selection (Hefny 2011).

Correlation coefficient: The genotypic and phenotypic correlation coefficients were presented in Tab. 3 and 4. In general, genotypic correlation coefficients were higher than corresponding phenotypic ones for most of the characters in both years. 17% of genotypic correlations and 6% of phenotypic correlations were statistically significant

(Tab. 3 and 4). The utilize selection based on genotypic correlation is an effective instrument for examining degree of relationships among traits due to phenotypic correlation obtain from genotype and environment interaction (Eşiyok *et al.*, 2011).

Significant correlations were estimated between yield and yield components for both years in this study (Tab. 3 and 4). In the first year of the experiment, husked ear yield was positively and significantly correlated with number of leaves per plant, husked ear weight and ear length (with correlation coefficients of 0.79, 1.00, and 1.00, respectively). Dehusked ear yield was positively and significantly correlated with fresh kernel yield, dehusked ear weight and ear length (with correlation coefficients of 0.89, 1.00 and 0.82, respectively). Fresh kernel yield had significant positive correlations with dehusked ear weight (0.89) and ear diameter (0.79). Number of leaves per plant was positively and highly correlated with husked ear weight. Husked ear weight and dehusked ear weight had positive significant correlations with ear length. Ear length was significantly and positively correlated with number of kernels per row (Tab. 3).

In the second year of the experiment, the relationships among yield and yield related traits were slightly different (Tab. 4). Husked ear yield had positive and significant correlation with dehusked ear yield, husked ear weight, dehusked ear weight and number of kernels per row (with correlation coefficients of 0.82, 1.00, 0.82 and 0.82, respectively).

Dehusked ear yield was positively and significantly correlated with fresh kernel yield, husked ear weight, dehusked ear weight, ear diameter, number of kernels per row (with correlation coefficients of 0.97, 0.88, 1.00, 0.82 and 0.79, respectively). Fresh kernel yield was positively and significantly correlated with husked ear weight, dehusked ear weight, ear diameter, number of rows per ear (with correlation coefficients of 0.75, 0.98, 0.92 and 0.82, respectively). Plant height had positive and significant association with ear length. Positive and significant correlation was existed between husked ear weight and dehusked ear weight. Ear diameter and number of rows per ear had positive significant correlations with dehusked ear weight (Tab. 4).

Yield and yield related traits revealed that husked ear yield was positively correlated with dehusked ear yield and dehusked ear yield was positively correlated with fresh kernel yield for both years (Tab. 3 and 4). These results are in close agreement with earlier report of Kashiani et al. (2010). Positive correlations were found between yield (husked, dehusked and fresh kernel) and yield components except plant height and 1000 seed weight (Tab.3 and 4). This indicates that high measurements of these traits had direct positive contribution to sweet corn yield and its possible to simultaneous improvement husked, dehusked or fresh kernel yield with these characters. Similar findings reported by Wong *et al.* (1994); Saleh *et al.* (2002); Kashiani *et al.* (2008); Öktem (2008); Kashiani *et al.* (2010); Khazaei *et al.* (2010); İlker (2011) in sweet corn and Kabdal *et al.* (2003) and Hefny (2011) in corn.

In this study, no significant correlation was obtained between plant height and yield. These results are in agreement with the findings of Asghar and Mehdi (1999) and Öktem (2008). On the other hand, Saleh *et al.* (2002); Kashiani (2010) and İlker (2011) reported that plant height had positive and significant association with yield.

Also, significant correlations were estimated between yield-yield components and kernel quality for both years in this study (Tab. 3 and 4). Negative correlations were found between number of leaves per plant and dry matter content; ear length and dry matter content; number of kernels per row and dry matter content in 2009; fresh kernel yield and sugar content; plant height and sugar content in 2010. Positive correlations were existed between plant height and starch content; 1000 seed weight and dry matter content in 2009; 1000 seed weight and dry matter content; 1000 seed weight and soluble solid concentration; plant height and starch content in 2010. Similar negative correlations reported between yield and yield components and kernel quality traits (Wong *et al.*, 1994; Saleh *et al.*, 2002; Solomon, 2011). This indicates that the difficulty of simultaneous improvement of sweet corn for both yield ability and kernel quality (Ha, 1999; Saleh *et al.*, 2002). On the other hand, non significant correlations between kernel quality and yield related traits reported by Khanduri *et al.* (2010), suggesting the potential for improvement of kernel sugar concentration independent of yield.

Kernel quality traits revealed that they were correlated among themselves for both years (Tab. 3 and 4). Positive and significant correlations were found between soluble solid concentration and starch content; dry matter content and soluble solid concentration. On the other hand, non-significant correlations between soluble solid concentration and starch content were reported by Kumari *et al.* (2006). Negative associations were existed between sugar content and starch content; protein content and starch content; soluble solid concentration and protein content. This kind of associations were also observed by Dudley and Lambert (1992); Ha (1999), Kumari *et al.* (2006). Many authors stated that soluble solid concentration has been utilised as a rapid, pre-harvest method to determine fresh sweet corn sugar content (Kleinhenz, 2003; Zhu *et al.*, 1992). Kumari *et al.* (2006) also reported total sugar, reducing sugar and non-reducing sugar content were positively and significantly correlated with soluble solid concentration. In contrast, Hale *et al.* (2005) reported that an overall R^2 of -0.99 between soluble solid concentration and total sugars; soluble solid cannot be used as a reliable indicator of sweet corn total sugar concentrations. In this study, non-significant but negative correlations were estimated between soluble solid concentration and sugar content.

Tab. 1. Genetic parameters of yield, yield components and kernel quality in sweet corn (2009)

	Range	Mean square	Mean	SE (±)	CD	CV (%)	σ^2g	σ^2p	GCV (%)	PCV (%)	h^2 (%)	GA (%)
Yield and Yield Components												
Husked ear yield (ton ha^{-1})	19.4 ± 29.3	51.60**	22.8	0.42	0.87[1] 1.20[2]	0.23	4.38	4.72	0.33	0.34	92.8	0.65
Dehusked ear yield (ton ha^{-1})	19.4 ± 24.6	403.08**	21.2	0.12	0.26 0.35	0.07	2.99	3.02	0.29	0.29	99.0	0.60
Fresh kernel yield (ton ha^{-1})	14.2 ± 20.9	6.08**	16.0	0.78	1.65 2.26	0.52	1.56	2.79	0.28	0.37	55.9	0.43
Plant height (cm)	176 ± 232	8.00**	206.6	6.68	14.03 19.21	1.24	156.07	245.20	0.22	0.27	63.7	0.35
No. of leaves/plant	5.4 ± 11.7	13.14**	8.0	0.56	1.17 1.60	0.53	1.87	2.49	0.62	0.71	75.1	1.10
Husked ear weight (g)	418 ± 525	53.08**	481.4	7.15	15.02 20.58	0.87	1331.49	1433.76	0.27	0.28	92.9	0.54
Dehusked ear weight (g)	340 ± 430	404.66**	371.3	2.12	4.45 6.10	0.29	906.23	915.21	0.29	0.29	99.0	0.59
Ear length (cm)	20 ± 25	3.80*	22.5	0.63	1.32 1.80	0.35	0.55	1.33	0.12	0.18	41.4	0.16
Ear diameter (mm)	44.9 ± 55.0	8.38**	48.7	1.12	2.36 3.23	0.43	4.63	7.15	0.16	0.20	64.8	0.26
No. of rows/ear	16 ± 22	6.15**	18.5	0.77	1.61 2.20	0.48	1.51	2.68	0.24	0.32	56.3	0.37
No. of kernels/row	38 ± 47	15.35**	42.3	0.81	1.70 2.33	0.33	4.68	5.99	0.18	0.21	78.1	0.33
1000-seed weight (g)	120 ±178	4.13**	139.9	8.72	18.32 25.09	1.97	122.27	274.29	0.28	0.42	44.6	0.39
Kernel Quality												
Dry matter content (%)	32.4 ± 40.0	11.26**	35.5	0.82	1.72 2.35	0.37	3.42	4.75	0.19	0.22	72.0	0.33
Soluble solid concentration (%)	14.4 ± 28.5	62.83**	21.7	0.73	1.53 2.10	0.42	16.40	17.46	0.67	0.69	93.9	1.33
Protein content (%)	9.6 ± 13.5	79.89**	11.1	0.19	0.39 0.53	0.15	1.37	1.44	0.38	0.39	95.1	0.76
Sugar content (%)	3.1 ± 10.0	77.39**	5.4	0.34	0.71 0.97	0.39	4.32	4.55	1.37	1.41	94.9	2.76
Starch content (%)	13 ± 20	20.39**	17.5	0.61	1.29 1.76	0.39	3.64	4.40	0.39	0.43	82.7	0.73

SE: Standard error, CD: critical difference, [1]5% and [2]1%, CV:coefficient of variation, σ^2 g: genotypic variances, σ^2p: phenotypic variances, GCV: genotypic coefficient of variation, PCV: phenotypic coefficient of variation, h^2: broad-sense heritability, GA: genetic advance, *,**significant at 5% and 1% probability level respectively

Tab. 2. Genetic parameters of yield, yield components and kernel quality in sweet corn (2010)

	Range	Mean square	Mean	SE (±)	CD	CV(%)	$\sigma^2 g$	$\sigma^2 p$	GCV (%)	PCV (%)	h^2 (%)	GA (%)
Yield and Yield Components												
Husked ear yield (ton ha^{-1})	21.5 ± 30.8	7.84**	26.3	1.15	2.42[1] 3.32[2]	0.60	4.54	7.19	0.29	0.36	63.1	0.47
Dehusked ear yield (ton ha^{-1})	18.4 ± 23.1	20.11**	20.6	0.51	1.08 1.48	0.30	2.53	3.06	0.28	0.30	82.7	0.52
Fresh kernel yield (ton ha^{-1})	13.3 ± 17.8	21.66**	15.5	0.44	0.91 1.15	0.30	1.96	2.33	0.32	0.35	84.1	0.61
Plant height (cm)	182 ± 237	16.14**	214.6	5.06	10.63 14.57	0.92	193.96	245.19	0.23	0.26	79.1	0.42
No. of leaves/plant	8.8 ± 12.2	11.13**	10.3	0.38	0.80 1.09	0.32	0.73	1.01	0.30	0.35	72.3	0.52
Husked ear weight (g)	393 ± 539	28.64**	464.3	11.02	23.14 31.70	1.37	1676.63	1919.31	0.31	0.34	87.4	0.61
Dehusked ear weight (g)	321 ± 403	20.03**	359.8	8.98	18.88 25.86	1.27	767.85	929.29	0.28	0.30	82.6	0.52
Ear length (cm)	21.3 ± 25.1	11.17**	23.1	0.39	0.82 1.13	0.22	0.78	1.09	0.14	0.16	71.6	0.24
Ear diameter (mm)	49 ± 56	19.21**	52.3	0.65	1.37 1.88	0.24	3.87	4.71	0.13	0.15	82.2	0.25
No. of rows/ear	15 ± 21	21.45**	17.6	0.51	1.08 1.48	0.33	2.70	3.23	0.33	0.36	83.6	0.63
No. of kernels/row	38 ± 46	7.46**	41.6	1.11	2.33 3.20	0.46	3.98	6.45	0.17	0.22	61.7	0.28
1000-seed weight (g)	127.3 ± 189.8	5.05**	158.5	9.83	20.65 28.29	2.09	195.68	388.87	0.32	0.44	50.3	0.46
Kernel Quality												
Dry matter content (%)	29.5 ± 44.2	2.93*	36.7	1.67	3.51 4.81	0.74	2.69	8.27	0.16	0.28	32.5	0.19
Soluble solid concentration (%)	11.6 ± 31.0	58.21**	22.2	1.11	2.32 3.18	0.63	35.03	37.48	0.95	0.98	93.5	1.89
Protein content (%)	8.9 ± 14.7	8.11**	11.1	0.57	1.19 1.63	0.46	1.14	1.78	0.34	0.43	64.0	0.57
Sugar content (%)	3.2 ± 8.6	201.62**	4.9	0.18	0.38 0.52	0.22	3.25	3.32	1.33	1.34	97.9	2.70
Starch content (%)	11.6 ± 19.1	24.53**	15.9	0.63	1.32 1.80	0.42	4.62	5.41	0.48	0.52	85.4	0.92

SE: Standard error, CD: critical difference, [1]5% and [2]1%, CV:coefficient of variation, σ^2 g: genotypic variances, σ^2p: phenotypic variances, GCV: genotypic coefficient of variation, PCV: phenotypic coefficient of variation, h^2: broad-sense heritability, GA: genetic advance, *,**significant at 5% and 1% probability level respectively

Table 3. Genotypic and phenotypic correlation coefficients of yield, yield components and kernel quality traits among sweet corn genotypes in 2009.

Traits		Husked ear yield	Dehusked ear yield	Fresh kernel yield	Plant height	No. of leaves/ plant	Husked ear weight	Dehusked ear weight	Ear length	Ear diameter	No. of rows/ear	No. of kernels/ row	1000-seed weight	Dry matter content	Soluble solid concentration	Protein content	Sugar content
Yield and Yield Components																	
Dehusked ear yield	G*	0.69															
	P	0.67															
Fresh kernel yield	G	0.54	0.89**														
	P	0.39	0.68														
Plant height	G	0.42	0.21	0.25													
	P	0.43	0.18	0.33													
No. of leaves/ plant	G	0.79*	0.36	0.55	0.62												
	P	0.68	0.32	0.48	0.65												
Husked ear weight	G	1.00**	0.69	0.54	0.42	0.79**											
	P	1.00**	0.67	0.40	0.33	0.68											
Dehusked ear weight	G	0.70	1.00**	0.89**	0.21	0.36	0.69										
	P	0.67	1.00**	0.69	0.18	0.32	0.67										
Ear length	G	1.00**	0.82*	0.71	0.43	0.71	1.00**	0.82*									
	P	0.62	0.52	0.23	0.13	0.33	0.62	0.52									
Ear diameter	G	0.14	0.62	0.79*	0.41	0.11	0.14	0.62	0.04								
	P	0.09	0.51	0.53	0.24	0.08	0.09	0.50	-0.02								
No. of rows/ear	G	0.24	0.57	0.71	0.08	0.38	0.24	0.58	0.33	0.14							
	P	0.18	0.41	0.34	-0.07	0.18	0.18	0.41	0.26	-0.02							
No. of kernels/row	G	0.70	0.45	0.30	0.61	0.60	0.74	0.45	1.00**	-0.07	0.13						
	P	0.64	0.42	0.29	0.51	0.48	0.64	0.42	0.58	-0.08	0.01						
1000-seed weight	G	-0.17	0.31	-0.04	-0.29	-0.51	-0.17	0.30	-0.40	0.55	0.12	-0.58					
	P	-0.09	0.21	0.09	-0.22	-0.30	-0.09	0.21	-0.23	0.23	-0.02	-0.41					
Kernel Quality																	
Dry matter content	G	-0.70	-0.21	-0.41	-0.56	-0.88**	-0.70	-0.22	-0.79*	0.07	0.07	-0.79*	0.86*				
	P	-0.60	-0.21	-0.38	-0.34	-0.61	-0.60	-0.21	-0.53	-0.06	0.08	-0.62	0.42				
Soluble solid concentration	G	-0.18	0.009	-0.27	0.37	-0.21	-0.19	0.01	-0.08	-0.06	0.39	0.12	0.26	0.51			
	P	-0.16	0.012	0.17	0.30	-0.14	-0.16	0.01	-0.12	-0.04	0.26	0.08	0.19	0.40			
Protein content	G	0.12	-0.08	0.01	-0.69	-0.01	0.12	-0.08	-0.03	-0.19	-0.33	-0.30	-0.04	-0.20	-0.93**		
	P	0.10	-0.08	0.01	-0.56	-0.36	0.10	-0.08	0.02	-0.11	-0.23	-0.26	-0.05	-0.19	-0.88**		
Sugar content	G	0.19	-0.36	-0.58	-0.44	0.02	0.20	-0.36	-0.06	-0.61	-0.56	-0.15	-0.05	-0.12	-0.49	0.71	
	P	0.20	-0.35	-0.44	-0.36	0.02	0.20	-0.35	-0.04	-0.47	-0.35	-0.13	-0.07	-0.13	-0.44	0.69	
Starch content	G	0.16	0.15	0.06	0.92**	0.25	0.13	0.15	0.18	0.26	0.16	0.49	-0.04	-0.05	0.81*	-0.98**	-0.59
	P	0.07	0.15	0.01	0.66	0.12	0.11	0.15	0.17	0.22	0.06	0.44	-0.05	-0.06	0.70	-0.86*	-0.56

G: genetic correlation coefficient, P: phenotypic correlation cofficient.,** Significant at 5% and 1% probability level, respectively

Tab. 4. Genotypic and phenotypic correlation coefficients of yield, yield components and kernel quality traits among sweet corn genotypes in 2010.

Traits		Husked ear yield	Dehusked ear yield	Fresh kernel yield	Plant height	No. of leaves/ plant	Husked ear weight	Dehusked ear weight	Ear length	Ear diameter	No. of rows/ ear	No. of kernels/ row	1000-seed weight	Dry matter content	Soluble solid concentration	Protein content	Sugar content
Yield and Yield Components																	
Dehusked ear yield	G*	0.82**															
	P	0.74															
Fresh kernel yield	G	0.65	0.97**														
	P	0.62	0.93**														
Plant height	G	0.17	0.33	0.44													
	P	0.16	0.34	0.43													
No. of leaves/ plant	G	0.39	0.15	0.19	0.44												
	P	0.32	0.14	0.16	0.35												
Husked ear weight	G	1.00**	0.88**	0.75*	0.12	0.27											
	P	0.89**	0.85*	0.71	0.08	0.21											
Dehusked ear weight	G	0.82*	1.00**	0.98**	0.33	0.15	0.88**										
	P	0.74	1.00**	0.93**	0.34	0.14	0.86**										
Ear length	G	0.32	-0.08	-0.31	-0.77*	-0.44	0.32	-0.08									
	P	0.18	-0.08	-0.22	-0.62	-0.35	0.20	-0.08									
Ear diameter	G	0.36	0.82*	0.92**	0.27	0.40	0.44	0.83*	-0.45								
	P	0.37	0.80*	0.86*	0.28	0.21	0.49	0.81*	-0.37								
No. of rows/ear	G	0.53	0.79*	0.82*	0.28	-0.13	0.53	0.80*	-0.14	0.52							
	P	0.46	0.77*	0.73	0.29	-0.12	0.52	0.77*	-0.15	0.54							
No. of kernels/row	G	0.82*	0.22	0.07	0.22	0.53	0.62	0.22	0.45	-0.36	0.02						
	P	0.52	0.26	0.11	0.12	0.44	0.51	0.26	0.29	-021	0.07						
1000-seed weight	G	-0.01	0.53	0.59	0.36	-0.36	0.13	0.53	-0.50	0.74	0.50	-0.64					
	P	-0.08	0.30	0.37	0.24	-0.26	0.08	0.30	-0.30	0.47	0.41	-0.35					
Kernel Quality																	
Dry matter content	G	0.08	0.37	0.30	-0.03	-0.66	0.13	0.37	-0.01	0.47	0.19	-0.61	0.79*				
	P	-0.19	0.07	0.03	-0.04	-0.32	0.04	0.07	-0.18	0.17	0.17	-0.36	0.66				
Soluble solid concentration	G	0.30	0.50	0.44	0.57	-0.19	0.35	0.50	-0.27	0.35	0.43	-0.01	0.77*	0.93**			
	P	0.24	0.46	0.42	0.52	-0.16	0.32	0.46	-0.19	0.33	0.37	-0.03	0.52	0.51			
Protein content	G	-0.28	-0.22	-0.11	-0.69	-0.09	-0.19	-0.22	0.19	0.14	-0.07	-0.43	-0.26	-0.44	-0.89**		
	P	-0.20	-0.15	-0.11	-0.55	-0.04	-0.16	-0.15	0.13	0.07	-0.09	-0.19	-0.33	-0.34	-0.73		
Sugar content	G	-0.33	-0.61	-0.75*	-0.81*	-0.40	-0.37	-0.62	0.53	-0.56	-0.50	-0.20	-0.33	0.26	-0.30	0.32	
	P	-0.34	-0.55	-0.66	-0.70	-0.34	-0.34	-0.55	0.49	-0.49	-0.46	-0.18	-0.25	0.11	-0.30	0.24	
Starch content	G	0.50	0.59	0.60	0.85*	0.21	0.50	0.60	-0.27	0.24	0.38	0.43	0.35	0.11	0.76*	-0.87**	-0.79*
	P	0.31	0.47	0.43	0.62	0.11	0.44	0.47	-0.23	0.19	0.39	0.36	0.34	0.20	0.65	-0.65	-0.74

G: genetic correlation coefficient, P: phenotypic correlation cofficient.,** Significant at 5% and 1% probability level, respectively.

Conclusions

The selection of hybrids for sources of favorable alleles in a breeding program for sweet corn improvement will depend on the consistency of a cultivar's performance over environments. Our replicated study in two years with seven hybrids was conducted to provide information concerning the effect of environmental factors on the performance of these hybrids. It can be conclude that sugar content, soluble solid concentration (for both years) and number of leaves per plant (in 2009) revealed the highest GCV and PCV values.

Sugar content, soluble solid concentration (for both years) and starch content (in 2010) showed high heritability with high genetic advance. These characters can be improved efficiently by individual selection or breeding strategies in the examined populations. Positive correlations were found between yield (husked, dehusked and fresh kernel) and yield components except 1000 seed weight. Husked ear weight and dehusked ear weight (for both years) could be used as the main criteria for all three types of yield improvement when selecting in the studied materials.

Significant negative correlations between kernel quality and yield and yield related traits, suggests that it should be unfeasible to develop sweet corn varieties with satisfactory yield potential and kernel quality for the different sweet corn markets.

Acknowledgement

The authors thanks Eskişehir Osmangazi University, The Scientific Research Projects Commission for supporting this study as the project number 200823026.

References

Asghar JA, Mehdi SS (1999). Variability for grain yield, its components and quality traits in a sweet corn population. Pakistan J Biol Sci 2(4):1336-1370.

Aydın İS (2011). Frozen fruits and vegetables in Turkey. Available at: http://tcp.gov.tr (Accessed:February 23, 2011).

Dudley JW, Lambert RJ (1992). Ninety generations of selection for oil and protein in maize. Maydica 37:1-7.

Earle FR, Milner RT (1944). Improvements in the determination of starch in corn and wheat. Cereal Chem 21:567-575.

Erdal Ş, Pamukçu M, Savur O, Tezel M (2011). Evaluation of developed standard sweet corn (*Zea mays sacharata* L.) hybrids for fresh yield, yield components and quality parameters. Turk J Field Crops 16(2):153-156.

Eşiyok D, Bozokalfa MK, Aşçıoğul TK (2011). Variability, heritability and association analysis in plant traits of Swiss chard (*Beta vulgaris* subsp. *cicla*). Genetica 43(2):239-252.

Ha V (1999). Genetic analysis of some yield components and kernel quality in sweet corn. Rom Agrıc Res 11-12:9-20.

Hale TA, Hassell RL, Phillips T (2005). Refractometer measurements of soluble solid concentration do not reliably predict sugar content in sweet corn. HortTechnology 15(3):668-672.

Hanson CH, Robinson HF, Comstock RE (1956). Biometrical studies of yield segragating populations of Korean Lespedesa. Agron J 45:268-272.

Hefny M (2011). Genetic paramaters and path analysis of yield and its components in corn inbred lines (*Zea mays* L.) at different sowing dates. Asian J Crop Sci 3(3):106-117.

İlker E (2011). Correlation and path coefficient analyses in sweet corn. Turk J Field Crops 16(2):105-107.

ISTA (1999). International rules for seed testing. Seed science and technology 24, Supplement, Zürich, Switzerland.

Johnson HW, Robinson HF, Comstock RE (1955). Estimation of genetic and environmental variability in soybeans. Agron J 47:314-318.

Kabdal MK, Verma SS, Ahmad N, Panwar UBS (2003). Genetic variability and correlation studies of yield and its attributing characters in maize (*Zea mays* L.). Agric Sci Digest 23(2): 137-139.

Kashiani P, Saleh G, Abdullah NAP, Abdullah SN (2008). Performance, heritability and correlation studies on nine advanced sweet corn inbred lines. Proceedings of the 10th Symposium of Malaysian Society of Applied Biology, November. 6-8, 2008. Malaysia 48.

Kashiani P, Saleh G, Abdullah NAP, Abdullah SN (2010). Variation and genetic studies on selected sweet corn inbred lines. Asian J Crop Sci 2(2):78-84.

Kempthorne O (1973). An introduction to genetic statistics. Iowa State University Press. Amec.

Khanduri A, Prasanna BM, Hossain F, Lakhera PC (2010). Genetic analyses and association studies of yield components and kernel sugar concentration in sweet corn. Indian J Genet Plant Breed 70(3):257-263.

Khazaei F, Alikhani MA, Yari L, Khandan A (2010). Study the correlation, regression and path coefficient analysis in sweet corn (*Zea mays* var. *saccharata*) under different levels of plant density and nitrogen rate. ARPN J Agric Biol Sci 5:14-19.

Kirk RS, Sawyer R (1991). Pearson's composition and chemical analysis of foods, 9th Ed. Longman Scientific & Technical, Essex, England.

Kleinhenz MD (2003). Sweet corn variety trials in Ohio: recent top performers and suggestions for future evaluations. HortTechnology 13(4):711-718.

Kleinhenz MD (2008). Sweet corn quality-What is it?. Available at: http://www.oardc.ohio-state.edu/kleinhenz (Accessed: May 15, 2008).

Kumari J, Gadag RN, Jha GK (2006). Heritability and correlation studies in sweet corn for quality traits, field emergence and grain yield. MNL 80:18-19.

Nanjeeb S, Sheikh FA, Ahangar MA, Teli NA (2011). Popularization of sweet corn (*Zea mays* L. *Saccharata*) under temperate conditions to boost the socieconomic conditions.

Maize Genetic Cooperation Newsletter 85.

Öktem A (2008). Determination of selection criterions for sweet corn using path coefficients analyses. Cereal Res Commun 36(4):561-570.

Panse VG (1957). Genetic of quantitative characters in relation to plant breeding. Ind J Genet 17:318-328.

Rangarajan A, Ingall B, Orfanedes M, Wolfe D (2002). In-row spacing and cultivar effects ear yield and quality of early-planted sweet corn. HortTechnology 12:410-415.

Rosenbrook RW, Andrew RH (1971). Diallel analyses of kernel carbonhydrates in sweet corn. Crop Sci 11:536-540.

Saleh GB, Alawi SAS, Panjaitan K (2002). Performance, correlation and heritability studies on selected sweet corn synthetic populations. Pakistan J Biol Sci 5(3):251-254.

Singh RK, Chaudhary BD (1985). Biometrical methods in quantitative genetic analysis. Kalyani Publisher, New Delhi, India.

Solomon KF, Martin I, Zeppa A (2011). Genetic effects and genetic relationships among shrunken (sh2) sweet corn lines and F1 hybrids. Euphytica 185(3):385-394.

Tracy WF (1993). Sweet corn: 777-807. In: G Kalloo and B.O. Bergh (Eds.) Genetic improvement of vegetable crops. Pergamon, Oxford. U.K.

Turgut I (2000). Effects of plant population and nitrogen doses on fresh ear yield component of sweet corn (*Zea mays* L. var. *saccharata Sturt.*) grown under Bursa conditions. Turk. J. Agric For 24:341-347.

Williams II MM (2006). Functional relationships between giant ragweed (*Ambrosia trifida*) interference and sweet corn yield and ear traits. Weed Sci 54:948-953.

Wong AD, Juvik JA, Breeden DC, Swiader JM (1994). Shrunken2 sweet corn yield and the chemical components of quality. J Amer Soc Hort Sci 119(4):747-755.

Wricke G, Weber WE (1986). Quantitative genetics and selection in plant breeding. Walter de Gruyter & Co. Berlin, Germany.

Znidarcic D (2012). Performance and characterization of five sweet corn cultivars as influenced by soil properties. J Food Agric Environ 10(1):495-500.

Zhu S, Mount JR, Collins JL (1992). Sugar and soluble solids changes in refrigerated sweet corn (*Zea mays* L.). J Food Sci 57:454-457.

Growth and Reproductive Success under Saline Conditions of Three *Plantago* Species with Different Levels of Stress Tolerance

Mohamad Al HASSAN[1], Andrea PACURAR[1,3], Alexandra GASPAR[1,3], Oscar VICENTE[1], Monica BOSCAIU[2*]

[1]*Instituto de Biología Molecular y Celular de Plantas (UPV-CSIC), Universitat Politècnica de València, CPI, edificio 8E, Camino de Vera s/n, 46022 Valencia, Spain; mohamad_al_hassan@live.com , ovicente@ibmcp.upv.es*

[2]*Instituto Agroforestal Mediterráneo (UPV), Universitat Politècnica de València, CPI, edificio 8E, Camino de Vera s/n, 46022 Valencia, Spain; mobosnea@eaf.upv.es (*corresponding author)*

[3]*Permanent address: University of Agricultural Sciences and Veterinary Medicine, Faculty of Horticulture, 3-5 Manastur Str., 400372 Cluj-Napoca, Romania; pacurar.andrea@yahoo.com , andragaspar@yahoo.com*

Abstract

Salt stress responses were studied in three *Plantago* species (*P. crassifolia, P. coronopus* and *P. major*), with different levels of salt stress tolerance, at growth and reproductive stages of the life cycle. Plants were treated with increasing salt concentrations (0, 100, 200 and 400 mM NaCl) under controlled conditions in the greenhouse. Fresh biomass, leaf length and reproductive parameters related to flowering were analysed. To evaluate the reproductive success of salt-stressed plants, seeds produced after two months of treatment were sampled, and the mean number of seeds per plant and mean seed weight per treatment were determined. Seeds were germinated *in vitro* and seed and seedling quality were assessed by determining germination percentages and rates, length of radicle, hypocotyl, cotyledonary leaves, and the angle of cotyledonary leaves. The 'seedling vigour index' was also calculated. Vegetative (fresh weight and leaf length) and reproductive traits (number of inflorescences, scape and spike length, mean number of seeds per spike and mean seed weight) were less affected by salinity in the more stress tolerant species *P. crassifolia* and *P. coronopus* than in *P. major*. Percentage and velocity of germination of seeds produced by plants grown under saline treatments were similar in all species; however seedling development was less successful in *P. major*, as shown by length of the radicle, hypocotyl and cotyledonary leaves. The relative sensitivity to salt of the investigated *Plantago* species correlated with their distribution in nature: *P. major*, never present in saline environments, was much more affected by salt than its halophytic counterparts, while *P. crassifolia*, which grows in saline habitats, exhibited the most efficient responses to salt stress, especially at the early seedling development stage.

Keywords: flowering, glycophytes, halophytes, reproductive success, seed germination, seedling development, seedling vigour

Introduction

Soil salinity is an environmental factor that significantly affects the distribution of plants in their natural ecosystems and it is one of the major threats for agriculture. According to FAO (2008), more than 800 million ha of land worldwide are affected by salinity. The extension of saline soils is increasing by approximately 1% each year due to poor rainfall, increasing temperatures with higher evaporation of water from soil, weathering of rocks, and anthropogenic alterations, such as irrigation with saline water, or the abusive and indiscriminate use of fertilizers (Flowers and Yeo, 1995; Munns and Tester, 2008; Rengasamy, 2010).

The immense majority of terrestrial plant species and all major crops belong to the category of glycophytes, or plants that cannot survive in conditions of (relatively) high soil salinity. On the contrary, halophytes include species that are able to survive and complete their life cycle in the presence of salt concentrations of at least 200 mM NaCl (Flowers *et al.,* 1986; Flowers and Colmer, 2008). Halophytes represent a small group of plants of only about 600 species that belongs to a few genera and families, which have adapted to saline ecosystems under harsh environmental conditions, thus avoiding competition with glycophytes (Flowers *et al.,* 1986).

Tolerance to salt stress involves complex mechanisms since numerous interactions take place between stress factors and the molecular, biochemical and physiological processes affecting plant development (Ashraf and Foolad, 2007). Halophytes would seem to be the most appropriate biological material to investigate these interactions; yet most experimental studies on plant responses to salt stress and salt tolerance mechanisms have been carried out on glycophytes, especially on model plants such as *Arabidopsis thaliana* (Zhu, 2000). The limitations of this approach and the necessity to increase ecophysiological research on halophytes have been recently discussed (Gil *et al.*, 2013; Grigore *et al.*, 2011). Most mechanisms of response to abiotic stress appear to be conserved in all plants, but their efficiency must largely vary among species so that they are effective enough to confer salt tolerance only in halophytes. Differences between glycophytes and halophytes seem to be the result of changes in the expression of the same basic packages of genes (Xiong and Zhu, 2002; Zhu, 2000, 2001); thus such differences are quantitative in nature rather than qualitative (Flowers *et al.*, 1986; Greenway and Munns, 1980; Vicente *et al.*, 2004; Zhu, 2001).

A very useful approach to gain a better comprehension of the concepts of 'stress responses' *vs.* 'stress tolerance' or 'stress adaptation', which are often confused, would be to carry out comparative studies on species closely related taxonomically, hence with a similar genetic background, but with different degrees of stress tolerance (Gil *et al.*, 2013). The genus *Plantago* with more than 200 species and 483 taxa in all (Tutel *et al.*, 2005), of which at least 20 species are considered halophytes, is one of the most suitable for such studies. Despite the relatively large number of papers that have dealt with one or several *Plantago* species, there are no comparative studies on their responses to salt stress at a crucial stage in the plant's life cycle, which is reproduction. Flowering, seed set, but mostly seed germination and seedling development, represents a bottleneck in the life cycle of those plants confronted by adverse environmental conditions. Germination is strongly inhibited by high salinity, even in most halophyte species, so many plants that have adapted to saline environments have developed strategies to postpone germination until the rain season, when soil salinity decreases. Therefore, salt tolerance at the specific stages of reproductive development and, mostly, seed germination, is probably one of the critical factors controlling plant distribution in habitats of different soil salinity.

A study into the effects of salt stress on germination in three *Plantago* halophytes (*P. coronopus, P. crassifolia*, and *P. microrrhiza*) showed that it was affected strongly by salinity and that, in all three species, there is an initial dormancy period, which is longer at higher temperatures (Luciani *et al.*, 2001). Dormancy is a common trend in halophytes, and seeds of numerous Mediterranean halophytes actually germinate in autumn when salinity is alleviated by abundant rainfall that washes out salts from the upper soil layers.

Although there are numerous publications on the physiological and biochemical responses to salt stress in species of the genus *Plantago*, the flowering and quality of the seeds produced by stressed plants have been analysed only in *P. crassifolia* (Boscaiu *et al.*, 2005). The present work analyses responses to saline stress at vegetative growth and reproductive stages in three *Plantago* species. *Plantago coronopus* L. is a cosmopolitan annual or biennial present in a wide range of variable habitats, mostly on disturbed lands, waste places, or littoral sand dunes. It is considered an indicator of the salinisation of marginal soils since it tolerates moderate salt stress; its growth appears to be even slightly stimulated by low salt concentrations (Koyro, 2006). Yet Luciani *et al.* (2001) found that *P. coronopus* seed germination was the most salt-sensitive among the *Plantago* halophytic species they studied. Taxonomically, it belongs to the section *Coronopus* (Lam & DC) Dietr. of the subgenus *Coronopus* (Lam. & DC.) Rahn, which includes 11 Mediterranean species (Rahn, 1996).

Plantago crassifolia Forssk. is a common perennial halophyte in littoral salt marshes and other saline habitats in the Mediterranean region. Previous studies (Boscaiu *et al.*, 2005; Vicente *et al.*, 2004) have shown the moderate salt tolerance of this species: it grows optimally in the absence of salt, and a drastic reduction in seed quantity and quality was observed under high salinity conditions. In natural environments, *P. crassifolia* appears only in areas with moderate salinity, on salt marsh borders, but never in strongly saline areas dominated by highly tolerant halophytes. From the taxonomic point of view, *P. crassifolia* is closely related to *P. coronopus*, which belongs to the same section.

Plantago major L. is a perennial (seldom annual) glycophyte, with origins in Eurasia and North Africa but naturalised throughout the world and considered as an important medicinal plant. It is much less salt tolerant than the two previous species and requires moderate humidity. Taxonomically, it is not so closely related to *P. crassifolia* and *P. coronopus*, since it is included in the section *Plantago* of the subgenus *Plantago*, the largest within the genus, with 130 species worldwide (Rønsted *et al.*, 2003).

The aim of this study was twofold. Firstly, to gain a better comprehension of the differences between mere *responses* and *adaptations* to stress in the three species. Secondly, to attempt and correlate the 'reproductive success' under high salinity conditions of these taxa with their distribution in nature. The paper compares the responses to salt of the three *Plantago* species during the growth and the reproductive stages, but it focuses mostly on seed production and seed and seedling quality as an expression of their relative reproductive success under salt stress conditions.

Material and methods

Plant growth and salt treatments

Seeds of *P. crassifolia* and *P. coronopus* were harvested in a salt marsh located in the Natural Park of La Albufera (Province of Valencia, Spain), and seeds of *P. major* were purchased from the company Spicegarden, EU.

The seeds of the three species under study were sown in seed trays containing a mixture of commercial peat and vermiculite (3:1), and were placed in a greenhouse with regulated temperatures of 17-23 °C, a photoperiod of 16 hours, and 85% relative humidity. Twenty days after sowing, young plants were transferred to individual pots on the same substrate in which they were grown for four weeks; then treatments with increasing salt concentrations (0, 100, 200 and 400 mM NaCl) were applied. The number of plants for each treatment and species was 3-5, and irrigation was carried out twice a week by adding 1.5 L of salt solutions (or distilled water, for the non-stressed control plants) to each tray (55 x 40 cm), which contained 12 standard pots (11 cm in diameter). After two months, all the plants were harvested and the vegetative parts were weighed individually on a precision balance. The length of the longest leaf was also measured. The reproductive parts of these plants were separated and used for further analysis.

Analysis of the plants' reproductive traits

The total number of spikes was recorded for each plant, and the lengths of floral scapes and spikes were measured. Seed set was estimated by counting and weighing seeds from approximately half the fertile spikes per treatment and species.

The quality of the seeds produced by the plants submitted to salt and control treatments was assessed by determining their germination capacity. Four replicas of 25 seeds per species and per treatment were placed in standard Petri dishes (9 cm in diameter), with cotton and filter paper moistened with 15 mL of distilled water; plates were sealed with parafilm to avoid evaporation and were incubated in a growth chamber at 25 °C under a 16-hour photoperiod. The number of germinated seeds, considered upon radicle emergence, was counted daily over a two-week period. Germination capacity was expressed as the final percentage of germination and the germination rate as MGT (mean germination time) according to the formula provided below (Ellis and Roberts, 1981):

$MGT = \Sigma D n / \Sigma n$,

Where D are the days from the beginning of the germination test, and n is the number of seeds newly germinated on day D.

The seedlings were analysed using ImageJ software (Rasband 1997-2012). The length of the radicle, hypocotyl and cotyledonary leaves, and the angle between cotyledons, were measured in 60 randomly selected, 12 days-old seedlings per species and per treatment. In addition, seedlings were weighed on a precision balance. Seedling Vigour Index (SVI) was calculated by the following formula (Abdul-Baki and Anderson, 1973):

SVI = Germination percentage x [Mean root length (mm) + Mean hypocotyl length (mm)].

Since *P. coronopus* plants that were watered without salt did not flower over the two-month treatment, the seeds from the initial stock were used as a control for the germination assay and seedlings measurements.

Soil analysis

Electrical conductivity of the substrate was checked at the end of treatments. Soil samples were taken from 3 pots per treatment, air-dried and then passed through a 2-mm sieve. A soil: water (1:5) suspension was prepared in distilled water and stirred for 60 min at 600 u/min at 21 °C. Electric conductivity was measured with a Crison Conductivity meter 522 and expressed in dS/m.

Statistical analysis

Data were analysed using SPSS, v. 16 and SYSTAT v. 12. A Levene test was applied to check whether the requirements of the analysis of variance (ANOVA) were accomplished. Germination percentages were normalized by arcsine transformation prior to the analysis of variance. Radicle, hypocotyl and cotyledonary leaves' lengths were transformed into percentages, considering values in control treatments as 100% in each species. The significance of the differences among treatments and among species was tested by applying a one-way ANOVA, at a confidence level of 95%. When the ANOVA null hypothesis was rejected, *post-hoc* comparisons were performed by the Tukey test. A linear fitting was applied to correlate the soil EC with the concentration of the saline solutions applied to the pots.

Results and discussion

Electrical conductivity of the substrate increased with the salt treatments in the three species (Fig. 1). The $EC_{1:5}$ of substrate in the treatment with 400 mM NaCl was about 13-fold higher than that registered in the control, for the three species. This increase was highly correlated with the

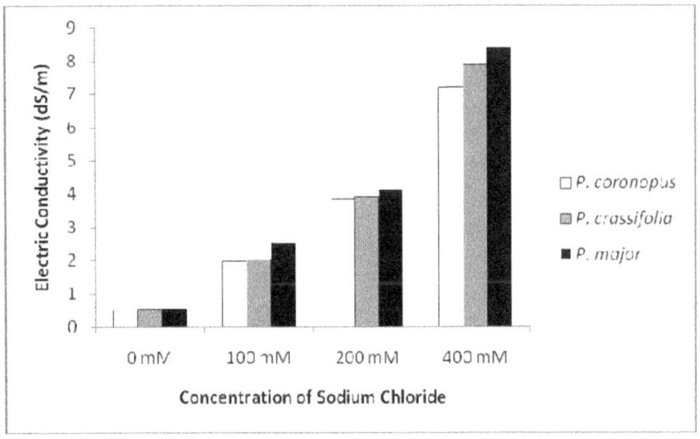

Fig. 1. Electric conductivity ($EC_{1:5}$) of the substrate after two months of salt treatments

saline solution applied, with high R^2 in the linear fitting in all species (0.92 in *P. coronopus*, 0.98 in *P. crassifolia* and 0.91 in *P. major*).

Sodium chloride treatments inhibited growth in a concentration-dependent manner in the three species under study. Maximal fresh weight (FW) was recorded in the control plants of the most vigorous species, *P. major*, which drastically decreased about 10-fold in the plants treated with 400 mM NaCl. The fresh weight of *P. coronopus* and *P. crassifolia* also decreased, but only by about 4- and 5.5-fold, respectively, at the same NaCl concentration (Fig. 2). These observations are consistent with the ecological requirements

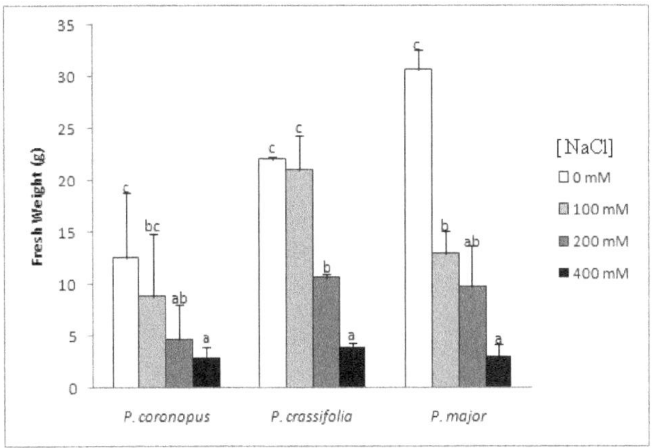

Fig. 2. Fresh weight of plants of *Plantago coronopus, P. crassifolia* and *P. major* treated for two months with the indicated NaCl concentrations. Same letters indicate homogeneous groups among treatments for each species

of the three species since *P. major,* the most affected by salt, is precisely the species that is never confronted with salinity in its natural environment.

Another parameter that is used to assess the effect of salinity on plant growth is the length of the longest leaf, which is generally affected by salt (Fig. 3). Indeed, a reduction in the longest leaf length was observed with increasing NaCl concentrations in the three taxa, but no significant differences were found among them (1.9-fold in *P. major*, 1.5-fold in *P. crassifolia* and 1.3-fold in *P. coronopus*). These data, related to vegetative growth, are similar to those previously reported in *P. crassifolia* (Vicente *et al.*, 2004), but no stimulation of growth at low salinity levels was detected in *P. coronopus*, contrarily to what Koyro described (2006).

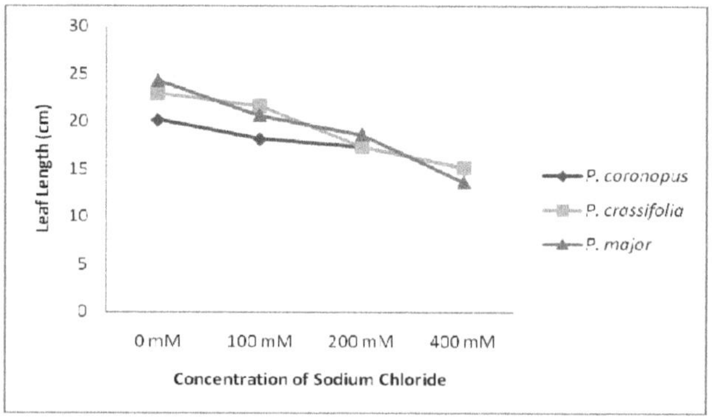

Fig. 3. Length of the longest leaf in the plants of *Plantago coronopus, P. crassifolia* and *P. major* treated for two months at the indicated NaCl concentrations

Effects of salinity on the floral characters were diverse among species. In *P. coronopus*, only the plants treated with 200 and 400 mM NaCl produced flowers (Tab. 1).

In *P. crassifolia,* the mean number of spikes per plant increased in the treatments with 100 and 200 mM NaCl, as reported in previous studies (Boscaiu *et al.*, 2005); at 400 mM NaCl the average number of spikes was smaller, similarly to that of the control (Tab. 1), whereas in a previous study (Boscaiu *et al.*, 2005) plants did not flower at all after five months of treatment at the same salt concentration. Such differences may be explained by the differences in the experimental conditions of plant growth, mostly regarding temperature and photoperiod. Salinity can induce inhibition or stimulation of flowering, and early or delayed flowering, depending on the plant species (Grigore *et al.*, 2011). The 'survival through the next generation' strategy can occur by both early flowering (Blanvillain *et al.*,

Tab. 1. Effects of increasing saline solutions on the traits related to reproductive success in *Plantago crassifolia, P. coronopus* and *P. major*. Mean ± SD values are shown. Same letters indicate homogeneous groups among treatments for each species, n.d. not determined since plants did not flower, and * seeds from the initial stock

Character	Species	Control	100 mM NaCl	200 mM NaCl	400 mM NaCl
Spike number/plant	*P. crassifolia*	3.66 ± 1.15^{a}	4.33 ± 1.52^{ab}	6.66 ± 1.52^{b}	3.33 ± 0.57^{a}
	P. coronopus	n.d.	n.d.	1.66 ± 0.57^{b}	1.33 ± 0.57^{b}
	P. major	7.66 ± 3.21^{b}	5.66 ± 1.15^{b}	4.66 ± 2.08^{ab}	1.33 ± 0.57^{a}
Scape length (cm)	*P. crassifolia*	24.55± 9.02^{c}	18.89 ± 7.52^{b}	11.32 ±3.18^{a}	7.09 ± 1.41^{a}
	P. coronopus	n.d.	n.d.	34.30 ± 8.04^{b}	10. 00 ± 3.08^{a}
	P. major	28.34 ± 4.19^{d}	23.33 ± 1.72^{c}	17.54 ± 1.35^{b}	8.81 ± 1.41^{a}
Spike length (cm)	*P. crassifolia*	5.15 ± 2.42^{c}	4.84 ± 1.5^{c}	3.00 ± 0.63^{b}	2.00 ± 0.38^{a}
	P. coronopus	6.70 ± 1.78^{b}	n.d,.	6.70 ± 1.78^{b}	1.62 ± 0.47^{a}
	P. major	10.42 ± 2.94^{c}	8.12 ± 1.38^{b}	7.37 ± 0.9^{b}	3.50 ± 0.75^{a}
Seed number/spike	*P. crassifolia*	88.71 ± 15.33^{c}	72.83 ± 2.75^{b}	70 ± 9.05^{b}	48.66 ± 3.05^{a}
	P. coronopus	n.d.	n.d.	289.5±27.57^{b}	41±29.69^{a}
	P. major	1369.20 ± 439^{b}	1345.83 ± 230 ^{b}	1358.33 ± 281^{b}	512.85 ± 291.0^{a}
Seed weight (mg)	*P. crassifolia*	0.49 ± 0.10^{c}	0.49 ± 0.07^{c}	0.41 ± 0.10^{b}	0.39 ± 0.11^{a}
	P. coronopus	0.34 ± 0.02^{c*}	n.d.	0.27 ± 0.02^{b}	0.17 ± 0.01^{a}
	P. major	0.39 ± 0.02^{c}	0.33 ± 0.02^{b}	0.28 ± 0.01^{a}	0.26 ± 0.01^{a}
Percentage of germination	*P. crassifolia*	98.40 ± 4.00^{b}	84.00 ± 0.00^{a}	81.00 ± 2.00^{a}	89.00 ± 11.00^{a}
	P. coronopus	96.00 ± 4.60^{a*}	n.d.	99.00 ± 2.00^{ab}	92.00 ± 0.00^{a}
	P. major	99.00 ± 2.00^{a}	98.00 ± 4.00^{a}	97.00 ± 6.00^{a}	98.00 ± 2.30^{a}
MGT (days)	*P. crassifolia*	4.79 ± 0.60^{a}	4.94 ± 0.20^{a}	4.66 ± 0.40^{a}	5.14 ± 1.10^{a}
	P. coronopus	2.92 ± 0.30^{a*}	n.d.	2.76 ± 0.40^{a}	3.02 ± 0.10^{a}
	P. major	3.90 ± 0.20^{a}	4.01 ± 0.10^{a}	4.25 ± 0.14^{a}	4.19 ± 0.20^{a}
Seedling weight (mg)	*P. crassifolia*	6.45 ± 0.22^{b}	6.32 ± 0.01^{b}	5.94 ± 0.12^{b}	5.22 ± 0.26^{a}
	P. coronopus	5.07 ± 0.06^{c*}	n.d.	3.57 ± 0.01^{b}	1.75 ± 0.02^{a}
	P. major	3.01 ± 0.39^{b}	2.18 ± 0.77^{ab}	0.98 ± 0.36^{a}	0.94 ± 0.38^{a}
SVI	*P. crassifolia*	2438.352	1176.84	1109.70	954.08
	P. coronopus	3901.44*	n.d.	3027.42	2650.52
	P. major	2677.95	2132.48	1428.81	1209.32
Cotyledonary leaves angle (°)	*P. crassifolia*	69.54 ± 17.81^{a}	94.20 ± 31.98^{b}	103.98 ± 24.78^{b}	98.15 ± 30.50^{b}
	P. coronopus	67.58 ± 16.45^{a*}	n.d.	76.53 ± 13.86^{ab}	79.87 ± 21.73^{b}
	P. major	45.41 ± 13.99^{a}	53.42 ± 18.72^{a}	72.24 ± 14.72^{b}	75.10 ± 22.90^{b}

2011) and an increase in the number of flowers produced, which is what we found in *P. crassifolia* and *P. coronopus*. On the contrary, the glycophyte *P. major* responded differently when exposed to salt stress. In this species, a concentration-dependent reduction of spike numbers was observed which was almost 6-fold at the highest salt concentration tested (Tab. 1).

The number of seeds produced by each species under different salinity conditions correlated well with the length of the spikes measured for the corresponding treatments, and decreased progressively with increasing external NaCl concentrations (Tab. 1). It is worth mentioning that in all cases, *P. major,* which is the most vigorous of the studied species, produced a considerably larger number of seeds in comparison with the other two. The mean mass of the seeds was also determined, and it became evident that those produced by control, non-stressed plants had the highest mass, which decreased in all three species with increasing salt concentrations (Tab. 1).

These findings show that, even though salinity may stimulate flowering, it clearly has a negative, concentration-dependent effect on reproduction in the three species under study: NaCl diminished the quality of the flowers produced, which were smaller and produced a reduced number and mass of seeds as compared with the controls grown in the absence of salt.

The quality of the seeds produced by the plants grown under different salinity conditions was assessed in *in vitro* germination tests, showing very high germination percentages (> 96%) in the control seeds of all three *Plantago* species, and the salt treatments of the mother plants did not lead to statistically significant changes in these values (Tab. 1). 'Mean germination time' (MGT) values differed for the seeds of the three species but, for each species, no significant differences were observed among the NaCl treatments (Tab. 1). On the contrary, the quality of the seedlings deriving from these seeds was clearly affected by the previous salt treatments of the mother plants. A significant loss of vigour of the seedlings with increasing salt concentrations, as estimated from the Seedling Vigour Index (SVI), was observed for the three *Plantago* species: when comparing the seeds produced by control plants and those from the plants watered with 400 mM NaCl, SVI values decreased by about 1.5-fold in *P. coronopus*, 2.5-fold in *P. crassifolia*, and 2.2-fold in *P. major* (Tab. 1). The weight of the seedlings followed a similar pattern, that is, decreasing with increasing salt concentrations, but with clear differences between the investigated species (Tab. 1). The treatment of the mother plants with the highest salt concentration tested, 400 mM NaCl, resulted in a reduction in seedling weight of about 20% (*P. crassifolia*), 65% (*P. coronopus*), and 70% (*P. major*) of the corresponding controls. These results indicate that *P. crassifolia* is minimally affected by high salinity at the seedling stage when compared with the other two species. This finding supports the idea that it is the most salt-tolerant of the three, which is in agreement with the distribution observed in the nature of these *Plantago* taxa.

All the growth parameters analysed by the ImageJ programme (length of radicle, hypocotyl and cotyledonary leaves' angle and length) showed a clear pattern of variation,

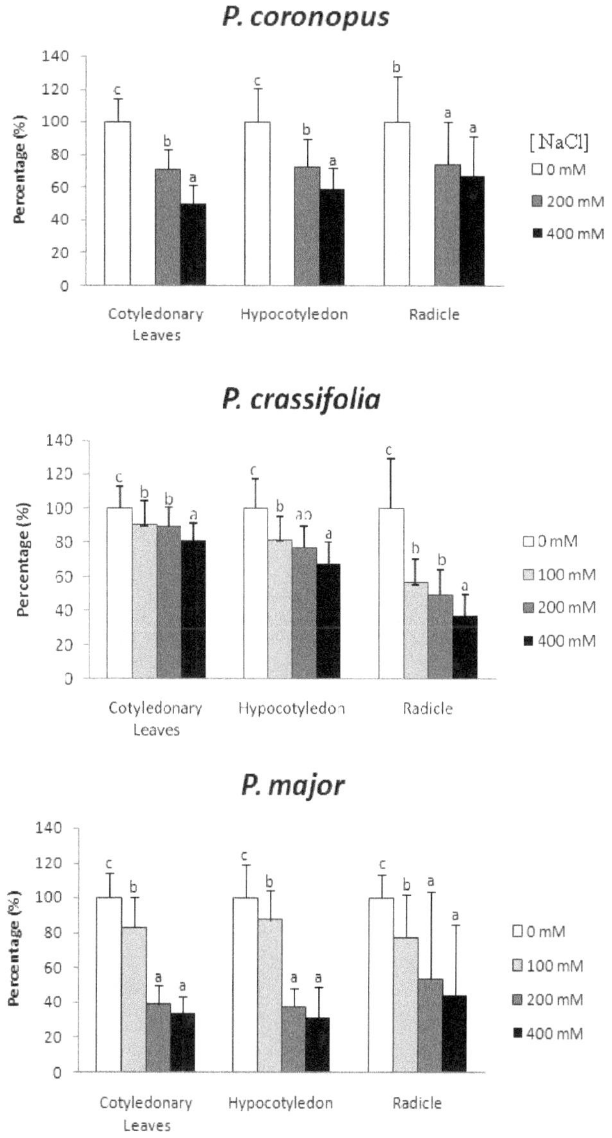

Fig. 4. Radicle, hypocotyle and cotyledon length reduction in 12 days-old seedlings obtained from seeds sampled from the plants treated for two months at the indicated NaCl concentrations. Percentages were calculated in relation to the controls' mean values which were considered 100%. Same letters indicate homogeneous groups among treatments for each species

which decreased in parallel to the increase in the NaCl concentration of the treatments applied to the mother plants (Fig. 4 and Tab. 1).

The most marked reduction in hypocotyl and cotyledon length was observed in *P. major*, and the least marked one in *P. crassifolia*, which once again is in agreement with the relative salt tolerance of these taxa. However, it was not possible to find a similar correlation in the salt-dependent decrease in radicle size (Fig. 4). Although the germination rates were very high in *P. major*, even for the seeds produced by those plants treated with 200 and 400 mM of NaCl, their seedlings were of poor quality; in almost half of them, cotyledons did not emerge as their development was blocked after radicle emergence (Fig. 5).

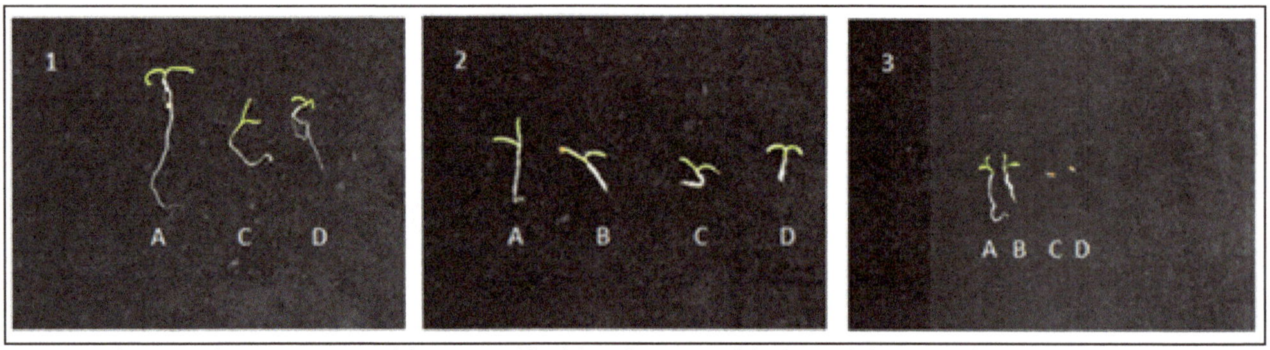

Fig. 5. Seedlings of *Plantago coronopus* (1), *P. crassifolia* (2) and *P. major* (3) obtained from the plants treated for two months with distilled water (A), 100 mM NaCl (B), 200 mM NaCl (C) or 400 mM NaCl (D)

The angle of cotyledonary leaves is a seedling characteristic that can be considered an early predictor of the growth rate. Lower *et al.* (1978) reported that cucumber (*Cucumis sativus*) seedlings which displayed their cotyledons in an upward position (smaller angle) grew faster than the plants which displayed their cotyledons in a downward position (larger angle). An upward cotyledon angle also correlated with the increase in plant height, node number and fresh weight at later developmental stages. In the analysed seedlings of the *Plantago* species, cotyledon angles (Tab. 1) increased in parallel to the increase in salinity, which affected mother plants, suggesting that stress conditions may negatively affect growth in the next generation.

In summary, salinity affected several growth and reproductive traits in the three *Plantago* species studied in the present work. The plants grown under salt stress conditions were less vigorous and their reproductive success was clearly affected by salinity, but two different patterns were detected: one corresponding to the salt-tolerant *P. coronopus* and *P. crassifolia* on the one hand, and to the glycophyte *P. major* on the other. Fresh weight decreased with increasing salinity in all three species, but this reduction was much more marked in *P. major* than in *P. coronopus* or *P. crassifolia*. Salinity stimulated flowering in halophytes, while the number of spikes lowered in *P. major* by salt treatments, although the absolute number of spikes was greater in the latter species, which is the most vigorous. Seed set and mass of seeds were also reduced in the three species in response to increasing salinity. Although no significant effect was observed on the germination capacity of seeds, which was very high under all the conditions tested, growth of mother plants in the presence of increasing salt concentrations negatively affected the quality of the corresponding seedlings. Seedling size and weight reduced upon salt treatment, but the observed reduction was much more evident in the glycophyte *P. major* than in the halophytes *P. crassifolia* and *P. coronopus*. Most importantly, upon germination of the *P. major* seeds obtained from plants treated with high NaCl concentrations, inhibition of seedling development was observed at an early stage, after radical emergence, and cotyledons did not emerge in almost half the seedlings. When comparing the two halophytes, *P. crassifolia* showed stronger resistance to salt stress than *P. coronopus*, as indicated by a slighter decrease in seedling weight or in hypocotyl and cotyledon lengths.

Conclusions

The results presented herein allow distinguishing how related species with different levels of salt tolerance respond to high soil salinity conditions. They also suggest that the reproductive success of a particular species under salt stress conditions depends largely on its tolerance at the specific early seedling development stage. The different responses of the studied species under the controlled greenhouse conditions correlate closely with their ecology and distribution in nature. *P. major* behaves as a typical glycophyte - although it is not extremely sensitive to salt – showing a very strong inhibition of vegetative growth and reproductive development, and low reproductive success, in conditions of salt stress; this species is never found in nature in saline habitats. When comparing the two *Plantago* taxa considered to be halophytes, our findings support the notion that *P. crassifolia* is more salt-tolerant than *P. coronopus* since its responses to salinity are more effective, especially at the early seedling development stage. This is also in agreement with their ecological characteristics, as only *P. crassifolia* is able to inhabit typical saline areas such as salt marshes.

Acknowledgements

Work in the UPV laboratories was funded by a grant to O.V. from the Spanish Ministry of Science and Innovation (Project CGL2008-00438/BOS), with contribution from the European Regional Development Fund. M.A.H. is a recipient of an Erasmus Mundus pre-doctoral scholarship financed by the European Commission (Welcome Consortium). A.P. and A.G. acknowledge the Erasmus fellowship programme for funding their stay in Valencia to carry out their Master Theses.

References

Abdul-Baki AA, Anderson JD (1973). Relationship between decarboxilation of glutamic acid and vigour in soybean seed. Crop Sci 13:222-226.

Ashraf M, Foolad MR (2007). Roles of glycine betaine and proline in improving plant abiotic stress resistance. Environ Exper Bot 59:206-216.

Blanvillain R, Wei S, Wei P, Kim JH, Ow DW (2011). Stress tolerance to stress escape in plants: role of the OXS2 zinc-finger transcription factor family. EMBO J 30:3812-3822.

Boscaiu M, Estrelles E, Soriano P, Vicente O (2005). Effects of salt stress on the reproductive biology of the halophyte *Plantago crassifolia*. Biol Plant 49:141-143.

Ellis RA, Roberts EH (1981). The quantification of aging and survival in orthodox seeds. Seed Sci Technol 9:373-409.

FAO (2008). Land and plant nutrition management service. <http://www.fao.org.ag/agl/agll/spush>.

Flowers TJ, Colmer TD (2008). Salinity tolerance in halophytes. New Phytol 179:945-963.

Flowers TJ, Hajibagheri MA, Clipson NJW (1986). Halophytes. Q Rev Biol 61:313-335.

Flowers TJ, Yeo AR (1995). Breeding for salinity resistance in crop plants: where next? Aust J Plant Physiol 2:875-884.

Gil R, Boscaiu M, Lull C, Bautista I, Lidón A, Vicente O (2013). Are soluble carbohydrates ecologically relevant for salt tolerance in halophytes? Funct Plant Biol 40:805-118.

Greenway H, Munns R (1980). Mechanisms of salt tolerance in non-halophytes. Annu Rev Plant Physiol 31:149-190.

Grigore MN, Boscaiu M, Vicente O (2011). Assessment of the relevance of osmolyte biosynthesis for salt tolerance of halophytes under natural conditions. Eur J Plant Sci Biotechnol 5:2-19.

Koyro H-W (2006). Effect of salinity on growth, photosynthesis, water relations and solute composition of the potential cash crop halophyte *Plantago coronopus* (L.). Environ Exp Bot 56:136-146.

Lower RL, Pharr DM, Smith OS, Sox HN (1978). Cotyledon angle as a predictor of growth rate of cucumber plant. Euphytica 27:701-706.

Luciani F, Cristaudo A, Aricó D. (2001). Germination ecology of three *Plantago* L. species living in a saline environment. Plant Biosyst 135:213-221.

Munns R, M. Tester (2008) Mechanisms of salinity tolerance. Annu Rev Plant Biol 59:651-681.

Rahn K (1996). A phylogenetic study of the Plantaginaceae. Bot J Linn Soc 120:145-198.

Rasband WS (1997-2012) ImageJ. U S National Institutes of Health. http://rsb.info.nih.gov/ij/, Bethesda, Maryland.

Rengasamy P (2010) Soil processes affecting crop production in salt-affected soils. Funct Plant Biol 37:613-620.

Rønsted N, Chase MW, Albach DC, Bello MA (2002). Phylogenetic relationships within *Plantago* (Plantaginaceae): evidence from nuclear ribosomal ITS and plastid *trnL-F* sequence data. Bot J Lin Soc 139:323-338.

Tutel B, Kandemir I, Kuş S, Kence A. (2005). Classification of Turkish *Plantago* L. species using numerical taxonomy. Turk J Bot 29:51-61.

Vicente O, Boscaiu M, Naranjo MA, Estrelles E, Bellés JM, Soriano P (2004). Responses to salt stress in the halophyte *Plantago crassifolia* (Plantaginaceae). J Arid Environ 58:463-481.

Xiong L, Zhu JK (2002). Salt-stress signal transduction, p. 165-197. In: Scheel D, Wasternack C (Eds.). Plant Signal Transduction, Frontiers in Molecular Biology Series. Oxford University Press, Oxford UK.

Zhu J-K (2000). Genetic analysis of plant salt tolerance using *Arabidopsis*. Plant Physiol 124:941-948.

Zhu J-K (2001). Plant salt tolerance. Trends Plant Sci 6:66-71.

Comparative Studies of the Phytoextraction Capacity of Five Aquatic Plants in Heavy Metal Contaminated Water

Erzsébet BUTA[1], Anamária TÖRÖK[2], Árpád CSOG[2], Bilassé ZONGO[3], Maria CANTOR[1], Mihai BUTA[1], Cornelia MAJDIK[2*]

[1]University of Agricultural Sciences and Veterinary Medicine, 3-5 Mănăştur Str., 400372, Cluj-Napoca, Romania
[2]University of Babeş Bolyai, Faculty of Chemistry and Chemical Engineering, Arany János 11, Cluj-Napoca, Romania; majdikc@yahoo.com (*corresponding author)
[3]University of Ouagadougou, Laboratory of Biology and Ecology of Plants, 09 BP 848 Ouagadougou 09, Burkina Faso, Western Africa

Abstract

The uptake capacity of the aquatic plants (*Salvinia natans* Kunth., *Eichhornia crassipes* Mart., *Lemna minor* L., *Elodea canadensis* Michx., *Pistia stratiotes* L.) was analyzed in phytoextraction of Cu^{2+}, Zn^{2+}, and Cd^{2+}. It was attend to study the plants capacity comparatively using mono and multimetallic systems. In particular, the chlorophyll, protein and carotenoids contents were studied during heavy metals uptake, in order to observe the stress effect on plants. The results obtained for the monometallic system showed that *Salvinia natans* Kunth. accumulated the highest quantity of Cu^{2+} (4.72 mg/g), Zn^{2+} (2.23 mg/g) and Cd^{2+} (1.90 mg/g). The leaves of *Lemna minor* L. accumulated the highest concentration of Cu^{2+} (10.80 mg/g) and Cd^{2+} (2.78 mg/g) in multimetallic system. The water lettuce (*Pistia stratiotes* L.) translocated the highest quantity of Zn^{2+} in its roots (4.80 mg/g). The chlorophyll and the carotenoids levels decreased under the stress of heavy metals in both systems, while protein content increased under the influence of Cu^{2+} and Cd^{2+}, but decreased for Zn^{2+}. The studied hydrophytes proved to be useful in the uptake of heavy metals in monometallic system and much more effective in the multimetallic system and showed great potential for further applications in the industrial and commercial wastewater treatments.

Keywords: hydrophytes, photosynthetic pigments, phytoaccumulation, pollution, proteins

Abbreviations: DW, dry wt – dry weight; chl – chlorophyll; chl *a* – chlorophyll *a*; chl *b* – chlorophyll *b*; total chl – total chlorophyll; car – carotenoids; FW – fresh weight; DW – dry weight; HM – heavy metals; NPK – nitrogen, phosphorus, potassium.

Introduction

The environmental contamination with toxic elements is a major issue on a global scale. It can damage or alter the biosphere reducing the agricultural productivity and can affect both animal and human bodies (Buszewski *et al.*, 2000; Cvjetko *et al.*, 2010; Kamal *et al.*, 2004; Kaur *et al.*, 2010; Lu *et al.*, 2004; Skinner *et al.*, 2007).

The remediation of pollutants in different environmental conditions can be achieved through plants (phytoremediation), being considered an alternative technology to clean up contaminated waters. This method was also used in the 18th century to treat wastewater, leading to extensive researches that allowed the study of metal phytoextraction processes. Moreover, the understanding of the physiological mechanisms of metal absorption, transportation and assimilation became possible (Mudgal *et al.*, 2010).

Phytoremediation includes: phytoextraction, phytostabilization, rhizofiltration, phytodegradation and phytovolatilization. Through phytoextraction, plants remove the pollutants from a contaminated site (Maczulak, 2009). Hydrophytes can translocate and concentrate many kinds of contaminants including heavy metals, pesticides, explosives, oils (Hasan *et al.*, 2007; Kamal *et al.*, 2004) and nutrients (nitrate, ammonium, phosphate).

The hyperaccumulator plants can gather high concentrations of heavy metals (Kachenko *et al.*, 2009) from contaminated sites, at a minimum percentage, which can vary depending on the pollutant involved (more than 1000 mg/kg of dry weight for copper, cobalt, nickel, chromium and lead; or more than 10,000 mg/kg for magnesium and zinc – Baker and Brooks, 1989). These plants are able to absorb heavy metals such as Pb, Hg, Zn, Co, Cd, and Cu via roots, stems and leaves, and deposit them in their organism. Some of these elements are essential for the metabolic function for a large class of organisms, e.g. Zn, Fe, Cu, Cr and Co (Hogan, 2010), while others are very toxic (Cd, Hg, Pb).

In the case of a high heavy metal concentration, the uptake inhibition and transportation can take place for essential metals as well (Fe, Zn and Mn), through antagonistic effects (Naaz and Pandey, 2010). Other authors analyzed the distribution, accumulation and selection of heavy metals on selective plants in the presence of chelating compounds (Csog *et al.*, 2012; Mihucz *et al.*, 2012).

The main source of these heavy metals usually find their path in the aquatic ecosystems, e.g. ponds, rivers and lakes (Kumar, 2010), and consist mostly of industrial residues, mining, military operations (Liang *et al.*, 2009), fertilizers and pesticide applications, fuel production and urban wastes. Recently, the use of aquatic plants for treating wastewater has been reported as an economical device for the treatment of heavy metal pollution in contaminated wastewaters (Kumar, 2010). Several studies proved that the cost of phytoextraction is lower than other conventional engineering technologies. Phytoremediation is not just environmentally friendly, cost-effective and non-intrusive, but it also conserves the ecosystem efficiently (Kamal *et al.*, 2004).

Many researches have been conducted regarding the ability of aquatics to remove heavy metals from contaminated waters. For a diversity of purposes, a number of studies have been made on several hydrophytes: *Salvinia sp.* (Dhir, 2009; Espinoza-Quinones *et al.*, 2005), *Potamogeton crispus* (Aysel *et al.*, 2010), *Potamogeton pectinatus* and *Potamogeton malaianus* (Peng *et al.*, 2008). Recent studies highlight the effectiveness of *Eichhornia crassipes* (Hogan, 2010; Naaz and Pandey, 2010; Singh and Kalamdhad, 2013; Skinner *et al.*, 2007), *Typha angustifolia* (Dilek and Aksoy, 2004), *Myriophyllum heterophyllum* (Aysel *et al.*, 2010), *Lemna minor* (Alka and Tripathi, 2007; Rahman and Hasegawa, 2011; Yeşim *et al.*, 2003) and *Azolla caroliniana* (Bennicelli *et al.*, 2004). Some research studies demonstrate that *Nymphaea* varieties can accumulate heavy metals, consequently, they can be used as hyperaccumulators (Lavid *et al.*, 2000).

The purpose of the present paper is to compare the phytoextraction capacity of five aquatic plants and monitor the accumulation and distribution of heavy metals into their morphological parts (leaves and roots). In order to evaluate the markers for stress responses in plants under heavy metal stress treatment, biochemical parameters such as chlorophyll, proteins and carotenoids were investigated.

Materials and methods

Plant material and growing conditions. During the experiment the following five hydrophytes were used: floating fern (*Salvinia natans* Kunth.), water hyacinth (*Eichhornia crassipes* Mart.), duckweed (*Lemna minor* L.), waterweed (*Elodea canadensis* Michx.) and water lettuce (*Pistia stratiotes* L.). The above mentioned species were grown in an educational greenhouse, for a period of 30 days at the University of Agricultural Sciences and Veterinary Medicine Cluj-Napoca, Romania. The plants were kept separately in 5 plastic tanks, filled with tap water, in addition of 0.5% Complex III (NPK) fertilizer.

After the growing period, the plants were used for Cu^{2+}, Zn^{2+} and Cd^{2+} phytoremediation experiments. They were kept for acclimatization for a period of 3 days in modified Hoagland nutrient solution in laboratory conditions. The Hoagland nutrient solution contained macro elements such as: 1.25 mM KNO_3 (Potassium nitrate), 1.25 mM $Ca(NO_3)_2$ (Calcium nitrate), 0.5 mM $MgSO_4$ (Magnesium sulfate), 0.25 mM KH_2PO_4 (Monopotassium sulphate) and micronutrients such as: 10 μM FeEDTA (Ethylenediaminetetraacetic acid) as iron source, 11.6 μM H_3BO_3 (Boric acid), 4.5 μM $MnCl_24H_2O$ (Manganese chloride tetrahydrate), 0.19 μM $ZnSO_47H_2O$ (Magnesium sulphate heptahydrate), 0.12 μM $Na_2MoO_42H_2O$ (Sodium molybdate), 0.08 μM $CuSO_45H_2O$ (Copper II sulphate pentahydrate). The dimension of the acquarium was 21 cm x 38 cm x 28 cm. During the phytoremediation experiments the plants were exposed to electric light for 14 hours then to darkness for 10 hours at a temperature of 24-28 °C.

The heavy metal stock solutions were made of salts such as $Cd(NO_3)_2$ (Cadmium nitrate), $ZnSO_4\times7H_2O$ (Zinc sulphate heptahydrate) and $CuSO_4\times5H_2O$ (Copper II sulphate pentahydrate). For accuracy purposes, the initial concentrations were analyzed by the Atomic Absorbance Spectrophotometer and the Inductively Coupled Plasma Mass Spectrometry. The concentrations of Cu^{2+}, Zn^{2+} and Cd^{2+}, used in all the aquatics studied, were of 10 mg/L.

The heavy metal contamination was investigated in the monometallic and multimetallic system, but the control plants were left without heavy metal treatment. It is essential to specify that the monometallic system can be defined as a process of each plant's separate absorption of every heavy metal one-by-one. Moreover, the multimetallic system uses the treatment of the plants with the mixture of the mentioned heavy metals. In order to determine the phytoextraction capacity, the results of the treated plant samples were compared with the results of the untreated plant samples. The untreated and treated plant samples were dried at 70-80 °C for a period of 48 hours. After digestion with HNO_3 (digestion of 1g plant material with 10 ml nitric acid - 69% HNO_3, left for 24 hours at 20 °C and for 8 hours at 100 °C) the Cu^{2+}, Zn^{2+} and Cd^{2+} concentrations were determined. Three replications were obtained of both systems.

In order to investigate the responses of the plants through the heavy metal treatments, the photosynthetic pigments (chlorophyll *a*–chl a; chlorophyll *b*–chl b, carotenoids-car) and the total content of protein were analyzed. The determination of chlorophyll in treated and untreated plant samples (250 mg) was achieved through the extraction of photosynthetic pigments with 96% chilled ethanol being kept in the dark (for 12 hours). After a ten-minute centrifugation, the supernatant absorbance (chl *a* A_{665}, chl *b* A_{649}, car A_{470}) was measured using the UV-Vis Spectrophotometer (GBC Cintra 202). The content of chl *a*, chl *b* and total chl and car was determined by the method of Arnon (1949) using the formula suggested by Lichtenthaler and Wellburn (1983) and Lichtenthaler (1987). The protein content was estimated following the method of Bradford (1976) using serum albumin as a standard protein and it was measured in both mono and multimetallic solutions.

In order to establish quantitative differences between values, statistical analyses were performed using Microsoft Office Excel. The data results were expressed as standard deviation (SD) from three replicates for each sample, consequently Duncan's multiple range test was undertaken.

Results and discussion

Based on the preliminary experiments and observations, the hydrophytes studied had a phytoremediation period of six days. In Romania, the permissible metal content in wastewaters in compliance with the law is: 0.1 mg/dm^3 for Cu^{2+}; 0.5 mg/dm^3 for Zn^{2+} and 0.2 mg/dm^3 for Cd^{2+} (http://www.epurare.euriteh.ro/legislatie/NTPA001%2028.02.2002.pdf). The phytoextraction capacity of the plants was determined by measuring the heavy metal concentrations before and after the experiments undertaken from the plant samples.

Monometallic system

The uptake of the plants studied and the bioaccumulation depend on the plant species and on the metal ion characteristics. Two hydrophytes *E. crassipes* Mart. and *P. stratiotes* L. were selected in order to analyze the heavy metal accumulation and distribution in the parts of the plant (root and leaves) and the obtained results are presented in Tab. 1. The controlled plant samples contain lower quantities of Cu^{2+} and Zn^{2+} ions (Tab. 1). The heavy metal distribution in the different morphological parts of the smaller plants studied was not investigated in monometallic condition.

Tab. 1. Accumulation and distribution of heavy metals (mg/g) in different parts of *E. crassipes* Mart. and *P. stratiotes* L., in monometallic system

Monometallic system					
Species	Treatment	Concentration (mg/g)		Accumulated concentration (mg/g)	
		leaves	roots	leaves	roots
E. crassipes Mart.	Cu^{2+}	0.005	0.003	0.36±0.02^d	1.46±0.3^b
	Zn^{2+}	0.075	0.041	0.32±0.03^d	0.81±0.07^c
	Cd^{2+}	0	0	0.13±0.01^d	1.72±0.2^a
P. stratiotes L.	Cu^{2+}	0.009	0.004	2.49±0.6^a	1.33±0.2^b
	Zn^{2+}	0.033	0.013	0.95±0.01^b	1.11±0.4^b
	Cd^{2+}	0	0	0.27±0.04^c	1.42±0.3^b

Values represent ± SD (n=3), small letters represent the statistical significant difference at $P<0.05$.

The results obtained showed that the concentration of the three heavy metal ions was higher in the root and lower in the leaves. One exception was noticed in the case of Cu^{2+} treatment for *P. stratiotes* L. which accumulated a higher concentration of Cu^{2+} ion (2.49 mg/g) in the leaves. These results demonstrate that *P. stratiotes* L. is the most effective plant for the removal of this metal from aqueous solutions. In diversion of the *E. crassipes* Mart., Cu^{2+} accumulation was measured in a lower quantity in the leaves. As for the Zn^{2+} monometallic treatment the highest concentration was accumulated in *P. stratiotes* L. roots (1.11 mg/g). This means that the phytoextraction period was not enough for the translocation of the Zn^{2+} ions from the root to the leaves. The results demonstrated that the highest Cd^{2+} ion accumulation was very significant in the case of *E. crassipes* Mart. in the plant roots (1.72 mg/g). In the case of Zn^{2+} ions, the Cd^{2+} heavy metal ions have been translocated only in lower concentrations in the leaves during the phytoremediation period. These results corresponded with Zhu *et al.* (1999) reports, regarding the phytoaccumulation and distribution process in the case of *E. crassipes* Mart.

The results concerning the accumulation of heavy metals in the entire plant of *El. canadensis* Michx., *S. natans* Kunth., *L. minor* L. are presented in the Tab. 2. The highest concentration of Cu^{2+} was registered in the leaves of *S. natans* Kunth. (4.71 mg/g), followed by *L. minor* L. (1.55 mg/g) with significant differences. Similar concentration in this element was removed by *El. canadensis* Michx. (1.34 mg/g). The highest amount of Zn^{2+} and Cd^{2+} ions was accumulated in *S. natans* Kunth. leaves (2.26 and 1.90 mg/g). The lowest concentration of Zn^{2+} (0.44 mg/g) was registered in *L. minor* L. The results showed that the greatest capacity to uptake and bind the heavy metals in monometallic systems was reached by *S. natans* Kunth. and *P. stratiotes* L. On the other hand, the ions distribution is representative only in the case of Cu^{2+} in the leaves of *P. stratiotes* L.

Tab. 2. The accumulated heavy metal concentration in leaves of different hydrophytes (*El. canadensis* Michx., *L. minor* L. and *S. natans* Kunth.), in monometallic system

Monometallic system			
Species	Treatment	Concentration in leaves (mg/g)	Accumulated concentration in leaves (mg/g)
El. canadensis Michx.	Cu^{2+}	0.044	1.34±0.1^a
	Zn^{2+}	0.086	0.58±0.02^b
	Cd^{2+}	0	0.44±0.01^b
L. minor L.	Cu^{2+}	0.003	1.55±0.6^c
	Zn^{2+}	0.044	0.44±0.01^c
	Cd^{2+}	0	0.06±0.06^{NS}
S. natans Kunth.	Cu^{2+}	0.007	4.71±0.5^a
	Zn^{2+}	0.003	2.23±0.6^c
	Cd^{2+}	0	1.90±0.2^c

Values represent ± SD (n=3), small letters represent the statistical significant difference at $P<0.05$.

Multimetallic system

An attempt was made to appraise the phytoextraction capacity of the plants, to present the competition between metals for a better understanding of metabolic pathways and to identify the phytoremediation mechanism of the plants.

Tab. 3 contains the results related to the accumulation and distribution of heavy metals in different morphological parts of *P. stratiotes* L. and *E. crassipes* Mart. The compared results showed that the presence of the three heavy metal ions has a major influence on the plants phytoextraction efficiency, and that the heavy metal bioaccumulation capacity has increased values. The heavy metal bioaccumulation of hydrophytes is not proportional in the plant body. A higher concentration of heavy metals was observed in the roots of the hydrophytes. An exception was noticed in the case of *P. stratiotes* L. where the highest quantity of Cu^{2+} ions was bioconcentrated in the leaves. The same effect was observed in the case of monometallic treatment. The highest quantity of Cu^{2+} was registered in the case of the *E. crassipes* Mart. The Cu^{2+} uptake of this plant increased; it was seven times higher in comparison with the monometallic system. The result obtained in the multimetallic experiment confirmed that *P. stratiotes* L. has removed the highest concentrations of Zn^{2+} ions from the synthetic, water in a concentration of 8.07 mg/g plant biomass (4.89 mg/g in the root and 3.17 mg/g in leaves). The two hydrophytes Cd^{2+} ions uptake had increased. The highest removed quantity of cadmium can be noticed in *P. stratiotes* L. The Cd^{2+} ions concentrations is three time higher than in the case of the monometallic treatment. Generally, the heavy metals are accumulated in the root and may be due to the process of rhizofiltration, which is commonly observed in aquatic plants (Vesely *et al.*, 2011).

Tab. 3. Accumulation and distribution of heavy metals (mg/g) in different parts of *E. crassipes Mart.* and *P. stratiotes* L., in multimetallic solutions

Multimetallic system					
Species	Treatment	Concentration (mg/g)		Accumulated concentration (mg/g)	
		leaves	roots	leaves	roots
E. crassipes Mart.	Cu^{2+}	0.004	0.005	1.76 ± 0.6^{c}	5.84 ± 0.4^{a}
	Zn^{2+}	0.029	0.006	1.73 ± 0.8^{c}	3.03 ± 0.2^{b}
	Cd^{2+}	0	0	0.39 ± 0.3^{d}	3.58 ± 0.5^{b}
P. stratiotes L.	Cu^{2+}	0.003	0.007	3.64 ± 0.06^{b}	2.98 ± 0.7^{b}
	Zn^{2+}	0.013	0.047	3.17 ± 0.3^{b}	4.89 ± 1.1^{a}
	Cd^{2+}	0	0	0.88 ± 0.02^{c}	3.29 ± 0.4^{b}

Values represent ± SD (n=3), small letters represent the statistical significant difference at $P<0.05$.

Results regarding the accumulation of HM (without distribution) in the smaller studied hydrophytes (*El. canadensis* Michx., *L. minor* L. and *S. natans* Kunth.) are presented in Tab. 4. The heavy metal uptake capacity was increased with the presence of three heavy metal ions in all the three aquatic plants. The highest Cu^{2+} ions bioaccumulation was measured in *L. minor* L. This plant Cu^{2+} ions uptake capacity has grown nine times more than in the monometallic system. Moreover, the obtained data showed that *L. minor* L. was the most efficient also in the uptake of Zn^{2+} and Cd^{2+} ions from the multimetallic systems. *El. canadensis* Michx. proved to be a weak HM accumulator, in both the systems, even though it has been seen, that in the monometallic system this plant accumulates high amounts of Cu^{2+} (1.34 mg/g), meanwhile registering low values of Zn^{2+} and Cd^{2+}. In the multimetallic system, this species accumulated high amounts of Cd^{2+} (1.51 mg/g), but low Cu^{2+} and Zn^{2+} quantities. The *S. natans* Kunth. Zn^{2+} ions uptake has increased (4.47 mg/g) in the presence of the Cu^{2+} and Cd^{2+} ions. The obtained results suggested that the poisoned plant system needs more Zn^{2+} ions for protection. The Zn^{2+} ion is known to be involved in the chlorophyll biosynthesis. The all-presence of these metals usually raises the phytoextraction capacities, showing non-concurrence between the metals during absorption.

Plants Biochemical Response (chl and car content). Heavy metal pollution causes certain physiological and biochemical changes in the plants. The most significant biochemical changes can be observed in the photosynthethic pigment concentrations and in the total protein content (Dhir *et al.*, 2008). Heavy metals stress can inhibit or stimulate the formation of new proteins in the plant samples. The photosynthetic pigments are considered as indicators of environmental stress (Liechtenthaler and Miehe, 1997) and their degradation can be observed as a response to HM toxicity (Bunea *et al.*, 2012).

El. canadensis Michx. was selected to analyze the biochemical changes after the monometallic treatment. Fig. 1 contains the results of chl and car content from the treated plant samples. The results showed that after the Zn^{2+} monometallic exposure the plants were the most affected.

Tab. 4. The accumulated heavy metal concentration in leaves of *El. canadensis* Michx., *L. minor* L. and *S. natans* Kunth., in multimetallic system

Multimetallic system			
Species	Treatment	Concentration in leaves (mg/g)	Accumulated concentration in leaves (mg/g)
El. canadensis Michx.	Cu^{2+}	0.036	0.48 ± 0.01^{b}
	Zn^{2+}	0.089	0.59 ± 0.03^{b}
	Cd^{2+}	0.003	1.51 ± 0.2^{a}
L. minor L.	Cu^{2+}	0.003	10.81 ± 1.3^{a}
	Zn^{2+}	0.046	4.89 ± 1.2^{b}
	Cd^{2+}	0.000	2.78 ± 0.9^{c}
S. natans Kunth.	Cu^{2+}	0.006	3.56 ± 0.1^{b}
	Zn^{2+}	0.002	4.47 ± 0.3^{b}
	Cd^{2+}	0.000	1.09 ± 0.4^{d}

Values represent ± SD (n=3), small letters represent the statistical significant difference at $P<0.05$.

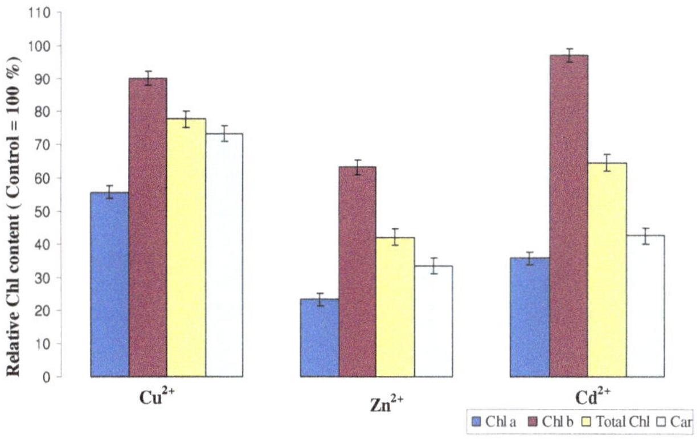

Fig. 1. The content of chl *a*, chl *b*, total chl and carotenoids of *El. canadensis* Michx. measured after the heavy metal treatment, in monometallic system

The multimetallic stress effects on the photosynthetic pigments content of the five studied aquatic plants are shown in Fig. 2 and 3. In this study, the chlorophyll concentrations decreased after six days of multimetallic exposure. A comparison of the relative sensitivity of the aquatic plants suggested that *S. natans* Kunth. is less sensitive to the three heavy metal ions than *L. minor* L., *P. stratiotes* L. and *E. crassipes* Mart. Hydrophytes showed the lowest chlorophyll content (Fig. 2) after the treatment. Moreover, the results confirmed that the three heavy metal ions had a major impact on the photosynthetic activity of the aquatic plants.

The carotenoids content in this experiment have shown a decreasing tendency in all the studied species. The lowest concentration was registered in the case of *El. canadensis* Michx. followed by *P. stratiotes* L. (Fig. 3). The obtained results are in agreement with previous literature reports about the monometallic systems. Naumann *et al.* (2007) showed that Cu^{2+} may inhibit the chl and car biosynthesis in case of *L. minor* L. The most sensitive parameters were observed on chl and car, followed by FW (fresh weight) and DW (dry weight). Hegazy *et al.* (2009) reported that chl and car content in *L. gibba* L. was altered by bioaccumulation of Zn^{2+} and Cu^{2+}, therefore the plants have shown a significant discoloration of leaves. In the same time, with the growth of concentration of HM, the content of chl *a* decreased, chl *b* increased and the car became higher than total chl. Singh *et al.* (2012) reported that the effect of combined heavy metal treatment on toxicity symptoms of aquatic plants depends on the HM concentration and also, on the exposure time.

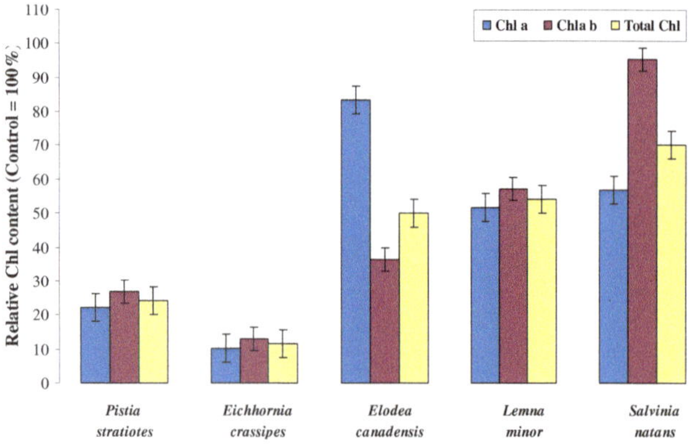

Fig. 2. The content of chl *a*, chl *b*, total chl of five aquatic plants measured after the multimetallic treatment

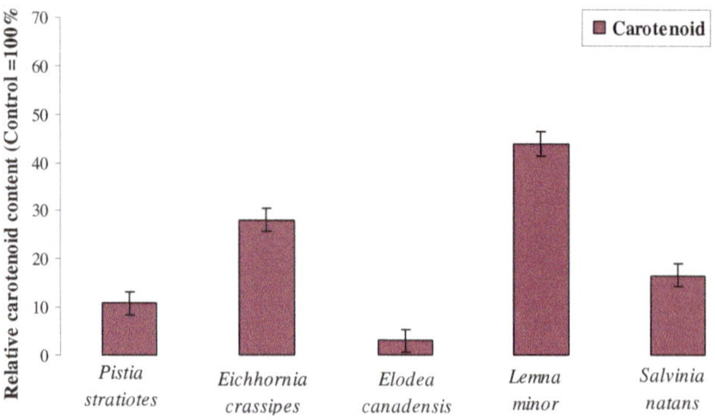

Fig. 3. The content of carotenoids of five aquatic plants measured after the multimetallic treatment

Protein Content

After the phytoremediation treatment with anorganic contaminants the plant metabolism is affected by lower or higher modifications in the protein content. The obtained results suggested that the correlation between removal rate and HM toxicity affected the protein content. The *El. canadensis* Michx. was selected for analysing the total protein content after the monometallic system. The results are presented in Fig. 4. Regarding these, the total protein content had increased after the Cu^{2+} and Cd^{2+} treatments. The total protein level decreased in the case of the Zn^{2+} exposure, which is in agreement with the report of Mishra and Tripathi (2008).

In the case of multimetallic systems the total protein content was measured in all five aquatic species. The results are shown in Fig. 5 in comparisons with the control plants content. The presence of the three toxic heavy metals in the aquatic environment has a main effect on the protein content in the case of *P. stratiotes* L., *E. crassipes* Mart. and *El. canadensis* Michx. The three mentioned aquatic plants protein concentration has increased after the multimetallic exposure. The plants possess a defence system, therefore the increased amount of protein could indicate the induction of stress proteins.

In the studied species with smaller morphological parts such as *L. minor* L. and *S. natans* Kunth., protein content has decreased after the heavy metal treatments. These results are in agreement with Singh's *et al.* (2012) reports concerning the decrease of total protein content in multimetallic system, which can generate higher toxicity effects on plant protein concentration.

The consequence of this study showed that the protein level in different aquatic species is in accordance with the HM accumulation. Phytoextraction results showed that the increasing protein content is involved in HM stress

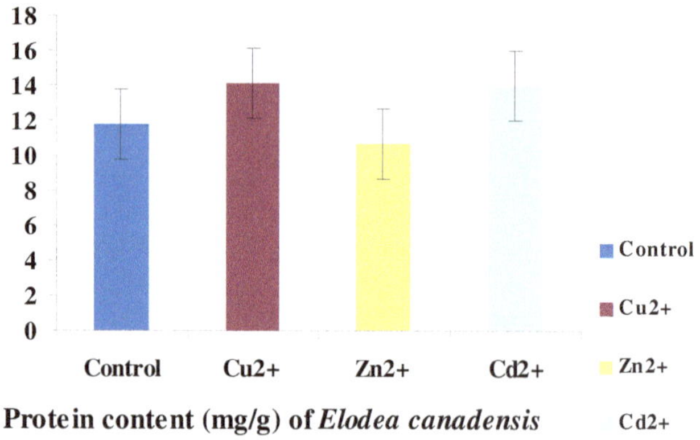

Fig. 4. The protein content of *El. canadensis* Michx. measured after the heavy metal treatment, in monometallic system (mg/g).

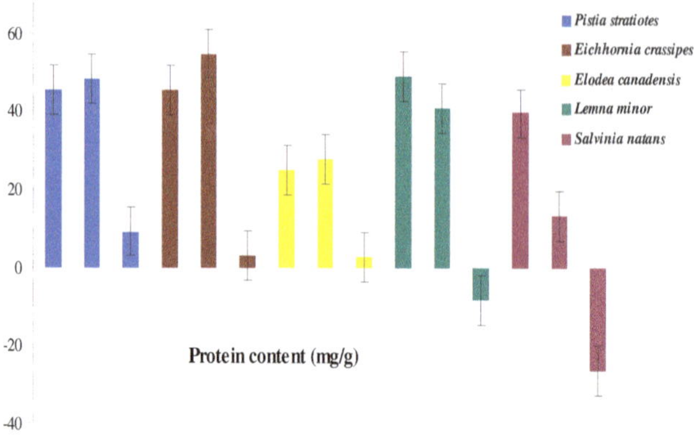

Fig. 5. Total protein content in control plants (mg/g), in treated plants and differences compared with control (untreated plants) and the studied species

response. It was also observed, that in case of *El. canadensis* Michx. the mixture of the three metal ions did not disturb the accumulation mechanism and the level of total proteins was higher than in the monometallic treatment.

Conclusions

The studied macrophytes proved to be useful in the uptake of heavy metals in monometallic system and much more effective in multimetallic system. The plant uptake and accumulation of toxic Cd^{2+} ions have been advantageously influenced by the presence of Cu^{2+} and Zn^{2+} ions. The responses of plants to heavy metal stress can vary depending on: the decrease in chl and car content, the inhibition and induction of soluble protein quantity. In the case of protein content two kinds of reactions were suggested: 1. the reduction of protein level as an effect of the metal ions toxicity, 2. the increase, from the production point of view, of new protein synthesis. The measurements of photosynthetic pigment concentration, revealed the highest heavy metal tolerance in the *L. minor* L. and *S. natans* Kunth species.

Acknowledgements

This study was financially supported by the University of Agricultural Sciences and Veterinary Medicine Cluj-Napoca, Romania, Grant No. 1215/18/2012 (Director: Lecturer Erzsebet BUTA, PhD).

References

Alka RU, Tripathi BD (2007). Principle and process of biofiltration of CD, Cr, Co, Ni and Pb from tropical opencast colamine effluent. Water Air Soil Pollut 180:213-223.

Arnon DI (1949). Cooper enzyme in isolated chloroplast polyphenol oxidase in *Beta vulgaris*. Plant Physiol 24:1-15.

Aysel S, Elmas E, Gümüş F, Ridvan ES (2010). Removal of cadmium by *Myriophyllum heterophyllum* Michx. and *Potamogeton crispus* L. and its effect on pigments and total phenolic compounds. Arch Environ Contam Toxicol 54:612-618.

Baker AJM, Brooks RR (1989). Terrestrial higher plants which hyperaccumulate metallic elements – A review of their distribution, ecology and phytochemistry. Biorecovery 181-126.

Bennicelli R, Pniewska ZSt, Banach A, Szajnocha K, Ostrowski J (2004). The ability of *Azolla caroliniana* to remove heavy metals (Hg (II), Cr (III), Cr (VI)) from municipal waste water. Chemosphere 55:141-146.

Bradford M (1976). A rapid and sensitive method for the quantitation of microgram quantities of protein utilizing the principle of protein-dye binding. Anal Biochem 72:248-254.

Bunea CI, Pop N, Babes AC, Matea C, Dulf FV, Bunea A (2012). Carotenoids, total polyphenols and antioxidant activity of grapes (*Vitis vinifera*) cultivated in organic and conventional system. Chem Cent J 6:1-9.

Buszewski BA, Jastrzębsk A, Kowalkowski T, Górna-Binkul A (2000). Monitoring of selected heavy metals uptake by plants and soils in the area of Toruń, Poland. Pol J Environ Stud 9:511-515.

Csog A, Mihucz VG, Tatar E, Fodor F, Virag I, Majdik C, Zaray Gy (2011). Accumulation and distribution of iron, cadmium, lead and nickel in cucumber plants grown in hydroponics containing two different chelated iron supplies. J Plant Physiol 168:1038-1044.

Cvjetko P, Tolić S, Šikić S, Balen B, Tkalec M, Vidaković-Cifrek Ž, Pavlica M (2010). Effect of copper on the toxicity and genotoxiticity of cadmium in duckweed (*Lemna minor* L.). Arh Hig Rada Toksikol 61:287-295.

Dhir B, Sharmila P, Saradhi PP (2008). Photosynthetic performance of *Salvinia natans* exposed to chromium and zinc rich wastewater. Braz J Plant Physiol 20:61-70.

Dhir B (2009). *Salvinia*: an aquatic fern with potential use in phytoremediation. Environ We Int J Sci Tech 4:23-27.

Dilek DR, Aksoy A (2004). Accumulation of heavy metals in *Typha angustifolia* (L.) and *Potamogeton pectinatus* (L.) living in Sultan Marsh (Kayseri, Turkey). Chemosphere 56:685-696.

Espinoza-Quinones FR, Zacarkim CE, Palacio SM, Obreg´on CL, Zenatti DC, Galante RM, Rossi N, Rossi FL, Pereira IRA, Welter RA (2005). Removal of heavy metal from polluted river water using aquatic macrophytes *Salvinia sp*. Braz J Phys 35:744-746.

Hasan SH, Talat M, Rai S (2007). Sorption of cadmium and zinc from aqueous solution by water hyacinth (*Eichhornia crassipes*). Bioresources Technol 98:918-928.

Hegazy AK, Kabiel HF, Fawzy M (2009). Duckweed as heavy metal accumulator and pollution indicator in industrial wastewater ponds. Desalin Water Treat 12:400-406.

Hogan CM (2010). Heavy metal. Encyclopedia of Earth. National Council for Science and the Environment. Eds E. Monosson & C. Cleveland, Washington DC, Retrieved from http://www.eoearth.org/view/article/153463.

Kachenko GA, Bhatia PN, Siegele R, Walsh KB, Singh B (2009). Nickel, Zn and Cd localization in seeds of metal hyperaccumulators using μ-PIXE spectroscopy. Nucl Instrum Meth B 267:2176-2180.

Kamal M, Ghaly AE, and Mahmoud N, Cote R (2004). Phytoaccumulation of heavy metals by aquatic plants. Environ Int 29:1029-1039.

Kaur L, Kasturi G, Sharma S (2010). Effect of pH and lead concentration on phytoremoval of lead from lead contaminated water by *Lemna minor*. Am-Euras J Agric and Environ Sci 7:542-550.

Kumar RP (2010). Phytoremediation of heavy metals in a tropical impoundment of industrial region. Environ Monit Assess 165:529-537.

Lavid N, Barkay Z, Tel-Or E (2000). Accumulation of heavy metals in epidermal glands of the waterlily (Nymphaeceae).

Plant J 212:313-322.

Liang HM, Lin TH, Chiou JM, Yeh KC (2009). Model evaluation of the phytoextraction potential of heavy metal hyperaccumulators and non – hyperaccumulators. Environ Pollut 157:1945-1952.

Lichtenthaler HK, Wellburn AR (1983). Determination of total carotenoids and chlorophylls a and b of leaf in different solvents. Biochem Soc Trans 11:591-592.

Lichtenthaler HK (1987). Chlorophylls and carotenoids: Pigments of photosynthetic biomembranes. Meth Enzymol 148:350-382.

Liechtenthaler HK, Miehe JA (1997). Flourescence imaging as a diagnostic tool for plant stress. Trends Plant Sci 2:316-320.

Lu X, Kruatrachue M, Pokethitiyook P, Homyok K (2004). Removal of cadmium and zinc by water hyacinth, *Eichhornia crassipes.* Science Asia 30:93-103.

Maczulak A (2009). Cleaning up the environment: Hazardous waste technology, Facts On File, Inc., NY.

Mihucz V, Csog A, Fodor F, Tatar E, Szoboszlai N, Silaghi-Dumitrescu L, Zaray Gy (2012). Impact of two iron (III) chelators on the iron, cadmium, lead and nickel accumulation in poplar grown under heavy metal stress in hydroponics. J Plant Physiol 169:561-566.

Mishra VK, Tripathi BD (2008). Accumulation of chromium and zinc from aqueous solution using water hyacinth (*Eichhornia crassipes*). Bioresource Technol 99:7091-7097.

Mudgal V, Madaan N, and Mudgal A (2010). Heavy metals in plants: phytoremediation: Plants used to remediate heavy metal pollution. Agr Biol J N Am 1:40-46.

Naaz S, Pandey SN (2010). Effects of industrial waste water on heavy metal accumulation, growth and biochemical responses of lettuce (*Lactuca sativa* L.). J Environ Biol 31:273-276.

Naumann B, Eberius M, Appenroth KJ (2007). Growth rate based dose-response relationships and EC – values of ten heavy metals using the duckweed growth inhibitions test (ISO 20079) with *Lemna minor* L. clone St. J Plant Physiol 164:1656-1664.

Peng K, Luo C, Lou L, Li X, Shen Z (2008). Bioaccumulation of heavy metals by the aquatic plants *Potamogeton pectinatus* L. and *Potamogeton malaianus* Miq. and their potential use for contamination indicators in wastewater treatment. Sci Total Environ 392:22-29.

Rahman MA, Hasegawa H (2011). Aquatic arsenic: Phytoremediation using floating macrophytes. Chemosphere 83:633-646.

Romanian Government (2002). Government decision 188/28/02/2002, concerning the approval of norms of waste water disposal in the environment. Retrieved from http://www.epurare.euriteh.ro/legislatie/NTPA.2002.pdf.

Singh A, Kumar CS, Agarwal A (2012). Physiological study of combined heavy metal stress on *Hydrilla verticillata* (l.f.) Royle. Int J Environ Sci 2(4):2234-2242.

Singh J, Kalamdhad AS (2013). Assessment of bioavailability and leachability of heavy metals during rotary drum composting of green waste (Water hyacinth). Ecol Eng 52:59-69.

Skinner K, Wright N, Porter-Goff E (2007). Mercury uptake and accumulation by four species of aquatic plants. Environ Pollut 145:234-237.

Vesely T, Tlustos P, Száková J (2011). The use of water lettuce (*Pistia stratiotes* L.) for rhizofiltration of a highly polluted solution by cadmium and lead. Int J Phytoremediat 13(9):859-72.

Zhu YL, Zayed AM, Qian JH, Souza M, Terry N (1999). Phytoaccumulation of trace elements by wetland plants: II. Water hyacinth. J Environ Qual 28:339-344.

Yeşim K, Davut B, Izzet K, Ali Z, Hasan G (2003). Bioaccumulation of nickel by aquatic macrophyta *Lemna minor* (Duckweed). J Agr Biol 3:281-283.

14

Photosynthesis-Involvement in Modulation of Ascorbate and Glutathione in *Euterpe oleracea* Plants Exposed to Drought

Maria Antonia Machado BARBOSA[1], Allan Klynger da Silva LOBATO[1*], Thaís Soares PEREIRA[1], Gélia Dinah Monteiro VIANA[1], José Ricardo Santos BARBOSA[1], Kelly Nayara Nascimento COELHO[1], Leila Sobral SAMPAIO[1], Benedito Gomes dos SANTOS FILHO[1], Joaquim Albenísio Gomes SILVEIRA[2]

[1]*Núcleo de Pesquisa Vegetal Básica e Aplicada, Universidade Federal Rural da Amazônia, Paragominas, Brazil*

[2]*Laboratório de Metabolismo de Plantas, Departamento de Bioquímica e Biologia Molecular, Universidade Federal do Ceara, Fortaleza, Brazil; allanllobato@yahoo.com.br (*corresponding author)*

Abstract

The present study aimed to determine if photosynthesis interferes with the modulation of antioxidant compounds in young *Euterpe oleracea* plants exposed to water deficiencies. A factorial, completely randomised experimental design was employed, and two water conditions (water deficit and control) and four evaluation points (0, 6, 12 and 18 days) were used, resulting in a total of eight measurements. The measured parameters included the water content and temperature of the leaf, gas exchange, electrolyte leakage, and antioxidant content. Compared to the control treatment, the net loss of photosynthesis due to water restriction increased by approximately 100% on the 18th day of drought. The ascorbate levels decreased due to water restriction, presenting significant differences on the 12th and 18th day. In some cases, the water deficit increased the glutathione content; however, significant effects were only observed on the 18th day after irrigation suspension. Water deficits had a negative impact on stomatal conductance, net photosynthesis rate, transpiration rate, and instantaneous carboxylation efficiency. Additionally, increases in the glutathione content, electrolyte leakage, and malondialdehyde content were observed; however, the ascorbate content decreased. Our results confirmed that the rate of photosynthesis interfered with the modulation of ascorbate and glutathione in young *Euterpe oleracea* plants exposed to drought.

Keywords: antioxidant compounds, *Euterpe oleracea,* gas exchange, reactive oxygen species, water deficiency

Introduction

Euterpe oleracea (Mart.) is a palm that grows in the Amazon forest, and this plant has nutritional and economical significance due to the commercialisation of energy drinks and the heart of the palm, which is produced from the fruit and stem (Rogez, 2000; Bobbio *et al.,* 2002). Regarding its nutritional proprieties, the fruit contains higher amounts of lipids and minerals such potassium, calcium, and magnesium (Menezes *et al.,* 2008; Yuyama *et al.,* 2011). Additionally, a study conducted by Bobbio *et al.* (2000) revealed that the fruit is a source of anthocyanins. Brazil is the main producer of *E. oleracea* and is a significant consumer and exporter of products derived from this palm (Menezes, 2005).

Water deficiency represents an abiotic stress and is considered one of the most important agricultural limitations due to its negative effects on growth and development (Kirnak *et al.,* 2001) and the corresponding yield losses (Endres *et al.,* 2010). Water supplies are essential in several metabolic and physiological processes such as the assimilation of nutrients (Wang *et al.,* 2011), gas exchange (Santos *et al.,* 2009), and translocation of organic solutes (Lobato *et al.,* 2008).

Plants exposed to water restrictions frequently exhibit negative responses linked to gas exchange (Yu *et al.,* 2009), and photosynthesis is strongly affected by drought (Isoda, 2005). Stomatal conductance has been used as indicator of water deficits (Oliveira *et al.,* 2005) because the regulation of this mechanism is influenced by the leaf water potential (Garnier and Berger, 1987) and water availability in the soil (Batista, 2011). In contrast, increases (Blum, 2005; Martins *et al.,* 2011) or decreases (Vu and Allen Jr., 2009) in water use efficiency occur during mild and severe stresses, respectively.

Recently, Sack and Holbrook (2006) reported that aquaporins present in bundle-sheath cells (BSCs) play a role in water regulation in the leaf; this finding was confirmed by

Shatil-Cohen *et al.* (2011), who showed that BSCs function as the first line of defence against drought.

Drought is directly related to the overproduction of reactive oxygen species (ROS) (Asada, 2006) such as hydrogen peroxide (H_2O_2) and superoxide (O_2^-) (Queiroz *et al.,* 2002), which are highly toxic compounds. ROS promote the oxidation of membranes and damage essential organelles such as chloroplasts (Carvalho, 2008) and mitochondria (Moller, 2001), which results in cell damage or death (Mittler, 2002).

Ascorbate and glutathione (GSH) have essential functions in antioxidant metabolism (Wang *et al.,* 2011) because ascorbate (ASC) is used as a substrate (Mehlhorn *et al.,* 1996). Additionally, GSH produces ascorbate and glutathione disulphide (GSSG), which is used to regenerate GSH via glutathione reductase (GR) (Creissen *et al.,* 1999).

The effects of gas exchange and ROS accumulation during water deprivation have been well studied; however, their effects have not been fully investigated in *E. oleracea.* Thus, the present study aimed to determine whether photosynthesis interferes in the modulation of antioxidant compounds in young *E. oleracea* plants exposed to water deficiencies.

Materials and Methods

Location and growth conditions

The experiment was performed on the Campus of Paragominas of the Universidade Federal Rural da Amazônia, Paragominas, Brazil (2°55'S and 47°34'W). The study was conducted in a greenhouse without environmental control, and the minimum, maximum, and median temperatures were 22 °C, 33 °C, and 26 °C, respectively.

The relative humidity during the experimental period varied between 65% and 93%, and the photoperiod was set to 12 h of light. During the measurement period (12:00 h), the amount of photosynthetically active radiation varied between 451 and 1,453 µmol m-2 s-1.

Plants, containers, and substrate

Seeds of *Euterpe oleracea* (Mart.) were germinated in 8-L pots (0.25 m in height and 0.20 m in diameter) equipped with holes for water drainage. The pots were filled with Plantmax® substrate, and the plants were irrigated daily with 1 L of distilled water. After 100 days, seedlings with similar aspects and sizes were selected. Subsequently, 120-day-old seedlings received 0.2 L of Hoagland and Arnon (1950) nutrient solution, which was modified for *E. oleracea.* Nutritional supplementation was applied at regular intervals (every 30 days) until the 12th month.

Experimental design

The experiment employed a factorial randomised design with two water conditions (water deficit and control) and four evaluation points (0, 6, 12 and 18 days), yielding a total of eight measurements. The experiment was assembled with five replicates for a total of 40 experimental units with one plant in each unit.

Water deficit application and harvest

All plants were grown until the 12th month, as described above. Plants receiving the water deficit treatment were submitted to 18 consecutive days without irrigation. Plants subjected to the control treatment were watered daily with distilled water. Parameters associated with gas exchange were measured during each evaluation period. The plants were harvested, and leaves present in the middle region of the plant were removed. The harvested leaves were frozen in liquid nitrogen and stored at −20 °C for subsequent biochemical determination.

Leaf gas exchange

The stomatal conductance, sub-stomatal CO2 concentration, leaf temperature, net photosynthetic rate, and transpiration rate were evaluated using an infrared gas analyser (ADC Bioscientific, model LCPro+). The parameters were measured using the adaxial surface of fully expanded leaves, which is localised in the middle of the plant.

The water use efficiency was estimated according to the method of Ma *et al.* (2004), and the instantaneous carboxylation efficiency was calculated using the formula described by Aragão *et al.* (2012). Gas exchange was evaluated in all plants between 9:00 and 12:00 h, and the irradiance was maintained at 800 µmol m-2 s-1 during the measurements.

Leaf relative water content

The leaf relative water content (LRWC) was evaluated in leaf disks with diameters of 10 mm. For each plant, 40 disks were removed, and the LRWC was calculated using the formula [(FM-DM)/(TM-DM)] × 100, as proposed by Slavick (1979).

Here, FM represents fresh matter, TM represents turgid matter evaluated after 24 h and saturated in deionised water at 4 °C in the dark, and DM represents dry matter determined after 48 h in an oven with forced air circulation at 80 °C.

Electrolyte leakage

Electrolyte leakage was measured according to the method described by Gong *et al.* (1998) with minor modifications. Fresh leaves (200 mg) were cut into pieces with a length of 1 cm and were placed in containers containing 8 mL of distilled deionised water. The containers were incubated in a water bath at 40 °C for 30 min, and the initial electrical conductivity of the medium (EC_1) was measured. The samples were boiled at 95 °C for 20 min to release the electrolytes. After the samples were cooled, the final electrical conductivity (EC_2) was measured (Gong *et al.,* 1998). The percentage of electrolyte leakage was calculated using the formula EL (%) = $EC_1 / EC_2 \times 100$.

Extraction of antioxidant compounds

Antioxidant compounds (malondialdehyde [MDA], ASC, and GSH) were extracted as described by Wu *et al.* (2006). Briefly, an extraction mixture was prepared by homogenising 500 mg of fresh leaf matter in 5 mL of 5% (w/v) trichloroacetic acid. Subsequently, the samples were centrifuged at 15,000 x g for 15 min at 3 °C, and the supernatant was collected.

Malondialdehyde quantification

MDA was determined by mixing 500 µL of supernatant with 1,000 µL of the reaction mixture, which contained 0.5% (w/v) thiobarbituric acid in 20% trichloroacetic acid. The mixture was incubated in boiling water at 95 °C for 20 min, and the reaction was terminated by placing the reaction container in an ice bath. The samples were centrifuged at 10,000 x g for 10 min, and the absorbance was measured at 532 nm. The amount of non-specific absorption at 600 nm was subtracted from the absorbance data. The amount of MDA–TBA complex (red pigment) was calculated based on the method of Cakmak and Horst (1991) with minor modifications, using an extinction coefficient of 155 mM^{-1} cm^{-1}.

Ascorbate quantification

The reaction for ASC detection contained 200 µL of supernatant and 1,800 µL of reaction mixture and was performed according to a method described by Cakmak and Marschner (1992) with minor modifications. The reaction mixture was prepared in 100 mM phosphate buffer (pH 7.6) composed of 400 µL of 10% trichloroacetic acid, 400 µL of 44% ortho-phosphoric acid, 400 µL of 4% 2,2'-dipyridyl in 70% ethyl alcohol, and 200 µL of 3% $FeCl_3$. The mixture was incubated at 40 °C for 40 min, and the resulting colour intensity was measured at 525 nm.

Glutathione quantification

For GSH detection, 200 µL of supernatant, 1,800 µL of reaction mixture (containing 100 mM phosphate buffer [pH 7.6] and 0.60 mM 2-nitrobenzoic acid) were combined, and the absorbance was measured at 412 nm (Wu *et al.*, 2006).

Data analysis

The data were subjected to an analysis of variance, and significant differences between the means were determined using the F test at a probability level of 5% (Steel *et al.*, 2006). Standard deviations were calculated for each treatment at all the evaluation points. Correlation analysis was performed using the Pearson parametric method. The statistical analyses were performed using SAS software.

Results

Relative water content, temperature, stomatal conductance, and substomatal CO_2 concentration in leaves

Exposure to water deficit occasionally reduced the LRWC; however, significant effects were only observed on the 12th and 18th day after stress application (Fig. 1). Compared with the control plants, a more severe reduction (60%) occurred on the 18th day. Water restriction reduced stomatal conductance, which showed significant changes at all points following water stress. A decrease of approximately 90% compared with the control plants was observed on the

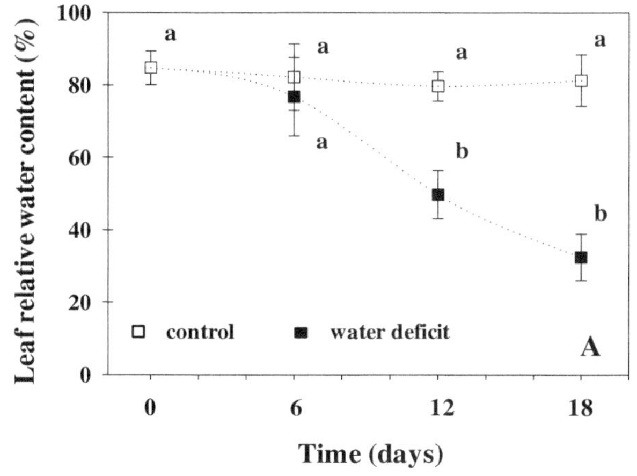

Fig. 1. Leaf relative water content in young *Euterpe oleracea* plants subjected to water deficit

Same letters do not show significant differences at F test (P < 0.05) to each time. Squares represent the mean values of 5 repetitions, and bars represent the standard deviations.

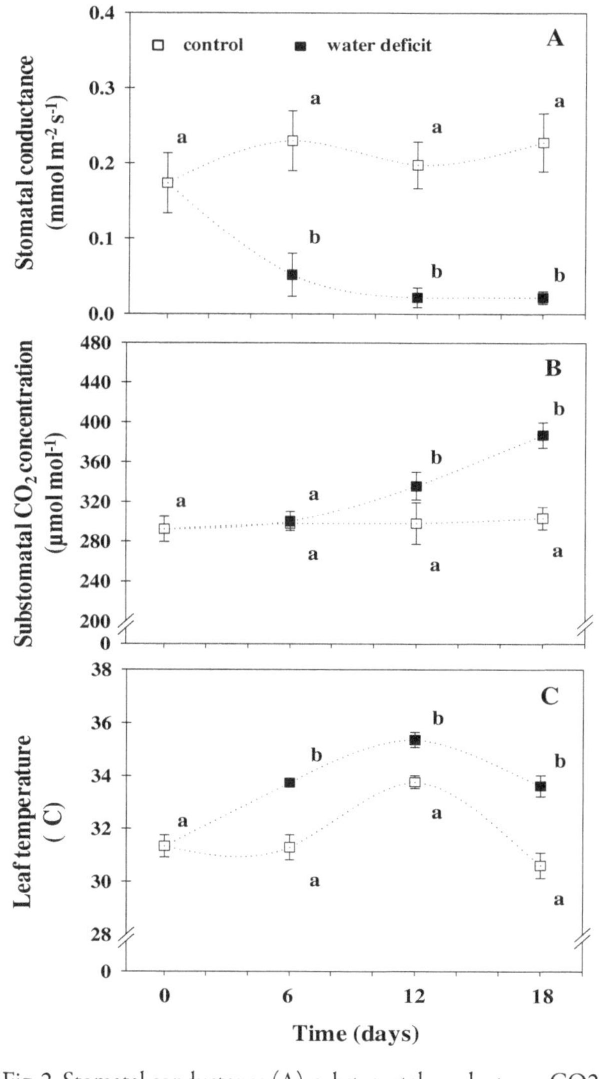

Fig. 2. Stomatal conductance (A), substomatal conductance CO2 concentration (B), and leaf temperature (C) in young *Euterpe oleracea* plants subjected to water deficit

Same letters do not show significant differences at F test (P < 0.05) to each time. Squares represent the mean values of 5 repetitions, and bars represent the standard deviations.

18th day (Fig. 2 A). A water stress-induced increase in the CO_2 concentration in the sub-stomatal cavity was observed on the 12th and 18th day after the application of drought (Fig. 2 B). The observed 27% increase in the sub-stomatal CO_2 concentration was higher than that of the control treatment on the 18th day. Water deficiency significantly increased the leaf temperature at all time points (Fig. 2 C). The observed increase was greater on the 18th day compared with the control plants.

Gas exchange

The net photosynthesis rate decreased due to water restriction, and this effect was significant on the 6th, 12th, and 18th days of water stress (Fig. 3 A). The greatest reduction compared with the control treatment was approximately 100% and occurred on the 18th day. Water stress decreased the transpiration rate, which showed significant reductions were observed on the 6th day (Fig. 3 D). By the 18th day, the instantaneous carboxylation efficiency had decreased by 100% compared with the control.

Electrolyte leakage and antioxidant compounds

Electrolyte leakage increased due to the water deficit, and significant increases were observed on the 12th and 18th days (Fig. 4 A). The increase in electrolyte leakage was more intense on the 12th day and was approximately 14% greater than that of the control plants. The water deficit increased the level of MDA; however, significant increases were only observed on the 18th day after stress application (Fig. 4 B). An increase of 26% compared with the control was observed on the 18th day. The ascorbate levels decreased after water restriction in the present study, and significant differences were noted at all time points (Fig. 4

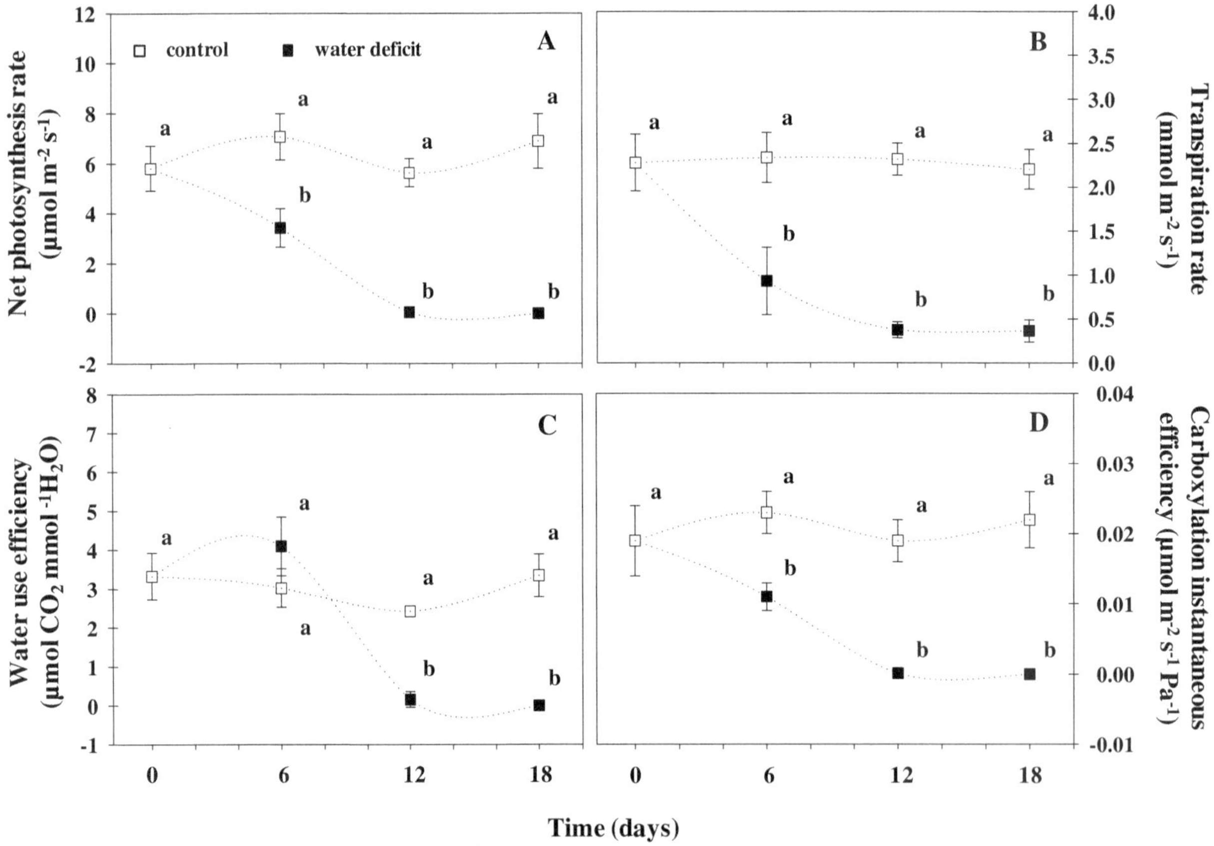

Fig. 3. Net photosynthetic rate (A), transpiration rate (B), water use efficiency (C), and carboxylation instantaneous efficiency (D) in young *Euterpe oleracea* plants subjected to water deficit

Same letters do not show significant differences at F test ($P < 0.05$) to each time. Squares represent the mean values of 5 repetitions, and bars represent the standard deviations.

significant differences from the 6th through the 18th day. The most intense reduction compared with the control was approximately 83% and occurred on the 12th day (Fig. 3 B). Water stress initially increased the water use efficiency, although the effect was not significant early in the experiment.

However, on the 12th and 18th day, the water use efficiency decreased significantly (Fig. 3 C). A more pronounced reduction was observed on the 18th day, when a 100% decrease was observed compared with the control plants. The efficiency of instantaneous carboxylation decreased due to the water deficit, and C). The greatest reduction (approximately 35%) was observed on the 18th day. Water stress increased the GSH concentration, showing significant effects on the 6th, 12th, and 18th day after irrigation suspension (Fig. 4 D). The greatest increase was observed on the 18th day, when an increase of approximately 62% was observed compared with the control treatment. The correlation analyses revealed two significant relationships between the net photosynthesis rate and the ascorbate content ($r = 0.91$; $P < 0.01$) (Fig. 5 A). Additionally, other relationships related to the net photosynthesis rate and GSH content were observed ($r = -0.81$; $P < 0.05$) (Fig. 5 B).

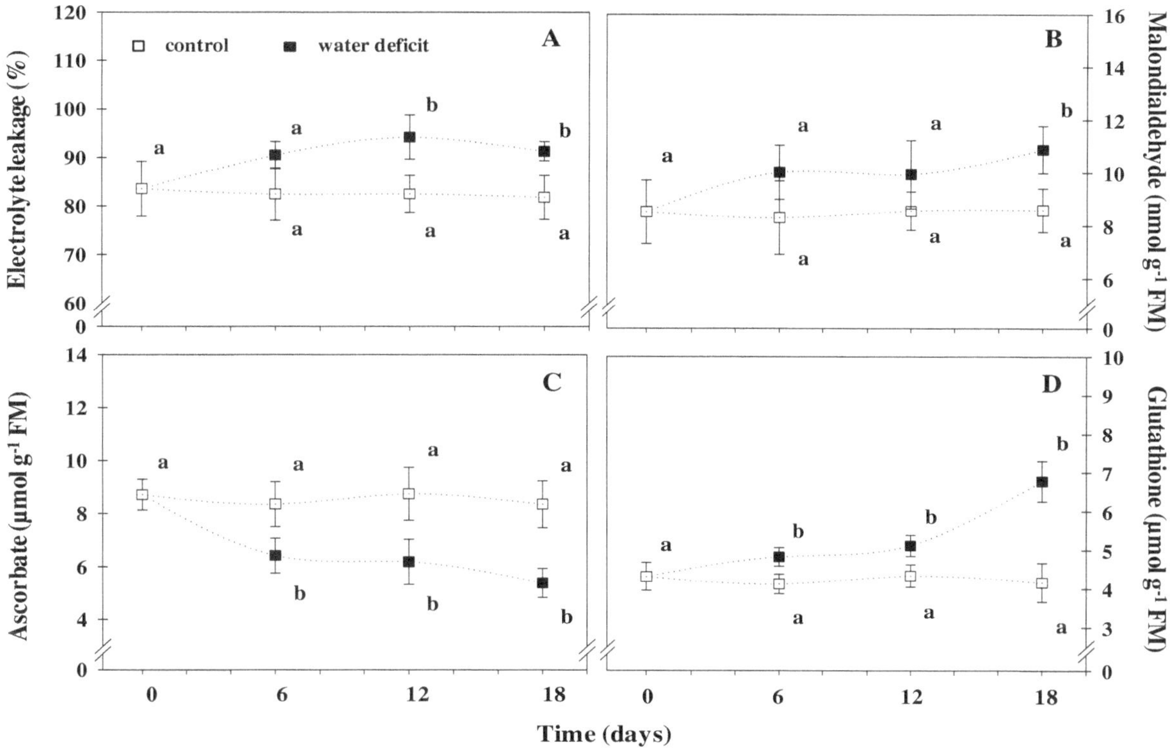

Fig. 4. Electrolyte leakage (A), malondialdehyde (B), ascorbate (C), and glutathione (D) in young *Euterpe oleracea* plants subjected to water deficit

Same letters do not show significant differences at F test ($P < 0.05$) to each time. Squares represent the mean values of 5 repetitions, and bars represent the standard deviations.

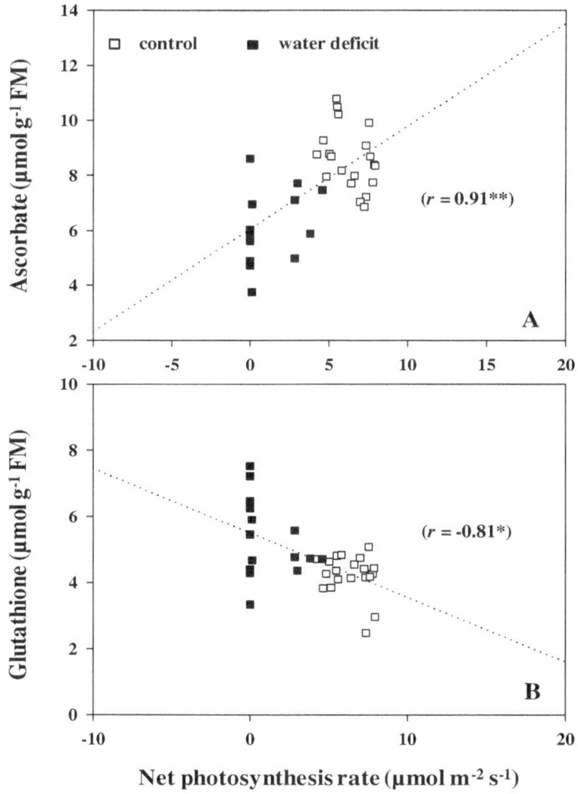

Fig. 5. Relationships between net photosynthesis rate and ascorbate (A), as well as net photosynthesis rate and glutathione (B) in *Euterpe oleracea* plants exposed to drought

Squares represent the values of 5 repetitions and asterisks (* and **) indicate significance at 0.05 and 0.01 probability level, respectively.

Discussion

Water stress decreased the LRWC because there was less water available for transpiration, which caused the plant to lose water to the atmosphere, resulting in wilted leaves. In other words, these symptoms occurred because the plant did not obtain water from soil macropores (Freitas *et al.,* 2007). Similar results were obtained by Carvalho *et al.* (1998), who found that water deficits significantly reduced the LRWC of *E. oleracea* plants.

The observed reduction in stomatal conductance in response to water deficit conditions was related to a decrease in the relative water content and the accumulation of abscisic acid in leaves (Egipto, 2011; Costa *et al.,* 2011). Silva *et al.* (2012) suggested that plants use stomatal mechanisms to reduce stomatal conductance and minimise water losses during conditions of drought. Similar results were also obtained by Oliveira *et al.* (2002) with *Bactris gasipaes* plants under water restriction, and these results corroborate the results obtained in the present study.

The observed increase in the substomatal CO_2 concentration of plants exposed to water deficit conditions was attributed to a decrease in the activities of enzymes involved in the process of CO_2 fixation, such as RUBISCO and PEPCASE (Machado *et al.,* 1999). These results also indicate that a reduction in the rate of photosynthesis, as opposed to a change in the stomatal mechanism, is the limiting factor (Mafakheri *et al.,* 2010). Similar results were reported by Repellin *et al.* (1997) with *Cocos nucifera* plants exposed to drought, which confirm the present results.

The observed increase in the leaf temperature during water deficit was due to the low transpiration rate, which reduced the flow of heat from the plant to the atmosphere and reduced the capacity for leaf thermoregulation (Nascimento, 2009). Salinas *et al.* (1996) suggested that an increase in leaf temperature due to reduced heat dissipation could be attributed to decreases in transpiration. An increase in leaf temperature was also observed by Silva *et al.* (2003), who evaluated young *Mimosa caesalpiniifolia, Enterolobium contortisiliquum*, and *Tabebuia aurea* plants exposed to water deficits.

A decrease in the photosynthesis rate was observed during progressive water deficiency. This response was due to lower stomatal conductance combined with the negative effects on instantaneous carboxylation efficiency in plants exposed to water deprivation. Mota (2011) obtained similar results and found that water deficiency significantly reduced the photosynthesis rate of *Acrocomia aculeata* plants.

The observed reduction in the transpiration rate in plants subjected to water restriction was due to the lower turgescence of guard cells, which resulted in stomatal closing and decreased the stomatal conductance (Machado *et al.*, 2009). In agreement with Magalhães Filho *et al.* (2008), stomatal closing is the first reaction of the plant to prevent water losses. Decreases in the transpiration rate were also observed in *E. oleracea* (Calbo and Moraes, 2000) and *Mauritia vinifera* (Calbo and Moraes, 1997) under water deficiency.

With respect to the water use efficiency, significant increases did not occur until the 6th day of drought stress due to the efficient response of this species to dehydration, which is mainly a result of a rapid reduction in the transpiration rate and superior water utilisation in the leaf (Taiz and Zeiger, 2009). However, later in the experiment (i.e., on the 12th and 18th days), the observed decrease in the water use efficiency may have been caused by the dehydration of mesophyll cells, which negatively affected essential processes that depend on the water supply such as photosynthesis and transpiration (Medina *et al.*, 1999). Souza *et al.* (2001) studied *Vitis* plants under water deficiency and reported similar water use efficiency results.

The severe decrease in the instantaneous carboxylation efficiency was attributed to lower values of stomatal conductance. Additionally, the observed decrease in this parameter provoked the accumulation of carbon dioxide in the substomatal cavity (Magalhães Filho *et al.*, 2008). Similar results were obtained by Tezara *et al.* (2002), who investigated the effects of water deficit in *Helianthus annuus* plants.

An increase in electrolyte leakage was occasionally observed due to the water deficit and was attributed to an increase in permeability or membrane rupture (Godoy, 2008). If the integrity of these structures is compromised, organelles can be irreversibly damaged and increases in electrolyte leakage can occur (Fioreze *et al.*, 2011). Gonçalves *et al.* (2011) evaluated *Jatropha curcas* plants under water deficit conditions and found that electrolyte leakage increased. Jiang and Zhang (2001) simulated oxidative stress with different levels of abscisic acid and also reported an increase in electrolyte leakage in *Zea mays* plants.

The observed increase in MDA levels revealed that water stress resulted in lipid peroxidation. This effect is normally produced by free radicals such as hydrogen peroxide, which is one of the main sources of reactive oxygen species (Luis, 2009). Carneiro (2011) also suggested that an increase in MDA levels may be correlated with inadequate activity of antioxidant enzymes. Similar results were obtained by Queiroz *et al.* (2002), who evaluated *Myracrodruon urundeuva* plants subjected to water stress.

The observed decrease in ascorbate levels was likely related to an increase in ascorbate peroxidase (APX) activity because ascorbate (ASC) is used as substrate in this enzymatic reaction (Apel and Hirt, 2004). Moreover, the decreased APX activity was also attributed to monodehydroascorbate reductase (MDHAR), which limits ascorbate production. These results suggest that the reaction catalysed by MDHAR is the limiting step in the antioxidant mechanism of *E. oleracea* plants under drought stress. Similar results were reported by Moran *et al.* (1994), who evaluated *Pisum sativum* plants exposed to water deficits.

The progressive increase in the GSH content was attributed to the minor activity of dehydroascorbate reductase (DHAR), which consumes GSH and catalyses the formation of glutathione disulphide (GSSG) (Gill and Tuteja, 2010). Additionally, GSH plays a role in the plant's defence system against reactive oxygen species that are formed during abiotic stress (May *et al.*, 1998; Potters *et al.*, 2007; Gill and Tuteja, 2010). Similar results were reported by Sofo *et al.* (2005), who studied four *Prunus* hybrids.

Conclusion

Water deficits have significant negative impacts on stomatal conductance, net photosynthesis rate, transpiration rate, and instantaneous carboxylation efficiency. Additionally, increases in GSH content, electrolyte leakage, and MDA were observed; however, the ascorbate content decreased. Our results confirmed the hypothesis that the photosynthesis rate interferes with the modulation of ascorbate and GSH in young *E. oleracea* plants exposed to drought.

Acknowledgements

This research had financial support from Conselho Nacional de Desenvolvimento Científico e Tecnológico (CNPq/Brazil) to Lobato AKS, and Barbosa MAM, Viana GDM, Barbosa JRS, Coelho KNN, and Pereira TS were supported by undergraduate scholarships also from Conselho Nacional de Desenvolvimento Científico e Tecnológico (CNPq/Brazil).

References

Apel K, Hirt H (2004). Reactive Oxygen Species: Metabolism, oxidative stress, and signal transduction. Annu Rev Plant Biol 55:373-399.

Aragão RM, Silva EM, Vieira CF, Silveira JAG (2012). High supply of NO3- mitigates salinity effects through an enhancement in the efficiency of photosystem II and CO2 assimilation in Jatropha curcas plants. Acta Physiol Plant 34:2135-2143.

Asada K (2006). Production and Scavenging of Reactive Oxygen Species in Chloroplasts and Their Functions. Plant Physiol 141:391-396.

Batista TMV (2011). Photosynthesis and stomatal conductance of tomato SM-16 and Mariana cultivated with different mulching. MSc, Universidade Federal Rural do Semiárido, Brazil.

Blum A (2005). Drought resistance, water-use efficiency, and yield potential – are they compatible, dissonant, or mutually exclusive? Aust J Agr Res 56: 1159-1168. Bobbio FO, Bobbio PA, Oliveira PA, Fadelli S (2002). Stability and stabilization of the anthocynins from *Euterpe oleracea* Mart. Acta Aliment 31:371-377.

Bobbio FO, Druzian JI, Abrão PA, Bobbio PA, Fadelli S (2000). Identification and quantification of anthocyanins in açai fruit (*Euterpe oleracea*) Mart. Ciência Tecnol Alime 20:388-390.

Cakmak I, Horst WJ (1991). Effect of aluminium on lipid peroxidation, superoxide, dismutase, catalase, and peroxidase activities in root tips of soybean (*Glycine max*). Physiol Plantarum 83:463-468.

Cakmak I, Marschner H (1992). Magnesium deficiency and high light intensity anhance activities of superoxide dismutase, ascorbate peroxidase, and glutathione reductase in bean leaves. Plant Physiol 98:1222-1227.

Calbo MER, Moraes JAPV (1997). Photosynthesis, Stomatal Conductance, Transpiration and osmotic adjustment of buriti plants subjected to water stress. Rev Bras Fisiol Veg 9: 117-123.

Calbo MER, Moraes JAPV (2000). Effects of water stress on plants of *Euterpe oleracea* (Acai). Braz J Bot 23:225-230.

Carneiro MMLC (2011). Gas exchange and antioxidative metabolism in sunflower plants in response to water deficit. MSc, Universidade Federal de Pelotas, Brazil.

Carvalho CJR, Rombold J, Nepstad DC, Sá TDA (1998). Water relations of açai palm in tidal forests of the amazon estuary. Rev Bras Fisiol Veg 10:213-218.

Carvalho MHC (2008). Drought stress and reactive oxygen species. Plant Signal Behav 3:156-165.

Costa RCL, Lobato AKS, Silveira JAG, Laughinghouse HD (2011). ABA-mediated proline synthesis in cowpea leaves exposed to water deficiency and rehydration. Turk J Agric For 35:309-317.

Creissen G, Firmin J, Fryer M, Kular B, Leyland N, Reynolds H, Pastori G, Wellburn F, Baker N, Wellburn A *et al.* (1999). Elevated glutathione biosynthetic capacity in the chloroplasts of transgenic tobacco plants paradoxically causes increased oxidative stress. Plant Cell 11:1277-1291.

Egipto RJL (2011). Effect of water availability on vegetative growth, yield and quality of grape in Aragonez (*Vitis vinifera* L.). MSc, Instituto Superior de Agronomia, Portugal.

Endres L, Souza JL, Teodoro I, Marroquim PMG, Santos CM, Brito JED (2010). Gas exchange alteration caused by water déficit during the bean reproductive stage. Rev Bras Eng Agríc Amb 14:11-16.

Fioreze SL, Pivetta LG, Fano A, Machado FR, Guimarães VF (2011). Performance of soybean genotypes under high intensity drought stress in greenhouse conditions. Rev Ceres 58:342-349.

Freitas JMN, Carvalho KS, Lobato AKS, Castro DS, Maia PSP, Oliveira Neto CF, Costa RCL (2007). Nitrate reductase activity, relative water content and total soluble chlorophyll contents in leaves of Açai (*Euterpe edulis* Mart.) submitted to drought and flooding. Braz J Biosciences 5:924-926.

Garnier E, Berger A (1987). The influence of drought on stomatal conductance and water potencial of peach trees growing in the field. Scientia Hortic 32:249-263.

Gill SS, Tuteja N (2010). Reactive oxygen species and antioxidant machinery in abiotic stress tolerance in crop plants. Plant Physiol Bioch 48:909-930.

Godoy AE (2008). Mechanical injury and its effects on physiology and quality of Golden papaya. MSc, Escola Superior de Agricultura Luiz de Queiroz, Brazil.

Gonçalves WG, Costa AL, Lima DP, Silva AA, Megguer CA (2011). Membrane permeabilty and relative water content in physic nuts submitted to fast water deficit. Global Sci Technol 4:131-139.

Gong M, Li YJ, Chen SZ (1998). Abscisic acid-induced thermotolerance in maize seedilings is mediated by calcium and associated with antioxidant systems. J Plant Physiol 153:488-496.

Smith SE, Read DJ (2008). Mycorrhizal Symbiosis. 3rd ed. London, UK: Elsevier Ltd.

Hoagland DR & Arnon DI (1950). The water culture method for growing plants without soil. California Agricultural Experiment Station. 1st ed. San Francisco, USA: California Publisher.

Isoda A (2005). Adaptive responses of soybean and cotton to water stress I. Transpiration changes in relation to stomatal area and stomatal conductance. Plant Prod Sci 8:16-26.

Jiang M, Zhang J (2001). Effect of abscisic acid on active oxygen species, antioxidative defence system and oxidative damage in leaves of maize seedlings. Plant Cell Physiol 42:1265-1273.

Lobato AKS, Oliveira Neto CF, Santos Filho BG, Costa RCL, Cruz FJR, Neves HKB, Lopes MJS (2008). Physiological and biochemical behavior in soybean (*Glycine Max* cv. Sambaiba) plants under water deficit. Aust J Crop Sci 2:25-32.

Luis RMFCB (2009). Responses of *Jatropha curcas* L. to water deficit Biochemical characterization and ecophysiological. MSc, Universidade Técnica de Lisboa, Portugal.

Ma CC, Gao YB, Guo HY, Wang JL (2004). Photosynthesis, transpiration and water use efficiency of *Caragana microphylla, C. intermedia* and *C. korshinskii*. Photosynthetica 42:65-70.

Machado EC, Medina CL, Gomes MMA (1999). Water content the growth substrate and photosynthesis orange 'Valencia'. Bragantia 58:217-226.

Machado RS, Ribeiro RV, Marchiori PER, Machado DFSP, Machado EC, Landell MGA (2009). Biometric and physiological responses to water déficit in sugarcane at different phenological stages. Pesq Agropec Bras 44:1575-1582.

Mafakheri A, Siosemardeh A, Bahramnejad B, Struik PC, Sohrabi E (2010). Effect of drougt stress on yield, proline and chlorophyll contents in three chickpea cultivares. Aust J Crop Sc 4: 580-585.

Magalhães Filho JR, Amaral LR, Machado DFSP, Medina CL, Machado ECM (2008). Water deficit, gas exchange and root growth in 'Valencia' Orange Tree budded on two rootstocks. Bragantia 67:75-82.

Martins VSR, Gonçalves-Vidigal MC, Moda-Cirino V, Vidigal Filho PS, Caixeta MP, Costa DV, Uchôa EB, Lacanallo GF, Lobato AKS (2011). Annu Rep Bean Improv Coop 54:164-165.

May M, Vemoux T, Leaver C, Van Montagu M, Inze D (1998). Glutathione homeostasis in plants: implications for environmental sensing and plant development. J Exp Bot 49:649-667.

Medina CL, Machado EC, Gomes MMA (1999). Stomatal conductance, transpiration and photosynthesis rates in 'Valencia' orange trees submitted to water stress. Ver Bras Fisiol Veg 11:29-34.

Mehlhorn H, Lelandais M, Korth HG, Foyer CH (1996). Ascorbate is the natural substrate for plant peroxidases. FEBS Letters 378:203-206.

Menezes EMS (2005). Effect of high hydrostatic pressure on açaí pre-frozen *(Euterpe oleracea* Mart.). MSc, Universidade Federal Rural do Rio de Janeiro, Brazil.

Menezes EMS, Torres AT, Srur AUS (2008). Nutritional value of açaí (*Euterpe oleracea* Mart) lyophilized. Acta Amazon 38:311-316.

Mittler R (2002). Oxidative stress, antioxidants and stress tolerance. Trends Plant Sci 7:405-410.

Moller IM (2001). Plant mitochondria and oxidative stress: Electron transport, NADPH turnover, and Metabolism of reactive oxygen species. Annu Rev Plant Physiol Plant Mol Biol 52:561-591.

Moran JF, Becana M, Iturbe-Ormaetxe I, Frechilla S, Klucas RV, Aparicio-Tejo P (1994). Drought induces oxidative stress in pea plants. Planta 194:346-352.

Mota CS (2011). Photosynthetic Characteristics of palm macaúba (*Acrocomia aculeata* (Jacq.) Lodd. Mart Ex.) Subjected to water deficiency. PhD, Universidade Federal de Viçosa, Brazil.

Nascimento SP (2009). Effect of water deficit in cowpea to identify genotypes with drought tolerance. MSc, Universidade Federal do Piauí, Brazil.

Oliveira AD, Fernandes EJ, Rodrigues TJD (2005). Stomatal Conductance as indicator of water stress in bean. Eng Agríc 25:86-95.

Oliveira MAJ, Bovi MLA, Machado EC, Gomes MMA, Habermann G, Rodrigues JD (2002). Photosynthesis, Stomatal conductance and Transpiration in peach palm under water stress. Sci Agric 59:59-63.

Potters G, Pasternak TP, Guisez Y, Palme KJ (2007). Stress-induced morphogenic responses: growing out of troble? Trends Plant Sci 12:98-105.

Queiroz CGS, Garcia QS, Lemos Filho JP (2002). Photosynthetic activity and membrane lipid peroxidation in plants mastic-the-wilderness under water stress and after rehydration. Braz J Plant Physiol 14:59-63.

Repellin A, Daniel L, Claude D, Braconnier S (1997). Water relations and gas exchange in ypirg coconut palm (*Cocos nucifera* L.) as influenced by water deficit. Can J Bot 75:18-27.

Rogez H (2000). Acai: Preparation, Composition and enhancement of conservation. 1st ed. Belém, BR: EDUFPA.

Sack L, Holbrook NM (2006). Leaf hydraulics. Annu Rev Plant Biol 57:361.

Salinas AR, Zelener N, Craviotto RM, Bisaro V (1996). Physiological responses that characterize behavior of different soybean cultivars a deficiency of water in the soil. Pesq Agropec Bras 31:331-338.

Santos MG, Ribeiro RV, Machado EC, Pimentel C (2009). Photosynthetic parameters and leaf water potencial of live common bean genotypes under mild water deficit. Biol Plantarum 53:229-236.

Shatil-Cohen A, Attia Z, Moshelion M (2011). Bundle-sheath cell regulation of xylem-mesophyll water transport via aquaporins under drought stress: a target of xylem-borne ABA? Plant J 67:72-80.

Silva EC, Nogueira RJMC, Neto ADA, Santos VF (2003). Stomatal behavior and leaf water potential in three woody species grown under water stress. Acta Bot Bras 17:231-246.

Silva ON, Lobato AKS, Ávila FW, Costa RCL, Oliveira Neto CF, Santos Filho BG, Martins Filho AP, Lemos RP, Pinho JM, Medeiros MBCL *et al.* (2012). Silicon-induced increase in chlorophyll is modulated by the leaf water potential in two water-deficient tomato cultivars. Plant Soil Env 58:481-486.

Slavick B (1979). Methods of studying plant water relations. 1st ed. New York, USA: Springer-Verlang.

Sofo A, Tuzio AC, Dichio B, Xiloyannis C (2005). Influence of water deficit and rewatering on the components of the ascorbate–glutathione cycle in four interspecific Prunus hybrids. Plant Sci 169:403-412.

Souza CR, Soares AM, Regina MA (2001). Gas exchange of vine cuttings obtained from two graftings submitted to water deficiency. Pesq Agropec Bras 36:1221-1230.

Steel RGD, Torrie JH, Dickey DA (2006). Principles and procedures of statistics: a biometrical approach. 1st ed. Moorpark, USA: Academic Internet Publishers.

Taiz L, Zeiger E (2009). Plant Physiology. 4rd ed. Porto Alegre, BR: Artmed.

Tezara W, Mitchell V, Driscoll SPE, Lawlor DW (2002). Effects of water deficit and its interaction with CO_2 supply on the biochemistry and physiology of photosynthesis in sunflower. J Exp Bot 53:1781-1791.

Vu JCV, Allen Jr LH (2009). Growth at elevated CO_2 delays the adverse effects of drought stress on leaf photosynthesis of the C_4sugarcane. J Plant Physiol 166:107-116.

Wang CQ, Liu TL, Xu HJ (2011). Ascorbate – Glutathione metabolism during PEG-Induced water deficit in *Trifolium repens.* Russ J Plant Physiol 58:597-602.

Wang Y, Liu F, Jensen CR (2012). Comparative effects of deficit irrigation and alternate partial root-zone irrigation on xylem pH, ABA and ionic concentrations in tomatoes. J Exp Bot 63:1907-1917.

Wu QS, Xia RX, Zou YN (2006). Reactive oxygen metabolism in mycorrhizal and non-mycorrhizal citrus (*Poncirus trifoliate*) seedlings subjected to water stress. J Plant Physiol 163:1101-1110.

Yu DJ, Kim SJ, Lee HJ (2009). Stomatal and non-stomatal limitations to photosynthesis in field-grown grapevine cultivars. Biol Plantarum 53:133-137

Yuyama, LKO, Aguiar JPL, Silva Filho DF, Yuyama K, Varejão MJ, Fávaro DIT, Vasconcellos MBA, Pimentel AS, Caruso MSF (2011). Physico-chemical characterization of acai juice from *Euterpe precatoria* Mart. coming from different ecosystems amazônicos. Acta Amazon 41:545-552.

Effect of Zinc and *Glomus intraradices* on Control of *Pythium deliense*, Plant Growth Parameters and Nutrient Concentrations of Cucumber

Zeliha KÜÇÜKYUMUK[1*], Hülya ÖZGÖNEN[2], İbrahim ERDAL[1], Figen ERASLAN[1]

[1]*Department of Soil Science and Plant Nutrition, Agricultural Faculty, Suleyman Demirel University, 32260, Isparta, Turkey; zelihakucukyumuk@sdu.edu.tr (*corresponding author)*

[2]*Department of Plant Protection, Agricultural Faculty, Suleyman Demirel University, 32260, Isparta, Turkey*

Abstract

Three levels of zinc fertilization (0, 5, 10 mg/kg) and an arbuscular mycorrhizal (AM) fungus *Glomus intraradices* were tested for their potential to control *Pythium deliense* on inoculated cucumber seedlings. Plant Zn, N, P, K, Mg, Ca, Fe, Mn, Cu contents, dry and fresh weights of plant and roots and disease severity were determined in the study. Resistance to Pythium rot was determined with the application of mycorrhiza with increasing doses of zinc. Zinc and mycorrhizal fungus applications had significant effects on plant nutrition except for K and Cu. While the highest N and P concentrations were noted under Zn_0 conditions, the values obtained under Zn_1 and Zn_2 conditions showed differences depending on *G. intraradices* and *P. deliense* treatments. Leaf Ca concentration reached up to highest level with $Zn_2GI_0Pd_1$ treatment and the lowest Ca content was recorded under GI_0Pd_0 for all Zn applications. Lower level of zinc together with $GIPd_0$ applications resulted in the highest leaf Mg concentration. The highest micronutrient concentrations were analysed on cucumber plants grown under Zn deficient conditions without GI but with *P. deliense.* Plant dry weight, root fresh and root dry weights were higher in cucumber plants challenged with AM fungus and *P. deliense* under zinc applied conditions. It was observed that certain rates of zinc and mycorrhiza based-treatments had positive effects on disease factors by suppressing Pythium rot and can be used for biological control.

Keywords: cucumber, *Glomus sp.*, mycorrhiza, *Pythium sp.*, zinc

Introduction

Pythium rot disease caused by *Pythium deliense* is one of the important diseases causing seedling losses after planting in vegetable crops. To solve this problem the use of zinc and mycorrhiza comes up.

Plant mineral nutrition level effects on the degree of disease factors. Especially, the micro nutrients decrease pathogen cell penetration and infection by affecting the cell wall rigidity and also the structural integrity of membranes as well as by their direct toxic effects to the pathogen (Marschner, 1995). Plants zinc nutrition and application of zinc with mycorrhiza had positive effects against diseases. Pythium seedling rot (*Pythium* sp.) is the most important diseases in vegetable seedlings.

Arbuscular mycorrhizal (AM) fungi were reported to be effective in controlling plant diseases (Agrios, 1997). Mycorrhizae are able to increase nutrient uptake in plants and to affect their development and were found to be effective against many fungal diseases (Marschner and Dell, 1994).

The disease severity reducing effects induced by mycorrhiza were reported in various pathosystems such as peanut - *Sclerotium rolfsii*, eggplant - *Verticillium dahlie*, tomato - *Phytophthora nicotianae* var. *parasitica*, pea-*Rhizoctonia solani*, tomato - *Fusarium oxysporum* f. sp. *lycopersici and F. oxysporum* f. sp. *radicis-lycopersici*, and pepper - *Phytophthora capsici* (Krishna and Bagyaraj, 1983; Caron *et al.*, 1986; Matsubara *et al.*, 1995; Trotta *et al.*, 1996; Karagiannidis *et al.*, 2002; Akküprü and Demir, 2006; Özgönen and Erkılıç, 2007).

Zinc has direct effect on fungal growth and secondary metabolism and indirect effects on host susceptibility (Duffy, 2009). Zinc-sufficient plants have been found to be more tolerant to root infection by Fusarium wilt than Zn-deficient plants (Marschner, 1995; Streeter *et al.*, 2001). This probably results from the fungitoxicity of Zn and the role of Zn in stabilizing the membranes of root cells (Marschner, 1995).

In view of the role of zinc in both generation and detoxification of oxygen radicals and hydrogen peroxide, plant resistance was reported to be markedly influenced by mineral nutrition during host plant-pathogen interactions (Marschner, 1995).

The resistance of plants to diseases is closely related to the nutrition levels. The deficiency of mineral elements influences the growth of the plant and can cause disease (Cai *et al.*, 2007).

There are some reports on zinc effect in controlling plant disease. In fact, *Fusarium solani* and *Rhizoctania solani* were reduced by increasing soil concentrations of zinc from 0 to 1.6 mg/kg (Siddiqui *et al.*, 2002). The potential of zinc in reducing disease severity has been signalled in various fungal diseases as is the case of Charcoal rot disease of maize caused by *Macrophomina phaseolina*, potato powdery scab, *Mycosphaeralla pinodes* and *Phoma medicaginis* in bean, *Rhizoctania* root rot of cowpea, alfalfa root rot, corn smut, Fusarium root rot of wheat (Braithwaite *et al.*, 1994; Kostandi *et al.*, 1997; Pareek and Pareek, 1999; Davidson and Ramsey, 2000; Grewal, 2001; Kalim *et al.*, 2003; Khoshgoftarmanesh *et al.*, 2010).

In a previous study, zinc and mycorrhiza were successfully used for the control of Phytophthora Blight of pepper. In fact, *Phytophothora capsici* infection and disease severity were reduced by individual or combined applications of Zn and AM fungus. These treatments also led to an increase in plant yield and nutrient concentration (Küçükyumuk *et al.*, 2013).

This study aimed to determine the effects of zinc and *mycorrhiza* on control of *Pythium deliense,* plant yield and mineral nutrient uptakes on cucumber plants.

Materials and methods

Cucumber (*Cucumis sativus* L.) cv. Silyon F1 was used as plant material. *Glomus intraradices* (GI) used as mycorrhizal inoculant produced on maize (*Zea mays* L.) (Harley and Smith, 1983). *Pythium deliense* (Pd) was isolated from naturally infected cucumber plants which showed post-emergence damping-off symptoms on corn meal agar (CMA) at 25 °C for 7 days. The soil mixture including soil: sand: pumice (1:1:1; v: v: v) was sterilized at 121 °C and 1 KPa two times for 1 hour. Plants were grown for 9 weeks in pots under greenhouse conditions with inoculation of *G. intraradices* at 1000 spores 10 g^{-1} soil and Zn fertilization was applied at three rates (0, 5 and 10 mg Zn kg^{-1} soil as $ZnSO_4$). *P. deliense* was inoculated 4 weeks after planting. For inoculation, Pd was cultured on CMA at 25 °C for 7 days and then cultures were added sterile distilled water and incubated at 4 °C for 30 min to induce zoospore release. Later, 10 ml at the concentration of 10^6 zoospore ml^{-1} were applied to each pot. In the experiment, there were 12 treatments in total with 8 replicates, giving a total of 96 pots in a completely randomized block design. The experimental soil was loam having pH 7.8 (1:2.5 water); $CaCO_3$ 17%; organic matter 1.5% (Jackson, 1973); 0.5 M $NaHCO_3^-$ extractable P 30 kg ha^{-1}; NH_4OAC-exchangeable K 550 kg ha^{-1} (Olsen *et al.*, 1954; Knudsen *et al.*, 1982); DTPA-extractable Zn 0.7 mg kg^{-1}(Lindsay and Norvell, 1978). As basal fertilization, 200, 120 and 150 mg kg^{-1} N, P and K was given to the pots as NH_4NO_3 and KH_2PO_4, respectively. For Zn fertilization, a stock solution was prepared using as $ZnSO_4 7H_2O$ given to the soil and mixed together with other fertilizers.

Plant samples were harvested 9 weeks after planting and the fresh weight of the above ground part of the plants was recorded. At first, samples were washed thoroughly with tap water and distilled water then they were kept at 65 ± 5 °C until stable weights reached. Finally, dry weights of plants were measured. After this, samples were grounded for nutrient analysis. In order to determine P, K, Ca, Mg, Fe, Zn and Mn concentrations, 0.4 g of grounded samples were wet digested at 180 °C for 15 min in microwave (CEM Mars X-press) and digestives were filled up with 50 ml with pure water. Phosphorus concentrations were determined by vanadate-molybdate colorimetric method. Zinc, K, Ca, Mg, Fe and Mn concentrations were determined using atomic absorption spectrophotometer as described by Kacar and İnal (2008). For fresh and dry weight measurement, shoots were separated from roots. The roots were washed with tap water and washed roots were left on two layers of filter paper to remove excess water. The roots in each treatment were weighed to determine fresh weight and then dried at 65°C to determine dry weight.

Disease severity was determined using 0-4 scale where 0 = healthy plant; 1 = 1-25% diseased of whole plant; 2 = 26-50% diseased plant; 3 = 51-75% diseased plant; 4 = 76-100% diseased plant (Zhang *et al.*, 1996). Disease severity and index values were calculated from scale values. In addition, the impact ratios were calculated according to the Abbot formula. Differences between treatments were analyzed by multiple comparison tests LSD (P = 0.01) using variance analyses to the results.

Results and discussion

Effect of zinc and *G. intraradices* on the dry weight of the above ground plant part was given in Tab. 1. As shown in this table, this parameter varied significantly depending on the treatments tested. Comparing to control conditions ($Zn_0GI_0Pd_0$) (1.49 g) dry weight of plants increased with zinc applications. The dry weight of plants inoculated with *P. deliense* was 0.64 g. Cucumber plants inoculated with *P. deliense* and treated with zinc and *G. intraradices* showed the highest records of dry weight (3.8 g).

Tab. 1. The effect of different zinc rates and *Glomus intraradices* applications above ground part of plant dry weight and root fresh and dry weight in cucumber

Applications		Plant dry weight (g)	Root fresh weight (g)	Root dry weight (g)
Zn_0	GI_0Pd_0	1.49 d	0.65 c	0.07 b
	GI_0Pd_1	0.64 e	0.13 d	0.03 c
	GI_1Pd_0	1.75 cd	0.85 bc	0.10 b
	GI_1Pd_1	1.76 cd	0.86 bc	0.08 b
Zn_1	GI_0Pd_0	1.66 cd	0.75 bc	0.08 b
	GI_0Pd_1	2.56 b	1.17 bc	0.09 b
	GI_1Pd_0	1.79 cd	0.87 bc	0.09 b
	GI_1Pd_1	3.8 a	1.68 a	0.17 a
Zn_2	GI_0Pd_0	2.06 c	0.87 bc	0.09 b
	GI_0Pd_1	3.15 b	1.27 ab	0.15 a
	GI_1Pd_0	2.11 c	0.87 bc	0.09 b
	GI_1Pd_1	3.8 a	1.70 a	0.17 a

For each column, means followed by the same letter are not significantly different according to LSD test (P = 0.05)

Effects of zinc and *G. intraradices* on cucumber root fresh and dry weights were given in Tab. 1. The results obtained revealed that the root fresh and dry weights varied significantly depending on the factors tested. In fact, results of fresh and dry weight of roots had a similar trend. Cucumber plants inoculated with *P. deliense* showed the lowest root fresh (0.13 g) and dry weights (0.03 g). Those inoculated with the pathogen and treated with zinc and *G. intraradices* presented the highest root fresh (1.7 g) and dry weights (0.17 g).

Effect of pathogen, zinc concentrations and *G. intraradices* on Zn contents in cucumber plants were given in Tab. 2. In fact, plants inoculated with *P. deliense* and treated with Zinc rates and *G. intraradices* showed significant variation in their zinc contents. This parameter was 42 mg kg^{-1} in control conditions, 46 mg kg^{-1} with Zn_1 application and 53 mg kg^{-1} with Zn_2 application.

According to control conditions, plant zinc content increased with the application of *G. intraradices*. As shown in Tab. 2, plants inoculated with *P. deliense* (73 mg kg^{-1}) had the highest zinc concentrations. Zn_2 application (53 mg kg^{-1}) had higher zinc concentrations compared to Zn_1 (46 mg kg^{-1}) applications.

As indicated in Tab. 2, copper concentrations did not vary significantly on different zinc rates and mycorrhizal applications tested. Plant contents ranged between 30 and 56 mg kg^{-1}.

Iron (Fe) and manganese (Mn) contents varied significantly depending on inoculation with pathogen, zinc rates and mycorrhizal applications. With zinc application, plants inoculated with *P. deliense* showed increased iron concentrations as compared to control (153 mg kg^{-1}) plants. *P. deliense* inoculated plants had the highest (314 mg kg^{-1}) iron concentrations. Mn concentrations shown in Tab. 2 were also significantly different depending on treatments tested. This parameter varied from 443 mg kg^{-1} in control plants whereas with zinc application, plant Mn concentrations were 483 and 488 mg kg^{-1} and they were in the same statistical group.

Tab. 2. The effect of different zinc rates and *Glomus intraradices* applications on Zn, Fe, Mn and Cu concentrations in cucumber plants

Applications		Zn (mg/kg)	Fe(mg/kg)	Mn (mg/kg)	Cu (mg/kg)
Zn_0	GI_0Pd_0	42 b*	153 b	443 ab	34 a
	GI_0Pd_1	73 a	314 a	534 ab	56 a
	GI_1Pd_0	40b	155 b	400 b	30 a
	GI_1Pd_1	43 b	195 ab	529 ab	34 a
Zn_1	GI_0Pd_0	46 b	164 b	483 ab	34 a
	GI_0Pd_1	50 ab	200 ab	492 ab	35 a
	GI_1Pd_0	47 ab	211 ab	500 ab	34 a
	GI_1Pd_1	50 ab	184 b	547 ab	36 a
Zn_2	GI_0Pd_0	53 ab	155 b	488 ab	34 a
	GI_0Pd_1	56 ab	186 b	575 a	35 a
	GI_1Pd_0	55 ab	190 b	548 ab	30 a
	GI_1Pd_1	55 ab	201 ab	605 a	32 a

*For each column, means followed by the same letter are not significantly different according to LSD test (P = 0.05)

Tab. 3 indicates that nitrogen and phosphorus concentrations of plants were significantly affected by the treatments tested while for potassium concentrations, no significant effect of different zinc doses and *G. intraradices* applications was recorded. Compared to control conditions, nitrogen concentrations were not affected by zinc sulphate applications. Nitrogen concentrations of host plants decreased according to control conditions. Plants inoculated with *P. deliense* and treated with Zn_2 applications showed the lowest nitrogen (4.21%) concentrations whereas their treatment with the mycorrhizal fungi led to highest nitrogen (4.65%) content.

Tab. 3. The effect of different zinc rates and *Glomus intraradices* applications on N, P and K concentrations

Applications		N (%)	P (%)	K (%)
Zn_0	GI_0Pd_0	4.34 bc*	0.43 b	5.0 a
	GI_0Pd_1	4.31 bc	0.89 a	5.5 a
	GI_1Pd_0	4.65 a	0.42 b	5.3 a
	GI_1Pd_1	4.54 ab	0.54 ab	5.1 a
Zn_1	GI_0Pd_0	4.31 bc	0.21 b	5.9 a
	GI_0Pd_1	4.37 abc	0.38 b	5.2 a
	GI_1Pd_0	4.49 abc	0.57 ab	5.1 a
	GI_1Pd_1	4.23 c	0.38 b	5.4 a
Zn_2	GI_0Pd_0	4.49 abc	0.39 b	6.0 a
	GI_0Pd_1	4.21 c	0.59 ab	5.2 a
	GI_1Pd_0	4.46 abc	0.50 ab	5.2 a
	GI_1Pd_1	4.47 abc	0.36 b	5.5 a

*For each column, means followed by the same letter are not significantly different according to LSD test (P = 0.05)

Tab. 4. The effect of different zinc rates and *Glomus intraradices* applications on Ca and Mg concentrations

Applications		Ca (%)	Mg (%)
Zn_0	GI_0Pd_0	4.9 c*	5.8 abcd
	GI_0Pd_1	5.3 bc	5.9 abcd
	GI_1Pd_0	4.9 c	5.6 cd
	GI_1Pd_1	5.4 a	6.0 abc
Zn_1	GI_0Pd_0	5.0 c	5.8 abcd
	GI_0Pd_1	5.9 abc	5.5 d
	GI_1Pd_0	5.5 bc	6.2 a
	GI_1Pd_1	6.4 ab	6.1 ab
Zn_2	GI_0Pd_0	5.0 c	5.9 abcd
	GI_0Pd_1	6.3 ab	5.7 bcd
	GI_1Pd_0	6.2 ab	6.1 ab
	GI_1Pd_1	6.8 a	6.1 ab

*For each column, means followed by the same letter are not significantly different according to LSD test (P = 0.05)

Tab. 5. The effects of disease severity of Pythium deliense on applications of mycorrhiza and zinc

Applications	Disease index
Zn_1+ Pd	2
Zn_2+ Pd	2
Zn_1+ Gi + Pd	2
Zn_2+ Gi +Pd	1
Pd +Gi	1
Pd	4
Control (-)	-

Tab. 6. Root colonization rates of mycorrhiza applications (%)

Applications	Root colonization rates (%)
G. intraradices (Gi)	71.3 a*
Zn_1 + Gi	70.2 a
Zn_2 + Gi	72.8 a
Pd +Gi	62.6 b

*Means followed by the same letter are not significantly different according to LSD test (P = 0.05)

Cucumber plants inoculated with *P. deliense* and untreated with zinc rates nor with *G. intraradices* exhibited the highest (0.89%) Phosphorus content but this parameter decreased to 0.43% on the uninoculated and untreated control plants, to 0.21% and 0.39% on uninoculated plants treated with Zn_1 and Zn_2 applications, respectively. In absence of pathogen inoculation, plants treated with Zn_1 and mycorrhiza showed a P content of about 0.57%. This parameter decreased to 0.38% and 0.36% on cucumber plants inoculated with *P. deliense* and treated with mycorrhiza and Zn_1 and Zn_2 applications, respectively (Tab. 3).

The Potassium concentrations in cucumber plants did not vary significantly depending on zinc and mycorrhizal fungi applications.

Calcium (Ca) and Magnesium (Mg) concentrations of cucumber plants varied significantly according to zinc doses and mycorrhizal treatments tested (Tab. 4). With zinc fertilization only, plant calcium concentrations (5.0%) were significantly similar to untreated control (4.9%). When combined treatments with zinc (Zn_1 and Zn_2) and mycorrhiza, plant Ca concentrations increased to 6.4% and 6.8%, respectively. Mycorrhiza based-treatments led to increased Ca concentrations.

Magnesium concentrations also varied depending on the applications tested.

Disease index results affected by mycorrhiza, pathogen inoculation and zinc rates were given in Tab. 5. This parameter ranged between 1.0 and 4.0. Disease index values were decreased by zinc and mycorrhiza based-treatments. Plants inoculated with *P. deliense* had the highest disease scale (4). Different zinc doses tested and mycorrhiza had protected cucumber plants from Pythium rot disease as compared to the control. Both of zinc doses showed typical damping off symptoms of *P. deliense* and got 2-scale value in cucumber plants untreated with *G. intraradices*. However, for plants treated with Zn_1 and Zn_2 together with mycorrhiza, disease scale was 2 and 1, respectively.

Root colonization

Root colonization results were given in Tab. 6. Colonization of cucumber roots ranged between 62.6 % and 72.8%. Plants treated with *G. intraradices* and Zn_1 or Zn_2 applications showed root colonization ranging between 70.2 and 72.8%. The lowest records (62.6%) were noted on cucumber plants inoculated with pathogen and challenged with AM fungi. Although mycorrhiza colonization decreased due to pathogen application, it was found to be successful in reducing disease severity.

The results recorded from this study indicated that mineral element concentrations in diseased inoculated plants were higher than the other plants. Some compounds such fitoalexins, fenols, flavonoids and oxins were reported to be effective in inducing disease resistance. These compounds were produced due to plant nutrient amounts and collected in the infection area (Uçgun and Gezgin, 2013).

When the results were evaluated, zinc and mycorrhiza applications increased plant dry weight. While plant zinc concentrations increased with zinc applications, phosphorus concentrations decreased. Mycorrhiza applications increased plant nitrogen and phosphorus concentrations. Similar results were determined by the other researchers (Kumar *et al.*, 1998; Çığşar *et al.*, 2000; Özcan and Taban, 2000; Tüfenkçi *et al.*, 2005; Küçükyumuk *et al.*, 2013) that zinc and mycorrhiza applications increased plant yield.

As a result, zinc and mycorrhiza had positive effects by controlling Pythium rot disease of cucumber seedlings and could, thus, be used for integrated control of this disease. Pythium seedling rot disease is harmful in nursery circuit for all vegetable crops. Results obtained in the cucumber-*Pythium* pathosystem could be extended to other vegetables. Joint application of Zinc rates with *G. intraradices* might be considered as a new approach in the struggle for Pythium seedling rot disease. Present study showed that the combination of *G. intraradices* and zinc concentration reduced disease severity under pot conditions.

Conclusion

It can be concluded that improving Zn nutrition and application of mycorrhizal fungus to cucumber seedlings will be an environmentally safe approach in controlling Pythium rot disease.

Acknowledgement

This study was financially supported by the TÜBİTAK (The Scientific and Technological Research Council of Turkey), Project No. 112O025.

References

Agrios GN (1997). Plant pathology, 4th ed. Academic Press, San Diego.

Akköprü A, Demir S (2006). Biological control of Fusarium wilts in tomato caused by *Fusarium oxysporum* f. sp. *lycopersici* by AMF *Glomus intraradices* and some rhizobacteria. J Phytopathol 153:544-550.

Braithwaite M, Falloon RE, Genet RA, Wallace AR, Fletcher JD, Braam, WF (1994). Control of powdery scab of patatoes with chemical seed tuber treatments. New Zeal J Crop Hort. 22: 121-128.

Cai J, Huang B, Chen J, Ji Q (2007). Effects of mineral elements on growth and resistance to Phytophthora blight of pepper. Jiangsu J Agric Sci 23:46-49.

Caron M, Fortin JA, Richard J (1986). Effect of *Glomus intraradices* on infection by *Fusarium oxysporum* f. sp. *radicis-lycopersici* tomatoes over a 12-week period. Can J Bot 64:552-556.

Çığşar S, Sarı N, Ortaş İ (2000). Hıyarda Vesiküler-Arbüsküler Mikorizanın Bitki Büyümesi ve Besin Maddeleri Alımı Üzerine Etkileri. Türk J Agric For 24:571-578.

Davidson JA, Ramsey MD (2000). Pea yield decline syndrome in South Australia: The role of diseases and impact of agronomic practices. Aust J Agr Res 51:347-354.

Duffy B (2009). *Zinc and Plant Disease.* Mineral Nutrition and Plant Disease. In: Mineral Nutrition and Plant Disease. The American Phytopathotological Society St. Paul, pp. 155-175. Minnesota, USA.

Grewal HS (2001). Zinc influences nodulation, disease severity, leaf drop and herbage yield of alfalfa cultivars. Plant and Soil 234:47-59.

Harley JL, Smith SE (1983). Mycorrhizal wheat under moisture stress condition. Indian J. Symbiosis. Academic Press, London.

Jackson ML (1973). Soil Chemical Analysis, Advanced Course: 2nd. ed., Publ. by the author, Dept. Soil Science, Univ. Wisc., Madison, Wisconsin, 895 pp.

Kacar B, İnal A (2008). Plant Analysis. Nobel, 1241 Ankara, Turkey.

Kalim S, Luthra YP, Gandhi SK (2003). Role of zinc and manganese in resistance of cowpea root rot. J Plant Dis Prot 110:235-243.

Karagiannidis N, Bletsos F, Stavropoulosi N (2002). Effect of Verticillium wilt (*Verticillium dahliae* Kleb.) and mycorrhizae (*Glomus mosseae*) on root colonization, growth and nutrient uptake in tomato and eggplant seedlings. Scientia Horticulturae 94:145-156.

Khoshgoftarmanesh A, Kabiri S, Shariatmadari HB, Sharifnabi B, Schulin R (2010). Zinc nutrition effect on the tolerance of wheat genotypes to Fusarium root rot disease in a solution culture experiment. Soil Sci Plant Nutr 56:234-243.

Knudsen D, Peterson GA, Pratt PF (1982). *Lithium, Sodium and Potassium.* In: 'Methods of Soil Analysis. Chemical and Microbiological Properties'. Agronomy no. 9, Part 2. 2nd edn (Ed. AL Page) pp. 225-246.

Kostandi SF, Soliman MF, Ghaly AA (1997). Smut disease and yield performance in corn (*Zea Glays* L.) as influenced by nitrapyrin, urea and zinc applications in coarse- textured soils. J. Agron. Crop Sci. 179:219-226.

Krishna KR, Bagyaraj DJ (1983). Interaction between *Glomus fasciculatum* and *Sclerotium rolfsii* in peanut. Can J Bot 61:2349-2351.

Kumar M, Yadav K, Thakur SK, Mandal K (1998). Effect of vasicular-arbuscular mycorrhizal fungi and *Rhizobium* inoculation on nodülation, root colonization, nitrogen fixation and yield to chick-pea. J Indian Soc Soil Sci 46:375-378.

Küçükyumuk Z, Özgönen H, Erdal İ (2013). Influence of zinc and mycorrhizal fungus *Glomus intraradices* on plant yield, nutrient concentrations and control of Phytophthora blight of pepper. J Food Agric Environ 11:807-812.

Lindsay WL, Norvell WA (1978). Development of a DTPA soil test for zinc, iron, manganese and copper. Soil Sci Soc Am J 42:421-428.

Marschner H, Dell B (1994). Nutrient uptake in mycorrhizal symbiosis. Plant and Soil 159: 89-102.

Marschner H (1995). Mineral Nutrition of Higher Plants, 2nd edn. AcadeGlic Press, London, United Kingdom.

Matsubara Y, Tamura H, Harada T (1995). Growth enhancement and Verticillium wilt control by vesicular-arbuscular mycorrhizal fungus inoculation in eggplant. J Jpn Soc Hortic Sci 64:555-561.

Olsen SR, Cole V, Watanabe FS, Dean LA (1954). Estimations of Available Phosphorus in Soils by Extractions with Sodium Bicarbonate. U.S. Dept. Of Agric. Cric. 939-941.

Özcan H, Taban S (2000). VA-Mycorrhiza'nın alkalin ve asit toprakla yetiştirilen mısır bitkisinin gelişimi ile fosfor, çinko, demir, bakır ve mangan konsantrasyonları üzerine etkisi. Türk J Agric Tübitak 24:629-635.

Özgönen H, Erkılıç A (2007). Growth enhancement and Phytophthora blight (*Phytophthora capsici* Leonian) control by arbuscular mycorrhizal fungal inoculation in pepper. Crop Prot. 26:1682-1688.

Pareek S, Pareek S (1999). Effect of macro- and micro-nutrients on charcoal rot disease development of maize induced by *Macrophomina phaseolina.* Annals of Agriculture. Research 20:129-131.

Siddiqui IA, Shaukat SS, Hamid M (2002). Role of zinc in rhizobacteria-mediated suppression of root-infecting fungi and root-knot nematode. J. Phytopathol 150:569-575.

Streeter TC, Rengel Z, Neate SM, Graham RD (2001). Zinc fertilization increases tolerance to *Rhizoctonia solani* (AG 8) in *Medicago truncatula.* Plant and Soil 228:233-242.

Trotta A, Varese GC, Gnavi E, Fusconi A, Sampo S, Berta G (1996). Interactions between the soilborne root pathogen *Phytophthora nicotianae* var. *parasitica* and the arbuscular mycorrhizal fungus *Glomus mosseae* in tomato plants. Plant and Soil 185:199-209.

Tüfenkçi S, Sönmez F, Sensoy R (2005). Effects of arbuscular mycorrhiza fungus inoculation and phosphorous and nitrogen fertilizations on some plant growth parameters and nutrients content of chickpea. Bull Pure Appl Sci 24:71-80.

Uçgun K, Gezgin S (2013). Relationship between disease and macro nutrients www.marim.gov.tr. Date of Access: 30.07.2013.

Zhang W, Dick WA, Hoitink HAJ (1996). Compost-induced systematic acquired resistance in cucumber to Pythium root rot and anthracnose . Phytopathology 86:1066-1070.

16

Chemical Composition of Celandine (*Chelidonium majus* L.) Extract and its Effects on *Botrytis tulipae* (Lib.) Lind Fungus and the Tulip

Marcel PARVU[1*], Laurian VLASE[2], Laszlo FODORPATAKI[3], Ovidiu PARVU[4], Oana ROSCA-CASIAN[5], Csaba BARTHA[3], Lucian BARBU-TUDORAN[6], Alina Elena PARVU[7]

[1]*Babes-Bolyai University, Faculty of Biology and Geology, Department of Biology, 42 Republicii St, 400015 Cluj-Napoca, Romania; marcel.parvu@ubbcluj.ro (*corresponding author)*

[2]*Iuliu Hatieganu University of Medicine and Pharmacy, Faculty of Pharmacy, Department of Pharmaceutical Technology and Biopharmaceutics, 12 Ion Creanga St, 400010 Cluj-Napoca, Romania; laurian.vlase@yahoo.com*

[3]*Babes-Bolyai University, Hungarian Department of Biology and Ecology, 1 Mihail Kogalniceanu St, 400084 Cluj-Napoca, Romania; lfodorp@gmail.com; barthacsabi@gmail.com*

[4]*Babes-Bolyai University, Faculty of Mathematics and Computer Science, 1 Mihail Kogalniceanu St, 400084 Cluj-Napoca, Romania; ovidiu.parvu@gmail.com*

[5]*Babes-Bolyai University, A. Borza Botanical Garden, 42 Republicii St, 400015 Cluj-Napoca, Romania; casioana@yahoo.com*

[6]*Babes-Bolyai University, Electron Microscopy Center, 5-7 Clinicilor St, 400006 Cluj-Napoca, Romania; lucianbarbu@yahoo.com*

[7]*Iuliu Hatieganu University of Medicine and Pharmacy, Faculty of Medicine, Department of Pathophysiology, 3 Victor Babes St, 400012 Cluj-Napoca, Romania; parvualinaelena@yahoo.com*

Abstract

In this study, the content of chelidonine and berberine alkaloids, and sterols and phenols in the *Chelidonium majus* plant extract were analyzed. Subsequently, the effects of the extract on the germination and growth of *Botrytis tulipae* fungus on nutritive medium were compared to the effects of fluconazole. The plant extract was used at the minimum inhibitory concentration on *B. tulipae* developed in tulip leaves and the *in vivo* effects were investigated. The influence of different concentrations of *C. majus* extract on the physiological processes of the tulip (gas exchange parameters, photosynthetic light use efficiency, and induced chlorophyll fluorescence) were also tested to assess the applicability of the extract for the protection of ornamental plants against fungal infection. Our results demonstrated that 2% celandine extract does not significantly change the gas exchange parameters (transpiration rate, carbon dioxide uptake, and stomatal conductivity) of leaves exposed for 2 h, and does not interfere with the photochemical processes in the leaves. However, in higher concentrations, it increases the transpiration rate and net carbon dioxide influx. At concentrations of 15% and 20%, the extract lowers the potential quantum yield efficiency of photosystem II and the vitality index of the photosynthetic apparatus. Therefore we recommend the use of lower concentrations (≤6%) of celandine extract for the biological protection of tulips against gray mold.

Keywords: alkaloids, antifungal action, chlorophyll fluorescence, electron microscopy, leaf gas exchange, tulip fire

Introduction

The *Botrytis* genus comprises over 20 species (Beever and Weeds, 2007), and *Botrytis* diseases are one of the most common and widely distributed; they have been identified on vegetables, ornamentals, fruits, and some field crops worldwide (Agrios, 2005). The members of this genus include *Botrytis cinerea* Pers., *Botrytis allii* Munn, *Botrytis fabae* Sardina, *Botrytis paeoniae* Oudem., and *Botrytis tulipae* (Lib.) Lind (Elad *et al.*, 2007).

Botrytis blight, which is also known as tulip fire, or tulip mold, is the most common and destructive disease to tulips, and is caused by the fungus *B. tulipae* (Hong *et al.*, 2002; Staats *et al.*, 2007). The fungus attacks all parts of the tulip and can rapidly kill its host's tissue and continue growing on the dead remains (Webster and Weber, 2007). *B. tulipae* produces abundant gray mycelium and long, branched conidiophores with one-celled, ovoid conidia. The conidiophores and clusters of conidia form a grape-like cluster (Agrios, 2005; Webster and Weber, 2007).

The ability of *B. tulipae* to infect living host plants may result from a combination of at least 4 factors: (1) possession of pathogenic factors (e.g., toxins and cell-wall degrading enzymes) that confer the ability to kill and invade plant tissue; (2) the ability to avoid or counteract plant resistance mechanisms; (3) the ability to survive outside host-plant tissue under less favorable conditions (e.g., low humidity and UV irradiation); and (4) the ability to reproduce and disperse (Staats *et al.*, 2005).

Because tulips occupy an important position among flowering plants cultivated worldwide and because gray mold is present each year in the crop, protection measures against tulip fire are vital (Agrios, 2005). The application of fungicides to control gray mold of plants is frequently used; however, the control of *Botrytis* in the field through chemical sprays is only partially successful, especially in cool, damp weather. Indeed, *Botrytis* strains resistant to several systemic fungicides, as well as some resistant to broad-spectrum fungicides have been found in various crops sprayed with these chemicals (Agrios, 2005; Webster and Weber, 2007). Plant fungicides based on synthetic chemicals cause severe and long-term environmental pollution, are highly and acutely toxic, and are carcinogenic to humans and animals (Strange and Scott, 2005). In addition, pathogens may become resistant to many of these chemicals. Consequently, the aim of new antifungal strategies is to develop drugs that combine low cost with sustainability, high efficacy, restricted toxicity, and increased safety for humans, animals, host plants and ecosystems. Biological control has become popular worldwide because fungicides of biological origin are biodegradable and have been demonstrated to be specifically effective against target organisms (Barker and Rogers, 2006; Carrillo-Munoz *et al.*, 2006; Fatehi *et al.*, 2005; Strange and Scott, 2005; Ienaşcu *et al.*, 2008).

Therefore, identifying new methods to control gray mold is an important requirement, in the protection of cultivated plants. In particular, the biological control of *Botrytis* species is very important and may be done via a variety of methods, which include the use of microbial antagonists (Elad and Stewart, 2004), and plant extracts (Choi *et al.*, 2004; Pârvu and Pârvu, 2011; Wilson *et al.*, 1997). Plants are rich in a wide range of bioactive secondary metabolites such as tannins, terpenoids, alkaloids, and flavonoids that are reported to have *in vitro* antifungal properties. In addition, a series of molecules that possess antifungal activity against different strains of fungus have been found in plants. These molecules may be directly used or exploited as models to develop better molecules (Arif *et al.*, 2011).

The *B. tulipae* fungus is found every year on tulips from Cluj-Napoca, Romania. We have studied the *in vitro* and *in vivo* effects of *C. majus* against gray mold on tulips because previous studies have shown that the *C. majus* extract has an antifungal effect (Matos *et al.*, 1999; Pârvu *et al.*, 2008) against phytopathogenic fungi. In brief, we determined the chemical composition of the *C. majus*, specifically, the content of chelidonine and berberine alkaloids, sterols, and polyphenols. In addition, the antifungal activity of *C. majus* on *B. tulipae* germination and growth was evaluated and the *in vivo* ultrastructural changes present in the tulip leaves attacked by tulip fire and treated with the *C. majus* plant extract at the minimum inhibitory concentration (MIC) for 2 h. Finally, we investigated the effects of different concentrations of *C. majus* extracts on the physiological processes of tulip plants, such as gas exchange parameters (transpiration rate, carbon dioxide uptake, and stomatal conductance) and efficiency indicators of photosynthetic light use revealed by induced *in vivo* chlorophyll fluorescence. The main aim of this study was to introduce the *C. majus* extract to the practice of biological disease management in tulip cultures.

Materials and methods

Plant material

Celandine (*C. majus* L.) was collected from the A. Borza Botanical Garden of Cluj-Napoca (46°45'36"N and 23°35'13"E) in April 2010 and was identified by Dr. M. Parvu, Babes-Bolyai University of Cluj-Napoca. A voucher specimen (CL 663 692) is deposited at the Herbarium of Babes-Bolyai University, Cluj-Napoca, Romania.

Preparation of fungal colony

B. tulipae (Lib.) Lind was isolated from a tulip (CL 663 693) and was identified in the Mycology Laboratory, Babes-Bolyai University, Cluj-Napoca, Romania, by Dr. M. Parvu. Colonies were obtained in Petri dishes on Czapek-agar medium (BD Difco, Budapest, Hungary), by inoculation in the central point with *B. tulipae* spore suspension ($1 \cdot 10^5$ conidia·mL^{-1}) and incubation at 22 °C for 5 days.

Preparation of alcoholic plant extract

Fresh *C. majus* herba (leaves, stems, and flowers fragments of 0.5 – 1.0 cm) was extracted with 70% ethanol (Merck, Bucuresti, Romania) in the Mycology Laboratory of Babes-Bolyai University, Cluj-Napoca, Romania by cold repercolation method (Mishra and Verma, 2009; Sundaram and Gurumoorthi, 2012), at room temperature, for 3 days (Sundaram and Gurumoorthi, 2012). The *C. majus* extract, containing 1 g plant material in 1 mL of 35% ethanol (w/v), was obtained by filtration. From this initial solution, dilutions were made with distilled water to obtain final concentrations of 2%, 6%, 10%, 15%, and 20%.

Chemical composition of the C. majus extract

Determinations of chelidonine and berberine alkaloids

A high-performance liquid chromatography method coupled with mass spectrometry (LC/MS) was accessed to quantify the amounts of berberine and chelidonine in the *C. majus* extract (Wu *et al.*, 2005).

The LC/MS system was an Agilent 1100 Series HPLC system (Agilent Technology Co., Ltd., USA) that consisted of a binary pump, degasser, autosampler, thermostat operating at 48 °C, VL ion trap detector, and a UV detector. Chromatographic separation was performed on a Zorbax SB-C18 column (100mm · 3.0mm i.d., 3.5μm; Agilent) proceeded by a 0.5 μm online filter.

The mobile phase consisted of acetonitrile and 0.1% (v/v) formic acid in water at 18:82 (v/v) and was delivered at a flow rate of 1 mL · min^{-1}. The autosampler injection volume was set at 10 µL. The mass spectrometer operated using an ESI source in the positive mode and was set for isolation and fragmentation of the berberine molecular ion with m/z = 336 and the chelidonine ion with m/z = 354.

The quantification of berberine was based on the sum of ions with m/z = 291.9 and 321.0 from the MS spectrum of the parent ion (Fig. 1a). Chelidonine was quantified based on the sum of the ions with m/z 275, 305, and 323 (Fig. 1b). The calibration curves were linear in the range of 6.8–68 ng · mL^{-1} for berberine and 14–140 ng · mL^{-1} for chelidonine, with a correlation coefficient greater than 0.997.

Identification and quantitative determinations of the polyphenols

A high-performance liquid chromatography method coupled with mass spectrometry (LC/MS) was used to analyze the polyphenolic compounds in the *C. majus* plant extract. The method used was a previously published HPLC method with minor changes (Nencu *et al.*, 2012; Compaore *et al.*, 2012; Meda *et al.*, 2011). The method is suitable for qualitative (18 compounds) and quantitative (14 compounds) analyses. In this study, 18 standards of the polyphenolic compounds were used, namely, caftaric acid, gentisic acid, caffeic acid, chlorogenic acid, paracoumaric acid, ferulic acid, sinapic acid, hyperoside, isoquercitrin, rutoside, myricetol, fisetin, quercitrin, quercetol, patuletine, luteolin, kaempferol, and apigenin.

Apparatus and chromatographic conditions

The experiments were performed using an Agilent 1100 HPLC Series system (Agilent) equipped with a degasser, binary gradient pump, column thermostat, autosampler, and UV detector. The HPLC system was coupled with an Agilent 1100 mass spectrometer (LC/MSD ion trap VL). For the separation, a reverse-phase analytical column was employed (Zorbax SB-C18 100 x 3.0 mm i.d., 3.5 µm particle); the temperature was 48 °C. The compounds were detected in both the UV and MS mode. The UV detector was set at 330 nm until 17.5 min, and then at 370 nm for the remainder of the experiment. The MS system used an electrospray ion source in the negative mode. The chromatographic data were processed using ChemStation and DataAnalysis software from Agilent. The mobile phase was a binary gradient prepared from methanol and a solution of 0.1% (v/v) acetic acid. The elution started with a linear gradient, beginning with 5% methanol and ending at 42% methanol, for 35 min; isocratic elution followed for the next 3 min with 42% methanol. The flow rate was 1 mL · min^{-1} and the injection volume was 5 µL.

Polyphenols

The MS signal was used only for qualitative analysis based on the specific mass spectra of each polyphenol. The MS spectra obtained from a standard solution of polyphenols were integrated in a mass spectra library. Subsequently, the MS traces/spectra of the analyzed samples were compared to spectra from the library, which allowed the positive identification of compounds, based on spectral matches. The UV trace was used for quantification of the identified compounds following MS detection. Using the chromatographic conditions described above, the polyphenols all eluted in less than 35 min (Tab. 1). Four polyphenols could not be quantified under the chromatographic conditions because of overlapping (caftaric acid with gentisic acid and caffeic acid with chlorogenic acid). However, all 4 compounds were selectively identified using MS detection (qualitative analysis) based on differences in their molecular mass and MS spectra. The detection limits were calculated as the minimal concentration required to produce a reproducible peak with a signal-to-noise ratio of >3. The quantitative determinations were performed using an external standard method. Calibration curves in the 0.5–50 µg · L^{-1} range with good linearity (R2 > 0.999) for a 5-point plot were used to determine the concentration of the polyphenols in the plant samples.

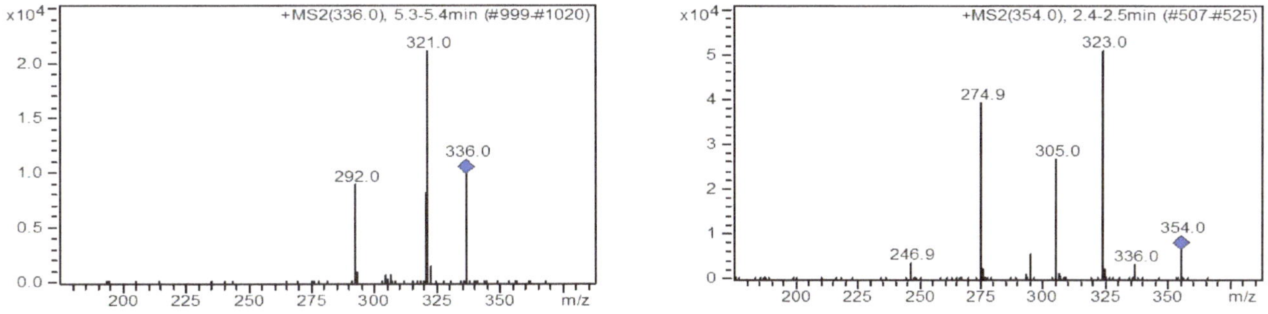

(a) (b)

Fig. 1. ESI/MS/MS spectra of berberine (a) and chelidonine (b)

Tab. 1. Retention times (tR) for the investigated polyphenols

Peak no.	Phenolic compound	$t_R \pm$ SD (min)	Peak no.	Phenolic compound	$t_R \pm$ SD (min)
1	Caftaric acid*	2.10 ± 0.06	11	Myricetin	20.70 ± 0.06
2	Gentisic acid*	2.15 ± 0.07	12	Fisetin	22.60 ± 0.15
3	Caffeic acid*	5.60 ± 0.04	13	Quercitrin	23.00 ± 0.13
4	Chlorogenic acid*	5.62 ± 0.05	14	Quercetol	26.80 ± 0.15
5	p-coumaric acid	8.7 ± 0.08	15	Patuletin	28.70 ± 0.12
6	Ferulic acid	12.2 ± 0.10	16	Luteolin	29.10 ± 0.19
7	Sinapic acid	14.3 ± 0.10	17	Kaempferol	31.60 ± 0.17
8	Hyperoside	18.60 ± 0.12	18	Apigenin	33.10 ± 0.15
9	Isoquercitrin	19.60 ± 0.10			
10	Rutoside	20.20 ± 0.15			

*overlapping UV peaks, qualitative analysis performed using MS detection

Identification and quantitative determinations of the sterols

The LC/MS technique was also used to analyze the sterols from the *C. majus* plant extract. The method used was a previously published HPLC method with minor changes (Sanchez-Machado *et al.*, 2004; Khalaf *et al.*, 2011). Three standards were used for the quantitative analysis, namely, beta-sitosterol, stigmasterol, and cholesterol.

Apparatus and chromatographic conditions

The analyses were performed using an Agilent 1100 HPLC Series system equipped with a G1322A degasser, G1311A binary pump, and G1313A autosampler. For the separation, we used a reverse-phased Zorbax SB-C18 analytical column (100 mm · 3.0 mm i.d., 5 μm particles) fitted with a precolumn Zorbax SB-C18, both operated at 40 °C. The mobile phase was prepared from methanol and acetonitrile 10:90 (v/v) isocratic elution. The flow rate was 1 mL · min^{-1} and the injection volume was 4 μL. All solvents were filtered through 0.5-mL Sartorius filters and degassed using ultrasound. MS/MS detection using multiple reaction monitoring (MRM) of specific daughter ions was used for each sterol. The HPLC was coupled with an Agilent ion trap 1100 VL mass detector, equipped with an atmospheric pressure chemical ionization (APCI) interface working in the positive ion mode. The operating conditions were: nitrogen gas, flow rate of 7 L · min^{-1}, heater at 400 °C, ion source temperature of 250 °C, nitrogen nebuliser at 50 psi, and capillary voltage of 4000 V. All chromatographic data were processed using ChemStation software and Data Analysis from Agilent.

Sterols

Under our chromatographic conditions, the retention times of the 5 analyzed sterols were 3.2 min for ergosterol, 3.9 min for brassicasterol, 4.9 min for both stigmasterol and campesterol (co-elution), and 5.7 min for beta-sitosterol. The ions monitored by the MRM method were ≤379 (253.3, 295.3, and 309.3) for ergosterol, ≤381 (201.3, 203.3, 215.2, and 217.3) for brassicasterol, ≤383 (147.3, 149.3, 161.3, and 175.3) for campesterol, ≤395 (163.3, 173.2, 187.3, 199.3, and 227.2) for stigmasterol and ≤397 (160.9, 174.9, 188.9, 202.9, and 214.9) for sitosterol. The quantitative experiments were performed using an external standard method. Calibration curves in the 60–3000 ng · mL^{-1} range with good linearity (R2 > 0.99) for a 7-point plot were used to determine the concentration of the sterols in the plant samples.

Determination of antifungal activity

The antifungal activity of the *C. majus* extract, expressed as the MIC, was determined by the agar-dilution assay, and was compared to the antimycotic drug fluconazole (2 mg · mL^{-1}, Krka, Novo Mesto, Slovenia) and a control (nutritive medium and 35% ethanol). The percentage of mycelial growth inhibition (P) at each concentration was calculated using the formula $P = (C\text{-}T) \times 100/C$, where C is the diameter of the control colony and T is the diameter of the treated colony (Nidiry and Babu, 2005).

Statistical analysis

Statistical analyses were performed using the program R environment, version 2.14.1. The results for each group were expressed as mean ± standard deviation. Data were evaluated by analysis of variance (ANOVA). A P value of ≤ 0.05 was considered statistically significant. The correlation analysis was performed by the Pearson test. Measurements of physiological processes in tulip leaves treated for 2 h with different concentrations of *C. majus* extract were performed in triplicate, and the post-ANOVA Tukey HSD test was used to analyze the significance of differences between treatments and control.

In vivo effect of C. majus extract against B. tulipae

Fresh tulip leaves from the field that were infected by *B. tulipae* were sprayed with the plant extract of *C. majus* at the MIC (6 %) and compared to the control tulips leaves after 2 h.

The samples of the treated leaves and control leaves were examined by transmission electron microscopy (TEM) with a JEOL JEM 1010 electron microscope (Japan Electron Optics Laboratory Co., Tokyo, Japan). Conidia controls of *B. tulipae* isolated from the tulip leaf surface were examined by scanning electron microscopy (SEM) with a JEOL JSM 5510 LV electron microscope (Hayat, 2000).

Measurement of leaf gas exchange

Specific gas exchange parameters were measured using a Ciras-2 leaf gas-exchange system (PP Systems) and a PLC6 automatic leaf cuvette. The photon flux density was set to 500 $\mu M \cdot m^{-2} \cdot s^{-1}$, the air temperature in the leaf cuvette was 26 °C, the reference carbon dioxide concentration was 340 ppm, and the reference relative air humidity was 75%. Measurements of transpiration rate, net carbon dioxide uptake, and stomatal conductivity were performed at midday, on the fully expanded leaves of the tulip (*Tulipa gesneriana* cv. 'Rococo'). Three leaves from each plant were examined at 2 h after different concentrations of celandine extracts were sprayed in a thin continuous layer on the leaves. The leaves were maintained for the 2 h under constant environmental conditions created in a vegetation chamber (Pinheiro *et al.*, 2008).

Measurement of induced chlorophyll fluorescence parameters

The parameters of the induced chlorophyll *a* fluorescence were measured using a pulse amplitude modulated chlorophyll fluorometer (PAM-FMS2, Hansatech), on 3 leaves of each plant. The leaves were left in the dark for 10 min prior to the measurements to terminate all previous photochemical reactions. The modulated light was sufficiently weak (0.04 $\mu M \cdot m^{-2} \cdot s^{-1}$) so as not to produce any significant variable fluorescence. A single saturating flash (2,000 $\mu mol \cdot m^{-2} \cdot s^{-1}$ for 0.5 s) was applied to reach the maximal fluorescence Fm. After the decline of the signal, the actinic light was turned on (100 $\mu mol \cdot m^{-2} \cdot s^{-1}$) to induce the kinetics. The determined parameters were the initial fluorescence (F_0), the maximal fluorescence (Fm), the F_v/F_m ratio (F_v or variable fluorescence, which is the difference between the maximal and initial fluorescence), the modulated maximal fluorescence (F_m'), the steady state fluorescence (F_s), the effective quantum use efficiency (Φ) representing the ratio (F_m' – F_s)/F_m', as well as the vitality index (relative fluorescence decrease, Rfd) expressed as the ratio (F_m – F_s)/F_s (Baker, 2008; Bartha and Fodorpataki, 2007; Horvath *et al.*, 1996). The experimental conditions were identical to those for the leaf gas exchange measurements, and the same leaves were used for the determination of the gas exchange parameters.

Results

The chelidonine and berberine alkaloids, polyphenols, and sterols present in the *C. majus* plant extract were determined.

The chromatograms of chelidonine (Fig. 2) and berberine (Fig. 3) from the *C. majus* extract revealed the following levels: 26.09 $\mu g \cdot mL^{-1}$ as berberine base or 31.65 $\mu g \cdot mL^{-1}$ as berberine · HCl · $2H_2O$, and 304.62 $\mu g \cdot mL^{-1}$ as chelidonine base. The retention time for chelidonine was 2.4 min and 5.3 min for berberine.

The non-hydrolyzed sample (Fig. 4) contains rutoside (31.8 $\mu g \cdot mL^{-1}$), whereas the hydrolyzed sample was determined to contain: p-coumaric acid (4.05 $\mu g \cdot mL^{-1}$), ferulic

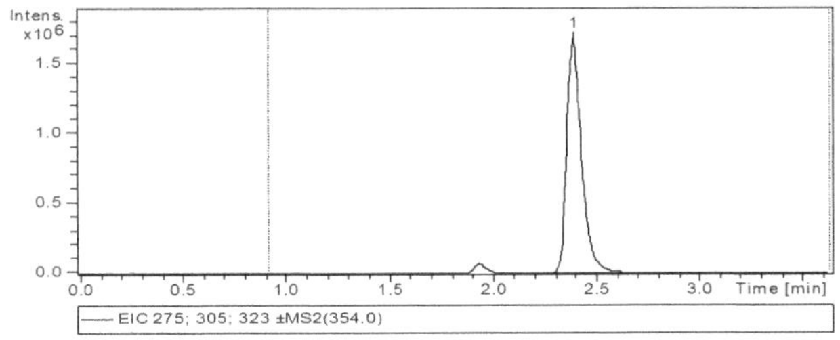

Fig. 2. Chromatogram of chelidonine from the *Chelidonium majus* extract. The chelidonine peak is marked "1"

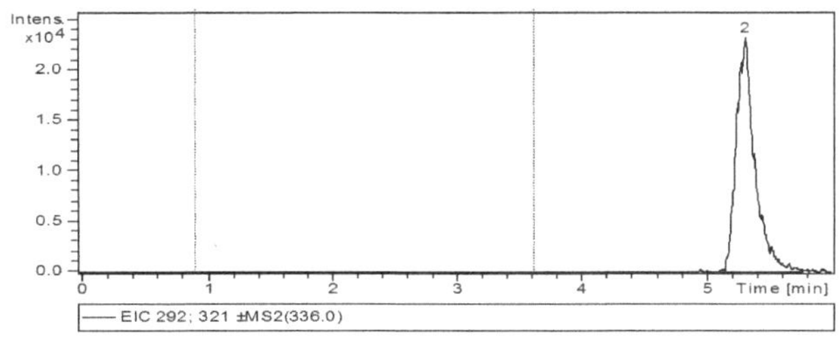

Fig. 3. Chromatogram of berberine from the *Chelidonium majus extract*. The berberine peak is marked "2"

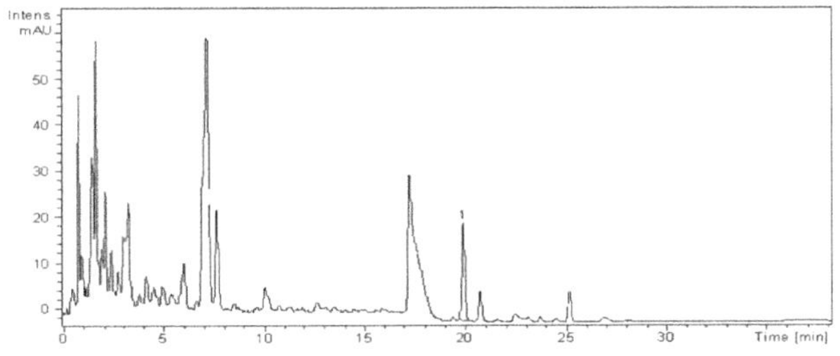

Fig. 4. Chromatogram of polyphenol rutoside from non-hydrolyzed sample of *Chelidonium majus* extract. The rutoside peak is marked "1"

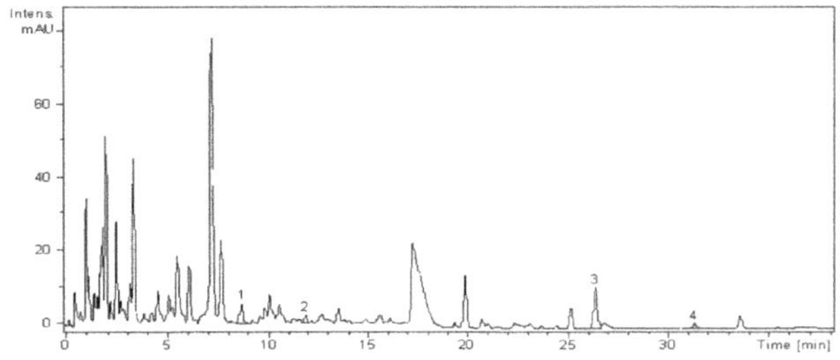

Fig. 5. Chromatogram of polyphenols from hydrolyzed sample of *Chelidonium majus* extract. The peaks are marked: "1" p-coumaric acid; "2" ferulic acid; "3" quercetol; "4" kaempherol

Tab. 2. *In vitro* effects of the *Chelidonium majus* extract on mycelial growth of *Botrytis tulipae* compared with the effects of the synthetic fungicide fluconazole

Chelidonium majus extract conc. (%)	*Botrytis tulipae*[a] Colony diameter (mm)	P[b](%)	Fluconazole conc. (%)	*Botrytis tulipae*[c] Colony diameter (mm)	P[d] (%)
C	62	0	C	62	0
1	60	3.22 ± 0.23	2	44	29.03 ± 0.23
2	46	25.80 ± 0.15	4	32	48.38 ± 0.15
3	30	51.61 ± 0.15	6	21	66.12 ±0.10
4	15	75.80 ± 0.15	8	12	80.64 ± 0.15
5	3	95.16 ± 0.16	10	4	93.54 ± 0.15
6	0	100 ± 0.21	12	0	100 ± 0.17

[a]Mycelial growth of *B. tulipae* at 5 days after inoculation in the presence of *C. majus*; [b]Inhibition % of radial growth in the presence of *C. majus*; [c]Mycelial growth of *B. tulipae* at 5 days after inoculation in the presence of fluconazole; [d]Inhibition % of radial growth in the presence of fluconazole;
C, 35% aq. EtOH;
Colony diameter is expressed as mean ± SE of 6 replicates

acid (0.81 · μgmL^{-1}), quercetol (7.88 μg · mL^{-1}), and kaempherol (1.21 μg · mL^{-1}) (Fig. 5).

The analyzed sample of *C. majus* extract contains stigmasterol (0.225 μg · mL^{-1}) and beta-sitosterol (0.191 μg · mL^{-1}).

The *C. majus* plant extract had a significant inhibitory effect on the mycelial growth of *B. tulipae* on culture medium. The *C. majus* MIC is 6% and the Fluconazole MIC is 12% (Tab. 2).

The *B. tulipae* control conidia were observed by electron microscopy. The SEM micrographs of the *B. tulipae* control revealed unicellular conidia with numerous randomly positioned protuberances (Fig. 6a). At the ultrastructural level, the *B. tulipae* control conidia contained a regular cell wall with a 2-layer structure, plasma membrane, cytoplasmic matrix with nucleus, and various cellular organelles, and lipids. The external cell wall layer was thin and electron dense, whereas the inner wall was thick, uniform, and less electron dense. The plasma membrane was tightly adhered to the cell wall. The cytoplasmic matrix (cytosol) was uniformly distributed, and the nucleus was ≤2 μm in diameter and ovoid or spherical (Fig. 6b)

The *B. tulipae* control hyphae were observed in the attacked tulip leaf below the cuticle (Fig. 7a), in the leaf mesophyll, and near the xylem vessel (Fig. 7b). At the ultrastructural level, *B. tulipae* appears to have septate hyphae with regular cell walls and plasma membranes, as well as cytoplasmic matrices with nuclei, mitochondria, endoplasmic reticulum, lipid bodies, and glycogen (Figs. 8a and b). The parasitic activity of *B. tulipae* destroyed the attacked leaf tissues (Figs. 9a and b).

When treated with *C. majus* plant extract at the MIC for 2 h, *B. tulipae* hyphae appeared damaged at the cellular level. Specifically, the organelles were partly and/or entirely destroyed, the cytoplasm was degenerated, and electron dense material appeared in the hyphal cells. In addition, the outside of the cell wall had an irregular shape and the plasma membrane was mostly destroyed and did not adhere to the cell wall. Furthermore, precipitation of the entire cytoplasm and destroyed organelles and nuclei were seen. Because of these effects, the morpho-functional relationship between the cell wall and cytoplasm was damaged and a less electron dense band was formed between the altered cytoplasm and cell wall (Figs. 9a and b).

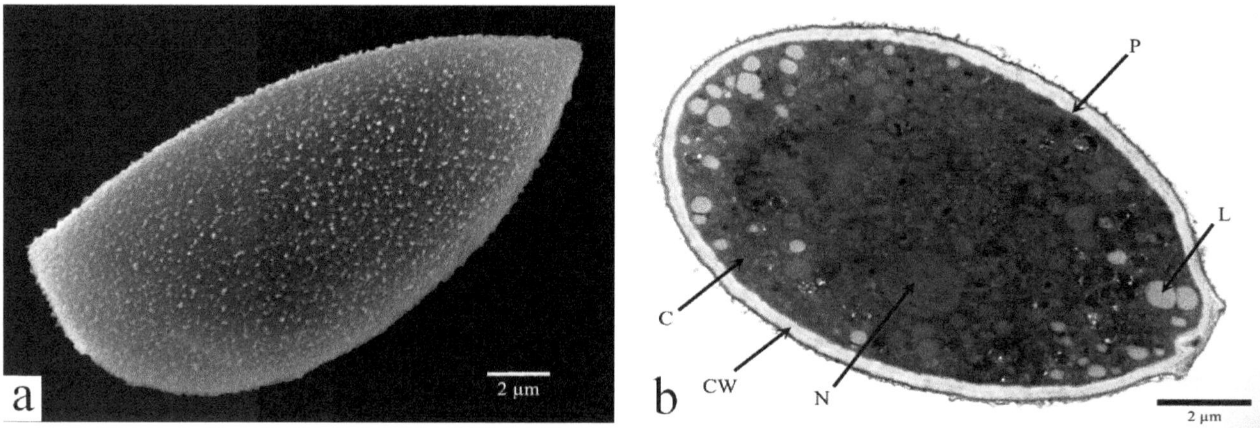

Fig. 6. Visualization of *Botrytis tulipae* conidium

a. Scanning electron micrograph showing protuberances on surface of cell wall. b. Transmission electron micrograph of an oblique section showing cell ultrastructure. CW cell wall; C cytoplasm; L lipids; N nucleus; P plasma membrane

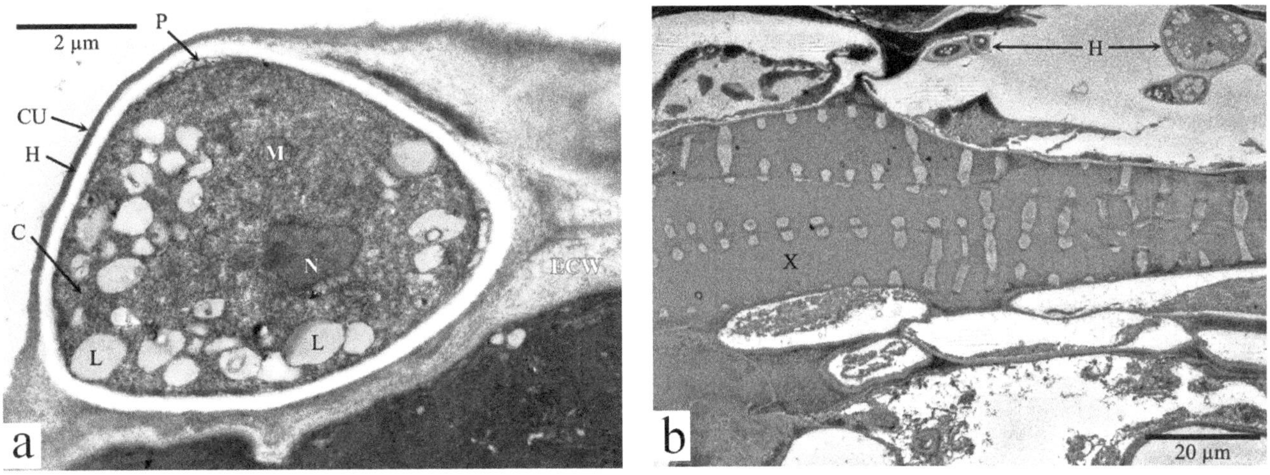

Fig. 7. Transmission electron micrograph of a tulip leaf cross section showing the *Botrytis tulipae* fungus

a. Hyphae (H) between the epidermal cell wall (ECW) and cuticle (CU) of the epidermis. b. Hyphae (H) in the leaf mesophyll, near the xylem (X). C cytoplasm; L lipids; M mitochondrion; N nucleus; P plasma membrane

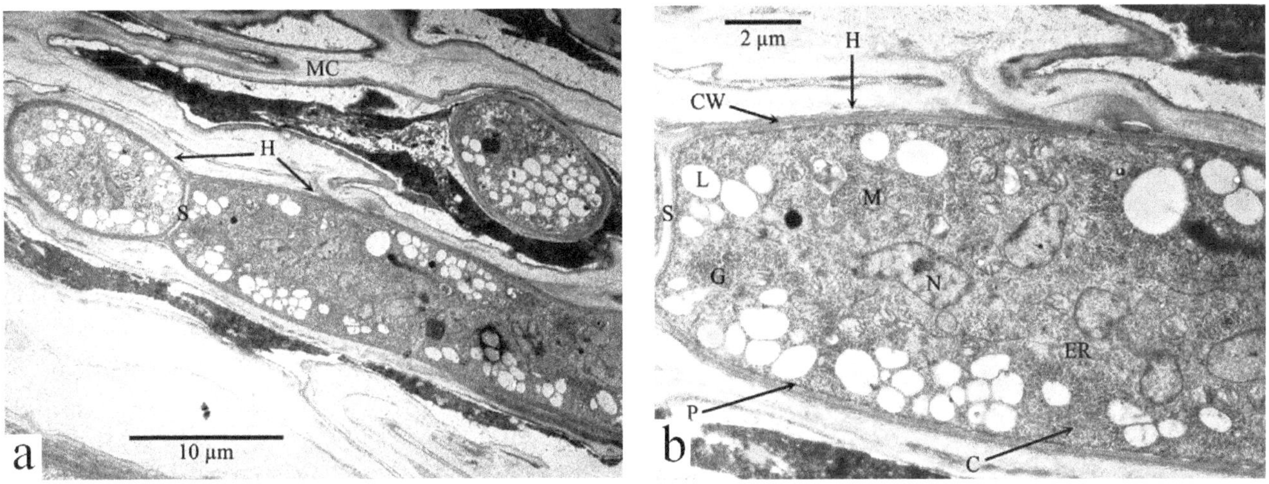

Fig. 8. Transmission electron micrograph of an oblique section of *Botrytis tulipae* control hyphae from the tulip leaf

a. Septate hyphae (H) in the mesophyll cells (MC). b. Detailed micrograph of septate hyphae (H) in the mesophyll. C cytoplasm; CW cell wall; ER endoplasmic reticulum; G glycogen; L lipids; M mitochondrion; N nucleus; P plasma membrane; S septum

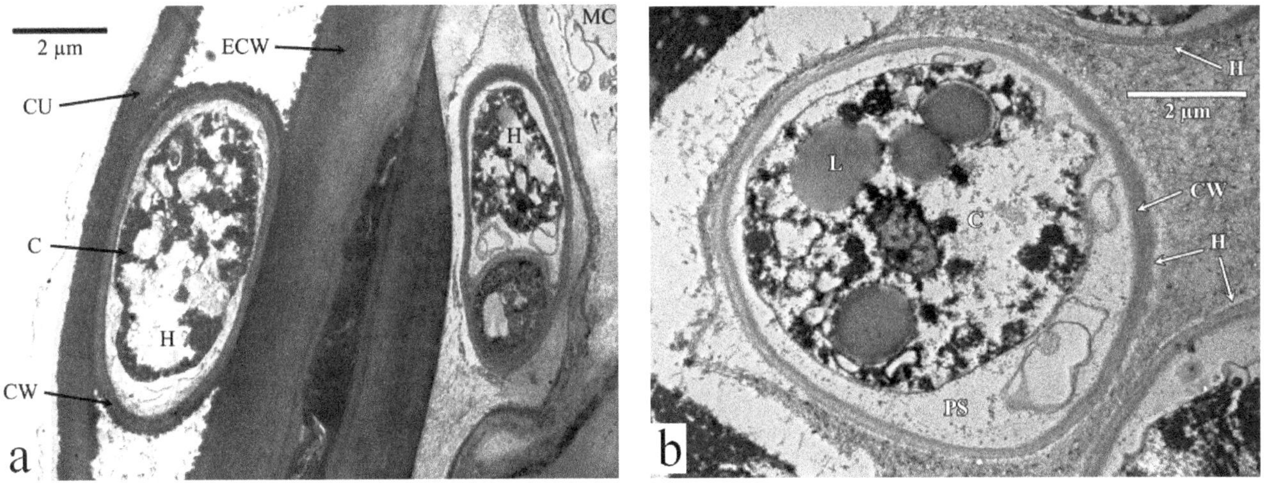

Fig. 9. Transmission electron micrograph of a tulip leaf cross section showing irreversible ultrastructural changes in *Botrytis tulipae* hyphae treated with *Chelidonium majus* plant extract at the MIC

a. Hyphae between the epidermal cell wall (ECW) and cuticle (CU) of the epidermis and hyphae (H) in the mesophyll cells (MC). b. hyphae (H) in the mesophyll cells (MC). C cytoplasm; CW cell wall; L lipids; PS. periplasmic space

The ground chlorophyll fluorescence was not significantly affected when the tulip leaves were covered for 2 h with solutions of 2%, 6%, and 10% extracts of celandine (Fig. 10).

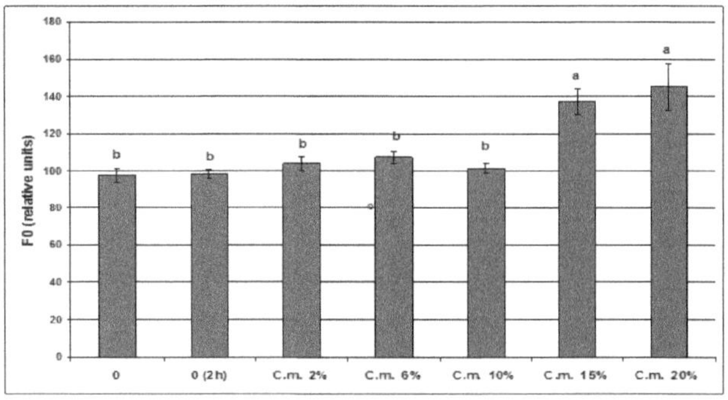

Fig. 10. Ground chlorophyll fluorescence (F_0) in the dark-adapted tulip leaves treated for 2 h with different concentrations of ce-landine (C.m.) extract

0, control leaves; 0 (2 h), control leaves after 2 h. Bars represent the standard error obtained from 3 independent experiments. The letters indicate significant differences at $p \leq 0.05$ according to the Tukey HSD test

The maximal chlorophyll fluorescence is more sensitive than the ground fluorescence, and its values were increased in the presence of 2% and 6% extracts, and decreased by 10%, 15%, and 20% celandine extracts (Fig. 11).

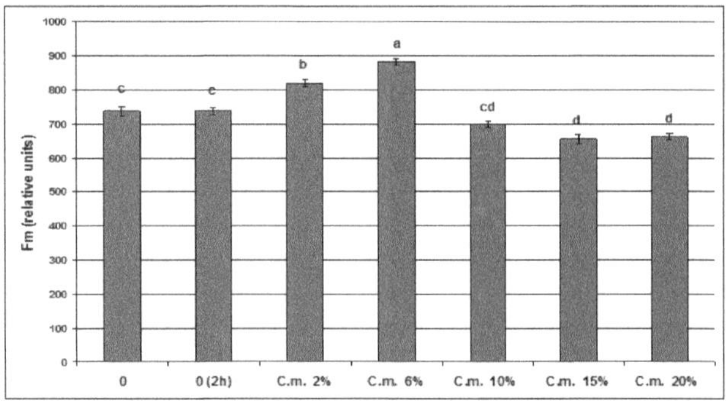

Fig. 11. Maximal chlorophyll fluorescence (F_m) in the dark-adapted tulip leaves treated for 2 h with different concentrations of celandine (C.m.) extract

0, control leaves; 0 (2 h), control leaves after 2 h. Bars represent the standard error obtained from 3 independent experiments. The letters indicate significant differences at $p \leq 0.05$ according to the Tukey HSD test

The potential quantum yield efficiency of photosystem II was not affected by the extract when it was sprayed on the tulip leaves at concentrations ≤ 15% (Fig. 12).

The vitality index of the photosynthetic apparatus starts to decrease significantly when the extract concentration is ≥ 10% (Fig. 13).

Progressive and statistically significant increases in all 3 gas exchange parameters (transpiration rate, net carbon dioxide uptake, and stomatal conductivity) were registered following treatment of the tulip leaves with concentrations of celandine extracts of ≥ 6% for 2 h. When the extract was used at concentrations of 2% no significant effects were noted on the investigated gas exchange parameters (Figs. 14-16).

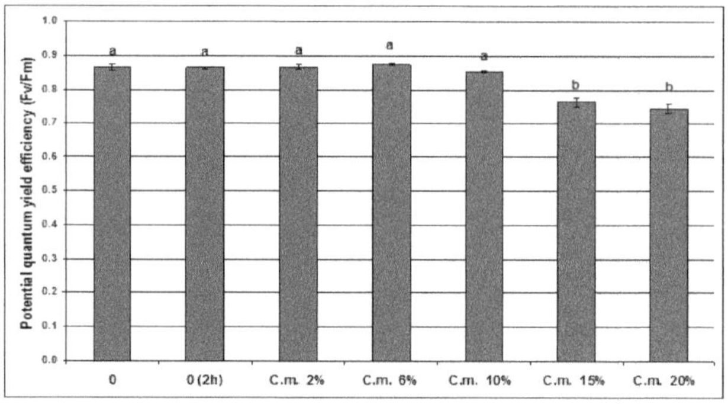

Fig. 12. Potential quantum yield efficiency of photosystem II

Values are based on the ratio between the variable and maximal chlorophyll fluorescence (F_v/F_m) in the dark-adapted tulip leaves treated for 2 h with different concentrations of celandine (C.m.) extracts. 0, control leaves; 0 (2 h), control leaves after 2 h. Bars represent the standard error obtained from 3 independent experiments. The letters indicate significant differences at $p \leq 0.05$ according to the Tukey HSD test

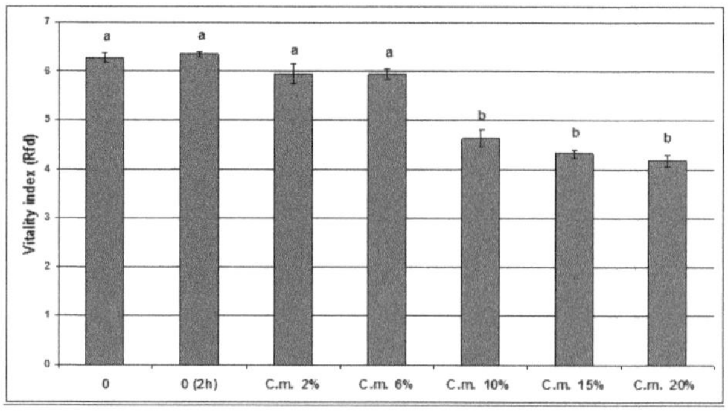

Fig. 13. Vitality index of the photosynthetic apparatus

Values are based on the relative chlorophyll fluorescence decrease (R_{fd}) in the tulip leaves treated for 2 h with different concentrations of celandine (C.m.) extracts. 0, control leaves; 0 (2 h), control leaves after 2 h. Bars represent the standard error obtained from 3 independent experiments. The letters indicate significant differences at $p \leq 0.05$ according to the Tukey HSD test

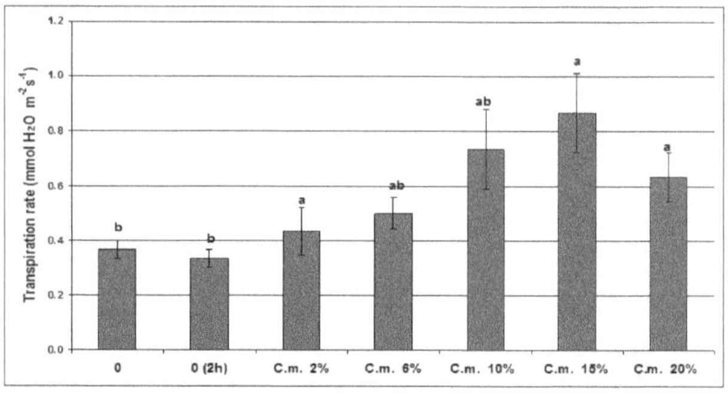

Fig. 14. Transpiration rate of the tulip leaves treated for 2 h with different concentrations of celandine (C.m.) extract

0, control leaves; 0 (2 h), control leaves after 2 h. Bars represent the standard error obtained from 3 independent experiments. The letters indicate significant differences at $p \leq 0.05$ according to the Tukey HSD test

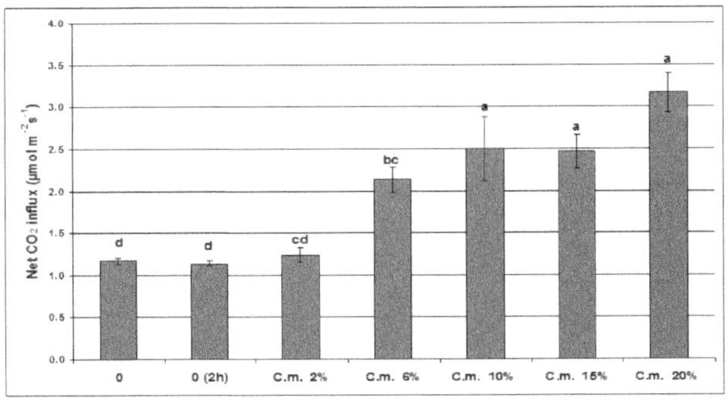

Fig. 15. Net carbon dioxide uptake by tulip leaves treated for 2 h with different concentrations of celandine (C.m.) extract

0, control leaves; 0 (2 h), control leaves after 2 h. Bars represent the standard error obtained from 3 independent experiments. The letters indicate significant differences at $p \le 0.05$ according to the Tukey HSD test

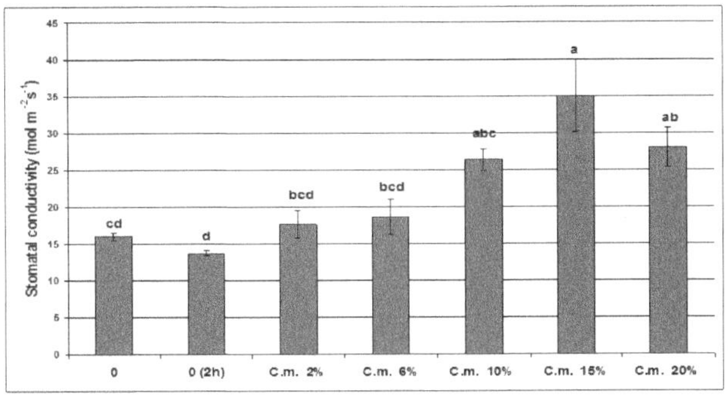

Fig. 16. Stomatal conductivity of the tulip leaves treated for 2 h with different concentrations of celandine (C.m.) extract

0, control leaves; 0 (2 h), control leaves after 2 h. Bars represent the standard error obtained from 3 independent experiments. The letters indicate significant differences at $p \le 0.05$ according to the Tukey HSD test

Discussion

B. tulipae is the only *Botrytis* species able to infect tulip (Staats *et al.*, 2005) and to abundantly produce sporulating gray mycelium on infected tissue (Yohalem *et al.*, 2003). The mitotically produced spores, macroconidia, can be transported long distances by wind (Agrios, 2005; Webster and Weber, 2007). This parasitic fungus overwinters in the soil as mycelium in decaying plant debris and as sclerotia, which are melanized mycelial survival structures (Agrios, 2005; Yohalem *et al.*, 2003).

The *B. tulipae* conidia are ellipsoidal or obovoid, unicellular (Hong *et al.*, 2002) and have numerous randomly positioned protuberances (Fig. 6a); however, these protuberances are fewer than those present in *B. cinerea* conidia (Pârvu *et al.*, 2008). Hydration and redrying causes these protuberances to disappear (Doss *et al.*, 1997). The cell wall of the conidia has 2 layers and appears dark (Fig. 6b) because of melanin, which protects the spores from enzyme action and probably UV radiation (Epton and Richmond, 1980).

B. tulipae fungus penetrates tulip leaves and produces irreversible ultrastructural changes in epidermal and mesophyll cells (Figs. 7 and 8). The ultrastructure of xylem cells (Fig. 7b) is not affected by the fungus and the fungus hyphae appear only in the leaf mesophyll tissues.

The infection of host plants by *B. tulipae* is mediated by numerous extracellular enzymes and metabolites. Each of these compounds plays a role in different stages of the infection process. Cutinases, lipases, and some cell wall-degrading enzymes facilitate penetration of the host surface, whereas toxins, oxalate, and reactive oxygen species enable host cell death. Several cell wall-degrading enzymes contribute to the conversion of host tissue into fungal biomass, and other enzymes such as laccases and proteases are involved in pathogenesis (Kars and van Kan, 2007).

Fungicide-resistant *Botrytis* strains have been identified in various crops (Agrios, 2005). In addition, plant fungicides based on synthetic chemicals are both pollutants and toxic (Barker and Rogers, 2006; Carrillo-Munoz *et al.*, 2006; Fatehi *et al.*, 2005; Strange and Scott, 2005; Ienaşcu *et al.*, 2008). Therefore, the biological control of *Botrytis* fungi with plant extracts is one of the important measures for enhancing farming techniques.

An important objective of our study was to test the *in vitro* action of *C. majus* extract on mycelium growth of *B. tulipae* and determine the MIC of the plant extract (Tab. 2). Previous studies suggested that *C. majus* possesses antifungal properties, and therefore, this extract is a promising source of active compounds against fungi such as *Fusarium* spp. (Matos *et al.*, 1999), *B. cinerea* (Pârvu *et al.*, 2008), and *Candida* species (Meng *et al.*, 2009).

Our study examined leaves attacked by gray mold and treated with *C. majus* extract at the MIC to demonstrate the *in vivo* inhibitory properties of the extract against *B. tulipae* (Fig. 9). The *C. majus* plant extract caused irreversible ultrastructural changes that abolished the cell wall's barrier function and its ability to activate cell wall-bound enzymes. The morpho-functional integrity of fungal cell components is required for viability and germination capacity (Isaac, 1992). The *B. tulipae* hyphae treated with *C. majus* extract revealed precipitation of the cytoplasm and destruction of organelles and nuclei that caused loss of viability and germination capacity.

In addition to the ultrastructural changes observed in *B. tulipae* hyphae treated with the plant extract (Fig. 9), the antimicrobial compounds from the *C. majus* extract induced important changes at the molecular level. The *C. majus* extract contains a large number of alkaloids and polyphenols, and is therefore, known for its antimicrobial activity (Meng *et al.*, 2009; Nawrot *et al.*, 2007; Zuo *et al.*, 2011). The main alkaloids identified in *C. majus* extracts are chelidonine, chelerythrine, sanguinarine, coptisine, and berberine (Sárközi *et al.*, 2006a; Zuo *et al.*, 2011). Formaldehyde formation due to demethylation is responsible for the antimicrobial activity of these alkaloids (Sárközi *et al.*, 2006b).

The alkaloids berberine, chelidonine, chelerythrine, sanguinarine, and coptisine (Wink, 1998) from the *C.*

majus plant extract induce changes at the molecular level in *B. tulipae* hyphae. Specifically, alterations involve disturbed DNA/RNA and related enzymes, as well as alterations to the cytoskeleton, ribosomal protein biosynthesis, and membrane permeability (Wink, 1998; Wink, 2008; Rosenkranz and Wink, 2008). The alkaloid berberine is present in *C. majus* extracts and *Berberis* extracts (Sárközi *et al.*, 2006a; Zuo *et al.*, 2011). Berberine inhibits esterases, DNA and RNA polymerases, cellular respiration, and acts in DNA intercalation (Aniszewski, 2007).

The alkaloids from *C. majus* are poisonous to *B. tulipae* fungus because they inhibit processes like DNA replication and RNA transcription that are vital for the microorganism (Wink, 1998).

Other antifungal compounds identified in the *C. majus* extract were the phenolic compounds rutoside, p-coumaric acid, ferulic acid, quercetol, and kaempherol (Tab.1). The mechanisms of action thought to be responsible for phenolic toxicity involve enzyme inhibition by the oxidized compounds, possibly through reaction with sulfhydryl groups or nonspecific interactions with the proteins (Arif *et al.*, 2009). In addition, antifungal phenolics from plants with action against phytopathogenic fungi *B. cinerea*, *Cercospora beticola*, *Colletotrichum circinans*, *Cladosporium herbarum*, *Fusarium oxysporum*, *Phytophthora infestans*, *Venturia inaequalis*, *Verticillium albo-atrum* have been identified (Lattanzio *et al.*, 2006).

The other antimicrobial compounds identified in the *C. majus* extract were the sterols stigmasterol (0.225 $\mu g \cdot mL^{-1}$) and beta-sitosterol (0.191 $\mu g \cdot mL^{-1}$). The most abundant plant sterols are sitosterol, campesterol, and stigmasterol (Moreau *et al.*, 2002), and the antifungal activity of plant sterols (Sharma and Kumar, 2009) and sterols from *Ganoderma annulare* mushroom have been described (Smania *et al.*, 2003). Moreover, free flavonoids and sterols of the *Tridax procumbens* plant extract completely inhibited spore germination of the *F. oxysporum* phytopathogenic fungus (Sharma and Kumar, 2009).

Our results reveal that the *C. majus* extract contains important antifungal compounds like alkaloids, phenols, and sterols. Importantly, these results clarify the antifungal activity of *C. majus* against phytopathogenic fungi (Matos *et al.*, 1999; Pârvu *et al.*, 2008) such as *B. tulipae*. The *C. majus* extract at MIC (6%) caused irreversible changes in *B. tulipae* hyphae, and therefore, *in vivo* studies examining the effect of the extract on tulips that have not been attacked by gray mold is required.

Whenever environmental stress factors directly or indirectly influence the energetic processes that occur during photosynthesis, they cause specific changes in the various parameters associated with induced chlorophyll fluorescence (Baker and Oxborough, 2004). This enables a quick *in situ* evaluation of alterations to photosynthesis in tulip leaves treated with antifungal celandine extracts. This is important because any agent used in pest management must be effective against parasites and minimally affect the vital processes of the host plant. *In vivo* induced chlorophyll fluorescence is a sensitive, non-destructive tool for the study of environmental impacts on the primary energy-conversion processes of photosynthesis. The potential quantum use efficiency of photosynthesis is reflected by the ratio between the variable and the temporary maximal fluorescence yield (F_v/F_m) in dark-adapted leaves. The F_v/F_m value is one of the most relevant functional markers of photosynthetic energy conversion, and therefore is used for detection of various stress factors that interfere with photochemical reactions in chloroplasts. Its drop below the value of 0.8 is directly related to the disturbed photochemical reactions that occur in photosystem II of thylakoid membranes in chloroplast, which leads to a less efficient photosynthetic use of the absorbed light energy. The initial fluorescence of the dark-adapted leaves (F_0), which is induced by a very weak red light flash, is related to the organization and energy transfer capacity of the light-harvesting antennae. The maximal fluorescence (F_m), which is generated by a flash of saturating red light, is related to the activity of the electron acceptors of photosystem II. One of the most sensitive parameters of induced chlorophyll fluorescence is the relative fluorescence decrease (R_{fd}), which is also known as the vitality index. This value is dependent on the difference between the temporary maximal fluorescence yield in dark-adapted samples and the steady state fluorescence level in constantly illuminated samples. The pulse amplitude modulation of chlorophyll fluorescence is obtained by regular saturating flashes on a background of a constant actinic light (Baker, 2008).

Because the ground chlorophyll fluorescence of the dark-adapted leaves was not significantly affected, one can deduce that the organization and energy transfer function of the light-harvesting pigment antennae of the leaves were not impaired by extract concentrations ≤15%. The registered values of maximal chlorophyll fluorescence indicate that small amounts of the celandine extract slightly stimulate photochemical reactions on the acceptor side of photosystem II (e.g., reduction of quinine acceptors), whereas higher extract concentrations moderately inhibit them, without causing a dramatic decline in the process. The fact that potential quantum yield efficiency of photosystem II was not affected by the extract implies that the overall conversion of light energy into storable chemical energy is not altered when plants are treated with dilute extracts of celandine. This is also valid for the vitality index of the photosynthetic apparatus, which is more sensitive than quantum efficiency, and decreases only upon treatment with extract concentrations reaching or exceeding 10% (Figs. 10-13). Based on the parameters of induced chlorophyll fluorescence, one can state that the photosynthetic light conversion capacity of tulip leaves is not reduced by the application of celandine extracts at concentrations of 2% or 6%.

Gas exchange processes through the leaf surfaces may be directly influenced by any substance that is sprayed onto the leaves, and may thus, penetrate the cuticle or enter the leaf through open stomata. This can lead to disturbances in carbon dioxide supply for photosynthetic carbon assimilation, or to an impaired regulation of stomatal movements that can threaten the water equilibrium and inorganic nutrient uptake of the whole plant. Therefore, it is important to investigate the effect of depositing celandine extracts on leaves on gas exchange processes. In the water economy of plants, the most important gas exchange parameter is transpiration rate, whereas for the photosynthetic carbon assimilation, net carbon dioxide uptake is a crucial prerequisite. The overall dynamics of gas exchange on the leaf surface is indicated by stomatal conductivity. For example, the intensity of the gas exchange processes per unit leaf area strongly decreases during drought and salt stress, as well as under the influence of different air pollutants (Allen and Pearcy, 2000; Medrano *et al.*, 2002; Hetherington and Woodward, 2003).

The results of the gas exchange measurements suggest that if the celandine extract is sprayed on tulip leaves at concentrations higher than 2%, it may cause increased transpirational water loss, which may also be beneficial if excess water is present in the soil and the higher suction force enhances the uptake of inorganic nutrients from the soil, and may ensure a better carbon dioxide supply to the leaves through the more widely open stomata. No inhibition of gas exchange processes were detected even when the celandine extract was applied at higher concentrations.

Conclusions

The *C. majus* plant extract exhibited strong *in vitro* and *in vivo* fungicidal activity against *B. tulipae*. The *C. majus* plant extract at the MIC caused severe ultrastructural changes in the tulip leaf hyphae that lead to loss of viability. The *Botrytis* strains have a high resistance to conventional fungicides, and therefore, we propose that *C. majus* is a good *in vivo* biological treatment against fungal infections like gray mold (Pârvu *et al.*, 2008), tulip mold, and other species (Matos *et al.*, 1999). After 2 h of surface treatment under growth chamber conditions, low concentrations (2%) of celandine extract did not affect the main physiological parameters associated with leaf gas exchange and photosynthetic light-use efficiency. The transpiration rate, net carbon dioxide influx, and overall stomatal conductivity increased at concentrations greater than 6%, thereby indicating stimulated stomatal opening, which favors carbon assimilation but may impair water economy. The ground fluorescence of chlorophyll *a* indicated that light harvesting by photosynthetic antenna pigments was affected only by higher concentrations (15% and 20%) of celandine extract. The maximal chlorophyll fluorescence yield of dark-adapted leaves, which is related to the photochemical processes on the acceptor side of photosystem II, increased upon treatment with lower concentrations (2% and 6%) of extract, but declined when the leaves were sprayed with higher concentrations (10-20%). The main efficiency parameters of photosynthesis, such as potential quantum yield efficiency and overall vitality index, were not affected by the extract when its concentration was ≤10%.

In conclusion, we recommend the use of the celandine extract in concentration of 6% for the efficient protection of tulips against the attack of gray mold.

Acknowledgment

These studies were financially supported by the Romanian Ministry of Education and Research from the CNCSIS grants 46/220/2006, 43/220/2007, and PNII–IDEI 2272/2009-2011.

References

Agrios GN (2005). Plant Pathology 5th ed. Elsevier, Amsterdam, 510 p.

Allen MT, Pearcy RW (2000). Stomatal behavior and photosynthetic performance under dynamic light regimes in a seasonally dry tropical rain forest. Oecologia 122:470-478.

Aniszewski T (2007). Alkaloids-secrets of Life Alkaloid Chemistry, Biological Significance, Applications and Ecological Role. Elsevier, Amsterdam, 181 p.

Arif T, Bhosale JD, Kumar N, Mandal TK, Bendre RS, Lavekar GS, Dabur R (2009). Natural products – antifungal agents derived from plants. J Asian Nat Prod Res 11:621-638.

Arif T, Mandal TK, Dabur R (2011). Natural products: Antifungal agents derived from plants, p. 283-311. In: Tiwari VK (Ed.). Opportunity, Challenge and Scope of Natural Products in Medicinal Chemistry. Research Signpost, Kerala (India).

Baker NR (2008). Chlorophyll fluorescence: a probe of photosynthesis *in vivo*. Annu Rev Plant Biol 59:89-113.

Baker NR, Oxborough K (2004). Chlorophyll fluorescence as a probe of photosynthetic productivity, p. 65-82. In: Papageorgiou GC, Govindjee (Eds.). Chlorophyll a Fluorescence: A Signature of Photosynthesis. Springer, Dordrecht.

Barker KS, Rogers PD (2006). Recent insights into the mechanisms of antifungal resistance. Curr Infect Dis Rep 8:449-456.

Bartha L, Fodorpataki L (2007). Physiological reactions of the succulent CAM plant *Bryophyllum daigremontianum* to increased salinity. Contrib Bot 42:47-56.

Beever RE, Weeds PL (2007). Taxonomy and genetic variation of *Botrytis* and *Botryotinia*, p. 29–52. In: Elad Y, Williamson B, Tudzynski P, Delen N (Eds.). *Botrytis*: Biology, Pathology and Control. Springer, Dordrecht.

Carrillo-Munoz AJ, Giusiano G, Ezkurra PA, Quindos G (2006). Antifungal agents: mode of action in yeast cells. Rev Esp Quimioter 19:130-139.

Choi GJ, Jang KS, Kim JS, Lee SW, Cho JY, Cho KY, Kim JC (2004). *In vivo* antifungal activities of 57 plant extracts against six plant pathogenic fungi. Plant Pathol J 20:184–191.

Compaore M, Lamien CE, Lamien-Meda A, Vlase L, Kiendrebeogo M, Ionescu C, Nacoulma OG (2012). Antioxidant, xanthine oxidase and lipoxygenase inhibitory activities and phenolics of *Bauhinia rufescens* Lam. (Caesalpiniaceae). Nat Prod Res 26:1069-1074.

Doss RP, Christian JK, Potter SW, Soeldner AH, Chastagner GA (1997). The conidial surface of *Botrytis cinerea* and several other *Botrytis* species. Can J Bot 75:612-617.

Elad Y, Stewart A (2004). Microbial control of *Botrytis* spp., p. 223–241. In: Elad Y, Williamson B, Tudzynski P, Delen N (Eds.). *Botrytis*: Biology, Pathology and Control. Springer, Dordrecht.

Elad Y, Williamson B, Tudzynski P, Delen N (2007). *Botrytis* spp. and diseases they cause in agricultural systems – an introduction, p. 1–8. In: Elad Y, Williamson B, Tudzynski P, Delen N (Eds.). *Botrytis*: Biology, Pathology and Control. Springer, Dordrecht.

Epton HAS, Richmond DV (1980). Formation, structure and germination of conidia, p. 41-83. In: Coley-Smith JR, Verhoeff K, Jarvis WR (Eds.). The Biology of *Botrytis*. Academic Press, London.

Fatehi M, Saleh TM, Fatehi-Hassanabad Z, Farrokhfal K, Davodi SJ (2005). A pharmacological study on *Berberis vulgaris* fruit extract. J Ethnopharmacol 102:46-52.

Hayat MA (2000). Principles and Techniques of Electron Microscopy: Biological Applications. University Press, Cambridge, 1 p.

Hetherington AM, Woodward FI (2003). The role of stomata in sensing and driving environmental change. Nature 424:901-908.

Hong SK, Kim WG, Cho WD, Kim HG (2002). Occurrence of tulip fire caused by *Botrytis tulipae* in Korea. Plant Pathol J 18:106-108.

Horvath G, Droppa M, Fodorpataki L, Istokovics A, Garab G, Oettmeier W (1996). Acridones: a chemically new group of protonophores. Proc Nat Acad Sci USA 93:3876-3880.

Ienaşcu IMC, Lupea AX, Hădăruga D, Hădăruga N, Popescu IM (2008). The antimicrobial activity and quantitative structure – biological activity relationships evaluation of some novel 2-hydroxybenzamide derivatives. Revista de Chimie 59:247-250.

Isaac S (1992). Fungal-Plant Interactions. Chapman & Hall, London, 10 p.

Kars I, van Kan JAL (2007). Extracellular enzymes and metabolites involved in pathogenesis of *Botrytis*, p. 99-118. In: Elad Y, Williamson B, Tudzynski P, Delen N (Eds.). *Botrytis*: Biology, Pathology and Control. Springer, Dordrecht.

Khalaf I, Corciova A, Vlase L, Ivănescu B, Lazăr D (2011). LC/MS analysis of sterolic compounds from *Glycyrrhiza glabra*. Studia UBB Chemia 56:97-102.

Lattanzio V, Lattanzio VMT, Cardinali A (2006). Role of phenolics in the resistance mechanisms of plants against fungal pathogens and insects, p. 23-67. In: Imperato F (Ed.). Phytochemistry: Advances in Research. Trivandrum, Kerala (India).

Matos OC, Baeta J, Silva MJ, Pinto Ricardo CP (1999). Sensitivity of *Fusarium* strains to *Chelidonium majus* L. extracts. J Ethnopharmacol 66:151-168.

Meda RNT, Vlase L, Lamien-Meda A, Lamien CE, Muntean D, Tiperciuc B, Oniga I, Nacoulma OG (2011). Identification and quantification of phenolic compounds from *Balanites aegyptiaca* (L) Del (Balanitaceae) galls and leaves by HPLC-MS. Nat Prod Res 25:93-99.

Medrano H, Escalona MH, Bota JM, Gulias J, Flexas J (2002). Regulation of photosynthesis of C_3 plants in response to progressive drought: stomatal conductance as a reference parameter. Ann Bot 89:895-905.

Meng FY, Zuo GY, Hao XY, Wang GC, Xiao HT, Zhang JQ, Xu GL (2009). Antifungal activity of the benzo[c]phenanthridine alkaloids from *Chelidonium majus* Linn against resistant clinical yeast isolates. J Ethnopharmacol 125:494-496.

Mishra R, Verma DL, (2009). Antifungal activity and flavonoid composition of *Wiesnerella denudata*. Steph Acad Arena 1:42-45.

Moreau R, Whitaker B, Hicks K (2002). Phytosterols, phytostanols, and their conjugates in foods: structural diversity, quantitative analysis, and health-promoting uses. Prog Lipid Res 41:457-500.

Nawrot R, Lesniewicz K, Pienkowska J, Gozdzicka-Jozefiak A (2007). A novel extracellular peroxidase and nucleases from a milky sap of *Chelidonium majus*. Fitoterapia 78:496-501.

Nencu I, Vlase L, Istudor V, Dutu LE, Gird CE (2012). Preliminary research regarding the therapeutic uses of *Urtica dioica* L. Note I. The polyphenols evaluation. Farmacia 60:493-500.

Nidiry ESJ, Babu CSB (2005). Antifungal activity of tuberose absolute and some of its constituents. Phytother Res 19:447-449.

Pârvu M, Pârvu AE (2011). Antifungal plant extracts, p.1055-1062. In: Méndez-Vilas A (Ed.). Science Against Microbial Pathogens: Communicating Current Research and Technological Advances. Formatex Research Center, Badajoz (Spain).

Pârvu M, Pârvu AE, Crăciun C, Barbu-Tudoran L, Tămaş M (2008). Antifungal activities of *Chelidonium majus* extract on *Botrytis cinerea in vitro* and ultrastructural changes in its conidia. J Phytopathol 156:550-552.

Pinheiro HA, Silva JV, Endres L, Ferreira VM, Camara CA, Cabral FF, Oliveira JF, Carvalho LWT, Santos JM, Filho BGS (2008). Leaf gas exchange, chloroplastic pigments and dry matter accumulation in castor bean (*Ricinus communis* L) seedlings subjected to salt stress conditions. Ind Crop

Prod 27:385-392.

Rosenkranz V, Wink M (2008). Alkaloids induce programmed cell death in bloodstream forms of trypanosomes (*Trypanosoma b. brucei*). Molecules 13:2462-2473.

Sanchez-Machado DI, Lopez-Hernandez J, Paseiro-Losada P, Lopez-Cervantes J (2004). An HPLC method for the quantification of sterols in edible seaweeds. Biomed Chromatogr 18:183-190.

Sárközi Á, Janicsák G, Kursinszki L, Kéry Á (2006a). Alkaloid composition of *Chelidonium majus* L. studied by different chromatographic techniques. Chromatographia 63:81-86.

Sárközi Á, Móricz ÁM, Ott PG, Tyihák E, Kéry Á (2006b). Investigation of *Chelidonium* alkaloids by use of a complex bioautographic system. J Planar Chromatogr 19:267-272.

Sharma B, Kumar P (2009). In vitro antifungal potency of some plant extracts against *Fusarium oxysporum*. Int J Green Pharm 3:63-65.

Smania EF, Delle Monache F, Smania A, Yunes RA, Cuneo RS (2003). Antifungal activity of sterols and triterpenes isolated from *Ganoderma annulare*. Fitoterapia 74:375-377.

Staats M, van Baarlen P, van Kan JA (2005). Molecular phylogeny of the plant pathogenic genus *Botrytis* and the evolution of host specificity. Mol Biol Evol 22:333-346.

Staats M, van Baarlen P, van Kan JAL (2007). AFLP analysis of genetic diversity in populations of *Botrytis elliptica* and *Botrytis tulipae* from the Netherlands. Eur J Plant Pathol 117:219-235.

Strange RN, Scott PR (2005). Plant disease: a threat to global food security. Annu Rev Phytopathol 43:83-116.

Sundaram U, Gurumoorthi P (2012). Validation of HPTLC method for quantitative estimation of L-Dopa from *Mucuna pruriens*. Int Res J Pharm 3:300-304.

Webster JW, Weber RWS (2007). Introduction to Fungi. University Press, Cambridge, 434 p.

Wilson CL, Solar JM, El Ghaouth A, Wisniewski ME (1997). Rapid evaluation of plant extracts and essential oils for antifungal activity against *Botrytis cinerea*. Plant Dis 81:204-210.

Wink M (1998). Modes of action of alkaloids, p. 301-326. In: Roberts MF, Wink M (Eds.). Alkaloids: Biochemistry, Ecology, and Medicinal Applications. Plenum Press, New York.

Wink M (2008). Ecological roles of alkaloids, p. 3-23. In: Fattorusso E, Taglialatela-Scafati O (Eds.). Modern Alkaloids Structure, Isolation, Synthesis and Biology. WILEY-VCH Verlag GmbH & Co. KGaA, Weinheim.

Wu W, Song F, Yan C, Liu Z, Liu S (2005). Structural analyses of protoberberine alkaloids in medicine herbs by using ESI-FT-ICR-MS and HPLC-ESI-MS(n). J Pharm Biomed Anal 37:437-46.

Yohalem DS, Nielsen K, Nicolaisen M (2003). Taxonomic and nomenclatural clarification of the onion neck rotting *Botrytis* species. Mycotaxon 85:175-182.

Zuo G-Y, Meng F-Y, Han J, Hao X-Y, Wang G-C, Zhang Y-L, Zhang Q (2011). *In vitro* activity of plant extracts and alkaloids against clinical isolates of extended-spectrum β-lactamase (ESBL)-producing strains. Molecules 16:5453-5459.

17

Pumpkin Fruit Flour as a Source for Food Enrichment in Dietary Fiber

Judita ČERNIAUSKIENĖ*, Jurgita KULAITIENĖ, Honorata DANILČENKO, Elvyra JARIENĖ, Edita JUKNEVIČIENĖ

*Institute of Agriculture and Food Sciences, Aleksandras Stulginskis University, Studentu str. 11, LT-53361 Akademija, Kaunas distr., Lithuania; judita.cerniauskiene@asu.lt (*corresponding author)*

Abstract

It is believed that fiber-rich diet reduces the risk of certain cancers, coronary heart and other diseases. Dietary fiber (DF) is not only used for its nutritional properties, but also for its functional and technological properties. Thus, to supplement daily diet, dietary fiber should be incorporated into frequently consumed foods. Pumpkins are a good source of carotenoids, mineral salts, vitamins, and other bioactive substances, such as phenol compounds. Also the pumpkin flour could be used for food enriching as fiber supplement. In the study standard methods were applied to determine and compare the neutral dietary fiber (NDF), acid dietary fiber (ADF), acid detergent lignin (ADL) and water-soluble carbohydrates (WSC) contents in pumpkins fruits flour obtained from different species and cultivars. The results imply that pumpkin fruit flour should be recommended as the component suitable for food production with high content of DF. The fiber content in the pumpkin fruit flour depends on the cultivar. The maximum insoluble fiber content (NDF, ADF, ADL) was accumulated in *C. maxima* 'Kroshka' pumpkin flour: 26.50%, 24.65% and 14.35%, respectively, and this cultivar is the most suitable one for enriching food with dietary fiber.

Keywords: acid dietary fiber, lignin, neutral dietary fiber, pumpkin fruit flour, water-soluble carbohydrates

Introduction

Dietary fiber is a group of food components that is resistant to hydrolysis by human digestive enzymes and is necessary for promoting good health (AACC Report, 2001). Although its health-promoting properties have been recognized for several decades, the dietary fiber itself has long been regarded as an indigestible ballast component of a edible plant (Jenkins *et al.*, 1998; Jiménez-Escrig and Sánchez-Muniz, 2000).

Many studies on dietary fiber have revealed that the components show the ability to bind numerous substances, including cholesterol and gastric juice (Jenkins *et al.*, 1998; Jiménez-Escrig and Sánchez-Muniz, 2000; Sangnark and Noomhorm, 2003; Mai *et al.* 2003; Correa Lima and Gomes da Silva, 2005; Rodrígez *et al.*, 2006). Also, dietary fiber plays an important role in the prevention and treatment of diabetes, obesity, atherosclerosis, heart diseases, colon cancer and colorectal cancer (Wang *et al.*, 2002; Ferguson and Harris, 2003; Ferguson, 2005; Figuerola *et al.*, 2005; Nawirska and Kwasniewska, 2005). Furthermore, increased consumption of DF improves serum lipid concentrations, lowers blood pressure, improves blood glucose, promotes regularity, aids in weight loss and appears to improve immune function (Anderson *et al.*, 2009). Papadina and Bloukas (1999) state that the diet characterized through an excess of energy-dense foods, rich in fats and sugar presents a deficiency of complex carbohydrates, which constitute the major portion of dietary fiber.

Dietary fiber, as a class of compounds, includes a mixture of plant carbohydrate polymers, both oligosaccharides and polysaccharides (e.g. cellulose, hemicelluloses, pectic substances, gums, resistant starch, inulin), that may be associated with lignin, and other non-carbohydrate components (e.g. polyphenols, waxes, saponins, cutin, phytates, resistant protein) (Elleuch *et al.*, 2011). The structural polysaccharides are the major part of plant cell walls. The types of plant material that are included within the definitions of dietary fiber may be divided into two forms, based on their water solubility: insoluble dietary fiber, which includes celluloses, some hemicelluloses and lignin; soluble dietary fiber, which include β-glucans, pectins, gums, mucilages and some hemicelluloses (Elleuch *et al.*, 2011).

In various countries, the pumpkin flesh is processed into jams, marmalades, sweetmeats, puree for children, ready-to-eat dried snacks (Gajc-Wolska *et al.*, 2005; Gliemmo *et al.*, 2009; Nawirska-Olszańska *et al.*, 2011; Konopacka *et al.*, 2010). Pumpkin can also be processed into flour, which has a longer shelf-life. This flour could be used due to its flavour, sweetness, deep yellow-orange color and considerable amount of dietary fiber. It could be use to supplement cereal flours in bakery products, soups, sauces, instant noodle and also as a natural coloring supplement for food (Noor Aziah and Komathi, 2009; Noor Aziah *et al.*, 2011). Bread made from wheat flour supplemented by pumpkin flour had good nutritional value and sensory characteristics (See *et al.*, 2007). With increasing the level of substitution with pumpkin flour from 5% to 15% the ash and crude fiber contents significantly increased. Pongjanta *et al.* (2006) obtained butter and chiffon cakes and cookies by adding pumpkin powder at the levels of 10, 20, 30, 40 and 50%. The supplementation significantly increased the β-carotene content in the samples by 2.5-9.0 times.

Foods prepared from pumpkin flour could have nutritional advantage in terms of vitamins, minerals, and dietary fiber content (Noor Aziah and Komathi, 2009). Pumpkin provides valuable source of carotenoids, provitamin A and ascorbic acid, which play major roles in human nutrition and have importance as antioxidants. Current scientific researches indicate that a diet rich in foods containing β-carotene may reduce the risk of developing certain types of cancer, offers protection against heart disease and helps prevent skin diseases and vision disorders (Fraserr and Bramley, 2004; Plaza *et al.*, 2006; Liu *et al.*, 2009).

Fiber-rich products present on the market are primarily those of natural origin (oat, peas, maize, grain bran) which may be added to food without modifications, as well as those that must be modified before they are added (Górecka *et al.*, 2010). Some reports state that fruit and vegetables DF concentrates have better nutritional quality than cereals, due to the presence of significant amount of bioactive compounds, such as flavonoids and carotenoids (Figuerola *et al.*, 2005). Pumpkins are a good source of carotenoids, pectin, mineral salts, vitamins and other bioactive substances, such as phenolic compounds, so we consider that it is possible to use the pumpkin flour for enriching food with fiber. There is shortage of information about the fiber fractions in the raw material for human nutrition. Usually, they are determined only in the raw materials for feed production.

Two pumpkin species, *Cucurbita maxima* and *Cucurbita pepo*, are commonly grown in Lithuania for fruit crops and are harvested at physiological maturity. The nutritional value of pumpkin fruits is high, but varies depending on the species and cultivars. The objective of our study was to determine and compare the amount of neutral dietary fiber (NDF), acid dietary fiber (ADF), lignin and water-soluble carbohydrates (WSC) contents in pumpkin fruits flour obtained from these two species and different cultivars.

Materials and methods

The following pumpkin cultivars were selected for the investigations: *Cucurbita pepo* L.: 'Miranda', 'Golosemiannaja', 'Herakles', 'Danaja', 'Olga'; *Cucurbita maxima* Duch: 'Chutorianka', 'Zalataja grusha', 'Arina', 'Chudo judo', 'Kroshka'.

The experiment was conducted during 2010-2012 at the Experimental Station of Institute of Agriculture and Food Sciences, Aleksandras Stulginskis University. Plants were grown in the soil with the following characteristics: limnoglacial loam on moraine loam, carbonate deeper gleyic luvisol (*Calearie Luvisol*), slightly neutral and neutral, medium humus content, phosphorus-rich and potassium-rich. The field was not additionally fertilized with mineral or organic fertilizers. The soil was drained by drainage, the relief was artificially levelled. Sprouts of pumpkins were planted in the third decade of May, in four repetitions. The crop was harvested in the third decade of September.

Flours making from pumpkin fruits

For the production of pumpkin flours ten fruits were taken from each cultivar. The pumpkin fruits were washed halved and the seeds were removed. The pulp was peeled, sliced and dried at 60 °C in the laboratory drying oven (Termaks, Norway). Dried slices of pumpkins were grinded using the grinder (Ultra Centrifugal Mill ZM 200, Retsch, Germany) to produce flours, which were kept, chilled in an air-tight container, until the analysis. The samples were evaluated in triplicate for each analysis.

Methods of samples preparation and chemical analyses

The study reported in this paper was carried out at the chemistry laboratory of Lithuanian Research Centre of Agriculture and Forestry (LAMMC) and laboratory of food technologies, department of Horticulture, Aleksandras Stulginskis University.

Dry matter (DM) content has been determined by drying samples to the constant weight at temperature of 105 °C (LST EN 12145:2001). The water-soluble carbohydrates (WSC) were determined using Anthrone method (Yemm and Willis, 1954). Samples were also subjected to the fiber components analyses: acid detergent fiber (ADF) (cellulose, lignin) and acid detergent lignin (ADL) and neutral detergent fiber (NDF) (cellulose, hemicellulose, lignin) using cell wall detergent fractionation method, according to Van Soest (Faithfull, 2002).

ADF extraction was done on an ANKOM220 Fiber Analyzer (ANKOM Technology Method 08-16-06) using F57 filter bags (25 - μm porosity). NDF was analyzed with sodium sulphite and the results were presented as ash-free. The content of cell wall structural carbohydrates hemicellulose and cellulose was calculated as it follows: cellulose = ADF - ADL and hemicellulose = NDF-ADF (Hindrichsen *et al.*, 2006).

The experimental data were statistically processed by the dispersion analysis method (ANOVA), software STATISTIKA 7.0 (StatSoft, USA). Dispersion analysis was applied to evaluate chemical composition of pumpkin fruit. Arithmetical means and standard errors of the experimental data were calculated. Tukey test ($p<0.05$) was applied to estimate statistical significance of differences.

Results and discussion

Dry matter content in fruits influences yield quality. This content depends on the cultivar characteristics and climatic conditions (Tarek *et al.*, 2001, Nawirska *et al.*, 2008). The results showed that dry matter content in *C. maxima* fruits was higher than in *C. pepo* fruits, the quantity was twice as high (Fig. 1).

Within *C. pepo* fruits, significantly more dry matter was found in 'Golosemiannaja' and 'Danaja' flesh, while within *C. maxima* fruits the highest level was observed in 'Arina', 'Zalataja grusha' and 'Kroshka' cultivars. A considerable low amount of dry matter was determined in the fruits of *C. pepo* 'Olga' and *C. maxima* 'Chutorianka'.

There are many references to DF and DF-related problems in literature, but there is shortage of information on the DF content in the raw material, which can be used

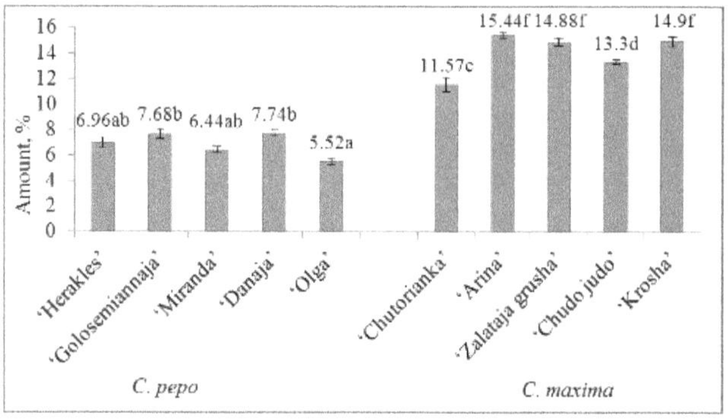

Fig. 1. Dry matter content in *C. maxima* and *C. pepo* pumpkin fruit flours; the same letters show no significant differences between means ($p<0.05$)

for the production of high-fiber content preparations (Borycka and Górecka, 2001; Nawirska *et al.*, 2008).

According to Nawirska (2008), the NDF fiber content in *C. pepo* fruits flesh ranges from 0.23% to 4.37% FW, in *C. maxima* fruits from-1.20% to 4.37% FW, ADF fiber- from 0.22% to 0.47% FW and 0.43-1.46 % FW, respectively. *C. maxima* fruits accumulated higher amounts of starch, pectin, hemicellulose and NDF fiber than *C. pepo* fruits. The amounts of other fiber fractions (lignin, ADF fiber.) in fruits of both species of pumpkin were similar (Nawirska *et al.*, 2008).

Our investigations of pumpkin fruits flours revealed that the content of ADF ranged between 10.12% and 24.65% DM. The largest amount of ADF was found in flour obtained from 'Kroshka', while the smallest amount in flour from 'Zalataja grusha' (Tab. 1).

According to the analysis of our data it can be concluded that the NDF fiber content is more dependent on the cultivar than on the specie. The NDF content ranged between 15.5% and 26.5% DM in the flour of ten pumpkin cultivars (Tab. 1). The NDF content is very high in the flour obtained from 'Kroshka' and 'Herakles' cultivars, slightly lower - in 'Chutorianka' and 'Miranda'.

The ADL content in the flour of all pumpkin cultivars was different. The ADL content varied between 0.86% and 14.3% DM depending on cultivar (Tab. 1). The highest ADL content was determined in 'Kroshka' flour and the lowest content in - 'Zalataja grusha' and 'Arina' flour.

It was observed that flour made of 'Kroshka' contained the highest amount of water-insoluble fiber. However, in 'Herakles', 'Olga' and 'Chudo judo' fruits had the highest number of NDF and ADF fiber was found, but less of lignin ADL, which decomposes more difficultly.

A big amount of lignin is an undesirable component in NDF fiber, since it reduces other fibers, hemicellulose and cellulose, degradation (Saha, 2003). In the tested pumpkin flour a moderate correlation between ADL and NDF fractions was established ($r = 0.490$, $p <0.05$). It might be thought that lower lignin content in pumpkin flour would make it easier for the human organism to absorb the cellulose and hemicellulose. The strong correlation between ADL and ADF fractions was established ($r = 0.972$, $p<0.05$). Lignin is more closely related to cellulose than to hemicellulose and has greater effect on its digestion (Van Soest *at al.*, 1991).

Water soluble components (WSC) are quickly digested energy-rich compounds and are the primary products of photosynthesis, so their content is highly dependent on temperature, sunlight and other environmental factors (Halling *et al.*, 2004).

The content of WSC varied, depending on cultivar. Pumpkin flours showed high content of total WSC (65.85% DM in 'Danaja'). The lowest content of WSC was characteristic to 'Kroshka' (26.45% DM).

Conclusions

The results imply that flour obtained from pumpkin fruits could be recommended as the component suitable for

Tab. 1. Fractional composition (mean ± s.d.) of dietary fiber in *C. maxima* and *C. pepo* pumpkin fruit flours samples

Cultivars	Amount* (% DM)			
	NDF	ADF	ADL	WSC
		C. maxima		
'Chutorianka'	$15.50\pm0.14^{b**}$	13.55 ± 0.07^{c}	2.91 ± 0.14^{ab}	61.20 ± 0.00^{a}
'Zalataja grusha'	24.30 ± 0.42^{c}	10.12 ± 0.39^{a}	0.86 ± 0.12^{a}	39.50 ± 0.28^{bc}
'Arina'	19.10 ± 0.00^{a}	11.50 ± 0.28^{ab}	1.31 ± 0.04^{a}	40.45 ± 0.64^{c}
'Chudo judo'	25.00 ± 0.28^{c}	21.95 ± 0.07^{c}	11.80 ± 0.42^{e}	38.30 ± 0.14^{b}
'Kroshka'	26.50 ± 0.00^{e}	24.65 ± 1.06^{f}	14.35 ± 0.21^{f}	26.45 ± 0.49^{e}
		C. pepo		
'Herakles'	26.50 ± 0.42^{e}	23.50 ± 0.28^{ef}	12.25 ± 0.64^{e}	51.05 ± 0.49^{f}
'Golosemiannaja'	19.10 ± 0.56^{a}	16.00 ± 0.56^{g}	5.66 ± 0.91^{c}	60.85 ± 0.07^{ad}
'Miranda'	16.40 ± 0.00^{b}	12.60 ± 0.14^{bc}	4.29 ± 0.95^{bc}	62.50 ± 0.00^{a}
'Danaja'	18.90 ± 0.14^{a}	18.20 ± 0.14^{d}	8.45 ± 0.42^{d}	65.85 ± 0.49^{g}
'Olga'	21.80 ± 0.42^{d}	19.35 ± 0.07^{d}	7.84 ± 0.39^{d}	59.20 ± 0.70^{d}

*Amount: NDF - neutral dietary fiber; ADF - acid dietary fiber; ADL - acid detergent lignin; WSC -water-soluble carbohydrates
**Means in column with different letters are significantly different ($p < 0.05$) for the different cultivars

food production with high content of dietary fiber (DF). The fiber content in pumpkin fruits flour depends on the cultivar. The highest insoluble fiber content (NDF, ADF, ADL) was characteristic to flour obtained from *C. maxima* 'Kroshka' fruits (26.50%, 24.65% and 14.35% DM, respectively), so the flour of this pumpkin cultivar is the most suitable for supplementation of food with dietary fiber.

Acknowledgements

This publication is funded by European Social Fund and the Budget of the Republic of Lithuania (Project „Eureka" E! 6855 – ECORAW "Higher functionality food products from organic vegetable raw materials").

References

AACC Report (2001). The definition of dietary fiber. Cereal Foods World 46, 3.

Anderson JW, Baird P, Davis RH Jr, Ferreri S, Knudtson M, Koraym A, Waters V, Williams CL (2009). Health benefits of dietary fiber. Nutr Rev 67(4):188-205.

Bingham SA, Day NE, Luben R, Ferrari P, Slimani N, Norat T, Clavel-Chapelon F, Kesse E, Nieters A, Boeing H, Tjønneland A, Overvad K, Martinez C, Dorronsoro M, Gonzalez CA, Key TJ, Trichopoulou A, Naska A, Vineis P, Tumino R, Krogh V, Bueno-de-Mesquita HB, Peeters PH, Berglund G, Hallmans G, Lund E, Skeie G, Kaaks R, Riboli E (2003). Dietary fiber in food and protection against colorectal cancer in the European Prospective Investigation into Cancer and Nutrition (EPIC): an observational study. The Lancet 361(5):1496-1501.

Borderías AJ, Sánchez-Alonsa I, Pérez-Mateos M (2005). New applications of fibres in foods: Addition to fishery products. Trends Food Sci Technol 16:458-465.

Borycka B, Górecka D (2001). Charakterystyka włókna pokarmowego porzeczkowych preparatów błonnikowych [Characterization of dietary fiber in the preparation of currant pomace]. Przem Ferm Owoc-Warz 2:30-32 [in Polish].

Corréa Lima MP, Gomes-da-Silva MHG (2005). Colorectal cancer: lifestyle and dietary factors. Nutr Hosp 20:235-241.

De Almada Costa GE, da Silva Queiroz-Monici K (2006). Chemical composition, dietary fiber and resistant starch contents of raw and cooked pea, common bean, chickpea and lentil legumes. Food Chem 94:327-330.

Dolreans M, Mandran N, Sauvant D (1995). Study the use of a protease with the van Soest procedure. Anim Feed Sci Technol 61:129-136.

El-Aldawy TA, Taha KM (2001). Characteristics and composition of different seeds oils and flours. Food Chem 74:47-54.

Elleucha M, Bedigian D, Roiseux O, Besbes S, Blecker Ch, Attia H (2011). Dietary fibre and fibre-rich by-products of food processing: Characterisation, technological functionality and commercial applications: A review. Food Chem 124:411-421.

Faithfull NT (2002). Methods in agricultural chemical analysis: a practical handbook. Wallingford, USA. 266 p.

Ferguson LR, Harris PJ (2003). The dietary fibre debate: more food for thought. The Lancet 361:1487-1488.

Ferguson LR (2005). Does a diet rich in dietary fibre really reduce risk of colon cancer? Digest Liver Dis 37:139-141.

Figuerola F, Hurtado ML, Estévaz AM, Chiffelle I, Asenjo F (2005). Fiber concentrates from apple pomace and citrus peel as potential fibre sources for food enrichment. Food Chem 91:395-401.

Fraser PD, Bramley PM (2004). The biosynthesis and nutritional uses of carotenoids. Prog in Lip Res 43:228-265.

Gajc-Wolska J, Gajewski M, Radzanowska J, Niemirowicz-Szczytt K, Korzeniewska A (2005). The fruit quality of chosen hybrids and cultivars of pumpkin (*Cucurbita maxima* Duch.). Veget Crops Res Bull 63:34-43.

Gliemmo MF, Latorre ME, Gerschenson LN, Campos CA (2009). Color stability of pumpkin (*Cucurbita moschata*, Duchesne ex Poiret) puree during storage at room temperature: Effect of pH, potassium sorbate, ascorbic acid and packaging material. Food Sci Technol 42:196-201.

Górecka D, Pachołek B, Dziedzic K, Górecka M (2010). Raspberry pomace as a potential fiber source for cookies enrichment. Acta Scientiarum Polonorum Technol Aliment 9(4):451-462.

Grigelmo-Miguel N, Martín-Belloso O (1999). Comparison of dietary fiber from by-products of processing fruits and greens and from cereals. Lebensm Wiss Technol 32:503-508.

Halling MA, Longland AC, Martens S, Nesheim L, O'Kiely P (2004). Accumulation of water soluble carbohydrates in two perennial ryegrass cultivars at nine European sites. Grassland Science in Europe 9:954-956.

Hindrichsen IK, Kreuzer M, Madsen J, Knudsen KEB (2006). Fiber and lignin analysis in concentrate, forage, and feces: Detergent versus enzymatic-chemical method. J Dairy Sci 89(6):2168-2176.

Jenkins DJA, Kendall CWC, Ransom TPP (1998). Dietary fiber, the evolution of the human diet and coronary heart disease. Nutr Res 18:633-652.

Jiménez-Escrig A, Sánchez-Muni FJ (2000). Dietary fiber from Edible seaweeds: chemical structure, physicochemical properties and effects on cholesterol metabolism. Nutr Res 20:585-598.

Konopacka D, Seroczyńska A, Korzeniewska A, Jesionkowsk K, Niemirowicz-Szczytt K, Plocharski W (2010). Studies on the usefulness of *Cucurbita maxima* for the production of ready-to-eat dried vegetable snacks with a high carotenoid content. Food Sci Technology 43(2):302-309.

Liu GN, Zhu YH, Jiang JG. (2009). The metabolomics of carotenoids in engineered cell factory. Appl Microbiol Biotechnol 83(6):989-999.

LST EN 12145:2001 (2001). Fruit and vegetable juices - Determination of total dry matter - Gravimetric method with loss of mass on drying.

Mai V, Flood A, Peters U, Lacey JV Jr, Schairer C, Schatzkin A (2003). Dietary fiber and risk of colorectal cancer in the Breast Cancer Detection Demonstration Project (BCDDP) follow-up cohort. Int J Epidemiol 32:234-239.

Nawirska A (2005). Binding of metal ions by selected fractions of fruit pomace. Food Chem 90:395-400.

Nawirska A, Kwasniewska M (2005). Dietary fiber fractions from fruit and vegetable processing waste. Food Chem 91:221-225.

Nawirska A, Sokół-Łętowska A, Kucharska AZ, Biesiada A, Bednarek M (2008). Porównanie zawartośći frakcji włókna pokarmowego w dmianach dyni z gatunku *Cucurbita maxima* i *Cucurbita pepo* (Comparing the contents of dietary fibre fractions in some varieties of *Cucurbita maxima* and *Cucurbita pepo*). Żywność Nauka Technologia Jakość 1(56):65-73 [in Polish].

Nawirska-Olszańska A, Biesiada A, Sokół-Łętowska A, Kucharsk AZ (2011). Content of bioactive compounds and antioxidant capacity of pumpkin puree enriched with japanese quince, cornelian Cherry, strawberry and apples. Acta Scientiarum Polonorum Technol Aliment 10(1):51-60.

Noor Aziah AA, Komathi CA (2009). Physicochemical and Functional Properties of Peeled and Unpeeled Pumpkin Flour. J Food Sci 74(7):328-333.

Noor Aziah AA, Komathi CA, Bhat R (2011). Evaluation of Resistant Starch in Creckers Incorporated with Unpeeld and Peeled Pumpkin Flour. Am J Food Technol 6(12):1054-1060.

Papadina SN, Bloukas JG (1999). Effect of fat level and storage conditions on quality characteristics of traditional Greek sausages. Meat Sci 51:103-113.

Plaza L, Sanchez-Moreno C, Elez-Martınez P, De Ancos B, Martın-Belloso O, Pilar Cano M (2006). Effect of refrigerated storage on vitamin C and antioxidant activity of orange juice processed by high-pressure or pulsed electric fields with regard to low pasteurization. Eur Food Res Technol 223:487-493.

Pongjanta J, Naulbunrang A, Kawngdang S, Manon T, Thepjaikat T (2006). Utilization of pumpkin powder in bakery products. J Sci Technol 28(1):71-9.

Prosky L (1999). What is fibre? Current controversies. Trends Food Sci Technol 10: 271-275.

Rahman ZU, Islam M, Shah WH (2002). Effect of microwave and conventional cooking on insoluble dietary fibre components of vegetables. Food Chem 80:237-240.

Rodrígez R, Jimenez A, Fernández-Bolaños J, Guillén R, Heredia A (2006). Dietary fibre from vegetable products as source of functional ingredients. Trends Food Sci Technol 17:3-15.

Saha BC (2003). Hemicelluloze bioconversion. J Ind Microbiol Biotechnol 30:279-291.

Sangnark A, Noomhorm A (2003). Effect of particle sizes on in-vitro calcium and magnesium binding capacity of prepared dietary fibers. Food Res Int 36:91-96.

See EF, Wan Nadiah WA, Noor Aziah AA (2007). Physico-Chemical and Sensory Evaluation of Breads Supplemented with Pumpkin Flour. ASEAN Food Journal 14(2):123-130.

Van Soest PJ, Robertson JB, Lewis BA (1991). Symposium: Carbohydrate methodology, metabolism, and nutritional implications in dairy cattle. J Dairy Sci 74:3583-3597.

Wang J, Rosell CM, de Barber CB (2002). Effect of the addition of different fibres on wheat dough performance and bread quality. Food Chem 79:221-226.

Yemm EW, Willis AJ (1954). The estimation of carbohydrates in plant extracts by Anthrone. Biochem J 57:508-514.

18

Effects of Various Mixed Salt-Alkaline Stress Conditions on Seed Germination and Early Seedling Growth of *Leymus chinensis* from Songnen Grassland of China

Jixiang LIN[1], Zhuolin LI[2], Shuai SHAO[1], Yingnan WANG[1], Chunsheng MU[2]*

[1]*Alkali Soil Natural Environmental Science Center, Northeast Forestry University/Key Laboratory of Saline-alkali Vegetation Ecology Restoration in Oil Field, Ministry of Education, Harbin, 150040, China; jixiang851012@gmail.com*
[2]*Key Laboratory of Vegetation Ecology of Ministry of Education, Institute of Grassland Science, Northeast Normal University, Changchun, 130024, China; mucs821@gmail.com (*corresponding author)*

Abstract

Soil salinization and alkalization always co-occur in grassland ecosystem, but little information exists concerning the mixed effects of salt-alkaline stresses on plants. *Leymus chinensis* is considered as one of the most promising grass species in Songnen Grassland of Northern China. In this study, we investigated the effects of 30 mixed salt-alkaline conditions (NaCl, Na_2SO_4, $NaHCO_3$ and Na_2CO_3; pH 7.10-10.18 and salinity 50-250 mM) on seed germination and seedling growth of *L. chinensis*. The results showed that germination percentage and rate were both decreased with increasing salinity and pH. Nongerminated seeds germinated well after being transferred to distilled water from treatment groups. Shoot and radicle growth were also affacted by salinity, pH and their interactions. However, radicle length decreased more markedly with increasing salinity and pH, and was strongly inhibited when pH reached 8.05. Stepwise regression analysis results showed that salinity was the dominant factor for seed germination under mixed salt-alkaline stress conditions. However, once radicle break through the seed coat, and pH changed into the dominant factor for seedling establishment. These results indicated that mixed salt-alkaline stresses had different impacts on germination and early seedling stages of *L. chinensis*. A better understanding of the germination and seedling processes should facilitate the effective utilization of this species under such complex environment.

Keywords: salt-alkaline stresses, seed germination, germination recovery, shoot and radicle growth

Introduction

Soil salinization and alkalization is a major environmental problem throughout the world, which not only limits growth and production of plant but also results in land degradation. For example, arable land acreage of the world is 1.5×10^9 ha, but 23% (0.34×10^9 ha) of the area is saline, and 37% (0.56×10^9 ha) is sodic. In Songnen Plains of Northeast China, approximately 70% of the natural grassland has been seriously degraded due to the impacts of salt and alkali soils, and this trend is still increasing (Zhang and Mu, 2009).

In most salt-alkaline soil of Northeast China, the main harmful salts are NaCl, Na_2SO_4, $NaHCO_3$ and Na_2CO_3, coming from neutral salts and alkaline salts (Yang *et al.*, 2008). Previous studies have proved that alkaline salt stress and neutral salt stress are differed greatly, and should be called alkali stress and salt stress, respectively (Shi and Yin, 1993). The deleterious effects of alkaline stress are more severe than salt stress (Yang *et al.*, 2008; Zhang and Mu, 2009; Guo *et al.*, 2009). In general, the impact of salt stress involves osmotic and ionic effects, alkali stress has same stress factors but added the influence of high pH, which can inhibit ion uptake and disrupt ionic balance of plant cells (Munns, 2002; Yang *et al.*, 2007). However, salinization and alkalization always co-occur in nature conditions, plant survive under salt and alkaline stresses are very complex (Shi and Wang, 2005; Li *et al.*, 2009). Therefore, mixed effect of salt stress and alkali stress should be further explored, not only that of salt stress or alkali stress.

Seed germination and seedling establishment are the most crucial stages for plant survival under saline-alkaline environment (Qu *et al.*, 2008). Seed germination can determine where and when seedling growth starts (Lin *et al.*, 2011). High salinities in the soil always result in both a delay and reduction of seed germination. It can also cause a complete inhibition of germination if salinity concentration is high enough. However, due to climatic factors such as rainfall or melted snow, salinity concentration in the soil always decreases, and thus provides a suitable condition to seed for germination. Consequently, most halophytes seeds remain viable at high salinity and germinate again until salinity concentration decreases (Ungar, 1991).

Leymus chinensis (Trin.) Tzvel. is a perennial rhizomatous forage grass of the family Poaceae. It is widely distributed in the eastern region of the Eurasian steppes, the Northern and Eastern parts of the People's Republic of Mongolia, the Northeastern China Plain, and the Inner Mongolia Plateau of China (Wang *et al.*, 2004). This plant is rich in proteins, carbohydrates, minerals and is also highly tolerant to saline-alkaline soils (Huang *et al.*, 2002). Due to its economic values and tolerant traits towards saline-alkaline stress, *Leymus chinensis* is considered as one of the most promising grass species for grassland rehabilitation and restoration in Northern China (Liu and Qi, 2004). However, to our knowledge, few studies focus on this species response to mixed salt-alkaline stresses especially for seed germination and early seedling stage. In this paper, mixtures of four salts NaCl, Na_2SO_4, $NaHCO_3$ and Na_2CO_3 were used in various proportions to simulate 30 treatments of mixed salt-alkaline conditions. The aims were (1) to test the effects of mixed salt-alkaline stresses on seed germination and early seedling growth of *L. chinensis*, (2) to analyze the key factor among all the stress factors for the two stages.

Materials and methods

Plant material

Mature seeds (Thousand-seed weight is 2.4 g) of *L. chinensis* were collected from the Grassland Ecosystem Field Station, Institute of Grassland Science, Jilin province, China (123°44'E, 44°44'N). This area is characterized by a semi-arid, continental monsoonal climate. Mean annual precipitation and temperature are 300-450 mm, and from 4.6 to 6.4 ℃. The soil type is mixed salt-alkali meadow soil (Zhang *et al.*, 2009). Seeds were then stored in paper bags at 4 °C until further use.

Mixed salt-alkaline stress conditions

To simulate mixed salt-alkali conditions, two neutral salts (NaCl and Na_2SO_4) and two alkaline salts ($NaHCO_3$ and Na_2CO_3) were chose according to the salt composition of the salt-alkaline soil in Northeast China (Peng *et al.*, 2008). Four salts were mixed in various proportions based on the tolerance of salt-alkaline stress and the salinity and pH in Songnen Grassland. Six treatment groups (A-F) were set with increasing alkalinity. Salt composition and molar ratio of various treatments were shown in Tab. 1. Within each group, five concentrations were used (50, 100, 150, 200 and 250 mM). Total 30 mixed stress treatments (A_1-F_5, pH 7.10-10.18, and salinity 50-250 mM) were shown in Tab. 2.

Tab. 1. Salt composition and molar ratio of various treatments

	Salt composition and molar proportions			
Treatments	NaCl	Na_2SO_4	$NaHCO_3$	Na_2CO_3
A	2	1	0	0
B	1	1	1	0
C	12	9	8	1
D	8	9	12	1
E	12	1	8	9
F	0	0	2	1

Tab. 2. Stress factors of various treatments

Treatment	Stress factors						
	pH value	Salinity(mM)	Na^+(mM)	Cl^-(mM)	SO_4^{2-}(mM)	HCO_3^-(mM)	CO_3^{2-}(mM)
A_1	7.10	50	66.7	33.3	16.7	0.0	0.0
A_2	7.14	100	133.4	66.6	33.4	0.0	0.0
A_3	7.20	150	200	100	50	0.0	0.0
A_4	7.24	200	266.8	133.2	66.8	0.0	0.0
A_5	7.28	250	333.5	166.5	83.5	0.0	0.0
B_1	7.80	50	66.7	16.7	16.7	16.7	0.0
B_2	8.05	100	133.4	33.4	33.4	33.4	0.0
B_3	8.22	150	200	50	50	50	0.0
B_4	8.28	200	266.8	66.8	66.8	66.8	0.0
B_5	8.35	250	333.5	83.5	83.5	83.5	0.0
C_1	8.90	50	66.7	20	15	13.3	1.7
C_2	8.92	100	133.4	40	30	26.6	3.4
C_3	8.96	150	200	60	45	39.9	5.1
C_4	8.98	200	266.8	80	60	53.2	6.8
C_5	9.02	250	333.5	100	75	66.5	8.5
D_1	9.05	50	66.7	13.3	15	20	1.7
D_2	9.10	100	133.4	26.6	30	40	3.4
D_3	9.15	150	200	39.9	45	60	5.1
D_4	9.18	200	266.8	53.2	60	80	6.8
D_5	9.24	250	333.5	66.5	75	100	8.5
E_1	9.56	50	66.7	20	1.7	13.3	15
E_2	9.60	100	133.4	40	3.4	26.6	30
E_3	9.64	150	200	60	5.1	39.9	45
E_4	9.69	200	266.8	80	6.8	53.2	60
E_5	9.76	250	333.5	100	8.5	66.5	75
F_1	9.90	50	66.7	0.0	0.0	33.3	16.7
F_2	9.94	100	133.4	0.0	0.0	66.6	33.4
F_3	9.98	150	200	0.0	0.0	100	50
F_4	10.12	200	266.8	0.0	0.0	133.2	66.8
F_5	10.18	250	333.5	0.0	0.0	166.5	83.5

Seed germination test

Seeds were surface sterilized in 0.1% mercury chloride for 10 min, and then washed with distilled water and air-dried to avoid fungus attack before being used in our experiments.

Seeds were placed in 11 cm Petri dishes on two layers of 12.5 cm filter paper moistened with 10 mL of the treatment solutions. Four replicates of 50 seeds were used for each treatment and distilled water was used as a control. The Petri dishes were placed in growth chambers and maintained at 30/20 °C with 12h photoperiod (Sylvania cool white fluorescent lamps, 200 $\mu molm^{-2}s^{-1}$,400-700 nm, HPG-400, Haerbin, China). Seeds were considered germinated with the emergence of the radicel. Germination percentage was recorded every 2 days for 20 d. Nongerminated seeds from all the treatments were then transferred to distilled water to study the recovery of germination, which was also recorded at 2 d intervals for 20 d.

Germination rate was estimated by using a modified Timson index of germination velocity, $\sum G/t$, where G is the percentage of seed germination and t is the germination time (Lin *et al.*, 2011). The maximum value with our data was 50 (i.e.1000/20). The recovery percentage was calculated according to the number of germinated seeds after being transferred to distilled water divided by the number of nongerminated seeds under saline and alkaline stresses.

Seedling growth test

For evaluation of the effects of mixed salt-alkaline stresses on early seedling growth, seeds were incubated initially in distilled water at 30/20 °C with 12 h photoperiod, when the coleoptile had just emerged, 20 of the early seedlings were then incubated with treatment solutions. Early seedling growth was ended after 15 d, and shoot lengths and radicle lengths were recorded.

Data analysis

All data were analyzed using SPSS 13.0 (SPSS Inc, Chicago, IL, USA). Tukey's tests were performed for multiple comparisons to determine significant ($p<0.05$) differences between individual treatments. A two-way ANOVA was used to test the effects of the factors (pH and salinity) and their interactions on seed germination and seedling growth. The data were expressed as mean ± S.E. Stepwise regression analysis was performed for the key factor.

Results

Germination of *L. chinensis* seeds was significantly affected by pH, salinity and their interactions of the two factors ($p<0.001$, Tab. 3). Maximum seed germination percentage was obtained in distilled water, and seed reached the highest final percentage of 72.5% under this condition. Germination decreased with increasing salinity under all treatment groups, and sharply decreased in group E and F (Fig.1, Tab. 4). At highest salt concentration (250 mM), germination percentage in group A was 6.5%, while seeds showed no capacity for germination in the other five treatment groups (B-F). In group F (highest pH/alkalinity), seed could not germinate when salinity was only 150 mM. However, in the other five treatment groups (A-E),

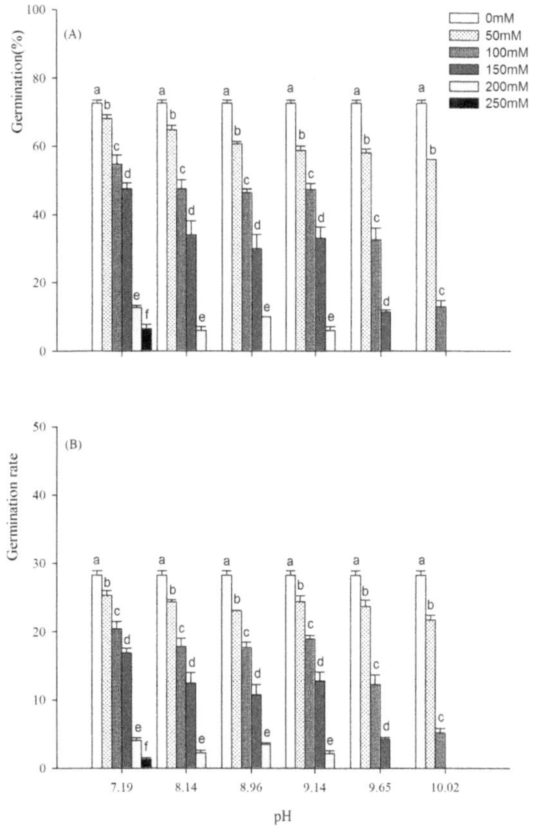

Fig. 1. Germination (A), germination rate (B) of *Leymus chinensis* in each treatment. The values are the means of four replicates. Means followed by different letters are significantly different at $p<0.05$ according to a least significant difference test

Tab. 3. Two-way ANOVA of effects of salinity, pH and their interactions on seed germination and seedling growth of *L. chinensis*

Source of variance	Germination (%)	Germination rate	Recovery percentage (%)	Shoot length (cm)	Root length (cm)
Salinity	1898.4***	1614.7***	104.6***	239.92***	637.48***
pH	113.5***	59.7***	137.9***	72.08***	589.48***
Salinity×pH	22.6***	16.8***	14.5***	5.54***	76.70***

Data represent F-values at 0.05 level. *** $p<0.001$

Tab. 4. Seed germination situations under various treatments

Treatment/Average pH	Salinity (mM)	Seed germination (%, mean±s.e., n=4)		
		Initial germination (%)	Recovery percentage(%)	Total germination (%)
Control	0	72.5 ± 1.0^{a}	0.0 ± 0.0	72.5 ± 1.0^{a}
A/ pH=7.19	50	68.0 ± 1.2^{b}	4.9 ± 0.6^{a}	69.6 ± 0.8^{a}
	100	54.7 ± 2.7^{c}	11.3 ± 0.7^{b}	59.9 ± 2.0^{b}
	150	47.5 ± 1.7^{d}	19.0 ± 1.0^{c}	56.7 ± 0.7^{b}
	200	12.7 ± 0.7^{e}	22.8 ± 0.8^{d}	32.7 ± 0.7^{c}
	250	6.5 ± 1.3^{f}	32.8 ± 1.3^{e}	36.0 ± 1.2^{c}
B/ pH=8.14	50	64.7 ± 1.3^{b}	5.0 ± 0.6^{a}	67.5 ± 2.3^{a}
	100	47.5 ± 2.6^{c}	18.3 ± 1.6^{b}	58.3 ± 2.4^{b}
	150	34.0 ± 4.2^{d}	37.0 ± 1.3^{c}	58.0 ± 2.0^{b}
	200	6.0 ± 1.2^{e}	37.7 ± 1.1^{c}	42.4 ± 1.6^{c}
	250	0.0 ± 0.0^{e}	40.7 ± 1.3^{c}	40.7 ± 1.3^{c}
C/ pH=8.96	50	60.7 ± 0.7^{b}	13.6 ± 1.8^{a}	64.7 ± 0.7^{b}
	100	46.5 ± 1.0^{c}	22.7 ± 1.9^{b}	59.3 ± 0.7^{c}
	150	30.0 ± 4.2^{d}	35.3 ± 0.6^{c}	54.7 ± 2.9^{d}
	200	10.0 ± 0.0^{e}	29.4 ± 1.4^{cd}	36.5 ± 1.3^{e}
	250	0.0 ± 0.0^{f}	32.7 ± 0.7^{d}	32.7 ± 0.7^{e}
D/ pH=9.14	50	58.7 ± 1.3^{b}	29.0 ± 2.0^{a}	70.7 ± 0.7^{a}
	100	47.3 ± 1.8^{c}	39.1 ± 2.0^{b}	68.0 ± 0.0^{ab}
	150	33.0 ± 3.4^{d}	43.4 ± 3.3^{b}	64.0 ± 3.8^{b}
	200	6.0 ± 1.2^{e}	40.9 ± 1.5^{b}	45.3 ± 1.3^{c}
	250	0.0 ± 0.0^{e}	44.5 ± 1.5^{b}	44.5 ± 1.5^{c}
E/ pH=9.65	50	58.0 ± 1.2^{b}	38.1 ± 2.4^{ab}	74.0 ± 1.2^{a}
	100	32.5 ± 3.6^{c}	35.3 ± 1.1^{a}	53.3 ± 1.3^{b}
	150	11.3 ± 0.7^{d}	35.5 ± 1.1^{a}	41.3 ± 1.3^{c}
	200	0.0 ± 0.0^{e}	38.0 ± 2.6^{ab}	38.0 ± 2.6^{c}
	250	0.0 ± 0.0^{e}	44.0 ± 3.0^{b}	44.0 ± 3.0^{c}
F/ pH=10.02	50	56.0 ± 0.1^{b}	41.9 ± 1.8^{b}	74.4 ± 1.3^{a}
	100	13.0 ± 1.7^{c}	34.2 ± 1.8^{a}	53.0 ± 4.7^{b}
	150	0.0 ± 0.0^{d}	32.5 ± 1.0^{a}	32.5 ± 1.0^{c}
	200	0.0 ± 0.0^{d}	39.0 ± 1.3^{b}	39.0 ± 1.3^{c}
	250	0.0 ± 0.0^{d}	47.5 ± 1.7^{c}	47.5 ± 1.7^{b}

Different letters indicate significant differences from different salinity concentrations ($p < 0.05$)

germination percentages were 47.5%, 34.0%, 30.0%, 33.0% and 11.3%, respectively.

Germination rate was also significantly affected by pH, salinity and their interactions ($p<0.001$, Tab. 3). The downtrend with increasing salinity and pH was similar to that of germination percentage. Germination rate in group A was higher than other groups, and group E and F had the greatest reduction, which were 4.2 and 0 at 150 mM salinity, significantly lower than the other treatment groups (A-D)($p<0.05$). Once salinity concentration ≥200 mM, germination rate were 0 in group E and F.

When nongerminated seeds were transferred from treatment solutions to distilled water, most of them germinated well. Recovery percentage was also affected by pH, salinity and their interactions ($p<0.001$, Tab. 3). Recovery percentages increased with increasing salinity in all the treatment groups, and reached highest at 250 mM salinity (Tab. 4).

Both shoot length and radicle length were affected by pH, salinity and their interactions ($p<0.001$, Tab. 3, Fig. 2). There was no shoot emerged at 250 mM salinity in all the treatment groups, and only for 200 mM in group E and F. Radicle length of *L. chinensis* seedlings under various mixed salt-alkaline stresses was significantly less than that of controls, especially under high pH groups. For example, radicle length was only 0.2 cm in group F even though salinity concentration was 50 mM.

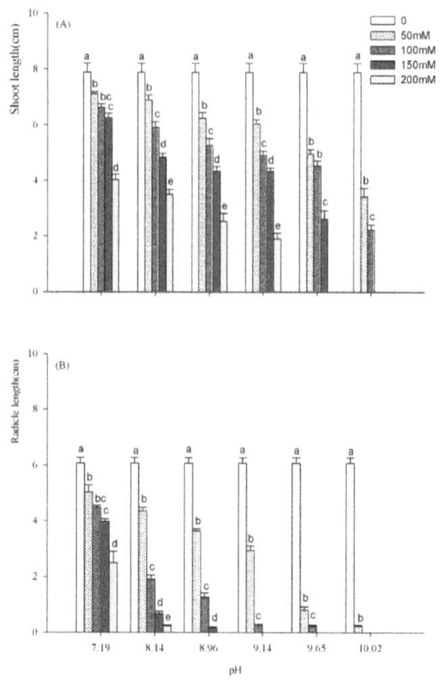

Fig. 2. Shoot length (A) and radicle length (B) of *Leymus chinensis* seedlings in each treatment. The values are the means of four replicates. Means followed by different letters are significantly different at $p<0.05$ according to a least significant difference test

Tab. 5. Results of stepwise regression between every index and stress factors

	Model	R2	ANOVA test	β1	β2	β3	β4	β5	β6
Germination (%)	Y=137.766-0.299X2-7.463X1	0.888	p < 0.001	-0.293	-0.873				
Germination rate	Y=49.546-0.118X2-2.452X1	0.889	p < 0.001	-0.248	-0.887				
Shoot length (cm)	Y=17.677-0.027X2-1.161X1	0.862	p < 0.001	-0.442	-0.776				
Root length (cm)	Y=19.339-1.891X1-0.029X5-0.011X4	0.818	p < 0.001	-1.107			-0.280	-0.497	

X_1=pH ; X_2=Salinity ; X_3= Cl^- ; X_4= SO_4^{2-} ; X_5= HCO_3^- ; X_6= CO_3^{2-}. β_1-β_6: Standardize regression coefficients corresponding X_1- X_6, the greater the absolute β value, the stronger effect of the stress factor on every index. R^2: the square of total correlation coefficient, n=30

Main stress factors of mixed salt-alkaline stresses are shown in Tab. 2. The results of stepwise regression analysis showed that there was a high linear correlation between each index and stress factors ($p<0.001$, Tab. 5). Among the absolute values of the regression coefficients in germination indexes, salinity (β_2) was highest for the indexes, indicating that salinity was the dominant factor among the stress factors. While among the absolute values of the regression coefficients in radicle growth, pH was highest for the indexes, indicating that pH changed into the key factor on seedling establishment of *L. chinensis*.

Discussion

Seed germination is the initial and one of the most critical stages over the life cycle of plants. Lower water potential caused by salinity is the determining factor inhibiting seed germination under saline environment (Debez *et al.*, 2004). Our results showed that high pH also significantly affected seeds germination. Germination percentage and rate were both affected by salinity, high pH, and their interactions (Fig. 1, Tab. 3). Furthermore, *L. chinensis* seeds were more tolerant to only salt stress (A group) than mixed salt-alkaline stresses (B-F groups). This mainly due to that mixed salt-alkaline stresses not only had the same stress characteristics with salt stress, but also added the impact of high pH stress, so created more harmful effects. However, higher germination percentages were observed at lower salinity (50 mM) in all the groups even though under higher pH treatments (D-F). This phenomenon indicated that germination was not inhibited by high pH when salinity concentration was low, which implied that high pH had a particular adjustment function on seeds at lower salinity. However, once salinity was higher, interactions of high pH and salinity might decrease seed germination, decompose seed structure and even result in seed death, the specific reason deserves further research. Most nongerminated halophyte seeds under high salinity environments will recover and germinate again after being transferred to distilled water (Song *et al.*, 2006; Qu *et al.*, 2008). Our result that recovery percentage in *L. chinensis* seeds showed increasing trends with salinity increment in all the treatment groups supports this phenomenon. This indicated an adaptive strategy of seed germination to high salinity stress, nongerminated seeds in a state of dormancy to escape from the rigorous environment (Debez *et al.*, 2004). This phenomenon also indicated that high salinities only delayed germination process for most *L. chinensis* seeds but did not cause them lose viability. In addition, there are several possible causes that nongerminated seeds at higher salinity stress recovery well. First, the lower salinity conditions always suitable to *L. chinensis* seeds and most of them can germinate quickly in order to avoid the competition on seedling stage. Second, high salinity may enhance the priming effect on the nongerminated seeds and increase seed vigor and germination ability.

Early seedling growth is another critical stage for plants survive in salt-alkaline conditions. Our results showed that although shoot growth well at lower salinity stress, but radicle growth was greatly retarded especially at high pH groups, indicating that the radicle of *L. chinensis* was more sensitive to salinity and pH stresses. The results agree with Li (2010) for *Spartina alterniflora*. In addition, we found that although seedling lengths decreased with increasing salinity in each treatment group, the treatment with larger proportion of alkalinity resulted in lower shoot and root length. Similar results have also been reported for other plants (Li *et al.*, 2010; Peng *et al.*, 2008). The reason of this phenomenon may be due to the effects of high pH and its interactions with ions stress. The high-pH environment surrounding the radicles of *L. chinensis* especially with high salinity stress can cause ions imbalance, metabolic disorders, and also destroys the structure and function of root cells, results in lower elongations of the seedlings (Shi and Yin, 1993; Shi and Wang, 2005).

The deleterious effects of mixed salt-alkaline stresses are more complex than that of single salt or alkali stress due to the various salt components and proportions (Liu and Zhu, 1997; Shi *et al.*, 2002). The stress characters of salt stress on plant contains osmotic and ionic effects, alkaline stress has the same characters with salt stress and exerts the high pH stress. However, when the neutral salts and alkali salts are mixed together, the negative influence on plant is much greater because of the interactions of salinity and high pH or various ions. For example, the stress on germination index and seedling growth index due to single salinity stress (from A1 to A5) or single high pH stress (A1, B1, C1, D1, E1 and F1) was smaller than that of salinity interacted with high pH (D4, D5, F4 and F5) (Figs. 1, 2). As a result, the reciprocal enhancement between salt and alkali stresses is characteristic of mixed salt-alkaline stress (Shi and Wang, 2005).

From the results of stepwise regression analysis among the stress factors, it was obvious that salinity was the dominant factor on germination stage under mixed salt-alkaline stress conditions (Tab. 5). This mainly due to that salinity concentration in the soil determines the water potential, and water absorption is the critical factor for seed germination. Therefore, with increasing salinity, the potential surrounding the seed (difference potential inside and outside of the cell) gradually decreases, and it is difficult for seeds to absorb water under such circumstances, then higher salinity strongly affects germination percentage and germination rate of *L. chinensis*. However, we found pH changed into the dominant factor for radicle growth, and strongly affacted seedling establishment. This means that the mixed salt-alkaline stresses have different impacts on *L. chinensis* during germination and early seedling stage. Compared with seed germination, seedling stage is much more complex, radicle should absorb great number of nutrient elements and also maintain osmotic and ionic balance, and high pH can strongly inhibit ion uptake and disrupt ionic balance of plant cells. Therefore, once radicles break through the seed coat, and pH changes into the dominant factor for seedling establishment.

Conclusion

In summary, this study clearly showed that seed germination and seedling growth of *L. chinensis* were affected by salinity and high pH and their interactions. Nongerminated seeds germinated well after being transferred to distilled water, indicating that seeds of this species have recovery ability under salt-alkaline environment. Although radicle and shoot length were both inhibited by mixed salt-alkaline stresses, radicle growth was more sensitive to the stress. Salinity was the key stress factor for germination, while pH was the key stress factor for seedling establishment under mixed salt-alkaline stresses. The better understanding of the germination and seedling processes of *L. chinensis* should facilitate the effective utilization of this species under such complex environment.

Acknowledgements

The research was supported by the Fundamental Research Funds for the Central Universities (DL12BA32, 2572014EA04), and was also funded by National Natural Science Foundation of China (31172259, 31370432).

References

Debez A, Hamed BK, Grignon C, Abdelly C (2004). Salinity effects on germination, growth, and seed production of the halophyte *Cakile maritime*. Plant Soil 262:179-189.

Grutterman, Y (1993). Seed germination in desert plants. Berlin: Springer-Verlag 169-206.

Guo R, Shi LX, Yang YF (2009). Germination, growth, osmotic adjustment and ionic balance of wheat in response to saline and alkaline stresses. Soil Sci Plant Nutri 55:667-679.

Huang ZH, Zhu JM, Mu XJ, Lin JX (2002). Advances on the mechanism of low sexual reproductivity of *Leymus chinensis*. Grassland China 24:55-60.

Li R, Shi F, Fukuda K (2010). Interactive effects of salt and alkali stresses on seed germination, germination recovery, and seedling growth of a halophyte *Spartina alterniflora* (Poaceae). South African J Bot 76:380-387.

Lin JX, Wang JF, Li XY, Zhang YT, Xu QT, Mu CS (2011). Effects of saline and alkaline stresses in varying temperature regimes on seed germination of *Leymus chinensis* from the Songnen Grassland of China. Grass For Sci 66:578-584.

Liu GS, Qi DM (2004). Research progress on the biology of *Leymus chinensis*. Acta Prataculturae Sinica 13:6-11.

Liu J, Zhu JK (1997). Proline accumulation and salt-stress-induced gene expression in a salt-hypersensitive mutant of arabidopsis. Plant Physiol 114:591-596.

Munns R (2002). Comparative physiology of salt and water stress. Plant Cell Envir 25:239-250.

Peng YL, Gao ZW, Gao Y, Liu GF, Sheng LX, Wang DL (2008). Eco-physiological characteristics of alfalfa seedlings in response to various mixed salt-alkaline stresses. J Integr Plant Biol 50: 29-39.

Qu XX, Huang ZY, Baskin JM, Baskin CC (2008). Effect of temperature, light and salinity on seed germination and radicle growth of the geographically widespread halophyte shrub *Halocnemum strobilaceum*. Ann Bot 101:293-299.

Shi DC, Wang DL (2005). Effects of various salt–alkali mixed stresses on *Aneurolepidium chinense* (Trin.) Kitag. Plant Soil 271:15-26.

Shi DC, Yin LJ (1993). Difference between salt (NaCl) and alkaline (Na_2CO_3) stresses on *Puccinellia tenuiflora* (Griseb.) Scribn.et Merr. Plants. Acta Bot Sin 35:144-149.

Shi DC, Yin SJ, Yang GH, Zhao, KF (2002). Citric acid accumulation in an alkali-tolerant plant *Puccinellia tenuiflora* under alkaline stress. Acta Bot Sin 44:537-540.

Song J, Feng G, Zhang FS (2006). Salinity and temperature effects on germination for three salt resistant euhalophytes, *Halostachys caspica*, *Kalidium foliatum* and *Halocnemum strobilaceum*. Plant Soil 279:201-207.

Ungar IA (1991). Ecophysiology of vascular halophytes. Boca Raton: CRC Press 9-48.

Wang ZW, Li LH, Han XG, Dong M (2004). Do rhizome severing and shoot defoliation affect clonal growth of *Leymus chinensis* at ramet population level? Acta Oecologica 26:255-260.

Yang CW, Chong J, Kim C, Li CY, Shi DC, Wang DL (2007). Osmotic Adjustment and ion balance traits of an alkali resistant halophyte *Kochia sieversiana* during adaptation to salt and alkali conditions. Plant Soil 294:263-276.

Yang CW, Shi DC, Wang DL (2008). Comparative effects of salt stress and alkali stress on growth, osmotic adjustment and ionic balance of an alkali-resistant halophyte *Suaeda glauca* (Bge.). Plant Growth Regul 56:179-190.

Zhang JT, Mu CS (2009). Effects of saline and alkaline stresses on the germination, growth, photosynthesis, ionic balance and anti-oxidant system in an alkali-tolerant leguminous forage *Lathyrus quinquenervius*. Soil Sci Plant Nutri 55:685-697.

Zhang JT, Mu CS, Wang DL, Wang JF, Chen GX (2009). Shoot population recruitment from a bud bank over two seasons of undisturbed growth of *Leymus chinensis*. Botany 87:1242-1249.

Evaluation of Drought Tolerance Indices for Selection of Confectionery Sunflower (*Helianthus anuus* L.) Landraces under Various Environmental Conditions

Esmaeil GHOLINEZHAD[1]*, Reza DARVISHZADEH[2], Iraj BERNOUSI[2]

[1]*Department of Agronomy, Payame Noor University, Tehran, Iran; Gholinezhad1358@yahoo.com (*corresponding author)*
[2]*Department of Plant Breeding and Biotechnology, Urmia University, Urmia, Iran; r.darvishzadeh@urmia.ac.ir, ibernousi@gmail.com*

Abstract

In order to study the response of 56 landraces of confectionery sunflower to drought stress, an experiment was conducted in the experimental field of Agricultural Research Center of West-Azerbaijan in 2012-2013 cropping seasons. The genotypes were evaluated in three separated rectangular 7 × 8 lattice design with two replications in three irrigation treatments including optimum irrigation, moderate stress and sever stress where irrigation was done after depletion of 50%, 70% and 90% of available water, respectively. Nine drought tolerance indices including mean productivity (MP), stress tolerance index (STI), geometric mean productivity (GMP), harmonic mean (HARM), stress non-stress production index (SNPI), yield index (YI), drought resistance index (DI), modified stress tolerance index in optimum irrigation (M_PSTI), and modified stress tolerance index in moderate and severe stress (M_sSTI) were calculated based on grain yield under well-watered (Y_P), moderate drought stress (Y_S-mild) and severe drought stress (Y_S-severe) conditions. Grain yield in well-watered (YP), mild and severe stress conditions was significantly and positively correlated with MP, STI, GMP, HARM, SNPI, YI, DI, M_PSTI and M_SSTI. Results of this study showed that in moderate drought stress conditions the indices MP, STI, GMP and HARM could be used as the most suitable indicators for screening drought tolerant genotypes. Based on cluster analysis, the studied genotypes were grouped in 3 clusters in each one of water treatment conditions. To select drought tolerant genotypes in severe drought stress conditions, it is suggested to use indices MP, STI, GMP, HARM, SNPI and M_SSTI because these indices supports stable and high yield in both non-stress and stress conditions.

Keywords: biplot presentation, confectionary sunflower, cluster analysis, landraces

Introduction

Sunflower (*Helianthus annuus* L.) is one of the 67 species in the genus *Helianthus*. Nowadays, two main types of sunflower are grown: for oilseed production and non-oilseed or confectionery types (Salunkhe *et al.,* 1999). Confectionery sunflower produces large seeds with low oil content and is used in baking and snack applications (Lu and Hoeft, 2007). Confectionery kernels are roasted and salted, or roasted and no salt added and marketed as edible chips. The confectionery type is one of the most popular and important crops in Iran. It is cultivated in all parts of the country especially in the northwest regions. Abiotic stresses such as cold, salinity, heat and water stress is the principal cause of crop failure worldwide which dipping average yields by more than 50% (Jaleel, 2007).

The different strategies have been suggested for selecting resistant and relatively resistant genotypes to drought stress. Fisher and Maurer (1978) stated that seed yield in drought environment could be considered as drought resistant index. Blum (1988) reported that selection of genotypes for drought resistant must be associated with high yield in non-stress environments. Drought indices based on loss of yield under stress conditions compared to normal state have been used for screening drought tolerant genotypes (Mitra, 2001). Several selection indices such as geometric mean productivity (GMP) and stress tolerance index (STI) by Fernandez (1992), mean productivity (MP) and tolerance index (TOL) by Rosielle and Hamblin (1981), harmonic mean (HARM) by Jafari et al. (2009), stress susceptibility index (SSI) by Fisher and Maurer (1978), yield stability index (YSI) by Bouslama and Schapaugh (1984) and yield index (YI) by (Gavuzzi *et al.,* 1997) were used for screening drought tolerant genotypes.

Khodarahmpour *et al.* (2011) showed that the STI and GMP indices were the more accurate criteria for selection of heat tolerant and high yielding genotypes. Eivazy *et al.* (2013) indicated that MP index was the best criterion for selecting genotypes with high grain yield at both well watered and drought stressed conditions. According to Ganjali *et al.* (2009), there were positive and highly significant correlations among MP, GMP, STI and HARM with yield in stressed and non-stressed conditions. Jabbari *et al.* (2008) studied drought tolerant indices in sunflower genotypes and found significant and positive correlation between STI and grain yield under non-stressed as well as mild and severe drought stresses conditions. Daneshian and Jonoubi (2008) reported that MP, GMP and STI are the most useful indices for evaluating tolerance to drought stress in sunflower. In some researches, YI led to ranking genotypes based on yield in stress conditions (Sio-Se Mardeh *et al.,* 2006). Drikvand *et al.* (2012) exhibited that the most suitable criteria for the identification of genotypes under irrigated and rain fed conditions were GMP, MP and STI. Anwar *et al.* (2011) observed that there were positive and significant correlations between grain yield and MP, HARM, GMP, STI and K1STI under irrigated and stressed conditions. Farshadfar *et al.* (2013) reported that K_1STI, K_2STI, SSPI, RDI, ATI, SNPI and DI could be used as the most suitable indicators for screening drought tolerant cultivars in wheat. Pourdad *et al.* (2008) indicated that cluster analysis based on STI, MP, GMP, YS and YP classify genotypes into three groups with different susceptibility to drought stress.

According to Fernandez (1992) report, genotypes can be classified into four groups based on their performance in stressed and non-stressed conditions: group A: genotypes with high yield under both stressed and non-stressed conditions, group B: genotypes with high yield under non-stressed conditions, group C: genotypes with high yield under stressed conditions, group D: genotypes with poor performance under both stressed and non-stressed conditions. A suitable index must be able to distinguish group A from the other groups. Biplot is an exploratory data visualization technique to display the multivariate data into a two dimensional scatter plot. The concept of biplot was first developed by Gabriel (1971). This technique has extensively been used in the analysis of multi-environmental trails (Ahmadi *et al.,* 2000; Farshadfar *et al.,* 2001; Golabadi *et al.,* 2006). The objectives of the present study were to evaluate several drought tolerance indices and identify drought- tolerant genotypes in confectionery sunflower.

Materials and methods

In order to evaluate the effect of drought stress on grain yield of confectionery sunflower, an experiment was carried out in the experimental field of Agricultural Research center of West-Azerbaijan in 2012-2013. 56 confectionery sunflower accessions were evaluated in three separated rectangular 7 × 8 lattice design with two replications under three irrigation treatments including optimum irrigation (irrigation after depletion of 50% of available water), moderate stress (irrigation after depletion of 70% of available water) and sever stress (irrigation after depletion of 90% of available water).

The soil physical and chemical characteristic of the experimental area is presented in Tab. 1. Mean monthly rainfall, humidity and temperature were recorded through the growing period (Tab. 2). Soil moisture samples were taken from 0-30 and 30-60 cm depth. Weight moisture percentage was determined by pressure plate (armfield CAT.REF: FEL13B-1 Serial Number: 6353 A 24S98). Field capacity and wilting point were 26 and 14, respectively. In order to obtain the irrigation times, soil samples were taken by auger from root development depth in each irrigation regimes 48 hours post irrigation. Irrigation intervals were determined from soil moisture percentage 20, 17.6 and 15.2 days for severe, moderate and well-watered conditions, respectively. Irrigation volume was calculated by equation 1 (Alizadeh, 2009):

$$V = \frac{(fc - \Theta m) \times \rho \times Droot \times A}{Ei}$$

where:

V= irrigation water volume (m^3), Θm = soil moisture weight percentage per replication, A= irrigated area (m^2), FC= field capacity, ρm = soil density (g cm^{-3}), Droot= developmental root depth (m).

Required water volume per irrigation regimes calculated and distributed equally by flume and chronometer with water distribution efficiency of 90 percent. Plants were harvested at maturity, and then the grain yield was recorded for every plot. Harvesting area was 3.6 m^2 from the two middle lines. The drought tolerance indices were calculated for every genotype using the corresponding well-watered and water-stressed plots. Drought tolerance indices were calculated by using the equations cited in Tab. 3.

Analysis of variance performed using PROC GLM in the SAS software (SAS Institute Inc., Cary, NC, USA). The comparison of the means was done by Tukey's test at 5% probability level. Correlations between grain yield per plot in each of the water regimes and drought tolerance indices were determined using SAS PROC CORR. The classification of genotypes using grain yield per plot in each of the water regimes and drought tolerance indices data was performed by Ward algorithm based on squared Euclidean distances. Biplot analysis was used to classify the drought tolerance indices as well as the confectionery sunflower landraces (Mohammadi et al., 2010). Data processing was performed using SPSS 20.00 statistical software (SPSS/PC-20, SPSS Inc., Chicago, IL, USA; http://www.spss.com).

Tab. 1. Chemical and physical properties of farm soil at depth of 0-30 cm

Soil texture	Soil density g cm^{-3}	Electrical conductivity (ds/m)	pH	Percentage of saturation (%)	Lime (%)	Clay (%)	Silt (%)	Sand (%)	Carbon organic (%)	Nitrogen (%)	Phosphor (ppm)	Potassium (ppm)
Clay loam	1.4	0.8	8	47	17	35	37	28	1.2	0.12	12	375

Tab. 2. Monthly climate data during sunflower growing season

Meteorological Parameters	Month					
	Mars	April	May	June	July	August
Maximum Temperature (°C)	16.6	23.3	28.2	30.1	33.1	28.6
Minimum Temperature (°C)	3.1	8.3	11.9	15.3	16.2	13.1
Mean Temperature (°C)	9.9	15.8	20	22.7	24.6	20.9
Total Rainfall (mm)	31.9	15	18.8	9.2	1.8	8.4
Total Evaporation (mm)	81.9	181.9	255.9	269.3	263.4	200.4
Mean Relative Humidity (%)	58	56	48	52	46	52

Tab. 3. Names of local landraces

Names of local landraces					
No.		No.		No.	
1	Saghez 1	20	Salmas 2	39	Hamadan 2
2	Anghane 4	21	Vaghaslou-Olya 4	40	Shabestar-Kouzeh-Kanan 3
3	Urmia-Barouj	22	Salmas-Gharaghashlagh-Pesteii	41	Saghez 4
4	Urmia-Maranghalou	23	Lalalou-Torab 2	42	Saghez 5
5	Marand-Dizaj- Ghalami	24	Shirabad 2	43	Saghez 3
6	Jabalkandi 2	25	Gharagoz 1	44	Shahroud 2
7	Salmas - Sadaghian	26	Vaghaslou-Sofla 1	45	Alibaglou 1
8	Babaghanje 6	27	Khanneshan 1	46	Baneh 2
9	Miyaneh-Basin	28	Heydarlou 1	47	Salmas-Gharagheshlagh- Ghalami
10	Boucan	29	Saribaglou 5	48	Marand-1389-2
11	Urmia - Nuoshinshahr	30	Chongharalou-Yekan 4	49	Salmas-Gharaghashlagh- Badami
12	Karimabad	31	Maranghalou 6	50	Shabestar-Kouzeh Kanan 1
13	Vaghaslou-Olya 1	32	Abajalou 1	51	Sanandaj
14	Vaghaslou-Olya 3	33	Hamadan 1	52	Shabestar-Kouzeh-Kanan 2
15	Ordoshahi 1	34	Saghez 2	53	Baneh 3
16	Marana-Yamchi-Pesteii	35	Piranshahr-Serokani	54	Piranshahr-Baleban
17	Mazandaran-Tirtash	36	Piranshahr Andizeh	55	Baneh 1
18	Sardasht	37	Mashhad	56	Marand-1389-1
19	Marana-Yamchi 4	38	Shahroud 1		

Results and discussion

Comparing landraces based on the resistance / tolerance indices

To investigate suitable drought tolerance indices for screening confectionery sunflower landraces under drought stress conditions, different indices were calculated based on grain yield under stressed and non-stressed environments (Tab. 6 and 7). A suitable index must have a significant correlation with grain yield under both non-stress and stress conditions (Mitra, 2001). Based on the stress tolerance index (STI) and grain yield, the landraces: 2, 12, 14, 16, 25, 28, 29, 31, 32, 33, 47, 49 and 56 were found drought tolerance with the highest STI and grain yield under mild stressed and non-stressed conditions while the landraces: 37 and 38 displayed the lowest amount of STI and grain yield under mild stressed and non-stressed conditions (Tab. 6). Landraces: 2, 7 and 12 had the most STI and grain yield under severe and non-stress conditions while the landraces 36 and 37 had the lowest amount of STI and grain yield under severe and non-stress conditions (Tab. 7). Khodarahmpour *et al.* (2011) and Khalili *et al.* (2012) reported that STI and GMP indices were the more accurate criteria for selection of heat tolerant and high yielding genotypes.

The highest GMP and MP were shown in landraces: 2, 12, 14, 25, 26, 28, 31, 32, 33, 49 and 56 under mild stressed and non-stressed conditions (Tab. 6). The lowest GMP and MP were observed in landraces: 36 and 37 under severe stressed and non-stressed conditions (Tab. 6). Under severe and non-stress conditions, the highest and lowest GMP and MP were observed in the accessions (2 and 12) and (36 and 37), respectively (Tab. 7). Based on Rosiele and Hamblin (1981) decline, the stress tolerance index (STI) and mean productivity (MP) are defined as the difference in yield and the average yield between stressed and non-stressed conditions, respectively.

The highest and lowest HARM were observed in the landraces: (12, 25, 26, 32, 33, 49 and 56) and (9, 10 and 41), respectively, in the mild stressed and non-stressed conditions (Tab. 6). Under severe stressed and non-stressed conditions, the highest and lowest HARM were related to landraces: (5, 7, 12 and 29) and (36, 37, 40 and 41), respectively (Tab. 7). Mevlut and Sait (2011) indicated that the genotypes with high STI usually present high difference in yield between two different conditions. They reported in general, common ranks for studied landraces by GMP, MP and STI, which suggests that these three indices are similar for screening drought tolerant landraces.

Based on the yield index (YI), the landraces 12, 25, 32 and 56 were the most tolerant landraces under mild stressed and non-stressed conditions (Tab. 6). Landraces: 5, 7, 12, 29, 31 and 43 had the highest YI under severe stressed

Tab. 4. Drought tolerance indices

Index name	Outcome	Equation	Reference
Mean Productivity (MP)	the landraces with high value of this index will be more desirable	$MP = \frac{Y_S + Y_P}{2}$	(Rosielle and Hamblin, 1981)
Geometric Mean Productivity (GMP)	the landraces with high value of this index will be more desirable	$GMP = \sqrt{(Y_S)(Y_P)}$	(Kristin *et al.*, 1997)
Stress Tolerance Index (STI)	The landraces with high STI values will be tolerant to drought stress	$STI = \frac{(Y_S)(Y_P)}{(\bar{Y}_P)^2}$	(Fernandez, 1992)
Harmonic Mean (HARM)	the landraces with high value of this index will be more desirable	$HARM = \frac{2(Yp \times Ys)}{Yp + Ys}$	(Jafari *et al.*, 2009)
Yield Index (YI)	the landraces with high value of this index will be suitable for drought stress condition	$YI = \frac{Y_S}{\bar{Y}_P}$	(Gavuzzi *et al.*, 1997)
Drought Resistance Index (DI)	the landraces with high value of this index will be suitable for drought stress condition	$DI = \frac{[Y_S \times (\frac{Y_S}{Y_P})]}{\bar{Y}_S}$	(Lan, 1998)
Stress Non-Stress Production Index	the landraces with high value of this index will be suitable for drought stress condition	$SNPI = \left[\frac{Yp + Ys}{Yp - Ys}\right]^{\frac{1}{3}} \times [Yp \times Ys \times Ys]^{\frac{1}{3}}$	(Moosavi *et al.*, 2008)
Modified Stress Tolerance Index in Optimum Irrigation (M_PSTI)	the landraces with high value of this index will be suitable for drought stress condition	$MpSTI = \frac{[(Y)]p)^2}{[(\bar{Y})]p)^2} \times STI$	(Farshadfar and Sutka, 2002)
Modified Stress Tolerance Index in Moderate and Severe Stress (M_SSTI)	the landraces with high value of this index will be suitable for drought stress condition	$MsSTI = \frac{[(Y)]s)^2}{[(\bar{Y})]s)^2} \times STI$	(Farshadfar and Sutka, 2002)

YS and *YP* are stress and optimal (potential) yield of a given genotype, respectively. $\bar{Y}_S$ and $\bar{Y}_P$ are average yield of all genotypes under stress and optimal conditions, respectively.

Tab. 5. Simple and combined analyses of variance for grain yield in 56 Iranian confectionery sunflower landraces under non-stressed, moderate and severe drought stressed conditions.

Source of variation	Individual				Combined	
	df	MS Optimum conditions	MS Mild stressed conditions	MS Severe stressed conditions	df	MS
(Environment)	-	-	-	-	2	46277099.58**
Replication	1	1293418.65*	707.52 ns	326538 **	-	-
Replication (Environment)	-	-	-	-	3	540221.39
Block (Replication)	14	195166.60 ns	250894.24 ns	119807.47 **	-	-
Block (Environment × Replication)	-	-	-	-	42	507269.85**
Genotype	55	878661.58 **	813047.52 **	286444.80 **	55	1380455.49**
(Environment × Genotype)	-	-	-	-	110	298849.20*
Error	41	326456.17	263024.93	51214.32	123	213565.1
C.V.(%)	-	21.32	24.84	16.23	-	22.58

** , * and Ns significant at the 1%, 5% probability levels and non significant respectively

conditions (Tab. 7). Based on M_PSTI and M_SSTI the landraces 2, 8, 12, 14, 25, 26, 32, 33 and 47 were the most tolerant under mild stressed conditions (Tab. 6). Under severe stressed conditions, the highest M_PSTI and M_SSTI were related to landraces 7, 12, 25, and 49 (Tab. 7). With regard to stress non-stress production index (SNPI) under mild and severe stressed conditions: landraces (2, 8, 14, 26 and 47) and (5, 7, 22, 29, 31, 34 and 43) were the most tolerant accessions, respectively (Tab. 6 and 7). Based on drought index (DI) under mild stressed conditions, the high tolerant landraces were 12, 16, 25, 31, 32 and 56 (Tab. 6). Based on drought index (DI) under severe stressed conditions, the most tolerant landraces were 5, 7, 9 and 43 (Tab. 7). Ilker et al. (2011) concluded that MP, GMP and STI are convenient indices to choose high yielding wheat genotypes in both stressed and non-stressed conditions.

Farshadfar et al. (2013) reported that K1STI, K2STI, SNPI and DI indices can be used as the most suitable indicators for screening drought tolerant cultivars in wheat.

Totally, based on MP, STI, GMP, HARM, YI, DI and MSSTI indices and ranking method the landraces: 2, 12, 25, 26, 31, 32, 33 and 56 were as the most drought tolerant and the landraces: 9, 10, 19, 35, 36, 37, 38, 40 and 41 were as the most susceptible under mild stressed conditions (Tab. 6). Under severe stressed conditions, based on MP, STI, GMP, HARM, YI, DI, SNPI and MSSTI indices and ranking method the landraces: 2, 5, 7, 12, 29 and 31 were as the most drought tolerant and the landraces: 9, 10, 17, 35, 36, 37, 40, 41 and 54 were as the most susceptible landraces (Tab. 7).

Cluster analysis

To determine the most desirable drought tolerance criteria, the correlation coefficient between Y_P, Y_S and other quantitative drought tolerance indices in mild and severe stressed conditions were calculated (Tab. 8 and 9). The results indicated that in moderate and severe drought stressed conditions there were positive and significant correlations among Y_P and MP, STI, GMP, HARM, SNPI, YI, DI, MPSTI and MSSTI. Also in moderate and severe drought stressed conditions, there were positive and significant correlations among Ys and MP, STI, GMP, HARM, YI, DI, MPSTI and MSSTI. However, indices

Tab. 6. Resistance/tolerance indices for 56 Iranian confectionery sunflower landraces under moderate drought stressed and non-stressed conditions.

No	Genotype	Rank	HARM	Rank	GMP	Rank	STI	Rank	MP	Rank	Ys-mild	Rank	Yp
1	Saghez 1	19	2.60	19.5	2.65	19.5	0.93	18.5	2.71	22	2169.72	15	3241.94
2	Anghane 4	13	2.81	4	3.38	4	1.51	1	4.06	38	1807.79	1	6310.77
3	Urmia-Barouj	25	2.41	24.5	2.51	24	0.84	23	2.62	35	1883.44	12	3355.65
4	Urmia-Maranghalou	39.5	2.10	38	2.20	38	0.64	34.5	2.31	42	1615.43	26	3001.82
5	Marand-Dizaj-Ghalami	27	2.36	27	2.42	27	0.78	28	2.49	33	1917.77	21	3055.23
6	Jabalkandi 2	29	2.35	31	2.37	31	0.74	31	2.39	23	2085.56	33	2695.38
7	Salmas - Sadaghian	18	2.61	19.5	2.65	19.5	0.93	21	2.69	19	2211.75	17	3177.21
8	Babaghanje 6	20	2.59	17.5	2.68	17.5	0.95	15	2.78	24	2048.18	7	3505.74
9	Miyaneh-Basin	54	1.25	54	1.25	54	0.21	54	1.25	52	1273.47	55	1221.97
10	Boucan	52	1.33	51	1.43	51	0.27	48.5	1.54	54	973.88	43	2103.75
11	Urmia - Nuoshinshahr	14.5	2.72	16	2.73	15.5	0.99	16.5	2.75	13	2482.86	25	3012.03
12	Karimabad	1	3.67	1	3.67	1	1.78	2	3.67	2	3611.45	3	3728.20
13	Vaghaslou-Olya 1	42	1.80	42	1.92	42	0.49	42	2.05	47	1328.74	31	2768.42
14	Vaghaslou-Olya 3	9	3.04	9	3.07	9	1.25	9	3.09	12	2686.43	9	3500.27
15	Ordoshahi 1	23	2.49	24.5	2.51	25	0.83	25.5	2.53	20	2208.24	29	2852.48
16	Marana-Yamchi-Pesteii	12	2.85	13	2.85	13.5	1.07	14	2.85	10	2889.97	30	2802.96
17	Mazandaran-Tirtash	43	1.75	43	1.75	43	0.41	43	1.75	36	1832.44	50	1675.26
18	Sardasht	41	2.05	41	2.07	41	0.57	41	2.09	37	1816.67	40	2356.11
19	Marana-Yamchi 4	48	1.50	48	1.52	48	0.31	48.5	1.54	50	1299.26	47	1783.27
20	Salmas 2	17	2.66	17.5	2.68	17.5	0.95	20	2.70	16	2367.56	22.5	3037.67
21	Vaghaslou-Olya 4	31	2.33	29.5	2.38	29.5	0.75	29	2.43	32	1927.74	28	2935.50
22	Salmas-Gharaghashlagh-Pesteii	22	2.52	23	2.53	23	0.85	25.5	2.53	15	2371.90	34	2691.73
23	Lalalou-Torab 2	21	2.53	21	2.62	21	0.91	18.5	2.71	28	2002.13	10	3418.40
24	Shirabad 2	35	2.27	33.5	2.29	33.5	0.70	34.5	2.31	26	2026.19	36	2591.75
25	Gharagoz 1	2	3.65	2	3.65	2	1.77	3	3.65	1	3778.83	5	3528.58
26	Vaghaslou-Sofla 1	4.5	3.27	5	3.28	5	1.43	5	3.30	7	2965.42	4	3635.39
27	Khanneshan 1	14.5	2.72	15	2.74	15.5	0.99	16.5	2.75	14	2467.21	22.5	3037.68
28	Heydarlou 1	10	3.00	11	3.00	11	1.20	12	3.01	9	2908.40	20	3103.55
29	Saribaglou 5	11	2.97	12	2.98	12	1.17	13	2.98	11	2853.26	19	3105.27
30	Chongharalou-Yekan 4	32	2.29	33.5	2.29	33.5	0.70	37	2.29	17	2344.98	42	2243.27
31	Maranghalou 6	8	3.05	10	3.05	10	1.23	10	3.05	5	3146.80	27	2961.98
32	Abajalou 1	3	3.55	3	3.55	3	1.67	4	3.55	3	3596.70	8	3505.17
33	Hamadan 1	6	3.26	7	3.26	6.5	1.41	6.5	3.27	6	3033.53	6	3513.70
34	Saghez 2	39.5	2.10	39.5	2.13	39.5	0.60	39	2.16	39	1799.07	37	2523.31
35	Piranshahr-Serokani	49.5	1.48	50	1.48	50	0.29	51	1.48	45	1439.76	51	1529.26
36	Piranshahr Andizeh	55	0.70	55	0.78	55	0.08	55	0.86	56	493.39	54	1222.34
37	Mashhad	56	0.63	56	0.65	56	0.06	56	0.66	55	515.86	56	807.34
38	Shahroud 1	49.5	1.48	49	1.50	49	0.30	50	1.51	49	1315.93	49	1700.66
39	Hamadan 2	47	1.59	47	1.63	46.5	0.35	46	1.68	51	1285.08	44	2071.94

40	Shabestar-Kouzeh-Kanan 3	51	1.37	52	1.37	52	0.25	52	1.38	48	1324.07	53	1426.11
41	Saghez 4	53	1.30	53	1.31	53	0.23	53	1.32	53	1177.20	52	1458.54
42	Saghez 5	37	2.13	37	2.24	37	0.66	33	2.34	41	1645.39	24	3036.84
43	Saghez 3	33.5	2.28	35.5	2.28	35.5	0.69	38	2.28	21	2183.06	39	2381.37
44	Shahroud 2	44	1.69	44	1.71	44	0.39	44.5	1.72	44	1502.04	46	1942.53
45	Alibaglou 1	38	2.12	39.5	2.13	39.5	0.60	40	2.14	31	1948.10	41	2331.95
46	Baneh 2	24	2.48	22	2.55	22	0.86	22	2.63	27	2009.26	16	3241.56
47	Salmas-Gharagheshlagh-Ghalami	16	2.67	14	2.85	13.5	1.07	11	3.03	29	1987.24	2	4076.50
48	Marand-1389-2	33.5	2.28	28	2.40	28	0.77	24	2.54	40	1721.08	11	3356.18
49	Salmas-Gharaghashlagh-Badami	7	3.13	8	3.14	8	1.30	8	3.14	8	2949.06	13	3340.35
50	Shabestar-Kouzeh Kanan 1	29	2.35	32	2.35	32	0.73	32	2.35	18	2305.50	38	2390.03
51	Sanandaj	29	2.35	29.5	2.38	29.5	0.75	30	2.40	25	2045.97	32	2756.99
52	Shabestar-Kouzeh-Kanan 2	45	1.66	45	1.69	45	0.38	44.5	1.72	46	1403.79	45	2030.41
53	Baneh 3	36	2.26	35.5	2.28	35.5	0.69	36	2.30	30	1979.44	35	2624.47
54	Piranshahr-Baleban	46	1.63	46	1.64	46.5	0.35	47	1.64	43	1522.93	48	1759.33
55	Baneh 1	26	2.37	26	2.44	26	0.79	27	2.52	34	1902.77	18	3141.23
56	Marand-1389-1	4.5	3.27	6	3.27	6.5	1.41	6.5	3.27	4	3228.56	14	3307.31

Tab. 6. (Continued)

No	Genotype	Rank	M$_S$STI	Rank	M$_P$STI	Rank	DI	Rank	YI	Rank	SNPI
1	Saghez 1	19	1.03	16	1.30	28	0.70	22	0.79	9	4.45
2	Anghane 4	16	1.16	1	7.97	53	0.25	38	0.66	1	9.08
3	Urmia-Barouj	30	0.70	17	1.25	42	0.51	34.5	0.69	8	4.50
4	Urmia-Maranghalou	41	0.39	31	0.77	50	0.42	42	0.59	21	3.68
5	Marand-Dizaj- Ghalami	32	0.67	26	0.96	37.5	0.58	32.5	0.70	15	3.99
6	Jabalkandi 2	28	0.76	33	0.72	23	0.78	23	0.76	26	3.24
7	Salmas - Sadaghian	18	1.07	18	1.24	25	0.75	19.5	0.81	12	4.32
8	Babaghanje 6	22	0.94	11	1.55	37.5	0.58	24.5	0.75	5	4.91
9	Miyaneh-Basin	52	0.08	54	0.04	31.5	0.64	52	0.46	51	-0.63
10	Boucan	54	0.06	46	0.16	54	0.22	54	0.36	39	1.91
11	Urmia - Nuoshinshahr	13	1.43	21	1.19	16	0.99	13	0.91	18	3.77
12	Karimabad	2	5.46	2	3.28	3	1.70	2	1.32	23	3.46
13	Vaghaslou-Olya 1	43.5	0.20	39	0.49	52	0.31	48	0.48	29	3.06
14	Vaghaslou-Olya 3	11	2.11	9	2.02	15	1.00	12	0.98	4	5.02
15	Ordoshahi 1	21	0.96	27	0.90	21	0.83	19.5	0.81	22	3.56
16	Marana-Yamchi-Pesteii	12	2.10	24	1.12	6	1.45	10	1.05	53	-2.29
17	Mazandaran-Tirtash	42	0.32	47.5	0.15	18	0.97	36	0.67	44.5	1.47
18	Sardasht	39	0.44	42	0.42	29	0.68	38	0.66	34	2.59
19	Marana-Yamchi 4	49.5	0.12	49	0.13	46.5	0.46	50.5	0.47	42	1.65
20	Salmas 2	15	1.25	22	1.16	20	0.90	16.5	0.86	16	3.95
21	Vaghaslou-Olya 4	34	0.65	28	0.86	34	0.61	32.5	0.70	19.5	3.76
22	Salmas-Gharaghashlagh-Pesteii	17	1.12	29.5	0.81	14	1.01	15	0.87	32	2.90
23	Lalalou-Torab 2	25	0.85	15	1.40	39	0.57	28	0.73	6	4.71
24	Shirabad 2	32	0.67	35	0.62	24	0.77	26	0.74	30	3.03
25	Gharagoz 1	1	5.92	3	2.91	1	1.96	1	1.38	56	-4.57
26	Vaghaslou-Sofla 1	6	2.95	5	2.50	12	1.17	7.5	1.08	3	5.20
27	Khanneshan 1	14	1.42	19	1.21	18	0.97	14	0.90	17	3.87
28	Heydarlou 1	9	2.37	12	1.53	7	1.32	9	1.06	27	3.13
29	Saribaglou 5	10	2.24	13	1.50	8.5	1.27	11	1.04	25	3.36
30	Chongharalou-Yekan 4	24	0.90	40	0.46	11	1.19	16.5	0.86	52	-1.78
31	Maranghalou 6	7	2.87	14	1.43	4	1.62	5	1.15	55	-3.18
32	Abajalou 1	3	5.07	4	2.72	2	1.79	3	1.31	54	-3.07
33	Hamadan 1	5	3.05	7	2.31	8.5	1.27	6	1.11	7	4.66

34	Saghez 2	38	0.46	38	0.51	33	0.62	38	0.66	31	2.94
35	Piranshahr-Serokani	47.5	0.14	51	0.09	30	0.66	45	0.53	47	0.94
36	Piranshahr Andizeh	55.5	0.00	55	0.02	56	0.10	56	0.18	49	0.72
37	Mashhad	55.5	0.00	56	0.00	55	0.16	55	0.19	50	0.44
38	Shahroud 1	49.5	0.12	50	0.11	43	0.49	48	0.48	43	1.51
39	Hamadan 2	47.5	0.14	44	0.20	51	0.39	50.5	0.47	37	2.08
40	Shabestar-Kouzeh-Kanan 3	51	0.10	52	0.07	35.5	0.60	48	0.48	48	0.89
41	Saghez 4	53	0.07	53	0.06	46.5	0.46	53	0.43	46	1.14
42	Saghez 5	40	0.42	29.5	0.81	48.5	0.43	41	0.60	19.5	3.76
43	Saghez 3	27	0.77	37	0.52	18	0.97	21	0.80	36	2.17
44	Shahroud 2	43.5	0.20	45	0.19	40.5	0.56	44	0.55	40	1.88
45	Alibaglou 1	36	0.54	41	0.43	22	0.79	31	0.71	35	2.44
46	Baneh 2	26	0.82	20	1.20	35.5	0.60	28	0.73	10	4.38
47	Salmas-Gharagheshlagh-Ghalami	20	0.99	6	2.36	44.5	0.47	28	0.73	2	5.88
48	Marand-1389-2	37	0.53	23	1.14	48.5	0.43	40	0.63	11	4.35
49	Salmas-Gharaghashlagh-Badami	8	2.66	10	1.93	10	1.26	7.5	1.08	13.5	4.14
50	Shabestar-Kouzeh Kanan 1	23	0.91	36	0.55	13	1.08	18	0.84	41	1.71
51	Sanandaj	29	0.73	32	0.75	26	0.74	24.5	0.75	24	3.40
52	Shabestar-Kouzeh-Kanan 2	46	0.17	43	0.21	44.5	0.47	46	0.51	38	2.05
53	Baneh 3	35	0.63	34	0.63	27	0.72	30	0.72	28	3.12
54	Piranshahr-Baleban	45	0.19	47.5	0.15	31.5	0.64	43	0.56	44.5	1.47
55	Baneh 1	32	0.67	25	1.03	40.5	0.56	34.5	0.69	13.5	4.14
56	Marand-1389-1	4	3.46	8	2.05	5	1.53	4	1.18	33	2.61

Tab. 7. Resistance/tolerance indices for 56 Iranian confectionery sunflower landraces under severe drought stressed and non-stressed conditions.

No	Genotype	Rank	HARM	Rank	GMP	Rank	STI	Rank	MP	Rank	Ys-mild	Rank	Yp
1	Saghez 1	19	2.60	19.5	2.65	19.5	0.93	18.5	2.71	22	2169.72	15	3241.94
2	Anghane 4	13	2.81	4	3.38	4	1.51	1	4.06	38	1807.79	1	6310.77
3	Urmia-Barouj	25	2.41	24.5	2.51	24	0.84	23	2.62	35	1883.44	12	3355.65
4	Urmia-Maranghalou	39.5	2.10	38	2.20	38	0.64	34.5	2.31	42	1615.43	26	3001.82
5	Marand-Dizaj-Ghalami	27	2.36	27	2.42	27	0.78	28	2.49	33	1917.77	21	3055.23
6	Jabalkandi 2	29	2.35	31	2.37	31	0.74	31	2.39	23	2085.56	33	2695.38
7	Salmas - Sadaghian	18	2.61	19.5	2.65	19.5	0.93	21	2.69	19	2211.75	17	3177.21
8	Babaghanje 6	20	2.59	17.5	2.68	17.5	0.95	15	2.78	24	2048.18	7	3505.74
9	Miyaneh-Basin	54	1.25	54	1.25	54	0.21	54	1.25	52	1273.47	55	1221.97
10	Boucan	52	1.33	51	1.43	51	0.27	48.5	1.54	54	973.88	43	2103.75
11	Urmia - Nuoshinshahr	14.5	2.72	16	2.73	15.5	0.99	16.5	2.75	13	2482.86	25	3012.03
12	Karimabad	1	3.67	1	3.67	1	1.78	2	3.67	2	3611.45	3	3728.20
13	Vaghaslou-Olya 1	42	1.80	42	1.92	42	0.49	42	2.05	47	1328.74	31	2768.42
14	Vaghaslou-Olya 3	9	3.04	9	3.07	9	1.25	9	3.09	12	2686.43	9	3500.27
15	Ordoshahi 1	23	2.49	24.5	2.51	25	0.83	25.5	2.53	20	2208.24	29	2852.48
16	Marana-Yamchi-Pesteii	12	2.85	13	2.85	13.5	1.07	14	2.85	10	2889.97	30	2802.96
17	Mazandaran-Tirtash	43	1.75	43	1.75	43	0.41	43	1.75	36	1832.44	50	1675.26
18	Sardasht	41	2.05	41	2.07	41	0.57	41	2.09	37	1816.67	40	2356.11
19	Marana-Yamchi 4	48	1.50	48	1.52	48	0.31	48.5	1.54	50	1299.26	47	1783.27
20	Salmas 2	17	2.66	17.5	2.68	17.5	0.95	20	2.70	16	2367.56	22.5	3037.67
21	Vaghaslou-Olya 4	31	2.33	29.5	2.38	29.5	0.75	29	2.43	32	1927.74	28	2935.50
22	Salmas-Gharaghashlagh-Pesteii	22	2.52	23	2.53	23	0.85	25.5	2.53	15	2371.90	34	2691.73
23	Lalalou-Torab 2	21	2.53	21	2.62	21	0.91	18.5	2.71	28	2002.13	10	3418.40
24	Shirabad 2	35	2.27	33.5	2.29	33.5	0.70	34.5	2.31	26	2026.19	36	2591.75

25	Gharagoz 1	2	3.65	2	3.65	2	1.77	3	3.65	1	3778.83	5	3528.58
26	Vaghaslou-Sofla 1	4.5	3.27	5	3.28	5	1.43	5	3.30	7	2965.42	4	3635.39
27	Khanneshan 1	14.5	2.72	15	2.74	15.5	0.99	16.5	2.75	14	2467.21	22.5	3037.68
28	Heydarlou 1	10	3.00	11	3.00	11	1.20	12	3.01	9	2908.40	20	3103.55
29	Saribaglou 5	11	2.97	12	2.98	12	1.17	13	2.98	11	2853.26	19	3105.27
30	Chongharalou-Yekan 4	32	2.29	33.5	2.29	33.5	0.70	37	2.29	17	2344.98	42	2243.27
31	Maranghalou 6	8	3.05	10	3.05	10	1.23	10	3.05	5	3146.80	27	2961.98
32	Abajalou 1	3	3.55	3	3.55	3	1.67	4	3.55	3	3596.70	8	3505.17
33	Hamadan 1	6	3.26	7	3.26	6.5	1.41	6.5	3.27	6	3033.53	6	3513.70
34	Saghez 2	39.5	2.10	39.5	2.13	39.5	0.60	39	2.16	39	1799.07	37	2523.31
35	Piranshahr-Serokani	49.5	1.48	50	1.48	50	0.29	51	1.48	45	1439.76	51	1529.26
36	Piranshahr Andizeh	55	0.70	55	0.78	55	0.08	55	0.86	56	493.39	54	1222.34
37	Mashhad	56	0.63	56	0.65	56	0.06	56	0.66	55	515.86	56	807.34
38	Shahroud 1	49.5	1.48	49	1.50	49	0.30	50	1.51	49	1315.93	49	1700.66
39	Hamadan 2	47	1.59	47	1.63	46.5	0.35	46	1.68	51	1285.08	44	2071.94
40	Shabestar-Kouzeh-Kanan 3	51	1.37	52	1.37	52	0.25	52	1.38	48	1324.07	53	1426.11
41	Saghez 4	53	1.30	53	1.31	53	0.23	53	1.32	53	1177.20	52	1458.54
42	Saghez 5	37	2.13	37	2.24	37	0.66	33	2.34	41	1645.39	24	3036.84
43	Saghez 3	33.5	2.28	35.5	2.28	35.5	0.69	38	2.28	21	2183.06	39	2381.37
44	Shahroud 2	44	1.69	44	1.71	44	0.39	44.5	1.72	44	1502.04	46	1942.53
45	Alibaglou 1	38	2.12	39.5	2.13	39.5	0.60	40	2.14	31	1948.10	41	2331.95
46	Baneh 2	24	2.48	22	2.55	22	0.86	22	2.63	27	2009.26	16	3241.56
47	Salmas-Gharagheshlagh-Ghalami	16	2.67	14	2.85	13.5	1.07	11	3.03	29	1987.24	2	4076.50
48	Marand-1389-2	33.5	2.28	28	2.40	28	0.77	24	2.54	40	1721.08	11	3356.18
49	Salmas-Gharaghashlagh-Badami	7	3.13	8	3.14	8	1.30	8	3.14	8	2949.06	13	3340.35
50	Shabestar-Kouzeh Kanan 1	29	2.35	32	2.35	32	0.73	32	2.35	18	2305.50	38	2390.03
51	Sanandaj	29	2.35	29.5	2.38	29.5	0.75	30	2.40	25	2045.97	32	2756.99
52	Shabestar-Kouzeh-Kanan 2	45	1.66	45	1.69	45	0.38	44.5	1.72	46	1403.79	45	2030.41
53	Baneh 3	36	2.26	35.5	2.28	35.5	0.69	36	2.30	30	1979.44	35	2624.47
54	Piranshahr-Baleban	46	1.63	46	1.64	46.5	0.35	47	1.64	43	1522.93	48	1759.33
55	Baneh 1	26	2.37	26	2.44	26	0.79	27	2.52	34	1902.77	18	3141.23
56	Marand-1389-1	4.5	3.27	6	3.27	6.5	1.41	6.5	3.27	4	3228.56	14	3307.31

Tab. 7. (Continued)

No	Genotype	Rank	M_SSTI	Rank	M_PSTI	Rank	DI	Rank	YI	Rank	SNPI
1	Saghez 1	31	0.46	22.5	0.76	40.5	0.36	35	0.46	36.5	2.29
2	Anghane 4	8	1.34	1	6.44	46.5	0.24	25	0.53	22	2.78
3	Urmia-Barouj	18	0.90	11	1.04	28.5	0.53	19.5	0.57	21	2.84
4	Urmia-Maranghalou	36.5	0.33	28	0.56	43	0.33	39.5	0.43	39	2.11
5	Marand-Dizaj- Ghalami	3	1.84	12.5	1.03	4	1.00	3	0.75	3	4.06
6	Jabalkandi 2	30	0.52	35	0.49	28.5	0.53	30.5	0.51	30	2.59
7	Salmas - Sadaghian	1	2.37	5	1.25	3	1.11	1	0.81	2	4.44
8	Babaghanje 6	43	0.24	22.5	0.76	52.5	0.21	47	0.36	46	1.85
9	Miyaneh-Basin	40.5	0.26	51.5	0.05	2	1.26	25	0.53	56	-3.08
10	Boucan	51	0.07	46	0.13	54	0.20	52	0.28	51	1.39
11	Urmia - Nuoshinshahr	13	1.17	19	0.85	12.5	0.76	11.5	0.65	11	3.34
12	Karimabad	2	1.98	2	1.80	12.5	0.76	5.5	0.72	8	3.63
13	Vaghaslou-Olya 1	27.5	0.56	30	0.54	26	0.54	28.5	0.52	25	2.63
14	Vaghaslou-Olya 3	23	0.73	9	1.09	35.5	0.43	25	0.53	28	2.61
15	Ordoshahi 1	19	0.82	26	0.66	19.5	0.66	17	0.59	19	2.98

16	Marana-Yamchi-Pesteii	27.5	0.56	29	0.55	28.5	0.53	28.5	0.52	28	2.61
17	Mazandaran-Tirtash	49	0.11	49.5	0.08	35.5	0.43	47	0.36	45	1.87
18	Sardasht	29	0.54	39	0.34	17	0.69	21	0.55	20	2.88
19	Marana-Yamchi 4	45.5	0.20	47	0.12	25	0.56	39.5	0.43	36.5	2.29
20	Salmas 2	47	0.19	37	0.47	50.5	0.22	49	0.35	48	1.76
21	Vaghaslou-Olya 4	17	0.97	24	0.75	14.5	0.70	15.5	0.62	15.5	3.15
22	Salmas-Gharaghashlagh-Pesteii	11	1.26	27	0.65	5	0.97	7	0.69	6	3.83
23	Lalalou-Torab 2	39	0.27	25	0.74	48.5	0.23	45	0.38	44	1.93
24	Shirabad 2	24	0.71	35	0.49	14.5	0.70	18	0.58	18	3.01
25	Gharagoz 1	10	1.28	3.5	1.35	22	0.62	13	0.64	13	3.17
26	Vaghaslou-Sofla 1	20	0.79	6	1.24	35.5	0.43	25	0.53	24	2.65
27	Khanneshan 1	15	1.03	20.5	0.84	17	0.69	15.5	0.62	15.5	3.15
28	Heydarlou 1	45.5	0.20	32	0.51	50.5	0.22	47	0.36	47	1.79
29	Saribaglou 5	4	1.78	10	1.07	6.5	0.96	4	0.74	4	3.94
30	Chongharalou-Yekan 4	26	0.59	41.5	0.31	9	0.79	19.5	0.57	17	3.14
31	Maranghalou 6	6	1.58	16.5	0.91	6.5	0.96	5.5	0.72	5	3.89
32	Abajalou 1	36.5	0.33	20.5	0.84	45	0.26	42.5	0.41	42	2.03
33	Hamadan 1	22	0.74	8	1.11	35.5	0.43	25	0.53	28	2.61
34	Saghez 2	16	1.02	32	0.51	8	0.93	9.5	0.66	7	3.68
35	Piranshahr-Serokani	51	0.07	51.5	0.05	42	0.35	50	0.31	49	1.59
36	Piranshahr Andizeh	55	0.01	55	0.02	55	0.15	55.5	0.18	55	0.90
37	Mashhad	55	0.01	56	0.00	52.5	0.21	55.5	0.18	54	0.93
38	Shahroud 1	44	0.22	48	0.11	21	0.65	36	0.45	31	2.56
39	Hamadan 2	40.5	0.26	43	0.19	31	0.51	37.5	0.44	38	2.28
40	Shabestar-Kouzeh-Kanan 3	53	0.03	53	0.04	46.5	0.24	53	0.25	52	1.25
41	Saghez 4	55	0.01	54	0.03	56	0.14	54	0.19	53	0.95
42	Saghez 5	12	1.21	18	0.89	11	0.77	9.5	0.66	10	3.38
43	Saghez 3	5	1.60	32	0.51	1	1.39	2	0.78	1	5.91
44	Shahroud 2	48	0.16	45	0.14	38	0.42	44	0.39	43	1.95
45	Alibaglou 1	33	0.43	41.5	0.31	23	0.60	30.5	0.51	26	2.62
46	Baneh 2	14	1.14	12.5	1.03	19.5	0.66	14	0.63	14	3.16
47	Salmas-Gharagheshlagh-Ghalami	35	0.41	3.5	1.35	48.5	0.23	42.5	0.41	40	2.10
48	Marand-1389-2	25	0.60	16.5	0.91	39	0.41	32.5	0.50	33	2.48
49	Salmas-Gharaghashlagh-Badami	9	1.30	7	1.17	17	0.69	11.5	0.65	12	3.28
50	Shabestar-Kouzeh Kanan 1	34	0.42	40	0.33	24	0.57	32.5	0.50	32	2.55
51	Sanandaj	32	0.45	35	0.49	33	0.47	34	0.49	34	2.42
52	Shabestar-Kouzeh-Kanan 2	42	0.25	44	0.18	28.5	0.53	37.5	0.44	35	2.30
53	Baneh 3	38	0.28	38	0.37	40.5	0.36	41	0.42	41	2.08
54	Piranshahr-Baleban	51	0.07	49.5	0.08	44	0.28	51	0.30	50	1.50
55	Baneh 1	7	1.35	14	1.00	10	0.78	8	0.67	9	3.45
56	Marand-1389-1	21	0.75	15	0.95	32	0.48	22	0.54	23	2.68

such as MP, STI, GMP, HARM, SNPI, YI, DI and M_SSTI were the best predictors of Y_P and Y_S than other indices. The observed relationship between YP, MP, STI, YS, MP and STI are in consistent with those reported by Fernansez (1992) in mungbean and Farshadfar *et al.* (2013) in wheat. Toorchi *et al.* (2012) showed that correlation between MP, GMP and YS and YP is positive. Ebrahimzadeh Banayjedi et al. (2012) introduced STI, MP and GMP as the best indices for yield predicting. Dehghani *et al.* (2009) reported that GMP, MP STI were significantly and positively correlated with stress yield. Farshadfar *et al.* (2001) showed that the most appropriate index for selecting stress tolerant cultivars is an index which has partly high correlation with grain yield under stress and non-stressed conditions. There were negative and significant correlations between YP and YSI in moderate and severe drought stressed conditions.

Correlation analysis

Based on cluster analysis, the studied landraces were grouped into 3 clusters in mild and severe drought stressed conditions. Therefore, there was considerable variation among landraces for drought tolerance. The dendrogram of accessions did not divide the accessions into distinct groups resembling the similar geographical distribution (Fig. 1 and 2). Therefore, it is suggested that selection of parents for hybridization does not need to be based on geographic diversity. Hybridization/crossing between any distantly related populations is expected to yield more heterosis and vigorous plants. According to the dendrogram 1, 30% of landraces situated in groups 1, 34% situated in groups 2 and 36% in group 3 under mild drought stressed conditions

(Fig. 1). In severe drought stressed conditions, 30% of landraces placed in groups 1, 21% situated in groups 2 and 45% in group 3 (Fig. 2). Cluster analysis has been used in drought tolerance studies by other researchers (Safahani Langroodi *et al.,* 2013; Dehbalaei *et al.,* 2013; Zaheri and Bahraminejad, 2012; Ajalli and Salehi, 2012; Zare, 2012; Tabatabaei, 2013). Golabadi *et al.* (2006) in characterizing durum wheat segregating populations with cluster analysis distinguished groups with superior lines in both drought stress and well-watered conditions, superior lines in only drought stressed conditions as well as superior lines only in well-watered conditions considering their grain yield. Sajad Bakaei et al. (2008) showed that based on cluster analysis, in normal conditions genotypes were divided into two groups while in mild and severe stressed conditions genotypes were divided into 4 and 5 groups, respectively. Based on results of Zahravi (2009) regarded to cluster analysis, genotypes were divided into three groups resistant, semi-resistant and susceptible.

Using the biplot diagram (Fig. 3) the landraces 2, 12, 14, 25, 26, 28, 31, 32, 33, 49 and 56 under mild stressed conditions were identified as tolerant and the landraces 9, 10, 35, 36, 37, 38, 40 and 41 were detected as sensitive to drought. In severe stressed conditions the landraces 2, 5, 7, 12, 25 and 29 were identified as tolerant and the landraces 10, 35, 36, 37, 40 and 41 were detected as sensitive to drought (Fig 4). According to the biplot presentation there is positive correlation between indices MP, GMP, HARM, STI and yield in three environments that, confirm the observed simple correlations between them. Accordingly, in this study four above indices are the most appropriate indices to screening drought tolerant genotypes. The results of this study are compatible with Fernandez (1992), Golabadi et al. (2006), Kaya *et al.* (2002) and Farshadfar *et al.* (2012).

Conclusion

Abiotic stress tolerance is a key component and in some cases the major factor in improving yield in crops (Tollenaar and Wu, 1999). Drought stress is considered as one of the most visible factors, which affect grain yield, and some of the constituents of the grain oil of sunflower (Razi and Assad, 1999; Ali et al., 2009). In this evaluation, based on cluster analysis, the studied genotypes were grouped in 3 clusters in each one of water treatment conditions. To select drought tolerant genotypes in severe drought stress conditions, it is suggested to use indices MP, STI, GMP, HARM, SNPI and M_SSTI because these indices supports stable and high yield in both non-stress and stress conditions.

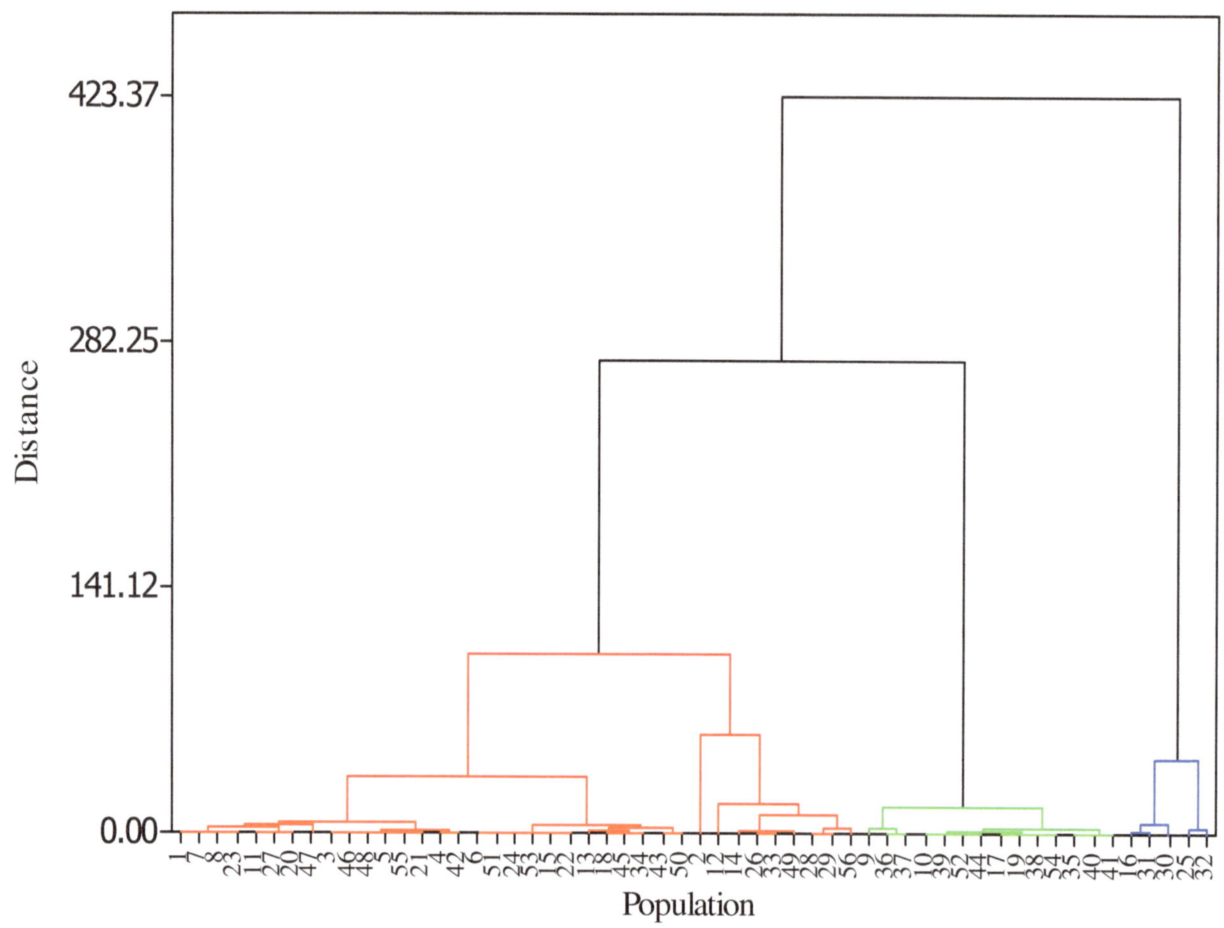

Fig. 1. Cluster diagram Dendrogram of 56 Iranian sunflower landraces generated by Wards clustering method by square Euclidean distance after data normalization based on drought tolerance indices (MP, STI, GMP, HARM, SNPI, YI, DI and M_SSTI) in moderate drought stress conditions

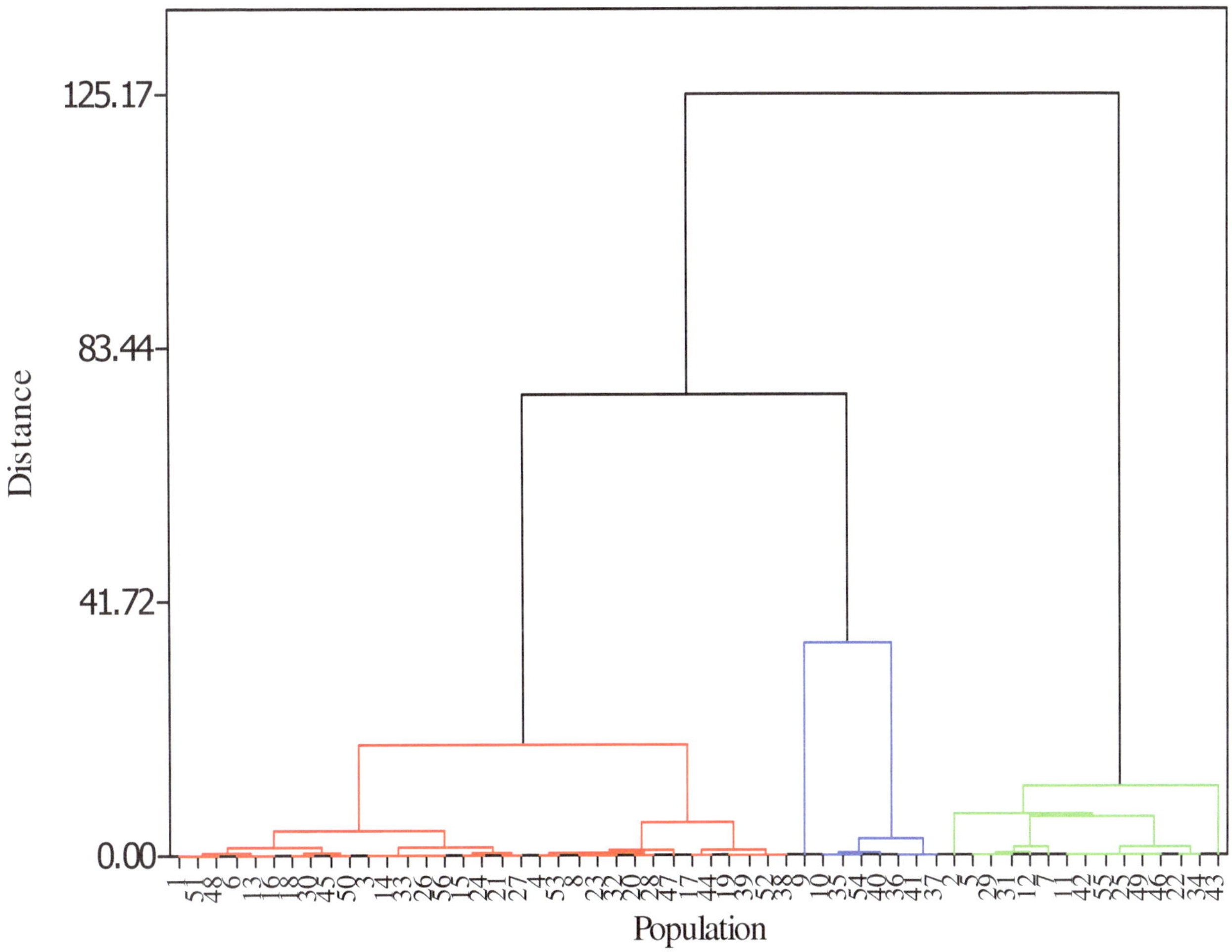

Fig. 2. Cluster diagram Dendrogram of 56 Iranian sunflower landraces generated by Wards clustering method by square Euclidean distance after data normalization based on drought tolerance indices (MP, STI, GMP, HARM, SNPI, YI, DI and M_SSTI) in severe drought stress conditions.

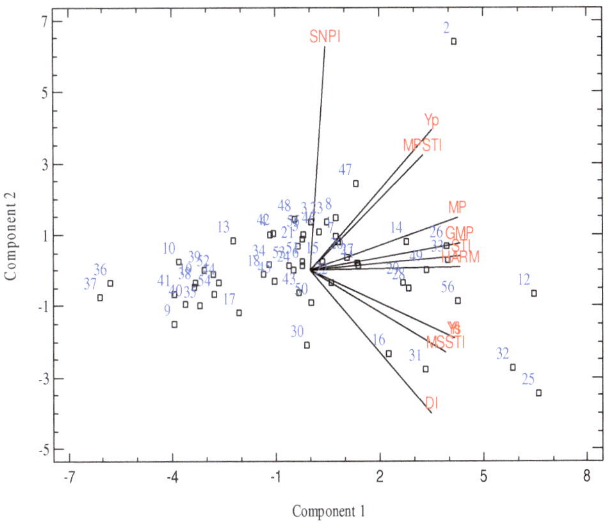

Fig. 3. Biplot for determination of the superior indices and genotypes in 56 Iranian sunflower landraces in moderate drought stress conditions.

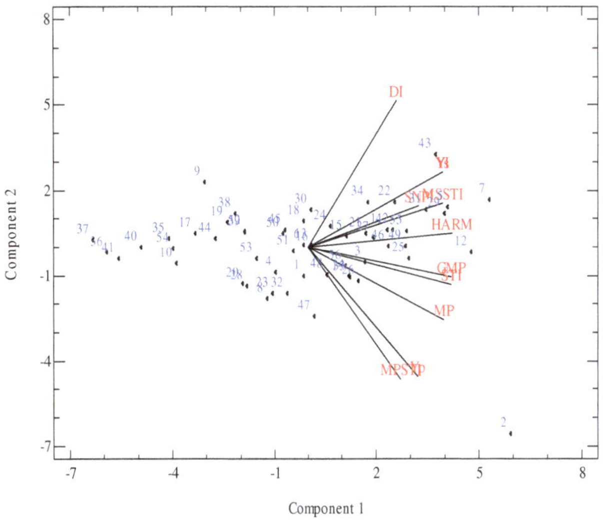

Fig. 4. Biplot for determination of the superior indices and genotypes in 56 Iranian sunflower landraces in severe drought stress conditions.

Tab. 8. Correlation coefficients between drought tolerance indices and grain yield in both moderate and sever drought stress conditions

	YP	YS	MP	STI	GMP	HARM	SNPI	YI	DI	MPSTI	MSSTI
YP	1										
YS	0.58**	1									
MP	0.91**	0.86**	1								
STI	0.82**	0.91**	0.97**	1							
GMP	0.85**	0.91**	0.99**	0.98**	1						
HARM	0.79**	0.95**	0.96**	0.97**	0.99**	1					
SNPI	0.53**	- 0.10^{ns}	0.28*	0.14^{ns}	0.21^{ns}	0.15^{ns}	1				
YI	0.58**	0.99**	0.86**	0.91**	0.91**	0.95**	- 0.09^{ns}	1			
DI	0.28*	0.93**	0.64*	0.75**	0.72**	0.78**	- 0.40**	0.93**	1		
MPSTI	0.88**	0.52**	0.80**	0.78**	0.73**	0.66**	0.37*	0.52**	0.31**	1	
MSSTI	0.51**	0.90**	0.77**	0.88**	0.81**	0.83**	- 0.23^{ns}	0.90**	0.87**	0.57**	1

Tab. 9. Correlation coefficients between drought tolerance indices and grain yield in both optimum and sever drought stress conditions

	YP	YS	MP	STI	GMP	HARM	SNPI	YI	DI	MPSTI	MSSTI
YP	1										
YS	0.45**	1									
MP	0.94**	0.72**	1								
STI	0.84**	0.83**	0.96**	1							
GMP	0.84**	0.85**	0.97**	0.98**	1						
HARM	0.70**	0.94**	0.89**	0.95**	0.97**	1					
SNPI	0.42**	0.72**	0.59**	0.67**	0.69**	0.74**	1				
YI	0.45**	0.99**	0.71**	0.83**	0.85**	0.94**	0.72**	1			
DI	- 0.02^{ns}	0.84**	0.29*	0.45**	0.47**	0.62**	0.46**	0.84**	1		
MPSTI	0.84**	0.33**	0.77**	0.74**	0.66**	0.52**	0.27*	0.33**	- 0.008^{ns}	1	
MSSTI	0.52**	0.91**	0.74**	0.88**	0.84**	0.89**	0.74**	0.91**	0.70**	0.48**	1

Tab 10. Ranking of 56 Iranian confectionery sunflower landraces based on the best indices in optimum and stress drought stressed conditions

No	Genotype	Yield			Rank	
		Ys- severe	YS - mild	YP	Normal-severe stress	Normal-mild stress
1	Saghez 1	1272.33	2169.72	3241.94	26.5	19.5
2	Anghane 4	1461.19	1807.79	6310.77	1	4
3	Urmia-Barouj	1575.41	1883.44	3355.65	14	24
4	Urmia-Maranghalou	1170.91	1615.43	3001.82	35.5	38
5	Marand-Dizaj- Ghalami	2065.43	1917.77	3055.23	4.5	27
6	Jabalkandi 2	1415.56	2085.56	2695.38	31	31
7	Salmas - Sadaghian	2217.95	2211.75	3177.21	3	19.5
8	Babaghanje 6	1002.17	2048.18	3505.74	35.5	17.5
9	Miyaneh-Basin	1462.42	1273.47	1221.97	48	54
10	Boucan	769.14	973.88	2103.75	50	51
11	Urmia - Nuoshinshahr	1783.14	2482.86	3012.03	12.5	15.5
12	Karimabad	1982.66	3611.45	3728.20	2	1
13	Vaghaslou-Olya 1	1439.29	1328.74	2768.42	28.5	42
14	Vaghaslou-Olya 3	1453.18	2686.43	3500.27	19	9
15	Ordoshahi 1	1611.69	2208.24	2852.48	23	25
16	Marana-Yamchi-Pesteii	1434.58	2889.97	2802.96	28.5	13.5
17	Mazandaran-Tirtash	996.16	1832.44	1675.26	49	43
18	Sardasht	1500.16	1816.67	2356.11	35.5	41
19	Marana-Yamchi 4	1177.18	1299.26	1783.27	45.5	48
20	Salmas 2	965.81	2367.56	3037.67	42	17.5
21	Vaghaslou-Olya 4	1693.70	1927.74	2935.50	20	29.5
22	Salmas-Gharaghashlagh- Pesteii	1901.66	2371.90	2691.73	17	23
23	Lalalou-Torab 2	1056.05	2002.13	3418.40	33	21
24	Shirabad 2	1588.41	2026.19	2591.75	26.5	33.5

25	Gharagoz 1	1746.63	3778.83	3528.58	6	2
26	Vaghaslou-Sofla 1	1471.15	2965.42	3635.39	12.5	5
27	Khanneshan 1	1703.78	2467.21	3037.68	15	15.5
28	Heydarlou 1	981.17	2908.40	3103.55	40.5	11
29	Saribaglou 5	2030.66	2853.26	3105.27	4.5	12
30	Chongharalou-Yekan 4	1567.05	2344.98	2243.27	35.5	33.5
31	Maranghalou 6	1985.41	3146.80	2961.98	8	10
32	Abajalou 1	1115.13	3596.70	3505.17	30	3
33	Hamadan 1	1453.90	3033.53	3513.70	17	6.5
34	Saghez 2	1807.28	1799.07	2523.31	25	39.5
35	Piranshahr-Serokani	860.42	1439.76	1529.26	52	50
36	Piranshahr Andizeh	497.26	493.39	1222.34	55	55
37	Mashhad	490.16	515.86	807.34	56	56
38	Shahroud 1	1242.51	1315.93	1700.66	45.5	49
39	Hamadan 2	1217.73	1285.08	2071.94	43.5	46.5
40	Shabestar-Kouzeh-Kanan 3	689.73	1324.07	1426.11	53	52
41	Saghez 4	525.13	1177.20	1458.54	54	53
42	Saghez 5	1801.78	1645.39	3036.84	11	37
43	Saghez 3	2143.35	2183.06	2381.37	17	35.5
44	Shahroud 2	1062.24	1502.04	1942.53	47	44
45	Alibaglou 1	1396.10	1948.10	2331.95	38.5	39.5
46	Baneh 2	1727.82	2009.26	3241.56	10	22
47	Salmas-Gharagheshlagh-Ghalami	1134.32	1987.24	4076.50	23	13.5
48	Marand-1389-2	1377.34	1721.08	3356.18	23	28
49	Salmas-Gharaghashlagh-Badami	1788.36	2949.06	3340.35	7	8
50	Shabestar-Kouzeh Kanan 1	1370.58	2305.50	2390.03	38.5	32
51	Sanandaj	1335.47	2045.97	2756.99	32	29.5
52	Shabestar-Kouzeh-Kanan 2	1222.83	1403.79	2030.41	43.5	45
53	Baneh 3	1153.55	1979.44	2624.47	40.5	35.5
54	Piranshahr-Baleban	828.05	1522.93	1759.33	51	46.5
55	Baneh 1	1844.57	1902.77	3141.23	9	26
56	Marand-1389-1	1489.70	3228.56	3307.31	21	6.5

References

Ahmadi G, Zeinaly Khane Ghah H, Rostamy MA, Chogan R (2000). The study of drought tolerance and biplot method in eight corn hybrids. Iranian Journal of Agriculture Science 31:513-523.

Ajalli J, Salehi M (2012). Evaluation of drought stress indices in barley (*Hordeum vulgare L.).* Annals of Biological Research 3 (12):5515-5520.

Ali Q, Ashraf M, Anwar F (2009). Physico-chemical attributes of seed oil from drought stressed sunflower (*Helianthus annuus* L.) plants. Grasas y Aceites 60:475-481.

Alizadeh A (2009). Soil and plant water relations. Press of Astan Ghods Razavi 484 p.

Anwar J, GHULAM Mahboob Subhani Genotype, Makhdoom H, Javed A, Mujahid H, Munir M (2011). Drought tolerance indices and their correlation with yield in exotic wheat genotypes. Pakistan J Bot 43(3):1527-1530.

Blum A (1988). Plant's breeding for Stress environments. CRC Press, Florida. P- 212.

Bouslama M, Schapaugh WT (1984). Stress tolerance in soybean. Part 1: evaluation of three screening techniques for heat and drought tolerance. Crop Sci 24:933-937.

Daneshian J, Jonoubi P (2008). Evaluation of sunflower new hybrids tolerance to water deficit stress. In proceeding of the 5th International Crop Science Congress. Jejo, Korea. Pp:189.

Dehbalaei S, Farshadfar E, Farshadfar M (2013). Assessment of drought tolerance in bread wheat genotypes based on resistance/ tolerance indices. International Journal of Agriculture and Crop Sciences 5(20):2352-2358.

Dehghani GH, Malekshhi F, Alizadeh B (2009). A study of drought tolerance indices in canola (*Brassica napus* L.) genotypes. Journal of Science and Technology of Agriculture and Natural Resources 13(48):77-90.

Drikvand R, Doosty B, Hosseinpour T (2012). Response of rainfed wheat genotypes to drought stress using drought tolerance indices. J Agr Sci 4(7):126-131.

Ebrahimzadeh Banayjedi M, Ehteshami SM, Delkhosh H, Abdi B, Rohami M (2012). Evaluation of drought tolerance indices and their relationship with grain yield of wheat genotypes. J Basic Appl Sci Res 2(2):1230-1232.

Eivazi AR, Mohammadi S, Rezaei M, Ashori S, Hossien Pour F (2013). Effective selection criteria for assessing drought tolerance indices in barley (*Hordeum vulgare* L.) accessions. International Journal of Agronomy and Plant Production 4 (4):813-821.

Farshadfar E, Ghannadha M, Zahravi M, Sutka J (2001). Genetic analysis of drought tolerance in wheat. Plant Breeding 114:542-544.

Farshadfar E, Moradi Z, Elyasi P, Jamshidi B, Chaghakabodi R (2012). Effective selection criteria for screening drought tolerant landraces of bread wheat (*Triticum aestivum* L.). Annals of Biological Research 3(5):2507-2516.

Farshadfar E, Poursiahbidi MM, Safavi SM (2013). Assessment of drought tolerance in land races of bread wheat based on resistance/tolerance indices. International Journal of Advanced Biological and Biomedical Research 1(2):143-158.

Farshadfar E, Sutka J (2002). Screening drought tolerance criteria in maize. Acta Agronomica Hungarica 50(4):411-416.

Fernandez GCJ (1992). Effective selection criteria for assessing stress tolerance. In: Kuo C.G. (Ed.), Proceedings of the International Symposium on Adaptation of Vegetables and Other Food Crops in Temperature and Water Stress, Publication, Tainan, Taiwan.

Fischer RA, Maurer R (1978). Drought resistance in spring wheat cultivars. I. Grain yield response. Aust J Agric Res 29:897-912.

Gabriel KR (1971). The biplot graphic display of matrics with application to principal component analysis. Biometrika 58:453-467.

Ganjali A, Bagheri A, Porsa H (2009). Evaluation of chickpea (*Cicer arietinum* L.) germplasm for drought resistance. Journal of Agronomic Research of Iran 7(1):183-194.

Gavuzzi P, Rizza F, Palumbo M, Campaline RG, Ricciardi GL, Borghi B (1997). Evaluation of field and laboratory predictors of drought and heat tolerance in winter cereals. Can J Plant Sci 77:523-531.

Golabadi M, Arzani A, Maibody SAM (2006). Assessment of drought tolerance in segregating populations in durum wheat. African Journal of Agriculture Research 5:162-171.

Ilker E, Tatar O, Aykut Tonk F, Tosun M, Turk J (2011). Determination of Tolerance Level of Some Wheat Genotypes to Post-Anthesis Drought. Turkish Journal of Field Crops 16(1):59-63.

Jabbari H, Akbari GA, Daneshian J, Alahdadi I, Shahbazian N (2008). Utilization ability of drought resistance indices in sunflower (*Heliantus annus* L.) hybrids. EJCP 1(4):1-17.

Jafari A, Paknejad F, Al-Ahmadi M (2009). Evaluation of selection indices for drought tolerance of corn (*Zea mays* L.) hybrids. Int J Plant Product 3:33-38.

Jaleel CA, Manivannan P, Sankar B, Kishorekumar A, Gopi R, Somasundaram R, Panneerselvam R (2007). Induction of drought stress tolerance by Ketoconazole in *Catharanthus roseus* is mediated by enhanced antioxidant potentials and secondary metabolite accumulation. Colloids Surf B Biointerfaces 60:201- 206.

Kaya Y, Plta C, Taner S (2002). Additive main effects and multiplicative interaction analysis of yield performance in bread wheat genotypes across environments. Turk J Agric 26:257-259.

Khalili M, Naghavi MR, Pour Aboughadareh AR, Talebzadeh SJ (2012). Evaluating of drought stress tolerance based on selection indices in spring canola cultivars (*Brassica napus* L.). Journal of Agricultural Science 4(11):78-85.

Khodarahmpour Z, Choukan R, Bihamta MR, Majidi Hervan E (2011). Determination of the best heat stress tolerance indices in maize (*Zea mays* L.) inbred lines and hybrids under Khuzestan province conditions. Iranian Journal of Crop Sciences 13:111-121.

Kristin AS, Serna RR, Perez FI, Enriquez BC, Gallegos JAA, Vallejo PR, Wassimi N, Kelly JD (1997). Improving common bean performance under drought stress. Crop Sci 37:51-60.

Lan J (1998). Comparison of evaluating methods for agronomic drought resistance in crops. Acta Agric Boreali-occidentalis Sinica 7:85-87.

Lu G, Hoeft E (2007). Sunflower. In Kole C, Hall TC (eds.), A compendium of transgenic crop plants Vol. 2, Oxford, Wiley-Blackwell.

Mevlut A, Sait C (2011). Evaluation of drought tolerance indices for selection of Turkish oat (*Avena sativa* L.) landraces under various environmental conditions. Zemdirbyste Agriculture 98(2):157-166.

Mitra J (2001). Genetics and genetic improvement of drought resistance in crop plants. Curr Sci India 80:758-762.

Moghadam A, Hadizadeh MH (2000). Study in density stress in selection of drought tolerant varieties in corn. Iranian Journal of Crop Sci 2(3):25-38.

Mohammadi R, Armion M, Kahrizi D, Amri A (2010). Efficiency of screening techniques for evaluating durum wheat genotypes under mild drought conditions. Int J Plant Product 4(1):1735-8043.

Moosavi SS, Yazdi Samadi B, Naghavi MR, Zali AA, Dashtid H, Pourshahbazi A (2008). Introduction of new indices to identify relative drought tolerance and resistance in wheat genotypes. Desert 12:165-178.

Pourdad SS, Alizadeh KH, Azizinezhad R, Shariati A, Askandari M, Khiyavi M, Ezatollahe N (2008). Evaluation of different safflowers in different regions. Sci Tech Agri Natur Resources 12 (45):403-415.

Razi H, Assad MT (1999). Comparison of seletion criteria in normal and limited irrigation in sunflower. Euphytica 105: 83-90.

Rosielle AA, Hamblin J (1981). Theoretical aspects of selection for yield in stress and non-stress environment. Crop Sci 21:943-946.

Safahani Langeroodi AR, Kamkar B, Ataei M, Teixeira Da Silva A (2013). Assessment of the response of sunflower cultivars to water shortage using various stress tolerance indices. Int J Agron Plant Prod 4 (7):1628-1636.

Sajad Bokaei A, Babaei H, Habibi D, Javidfar S, Mohammadi A (2008). Evaluation of different soybean (*Glycine max* L.) genotypes under drought stress conditions. Journal of Agronomy and Plant Breeding 4(1):28-38.

Salunkhe DK, Chavan JK, Adsule RN, Kadam S S (1999). World oil seeds: Chemistry, technology, and utilization. New York: Van Nostrand Reinhold.

Sio-Se Mardeh A, Ahmadi A, Poustini K, Mohammadi V (2006). Evaluation of drought resistance indices under various environmental conditions. Field Crop Res 98:222-229.

Tabatabaei SA (2013). Study relationship of drought tolerance indices in wheat (*Triticum aestivum*) genotypes. International Journal of Biosciences 3(7):15-22.

Tollenaar M, Wu J (1999). Yield improvement in temperate maize is attributable to greater stress tolerance. Crop Sci 39:1597-1604.

Toorchi M, Naderi R, Kanbar A, Shakiba MR (2012). Response of spring canola cultivars to sodium chloride stress. Annals of Biological Research 2(5):312-322.

Zaheri A, Bahraminejad Sunflower (2012). Assessment of drought tolerance in oat (*Avena sativa*) genotypes. Annals of Biological Research 3(5):2194-2201.

Zahravi M (2009). Evaluation of genotypes of wild barley (*Hordeum spontaneum*) based on drought tolerance indices. Seed and Plant 25(4):533-549.

Zare M (2012). Evaluation of drought tolerance indices for the selection of Iranian barley (*Hordeum vulgare*) cultivars. African Journal of Biotechnology 11(93):15975-15981.

Development of SCAR Marker Related to Summer Stress Tolerance in Tall Fescue (*Festuca arundinacea*)

Xiaojun YUAN[1a], Zhenjing BAO[2a], Yali HE[2*], Qun CHEN[2], Gang WANG[1*]

[1]*School of Life Science, Shanghai University, Shanghai 200444; Shanghai Key Laboratory of Bio-Energy Crops, Shanghai University, China*
[2]*School of Agriculture and Biology, Shanghai Jiaotong University, Shanghai 200240, China;*
*heyali@sjtu.edu.cn, wg@shu.edu.cn (*corresponding authors).* [a] *These authors contributed equally to this work*

Abstract

Summer stress tolerance (SST) is one of the most important breeding objectives in tall fescue (*Festuca arundinacea*), an important perennial cool-season grass. However, breeding for better SST is generally complicated by the many environmental factors involved during the growing season. Utilizing the bulked segregant analysis (BSA), we were able to identify one marker related to SST from 100 inter-simple sequence repeat (ISSR) markers and 800 random amplified polymorphic DNA (RAPD) markers, and successfully developed a dominant sequence characterized amplified region (SCAR) marker T_SC856 from the UBC856 sequence. Furthermore, the SCAR marker was tested in different clones of new populations, which were identified under complex summer stress (high temperature and humidity, Pythium blight, and brown patch), and it exhibited relatively high consistency (77%) with the phenotype. We believe that with more markers obtained in the future, better efficiency is likely to be achieved in breeding for improved SST in tall fescue and possibly other species as well. Further studies that analyze the factors relating to the SCAR marker are needed.

Keywords: *Festuca arundinacea*, summer stress tolerance, SCAR marker, marker-assisted selection

Introduction

Tall fescue (*Festuca arundinacea*) is a common perennial cool-season grass (Poaceae) spread throughout the temperate regions of the world. It is widely used in parks, home lawns, athletic fields, golf courses and soil conservation sites. The optimum temperature for the growth of tall fescus is 16 °C～24 °C (Emmons, 1994). When grown and managed in temperate and transition climatic zones, quality of tall fescue decreases often suffers greatly during summer months as evidenced by stunt growth, severe disease and insect pressure, and dead patches. The comprehensive resistance exhibited by tall fescue against multiple summer abiotic and biotic stresses is referred as summer stress tolerance (SST). Tall fescue genotypes with better SST have the potential to maintain better overall quality through summer time, especially in the transition zone. Therefore the improvement of tall fescue SST resistance is an important target in tall fescue breeding efforts.

Many previous studies evaluating summer performances of cool-season grasses haven focused on the impact of a single specific stress, such as heat (Du *et al.*, 2011; He and Huang, 2007; Kim *et al.*, 2010; Lefsrud *et al.*, 2010; Xu and Huang, 2008; Xu *et al.*, 2006; Xu *et al.*, 2010; Xu *et al.*, 2011), drought (Abraham *et al.*, 2004; Merewitz *et al.*, 2010), brown patch (*Rhizoctonia* ssp.) (Bonos *et al.*, 2006; Dong *et al.*, 2008; Watkins and Meyer, 2004), and Pythium blight (*Pythium* ssp.) (Allen *et al.*, 2005), etc. Among these commonly studies stresses, heat and drought are the two major ones limiting the summer growth and quality of cool-season turfgrasses. And reports have been made on the combined the effects of both heat and drought, as well as their impacts on the summer health of cool-season grass species (Abraham *et al.*, 2008; Su *et al.*, 2007; Wang and Huang, 2004). In a study done in New Jersey, USA, Bonos and Murphy (1999) evaluated growth and performance of Kentucky bluegrass (KBG, *Poa pratensis*) genotypes under summer stresses (heat and drought) and identified genotypes that were relatively more tolerant to these stresses. The growth of KBG is also reported to be limited by drought and heat stress during summer in North-western China (Liu *et al.*, 2008). However, there are limited reports on the impacts of combined stress from heat, high humidity, diseases and insects, as commonly encountered in field condition during summer months.

Commercial tall fescue is allohexaploid ($2n=6x=42$) with a large genome ($5.27\text{-}5.83\times10^9$ bp) (Seal, 1983). Due to a cross-pollinated habit, each seed or plant of the grass is genetically unique. Therefore, intraspecific genetic variation in tall fescue is commonly observed for individual selection breeding (i.e., recurrent selection and ecotype breeding). However, due to the heterogeneity of soil physiochemical characteristics and the variation in spreading rate of fungal pathogen in field, the selection efficiency of conventional breeding programs targeting at summer stress tolerance is relatively low.

Along with the development of molecular technology, transgenic tall fescue lines have been reported with improved stress resistance. Wu *et al.* (2006) obtained transgenic tall fescue plants with enhanced salt and osmosis tolerances, containing *CBF1* gene from *Arabidopsis thaliana* via *Agrobacterium tumefaciens* - mediated transformation. Dong *et al.* (2007) obtained transgenic plants exhibiting resistance to two major fungal diseases greay leaf spot (*Magnaporthe grisea*) and brown patch (*Rhizoctonia solani*). Meanwhile, molecular marker techniques were also applied in tall fescue researches, such as linkage maps construction, genetic diversity analysis, cultivar identification, and so on. The first tall fescue linkage map was reported by Xu *et al.* (1995), generated from the F2 population of HD28-56 and 'Kentucky-31'. It consisted of 108 restriction fragment length polymorphisms (RFLPs) and covered 1274 cM on 19 linkage groups with a mean marker interval of 17.9 cM. Ten years later, an amplified fragment length polymorphism (AFLP) and simple sequence repeat (SSR) based genetic linkage map of tall fescue was constructed by Saha *et al.* (2005), who combined the female (HD28-56) map and the male (R43-64) map into an integrated map, which covered 1841 cM on 17 LGs with an average marker interval of 2.0 cM. Furthermore, molecular mapping of locating quantitative trait loci (QTLs) related to forage digestibility in tall fescue was also reported by Saha *et al.* (2009). Base on RFLP data, the results of genetic diversity showed that variation within tall fescue cultivars was high (Xu *et al.*, 1994). Analyzed by SSR, random amplified polymorphic DNA (RAPD), and inter-simple sequence repeat (ISSR) markers, Mester *et al.* (1999) classified 30 somaclones into four distinct groups. Using 461 AFLPs from six primer combination, Mian *et al.* (2002) clustered 16 tall fescue plants in groups. In addition, identification of 12 commercial varieties of tall fescue was successful via RFLPs (Busti *et al.*, 2004), in which seven specific bands were distinguished.

As an effective breeding method, marker-assisted selection (MAS) has been applied in many species successfully. It is especially useful for some complex traits. Unfortunately, there are no reports on MAS of SST in tall fescue. Considering the hereditary characteristics of tall fescue, this study was designed to identify molecular markers (ISSR and RAPD) that are linked to SST trait via pseudo-bulked segregant analysis (BSA) method, using tall fescue clones with varying SST. Also, a sequence characterized amplified region (SCAR) marker developed from polymorphic random molecular marker was tested in other clones. Our purpose of this study is to seek and offer sequence specific markers that could be valuable in assisting breeding for summer stress tolerance in tall fescue, and hopefully render future breeding efforts of this kind more simplicity and reliability.

Materials and methods

Plant materials

Selection population: In the autumn of 2007, 130 tall fescue clone lines (genotypes) from different varieties and lines (Tab. 1) were planted in Funing County, Jiangsu Province, China. Each clone line included twenty-two plants (4-5 tillers per plant). Funing County is located at 33'N latitude and 119'E longitude, and it has four distinct seasons featuring high temperature and plenty of rainfall in summer. Soil type of this area is sandy loam. Till the end of the experiment in 2009, we only fertilized twice (late autumn: 88 kg N·ha^{-1}; early spring: 34 kg N·ha^{-1}). No pesticide was used for insects and diseases control and weeds were hand-picked twice a year in late autumn and early June.

Tab. 1: Identification of summer stress tolerance in different tall fescue germplasms

Clones	Cultivar[a]	Summer Stress Tolerance[b]	Clones	Cultivar[a]	Summer Stress Tolerance[b]
TF01	Southeast	T	TF34	Triple A	S
TF02	Paraso	T	TF35	Plantation	S
TF03	SHED	T	TF36	Plantation	S
TF04	Plantation	T	TF37	SND×98-19	S
TF05	Triple A	T	TF38	SND×98-19	S
TF06	Plantation	T	ST01	SHED	T
TF07	Plantation	T	ST02	SHED	T
TF08	Plantation	T	ST03	SHED	T
TF09	Plantation	T	ST04	SHED	T
TF10	Triple A	T	ST05	SHED	T
TF11	Plantation	T	ST06	SHED	T
TF12	Triple A	S	ST07	SHED	T
TF13	Plantation	S	ST08	SHED	T
TF14	Triple A	S	ST09	SHED	T
TF15	Plantation	S	ST10	SHED	T
TF16	Plantation	S	ST11	SHED	T
TF17	SND×98-19	S	ST12	SHED	T
TF18	SND×98-19	S	WT01	SHED	S
TF19	SND×98-19	S	WT02	SHED	S
TF20	SND×98-19	S	WT03	SHED	S
TF21	SND×98-19	S	WT04	SHED	S
TF22	SND×98-19	S	WT05	SHED	S
TF23	98-19×SND	S	WT06	SHED	S
TF24	98-19×SND	S	WT07	SHED	S
TF25	98-19×SND	S	WT08	SHED	S
TF26	98-19×SND	S	WT09	SHED	S
TF27	SND	S	WT10	SHED	S
TF28	98-19	S	WT11	SHED	S
TF31	Plantation	S	WT12	SHED	S
TF32	Triple A	S	WT13	SHED	S
TF33	Triple A	S			

[a]SHED is Shanghai Evergreen Dwarf Tall Fescue (registered by "Shanghai Crop Cultivars Examination and Approval Committee" under the code of 'Hu Nong Pin Ren Turf (2005) No. 1'). SND is Shangnong Dwarf Tall Fescue. SND and 98-19 are breeding lines from Grass Research Group of Shanghai Jiaotong University (He et al., 2001). SND×98-19 is the hybrid of SND and 98-19, maybe including the SND inbred seeds. 98-19×SND is the hybrid of 98-19 and SND, maybe including the 98-19 inbred seeds.

[b]T, Summer stress tolerant; S, Summer stress sensitive.

Testing population: In 2005, plants of tall fescue variety 'Shanghai Evergreen Dwarf' were grown in Sanjiagang area, Shanghai, China. The seeding rate was 20 $g{\cdot}m^{-2}$. Clone lines were constructed from fifty plants (including summer stress tolerant and sensitive plants) selected under summer stress (mainly insect and disease pressure), and transplanted to research farm at School of Agriculture and Biology, Shanghai Jiao Tong University, Shanghai, China. Subsequently, twelve clones performed well under rainy and high temperature conditions and thirteen other clones performed poorly under same conditions were selected in autumn of 2008 and 2009. All these plants were used as testing population.

Shanghai is located at 31'N latitude and 121'E longitude. The climate here is similar to Funing County, with an average temperature of 27.9 °C in July (recorded highest temperature of 40.7 °C) and an annual precipitation of 1087 mm (He *et al.*, 1997). The Soil type at testing location is loam. Management program was similar to that of selection population.

Identification of summer stress tolerance: We identified SST from clone lines of selection population and clones of testing population. The identification was operated after the full duration of summer stress and before recovery started to occur. Based on the status of plants (survival rate, green leaf index, degree of leaf curling, disease incidence, and so on), three breeders marked each clone by visual inspection independently. Scoring is based on a scale of 1 to 9, with 9 being the best growing state (He *et al.*, 2001). Summer stress tolerant and summer stress sensitive clone lines (Tab. 1) were selected, respectively, for future work. Some of the summer stress sensitive clones lost green shoots due to summer stress, but they were able to recover from crown in the next year.

DNA extraction and pseudo-BSA: The same amounts of leaves were collected from each plant per clone line, and genomic DNA mixture of each clone lines was extracted with the CTAB method (Clark 1998). Extracted DNA samples were visualized after electrophoresis on 0.8% agarose gels in 1×TAE. DNA concentration and purity was measured with a UV spectrophotometer. The DNA was adjusted to a final concentration of 20 $ng{\cdot}\mu L^{-1}$ with TE buffer (pH 8.0) and stored at -20 °C until use.

Equal amounts of DNA from eleven summer stress tolerant clone lines and eleven summer stress sensitive clone lines were pooled to construct two DNA bulks for BSA (Michelmore *et al.*, 1991), summer stress tolerant pool (STP) and summer stress sensitive pool (SSP). Obviously, the polymorphism between these two DNA pools was affected by multiple factors rather than only one. To indicate the difference with the traditional BSA approach, we refer our approach as pseudo-BSA.

RAPD and ISSR analyses: A total of 100 ISSR primers and 800 RAPD primers were used to screen polymorphisms in the two bulks. The primers that were able to amplify polymorphic bands between the two pools were tested in the 22 individual clone lines that made up the pools for further check of the polymorphism. RAPDs and ISSRs were all according to sequences of NAPS Unit standard primers (University of British Columbia, Canada). All the primers were synthesized by Sangon Biological Engineering Technology and Service Co. Ltd, Shanghai.

PCR reactions for ISSR and RAPD markers were carried out in 10-μl mixture [20 ng genomic DNA, 10 pmol primer, 100 $\mu mol{\cdot}L^{-1}$ dNTPs, 1×Taq Buffer, 1.5mmol·L^{-1} $MgCl_2$, and 0.5 unit of Taq DNA polymerase (TaKaRa, Talien, China)]. The reaction conditions were as follows: 94 °C for 3min, followed by 35 cycles at 94 °C for 10s, an annealing temperature depending different primers for 40s, and 72 °C for 45s and a final extension at 72 °C for 6 min. The amplification products were separated on 1.5% agarose gels, and then soaked in a 1.0 $\mu g{\cdot}mL^{-1}$ of ethidium bromide water solution for 10 min. Then, the gels were illuminated by UV light, and photographed with a Tanon3500 imaging system (Tanon Science & Technology Co., Ltd., Shanghai, China).

Cloning and sequencing of the ISSR fragment: Sequence-related amplified polymorphism fragment was retrieved with the Gel Extraction Kit (DV805A, TaKaRa, Talien), and the quality of fragment extracted was checked by 0.8% agarose gels. The target band was cloned into the PMD18-T vector (D101A, TaKaRa, Talien). Positive colonies bearing DNA of the expected size were sequenced with ABI 3700 Sequencer (Sangon, China).

Conversion of ISSR marker to SCAR (sequence characterized amplified region) marker: According to the sequence of the fragment, SCAR primers were then designed using the Primer Premier5.0. The SCAR primers were tested in two pools and 22 clones. The condition of the SCAR amplification was as follows: 94 °C for 3 min, followed by 29 cycles at 94 °C for 10s, 63 °C for 40s, and 72 °C for 30s and a final extension at 72 °C for 6 min. We obtained one polymorphic SCAR marker, named as T_SC856 (Forward primer sequence: 5'ACACACACACACACACCAATTG3'; Reverse primer sequence:5'ACACACACACACACACTAC CTC3').

Results

Identification of summer stress tolerant clones in selection population

Throughout the whole summer season, the 130 clone lines of tall fescue planted exhibited very distinct summer hardiness. Some clones (i.e., TF18) almost lost all above ground tissues (Fig. 1a), while some other clones (i.e., TF02) were still growing well, with abundant green and dense leaves (Fig. 1b). Obviously, the former type was summer stress sensitive clones, and the latter was summer stress tolerant ones. Careful examination of the dead plants revealed large populations of millipedes (*Spirobolus bungii*) nearby.

Finally, eleven summer stress tolerant clone lines (TF01-TF11) and twenty-five summer stress sensitive clone lines (TF12-TF28, TF31-TF38) were selected from the selection population (Tab. 1). All the summer stress tolerant ones and 11 of the summer stress sensitive ones (TF12-TF22) were selected randomly to construct the DNA bulks.

Identification of markers related to the summer stress tolerance trait

Of the 100 ISSR primers tested, 13 primers were polymorphic between the two bulks (Fig. 2). Unfortunately, none of the 800 RAPD primers showed polymorphism. From the successful reaction system, about 5.5 marker bands were amplified clearly per primer. Upon further testing with 22 clone lines DNA that made up the pools, only one ISSR marker (UBC856) was able to generate the same polymorphic bands in the clone lines and two bulks (Fig. 3a). It was a dominant marker appearing brightly in SSP, and the size of the band was about 1300bp.

Fig. 1. Identification of tall fescue clone lines in Funing County, Jiangsu, China. Summer stress sensitive clone line TF18 (a) and summer stress tolerant clone line TF02 (b)

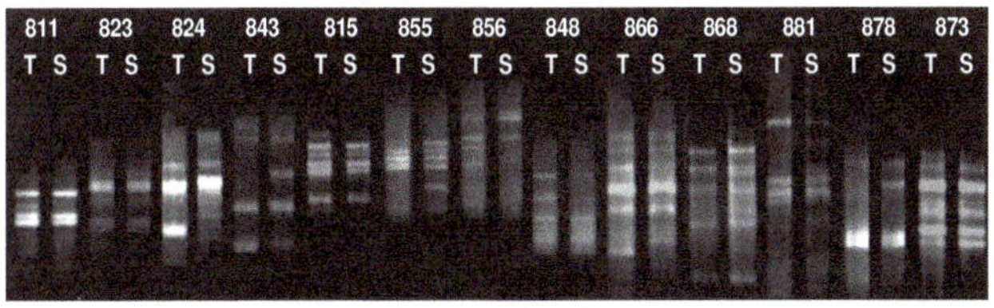

Fig. 2. Polymorphic bands amplified by 13 ISSR primers in two bulks. T, summer stress tolerant pool; S, summer stress sensitive pool. Number indicates ISSR primer names (i.e. 811 means UBC811)

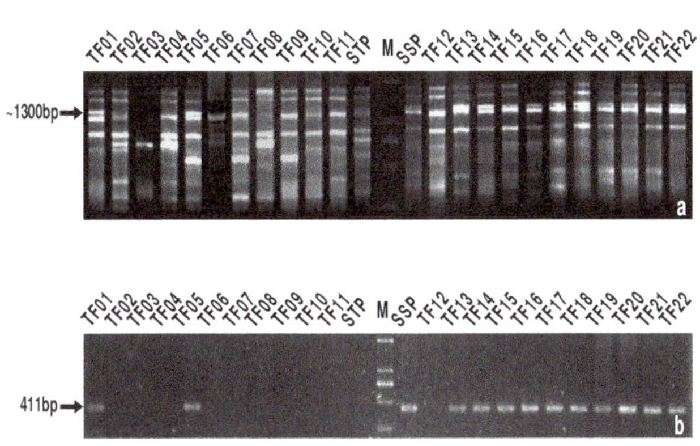

Fig. 3. PCR amplification of the UBC856 (a) and T_SC856 (b) in two bulks and 22 clone lines that comprising the two bulks. STP, summer stress tolerant pool; SSP, summer stress sensitive pool

Conversion of ISSR marker to SCAR marker

The ISSR marker related to summer stress tolerance were converted into SCAR marker (T_SC856). The polymorphic product was 411bp, appeared brightly in SSP and almost none in STP (Fig. 3b). Among the 22 individuals of the two pools, the polymorphic bands amplified by the SCAR marker were identical to those of its corresponding ISSR marker. In the future, this anchor marker could be used for large-scale screens for summer stress tolerance of MAS.

Anchor marker verified in the testing population

Summer stress tolerant clones and summer stress sensitive clones were selected from "Shanghai Evergreen Dwarf" testing population. Combined with the clone lines (TF23-TF28, TF31-TF38) identified in selection population (not include in the bulk construction), they were all tested by the SCAR maker. According to the comparison between polymorphism of SCAR amplification and phenotype of SST in the field, 4 of 13 WT clones (WT3, WT5, WT6, and WT11) were not consistent (Fig. 4b); 2 of 12 ST clones (ST08 and ST12) were not consistent (Fig. 4a). Besides, 3 of 14 clone lines (TF31, TF35, and TF36) were inconsistent (Fig. 4c). Taken together, accuracy of the molecular marker selection reached 77%.

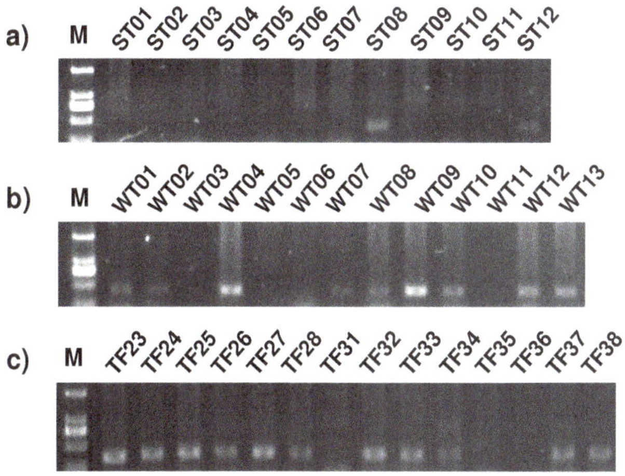

Fig. 4. Amplification of the SCAR marker T_SC856 in summer stress tolerant clones (ST) (a) and summer stress sensitive clones (WT) (b) from SHED testing population, and several summer stress sensitive clones from the selection population (c)

Discussion

It generally takes turfgrass breeders more than ten years to breed a new variety, from single plant selection and cloning to quality evaluation via field tests. Through efforts in conventional phenotypic selection, we have indeed obtained some summer stress tolerant varieties and lines of tall fescue in Shanghai, China (He *et al.*, 2001; He *et al.*, 2002). However, released varieties would be usually replaced by more recent ones after five years or so.

Combining phenotyping with the molecular marker polymorphism selection could certainly accelerate the breeding process, especially for some complex traits. In order to do so, abundant molecular markers closely linked to target traits have to be available to breeders first. Several applications of turfgrass SCAR markers have been reported. For example, SCAR markers created by RAPD bands were used to differentiate some species of bentgrass, and identify progenies derived from hybridization among bentgrass species (Scheef *et al.*, 2003).

Identification and detection of *Phoma sclerotioides* was operated by sequence-characterized DNA markers in the brown root rot of alfalfa (Larsen *et al.*, 2002). Humaid *et al.* (2004) detected the genetic variation and *Fusarium* resistance in seven tufgrass genotypes with RAPD and SCAR markers. Our work is the first report of the development of SCAR marker relating to summer stress tolerance trait. And it showed the relatively high accuracy of the molecular marker selection in different clones. It is undoubtedly a worthwhile effort using MAS for summer stress tolerance in tall fescue and could be potentially extended to other species.

With continued trend in global warming, grass breeders have been paying more attention to SST in tall fescue. Of course, this is not an easy breeding target to work with. SST in turfgrass species is a very practical trait, but is also very complex indeed. Each involving factor (i.e. heat, disease, insect, drought, water-logging tolerance, etc.) in SST is a complex trait itself controlled by multiple genes. For example, several QTLs related to these factors have been located in diploid ryegrass, which is a main research object in grasses. Pfender *et al.* (2011) detected three stem rust resistance QTLs by SSR, sequence-tagged site (STS), and restriction-site associated DNA (RAD). The most prominent QTL accounted for more than 30% of the phenotypic variance. Pearson *et al.* (2011) identified 37 QTLs for morphological traits influencing water logging tolerance. Certainly, heat tolerance is one of the most important SST components, and some studies on heat tolerance genes have been reported. In C3 *Agrostis* grass species, Xu *et al.* (2007) identified and characterized an expansin gene *AsEXP1* associated with heat tolerance. Kim *et al.* (2010) overexpressed an *Arabidopsis* 2-Cys Prx in transgenic tall fescue plants to enhance heat tolerance.

Factors involved with the combined summer stress could offer different levels of impact depending on the specific climate of a given region or a specific year. In a wet years or regions with ample summer precipitation, summer stress would be dominated by high temperature, high humidity, and brown patch/Pythium blight promoted by these environmental condition. In a dry years and regions with less scarce rainfall, the main stress would be heat and drought.

In this study, the polymorphism of SCAR markers corresponds well with the phenotypic identification in the field, between Funing County and Shanghai. As a result of the environmental factors analysis of the two locations in these years, they were both characterized by high temperature, ample rainfall, and Pythium blight promoting in summer. Later, we identified the polymorphism of the SCAR in another tall fescue population in 2011. However, the result was not ideal (data not shown). The analysis of 2011 found lower temperature, less rainfall, and better field management than those in 2008 and 2009. Under the summer stress of 2011, there were only some plants wilting in the afternoon, and of course, resulted in no brown shoots. It was determined that there were no significant correlations ($p > 0.05$) between T_SC856 and wilting trait. Instead, the SCAR marker maybe linked to heat, high humidity or Pythium blight tolerance. Yet, it is hard to single out which factor relates to the SCAR marker developed indeed. Further studies that specifically look at heat, high humidity tolerance, etc., under carefully controlled environment are needed to verify the factor.

Although this study reported a SST related SCAR marker with an accuracy closed to 80%, it is impractical to base MAS for SST on this one marker alone. Phenotype, affected by many uncertain factors, is unlikely to express consistently if plant materials were selected using one molecular marker, even though they are linked tightly. As a result, if locations and years change, MAS for SST based on one molecular maker is likely to fail. Currently, we are testing the SCAR marker in more clones from different locations and years. Next, many more types of molecular markers related to SST trait need to be developed. A combination use of two or more molecular markers for summer stress tolerance selection could be more accurate and stable. Development of multiple markers linked to SST will certainly be good news for MAS breeding in tall fescue and other species as well.

Acknowledgments

The authors greatly appreciate Drs. Chenxi Zhang and Daniel Bowman of North Carolina State University for proofreading and offering comments on the manuscript. This work was supported by the Natural Science Foundation of China (Grant No. 30972134).

References

Abraham EM, Huang BR, Bonos SA, Meyer WA (2004). Evaluation of drought resistance for Texas bluegrass, Kentucky bluegrass, and their hybrids. Crop Sci 44:1746-1753.

Abraham EM, Meyer WA, Bonos SA, Huang BR (2008). Differential responses of hybrid Bluegrass and Kentucky bluegrass to drought and heat stress. HortScience 43:2191-2195.

Allen TW, Han DY, Bowen KL (2005). Effect of environmental characteristics on *Pythium* and *Mesocriconema* spp. in golf course greens in Alabama. Can J Microbiol 51:287-293.

Bonos SA, Murphy JA (1999). Growth responses and performance of Kentucky bluegrass under summer stress. Crop Sci 39:770-774.

Bonos SA, Clarke BB, Meyer WA (2006). Breeding for disease resistance in the major cool-season turfgrasses. Annu Rev Phytopathol 44:213-234.

Busti A, Caceres ME, Calderini O, Arcioni S, Pupilli F (2004). RFLP markers for cultivar identification in tall fescue (*Festuca arundinacea* Schreb.). Genet Resour Crop Evolution 51:443-448.

Clark MS (1998). Plant Molecular Biology – A Laboratory Manual. Springer-Verlag, Heidelberg.

Dong S, Tredway LP, Shew HD, Wang GL, Sivamani E, Qu R (2007). Resistance of transgenic tall fescue to two major fungal diseases. Plant Sci 173:501-509.

Dong S, Shew HD, Tredway LP, Lu J, Sivamani E, Miller ES, Qu R (2008). Expression of the bacteriophage T4 lysozyme gene in tall fescue confers resistance to gray leaf spot and brown patch diseases. Transgenic Res 17:47-57.

Du H, Wang Z, Yu W, Liu Y, Huang B (2011). Differential metabolic responses of perennial grass *Cynodon transvaalensis*×*Cynodon dactylon* (C_4) and *Poa pratensis* (C_3) to heat stress. Physiol Plant 141:251-264.

Emmons RD (1994). Turfgrass Science and Management. 2nd ed. New York: Delmar Publishers TM 62～81.

He YL, Hu XH, Zhang AM, Xu R, Qi J (1997). A study on field experiment of introduced varieties of cool season turf grass. J Shanghai Agr Coll. 15:276-282 (in Chinese with English summary).

He YL, Hu XH, Tang J (2001). Breeding, comparison and extension of new lines of cool-season turf-type tall fescue-"98-8" and "Shangnong Dwarf Tall Fescue". Pratac Sci 18:60-64 (in Chinese with English summary).

He YL, Hu XH, Chen W, Xu H (2002). Breeding of new lines of turf-type tall fescue. Grassl China 24:33-39 (in Chinese with English summary).

He Y, Huang B (2007). Protein changes during heat stress in three Kentucky bluegrass cultivars differing in heat tolerance. Crop Sci 47:2513-2520.

Humaid AA, Motawei MI, Abdalla MY, Mana F (2004). Detection of genetic variation and Fusarium resistance in turfgrass genotypes using PCR-based markers (ISSR and SCAR). J Food Agric Environ 2:225-229.

Kim KH, Alam I, Lee KW, Sharmin SA, Kwak SS, Lee SY, Lee BH (2010). Enhanced tolerance of transgenic tall fescue plants overexpressing 2-Cys peroxiredoxin against methyl viologen and heat stresses. Biotechnol Lett 32:571-576.

Larsen RC, Hollingsworth CR, Vandemark GJ, Gritsenko MA, Gray FA (2002). A rapid method using PCR-based SCAR markers for detection and identification of *Phoma sclerotioides*: the case of brown root rot disease of alfalfa. Plant Dis 86:928-932.

Lefsrud MG, Sorochan JC, Kopsell DA, McElroy JS (2010). Pigment concentrations among heat-tolerant turfgrasses. HortScience 45:650-653.

Liu JR, Xie XR, Du JX, Sun JX, Bai XM (2008). Effects of simultaneous drought and heat stress on Kentucky bluegrass. Sci Hort-Amsterdam 115:190-195.

Merewitz E, Meyer W, Bonos S, Huang BR (2010). Drought stress responses and recovery of Texas x Kentucky hybrids and Kentucky bluegrass genotypes in temperate climate conditions. Agron J 102:258-268.

Mester Z, Murenyetz L, Gyulai G, Janowszky J, Hangyel TM, Bottka S, Heszky L (1999). Somaclonal variation of tall fescue (*Festuca arundinacea* L.) analyzed by RAPD-, ISSR-, and SSR-PCR. In Vitro Cell Dev Biol-Plant 35:175.

Mian MAR, Hopkins AA, Zwonitzer JC (2002). Determination of genetic diversity in tall fescue with AFLP markers. Crop Sci 42:944-950.

Michelmore RW, Paran I, Kesseli RV (1991). Identification of markers linked to disease-resistance genes by bulked segregant analysis: a rapid method to detect markers in specific genomic regions by using segregating populations. Proc Natl Acad Sci USA 88:9828-9832.

Pearson A, Coqan NO, Baillie RC, Hand ML, Bandaranayake CK, Erb S, Wang J, Kearney GA, Gendall AR, Smith KF, Forster JW (2011). Identification of QTLs for morphological traits influencing waterlogging tolerance in perennial ryegrass (*Lolim perenne* L.). Theor Appl Genet 122:609-622.

Pfender WF, Saha MC, Johnson EA, Slabaugh MB (2011). Mapping with RAD (restriction-site associated DNA) markers to rapidly identify QTL for stem rust resistance in *Lilium perenne*. Theor Appl Genet 122:1467-1480.

Saha MC, Mian R, Zwonitzer JC, Chekhovskiy K, Hopkins AA (2005). An SSR- and AFLP-based genetic linkage map of tall fescue (*Festuca arundinacea* Schreb.). Theor Appl Genet 110:323-336.

Saha M, Kirigwi F, Chekhovskiy K, Black J, Hopkins A (2009). Molecular Mapping of QTLs Associated with Important Forage Traits in Tall Fescue. In: Yamada T and G Spangenberg (eds). Molecular Breeding of Forage and Turf 251-258. Springer, New York.

Scheef EA, Casler MD, Jung G (2003). Development of species-specific SCAR markers in bentgrass. Crop Sci 43:345-349.

Seal AG (1983). DNA variation in *Festuca*. Heredity 50:225-236.

Su KM, Bremer DJ, Keeley SJ, Fry JD (2007). Effects of high temperature and drought on a hybrid bluegrass compared with Kentucky bluegrass and tall fescue. Crop Sci 47:2152-2161.

Wang ZL, Huang BR (2004). Physiological recovery of Kentucky bluegrass from simultaneous drought and heat stress. Crop Sci 44:1729-1736.

Watkins E, Meyer WA (2004). Morphological characterization of turf-type tall fescue genotypes. HortScience 39:615-619.

Wu GT, Chen JQ, Hu ZH, Lang CX, Chen XJ, Wang FL, Jin W, Xia YW (2006). Production of transgenic tall fescue plants with enhanced stress tolerances by *Agrobacterium tumefaciens*-mediated transformation. Sci Agric Sin 5:330-338.

Xu C, Huang B (2008). Root proteomic responses to heat stress in two *Agrostis* grass species contrasting in heat tolerance. J Exp Bot 59:4183-4194.

Xu JC, Tian J, Belanger FC, Huang B (2007). Identification and characterization of an expansin gene *AsEXP1* associated with heat tolerance in C3 *Agrostis* grass species. J Exp Bot 58:3789-3796.

Xu S, Li JL, Zhang XQ, Wei H, Cui LJ (2006). Effects of heat acclimation pretreatment on changes of membrane lipid peroxidation, antioxidant metabolites, and ultrastructure of chloroplasts in two cool-season turfgrass species under heat stress. Environ Exp Bot 56:274-285.

Xu WW, Sleper DA, Krause GF (1994). Genetic diversity of tall fescue germplasm based on RFLPs. Crop Sci 34:246-252.

Xu WW, Sleper DA, Chao S (1995). Genome mapping of polyploid tall fescue (*Festuca arundinacea* Schreb.) with RFLP markers. Theor Appl Genet 91:947-955.

Xu Y, Gianfagna T, Huang B (2010). Proteomic changes associated with expression of a gene (ipt) controlling cytokinin synthesis for improving heat tolerance in a perennial grass species. J Exp Bot 61:3273-3289.

Xu Y, Zhan C, Huang B (2011). Heat Shock Proteins in Association with Heat Tolerance in Grasses. Int J Proteomics doi:10.1155/2011/529648.

Alien Species of *Lepidium* in the Flora of Romania: Invasion History and Habitat Preference

Culiță SÎRBU[1*], Adrian OPREA[2], Cristian Valeriu PATRICHE[3], Costel SAMUIL[1], Vasile VÎNTU[1]

[1]*University of Agricultural Sciences and Veterinary Medicine Iași, Faculty of Agriculture, 3, Mihail Sadoveanu Alley, 700490, Iași, Romania; csirbu@uaiasi.ro (*corresponding author); vvintu@uaiasi.ro; csamuil@uaiasi.ro*
[2]*University "Alexandru Ioan Cuza", Botanical Garden, 7-9, Dumbrava Roșie Street, 700487, Iași, Romania; aoprea@uaic.ro*
[3]*Romanian Academy, Department of Iași, Geography Group, B-dul Carol I, 8, 700505, Iași, Romania; pvcristi@yahoo.com*

Abstract

The history of spread and habitat preferences during invasion of four alien species of *Lepidium* in Romania, namely *L. densiflorum*, *L. oblongum*, *L. sativum*, and *L. virginicum* are presented in this paper. The floristic records of these species, collected between 1816 and 2013, were used in the study. *L. sativum* was sporadically mentioned by earlier botanic references, as a casual alien plant in ruderal places. *L. oblongum* was reported from a single locality, but it may be considered a potential invader of ruderal habitats from southern regions. For *L. densiflorum* and *L. virginicum*, distribution maps were made in the Universal Transverse Mercator system, and exponential regression models were fitted to the cumulative number of records against time. Both *L. densiflorum* and *L. virginicum* are invasive in Romania, and their spread was initiated at the middle of last century. The slope of invasion curves, which was used as a measure of the invasion rate, did not differ significantly between the two invasive species. Although both *L. densiflorum* and *L. virginicum* were first reported in the North-West of the country, they followed different paths of migration in Romania. Ruderal habitats associated with railways played an important role in the invasion process of both species, but invaded habitats became more diversified over the time. Among the four species, only *L. densiflorum* tends to invade agricultural crops.

Keywords: distribution map, herbarium specimen, non-native weeds, transportation routes

Introduction

The genus *Lepidium* L. is one of the largest genera in the *Brassicaceae* family, being distributed on all continents, except Antarctica (Al-Shehbaz and Gaskin, 2010). In the alien flora of Europe, the genus *Lepidium* is currently represented by 20 species, of which 11 are from outside of Europe, and nine are native in certain regions of the continent, but alien outside of their native range (Drake, 2009). Among these, three species were recorded in Romania (Ciocârlan, 2009), namely: *L. virginicum* L., *L. densiflorum* Schrader (including *L. neglectum* Thell.), and *L. sativum* L. All these three species are listed among the most widespread alien plant species in Europe (Lambdon *et al.*, 2008). A fourth alien species, *L. oblongum* Small, was newly reported from Romania by Sîrbu and Oprea (2011).

In order to understand causes of invasions, and develop successful management strategies for invasive alien species, a useful tool is the reconstructing of the spread history of invasive plants, using data from herbaria (Chauvel *et al.*, 2006; Crawford and Hoagland, 2009; Csontos *et al.*, 2010; Delisle *et al.*, 2003; Lavoie *et al.*, 2007), monographs (Mihulka and Pyšek, 2001), or from multiple floristic sources in a given region (*i.e.* herbaria, literature and unpublished records) (Essl *et al.*, 2009; Pyšek and Prach, 1993; Williamson *et al.*, 2005). Maps based on historical floristic records can indicate points of introduction, invasion pathways, rates of spread and other data on the spreading of alien plants in the invaded areas (Csontos *et al.*, 2010; Lavoie *et al.*, 2007).

In this study, we aimed to reconstruct the history of spread and habitat preferences of alien species of *Lepidium* in Romania, using all the available data from herbaria and literature, as well as our own recent unpublished records.

Materials and methods

In order to reconstruct the history of invasion and habitat preferences of alien species of *Lepidium* in Romania, all occurrence data up to 2013 were registered. The information was gathered from literature (see Sîrbu and Oprea (2011) for primary sources of data), public herbaria, and recent records of the authors (2012-2013).

Herbarium data were collected from the following main herbaria of Romania (abbreviations according to Holmgren *et al.*, 1990): BUAG, BUC, BUCA, BUCM, CL, CLA, CRAI, I, IAGB, and IASI. In addition, some smaller collections were investigated, from Bacău, Oradea, and Tg. Mureş cities.

Each herbarium specimen was checked out for correct identification, sampling location, date of sampling, and habitat characteristics. The date of collection on the specimen label (for herbarium data), or the publication year (for the data from literature) unless other specific information from the fieldwork was given, were accepted as data of presence of a certain species in a given area. Repeated reports on the occurrence of a species from the same locality and the same year were treated as a single record. The nomenclature of *Lepidium* taxa follows Al-Shehbaz and Gaskin (2010).

The status of populations (naturalized or non-naturalized) was assessed for each record according to the terminology and definitions recommended by Richardson *et al.* (2000), and the pattern used by Essl *et al.* (2009). Populations were considered naturalized, if (i) this status has resulted unambiguously in the original data source, (ii) populations consisted of a large number of reproductive individuals (more than 100), (iii) populations smaller than 100 individuals were reported in the same location at least two consecutive years, or several small populations were recorded in the same locality, the same year. In all other cases, populations were considered non-naturalized.

The distribution maps of *L. densiflorum* and *L. virginicum* in Romania were made using the Universal Transverse Mercator (UTM) grid system (Lehrer and Lehrer, 1990). All records of these two species were marked on the map based on the locality code from UTM grid with cells of 10 × 10 km.

In order to obtain a global estimation of the invasion rate for *L. densiflorum* and *L. virginicum*, exponential regression models ($y = e^{a+bx}$) were fitted to the cumulative number of records against time and the slope *b* was used as a measure of the invasion rate (Delisle *et al.*, 2003; Mihulka and Pyšek, 2001; Pyšek and Prach, 1993). The significance of difference between the slopes was tested using the *t*-test of parallelism (Zar, 2010). We also tested, for these two species, whether the length of the railway network (km), as well as the rail traffic intensity of freight (tons per year) and people (thousand passengers per year), between 1950 and 2010, was positively related to the number of records by calculating the Spearman's correlation coefficients (r_s), with right-tail tests of significance (Zar, 2010). To assess habitat preferences of *L. densiflorum* and *L. virginicum*, a G^2-test of independence (McDonald, 2008) was performed on the observed frequencies of species in various habitat types. The program XLSTAT (version 2011) was used for statistical analyses.

Results

General result

A total number of 174 records of four alien species of *Lepidium* (*L. sativum* L., *L. densiflorum* Schrad., *L. virginicum* L., and *L. oblongum* Small) have been reported in Romania in a period of 197 years (between 1816 and 2013). Of these, 62 records were obtained from 92 herbarium sheets, 94 from literature sources, and 18 from field investigations (2012-2013). Herbarium data were recorded in 13 public herbaria, of which the most important was IASI (38 specimens). Four herbaria contained no specimens of alien species of *Lepidium*.

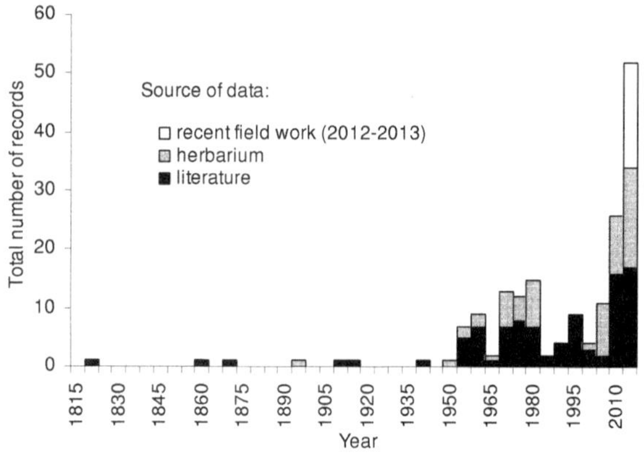

Fig. 1. Distribution of the total number of records of the alien species of *Lepidium* in Romania, between 1816 and 2013

Four periods of intensive recording of alien species of *Lepidium* can be distinguished (Fig. 1), of which the most important was between 2004 and 2013. Herbarium specimens were collected especially between 1966 and 1980 and between 2004 and 2013.

History of invasion and distribution

Lepidium sativum was the first alien species of *Lepidium* reported in Romania, in 1816. Until 1955, this species was reported only in 6 localities, and there are no reliable data on its naturalization in Romania. Since 1955 the species was no longer reported. All specimens of *L. sativum* from various herbaria were collected from vegetable gardens.

Historical data on the distribution of *L. densiflorum* and *L. virginicum* in Romania are represented on the maps in Fig. 2 and Fig. 3, respectively.

Lepidium densiflorum was first observed in Romania in the province of Crişana (1893), and later on in Transylvania (1958), Moldavia (1967), Muntenia (1968), Oltenia (1976), and Dobrudja (2011). Overall, *L. densiflorum* has been reported so far from 110 UTM cells of 100 km^2 (*i.e.*, 4.6% of the country area), corresponding to 89 localities, of which 1.1% before 1950, 41.6% between 1951 and 2000, and 57.3% after the year 2000. This species is widespread in the eastern half of Romania, mainly in Moldavia (60 localities) and Muntenia (11 localities), and it is less widespread in the western half of the country. In Maramureş and Banat it has not been reported so far. The first reliable data on naturalized populations of this species in Romania dates back to 1967, in Moldavia. Until 2013, *L. densiflorum* has been reported in Romania, as a

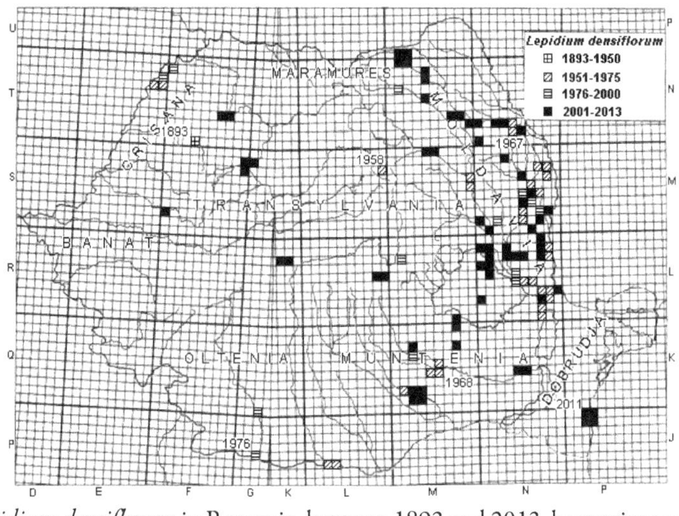

Fig. 2. Distribution of *Lepidium densiflorum* in Romania, between 1893 and 2013, by species occurrence within UTM cells of 10 x 10 km. The years of the oldest records for each historical region were indicated

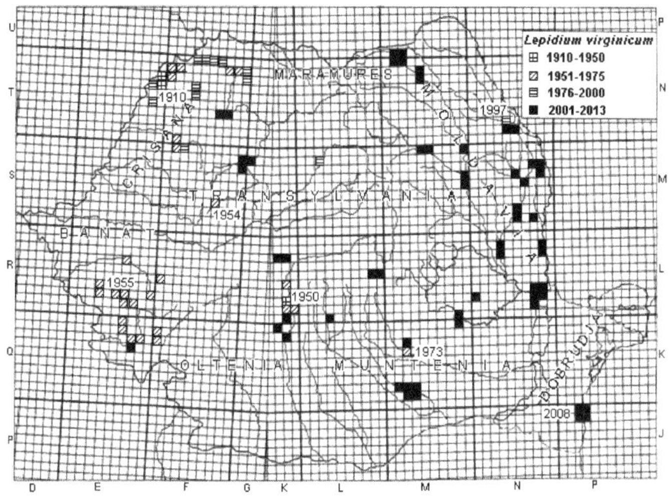

Fig. 3. Distribution of *Lepidium virginicum* in Romania, between 1910 and 2013, by species occurrence within UTM cells of 10 x 10 km. Meaning of numbers displayed as in Fig. 2

naturalized species, from 59 localities (which is 66.3% of all records), of which 51 localities (86.4%) after the year 2000.

Lepidium virginicum was first documented in 1910, in Crişana, and subsequently in Oltenia (1950), Banat (1955), Transylvania (1962), Muntenia (1973), Maramureş (1974), Moldavia (1997), and Dobrudja (2009). The occurrence of *L. virginicum* has been reported so far from 101 UTM cells of 100 km^2 (*i.e.* 4.2% of the country area), corresponding to 76 localities, of which 2.6% between 1910 and 1950, 48.7% between 1950 and 1999, and 48.7% after the year 2000. Nowadays, this species has spread to all historical provinces of Romania, being more frequent in Moldavia (22 localities), Crişana (16 localities), and Banat (12 localities). In Oltenia, Muntenia, and Transylvania it is sporadic, while in Dobrudja and Maramureş, it is rare. The first record of a naturalized population was in 1955, in Banat. Until 2013, this species has been reported as naturalized in 32 localities (which is 42.1% of all records), of which 27 localities (84.4%) after the year 2000.

The slopes of invasion curves of *L. densiflorum* and *L. virginicum* (Fig. 4) do not differ significantly from each

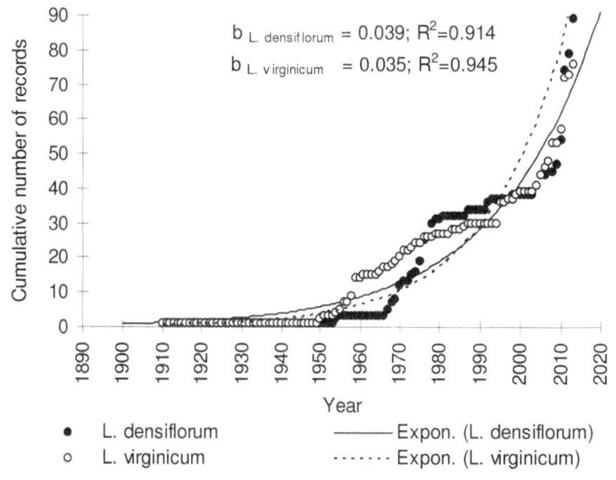

Fig. 4. Exponential regression curves of the cumulative number of records of *Lepidium densiflorum* and *L. virginicum* in Romania. The estimated invasion rates (slopes b) are both statistically different from zero at 5% probability level

other (test of parallelism: t = 0.207, P = 0.836). Two periods of invasiveness (*i.e.* sharp increase of the lines of distribution) can be distinguished for both *L. densiflorum* (*i.e.* between 1967-1979, and 2004-2013, respectively) and *L. virginicum* (*i.e.* 1950-1976, and 1995-2013, respectively).

No significant positive relationships were found between the length of railway network (km) and the number of records of *L. densiflorum* and *L. virginicum* (r_s = -0.003, p = 0.491; r_s = -0.438, p = 0.001, respectively). In addition, no significant positive relationships were found between the rail traffic of goods (thousand tonnes per year) or passengers (thousand passengers per year) for the period 1950-2010 and number of records of species *L. densiflorum* (r_s = 0.118, p = 0.183; r_s = -0.068, p = 0.301, respectively) and *L. virginicum* (r_s = -0.387, p = 0.001; r_s = -0.382, p = 0.001, respectively).

The most recent alien species of *Lepidium* recorded in Romania's flora is *L. oblongum* (Fig. 5). It was found in 2011, only in a single locality, in the railway station of Râmnicu Sărat, Buzău County, between the platform of the station and the rail lines.

Habitat preferences over time.

Two of the species considered in this paper were observed each in a single type of habitat, namely, ruderal habitats near gardens (*L. sativum*) or associated with railways (*L. oblongum*). The proportion of each habitat types invaded over time by the other two species (*L. densiflorum* and *L. virginicum*) is shown in Fig. 6. The difference in invaded habitat types between *L. densiflorum* and *L. virginicum* was significant (Wilks' G^2 = 14.07, d.f.=7; p=0.041).

Fig. 5. *Lepidium oblongum*: a) herbarium specimen, b) basal leaf, c) upper cauline leaves; d) fruits, e) seed. Scale bar: a = 20 mm, b = 10 mm, c = 5 mm, d, e = 1 mm

Since 1883 *L. densiflorum* has been mainly recorded in ruderal places associated with railways. Diversification of habitats invaded by this species began in the late 1960s, when it was collected from floodplains of some rivers. Since 1970s, it has been also observed in harbours, cereal crops or vineyards on sandy soils, along roadsides and in other ruderal habitats, as well as on sandy lands. However, in recent decades, a significant diversification of invaded habitats has not been further registered. The main habitats,

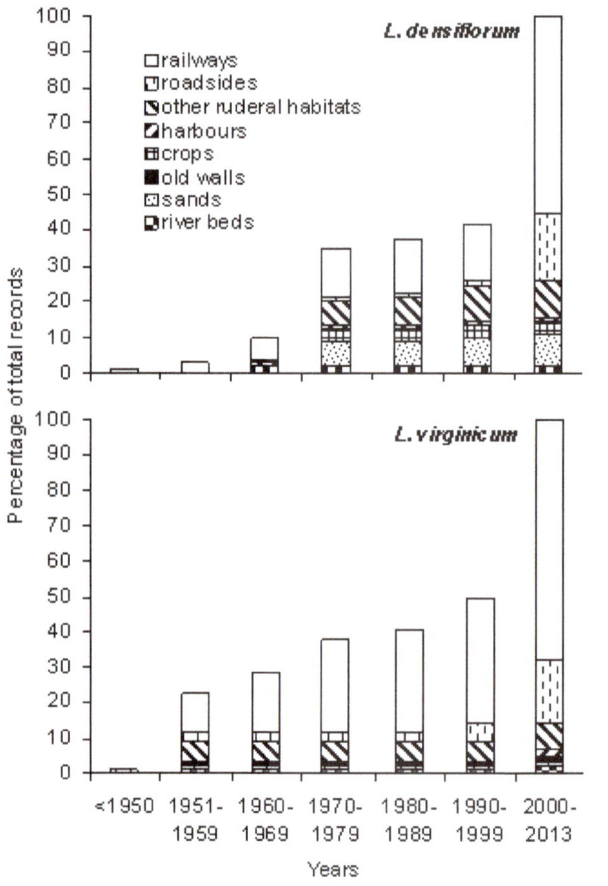

Fig. 6. Relative frequency of habitat types invaded over time in Romania by *Lepidium densiflorum* and *L. virginicum*

in which *L. densiflorum* has been observed since 1980 are represented by railways and, to a lesser extent, roadsides and other ruderal sites.

Lepidium virginicum has been mainly recorded along railways. To a lesser extent, this species invaded roadsides and other ruderal habitats. It was also observed on ruderalized sands, on old walls, on sandy gravels of river beds, as well as in harbours. It has not been recorded until now as a weed in crops.

Discussion

General distribution and invasion history

Lepidium sativum originated in North-East Africa and South-West Asia, being cultivated since ancient times as a vegetable or as a herb (Jansen, 2004; Markgraf, 1963; Thellung, 1906). According to Akeroyd and Rich (2010), it is frequent as a casual weed and locally naturalized in Europe, but it is impossible from the available data to distinguish between naturalized and casual distribution. However, Lambdon *et al.* (2008) listed this species among the most widespread 20 alien plant species in Europe. *L. sativum* was also reported as naturalized in North America, South America (Argentina), Australia (Al-Shehbaz, 2010), and Eastern Asia (Taiyan *et al.*, 2001).

We suppose *L. sativum* was introduced in Romania as a vegetable, before the nineteenth century, when it was first reported from Transylvania (Baumgarten, 1816). Up to the middle of the last century, this species was mentioned sporadically in the literature, as a subspontaneous plant, escaped from gardens (Nyárády, 1955), in Transylvania (Baumgarten, 1816; Schur, 1866), Banat (Heuffel, 1858), Muntenia (Panţu, 1912) and Bukovina (Prodan, 1939). However, according to Simonkai (1886), *L. sativum* was found in Transylvania only as a cultivated plant. After 1955, this species has not been reported from other localities in Romania, and nowadays it is rarely cultivated as a vegetable. All the examined specimens of *L. sativum*, from herbaria, were collected from gardens. Therefore *L. sativum* can be considered as a casual alien plant in Romania, based only on old data from the literature.

Lepidium densiflorum is a species native to North America, recorded as a casual alien after 1870 in Central Europe (Markgraf, 1963). Nowadays, it is naturalized in most of Europe, except in Great Britain, the Iberian Peninsula and the Balkan Peninsula (Akeroyd and Rich, 2010), being listed among the 100 most widespread alien plant species in the continent (Lambdon *et al.*, 2008). It is also naturalized in South America (Al-Shehbaz, 2010) and Asia (Taiyan *et al.*, 2001). In the neighbouring countries of Romania, *L. densiflorum* has been known since 1895 in Ukraine (Protopopova *et al.*, 2006), and since 1935, in Hungary (Pénzes, 1941).

The presence of *L. densiflorum* in Romania was first reported by Negrean (1980), based on herbarium specimens collected by the author in several localities, as well as on the revision of some earlier herbarium collections that had been erroneously identified as *L. virginicum*. The first record of this species in Romania is a specimen collected by Richter, in 1893, in the northwestern part of the country (Crişana), at Ciucea village. According to Negrean (1980), this specimen was first assigned to *L. virginicum*, and subsequently revised by Thellung (1906), as *L. neglectum* (today included in *L. densiflorum*). However, it seems that the plants that arrived here accidentally from the West, via rail transport, were not able to naturalize in that initial point of introduction, which thus remained isolated on the distribution map. Although in the 1950s new introductions appeared independently of the initial introduction both in Crişana, and in eastern Transylvania, a significant wave of migration of this species cannot be recognized yet in these two provinces.

In the eastern Romania (Moldavia), *L. densiflorum* was collected in 1967, in Iaşi and Socola railway stations (Iaşi county), but confused with *L. virginicum* and distributed under that name in *Flora Romaniae Exsiccata* (FRE), no 3060, and *Flora Moldaviae et Dobrogiae Exsiccata* (FMDE), no. 40 (Negrean, 1980), a clear evidence that it was abundant in these localities at the time. After that year, an important increase in records of this species occurred in Moldavia, especially along railways. Populations of *L. densiflorum* from Moldavia could have originated from Ukraine, where this species is quite widespread and was first registered in 1895 (Protopopova *et al.*, 2006). The fact that the first information about the occurrence of this species in Moldavia come from two cities situated near the eastern border of Romania, along the railways linking Ukraine and Romania, can support this hypothesis.

In the southern regions of Romania *L. densiflorum* was first collected in Muntenia (1968), then in Oltenia (1976) (Negrean, 1980) and much later in Dobrudja (2011). We suppose that populations from southern regions are linked with those of Moldavia from where they could migrate along railways. The expansion from Transylvania, through mountain passes, toward eastern and southern regions of the country, cannot be excluded, but it is less likely, given that *L. densiflorum* is even nowadays a quite rare species in Transylvania. Probably it could not arrive in these regions either from the South, because it has not been registered yet in the southern neighbouring countries (Bulgaria, Serbia) (Akeroyd and Rich, 2010).

Lepidium virginicum is another North American species, widely naturalized in Europe (Akeroyd and Rich, 2010), as well as in Asia (Taiyan *et al.*, 2001), South America, Africa, and Australia (Al-Shehbaz and Gaskin, 2010). According to Markgraf (1963), *L. virginicum* was introduced in the Botanic Garden of Montpellier, in 1697, and has been recorded in Central Europe as a subspontaneous plant, beginning with the eighteenth century (in 1786). Subsequently, it has gradually spread throughout the continent (Akeroyd and Rich, 2010; Markgraf, 1963; Thellung, 1906), mainly along railways. According to Lambdon *et al.* (2008), *L. virginicum* is currently one of the most widespread alien plant species in Europe, occurring in more than 80% of European regions.

In neighbouring countries of Romania, *L. virginicum* was reported in Hungary, in the beginning of the last century (Thellung, 1906), Serbia, in the 1970s (Ivković, 1978), the Caucasus region (Fedorov, 1979; Vinogradova *et al.* 2010), Ukraine (Protopopova *et al.*, 1994), and in Turkey (Terzioğlu and Anşin, 2001).

In Romania, *L. virginicum* has arrived most probably from Central Europe, along railways, being recorded in 1910, in the North-West of the country (Crişana), near the actual border with Hungary (the Pişcolt village) (Karácsonyi, 1995). The previous indication from Crişana (at Ciucea village, leg. 1901) made by Degen (1902), is not a certain one, since subsequently in the monograph of the genus *Lepidium*, Thellung (1906) mentioned only "*L. neglectum*" from that locality: "1893, L. Richter – Herb. Degen".

After these first indications, the occurrence of this species was reported, almost simultaneously, at the middle of the last century, both in Crişana (Vadu Crişului railway station, Bihor County, leg. Soran, 03.VI.1951, herbarium CL-216713), and in three other provinces in the western half of the country, namely: Oltenia (herbarium CLA-06400, Lotru railway station, leg. Nyárády, 19.VI.1950), Transylvania, in the Muşca railway canton, Alba county (leg. Soran, 1954) (Soran, 1962), and Banat, in the railway station of Câlnic, Caraş-Severin County (leg. Goga, 1955) (Goga, 1960). Reports on the presence of this species during such a short period of time (1950-1955) in localities so distant from one another (100-250 km), from four different provinces, suggests that in the first half of last century, an active spread of this species took place, along railways, starting from Crişana, the initial centre of introduction, southward, in Transylvania, Banat and Oltenia. However, we cannot exclude the hypothesis that new introductions occurred in all that time independently of the initial introduction. The introduction of this species in the southwestern Romania (Banat) from the neighbouring country (Serbia) is unlikely because there, *L. virginicum* was reported much later, in the 1970s (Ivković, 1978). In the subsequent decades, the spread of this species has continued along the railway network, especially in Banat, Crişana, and, to a lesser extent, in Oltenia (along the Olt River) and Transylvania. Also, *L. virginicum* migrated from Crişana to Maramureş, where it was first observed in 1974 (at railway station Ilba) (Negrean, 1980).

In Muntenia, *L. virginicum* was first collected in 1973, in the railway station Prahova, situated on the railway line Braşov-Bucharest. It was introduced there, probably, by rail transports, from Transylvania. The absence of this species in Bulgaria (Akeroyd and Rich, 2010) excludes its arrival in southern Romania from the Balkans.

In Moldavia, *L. virginicum* was recorded by error between 1967 and 1997, based on specimens of *L. densiflorum*, as mentioned above. Because of these wrong identifications, Negrean (1980, 1987) considered the presence of this species in Moldavia as uncertain. Indeed, upon examination of the herbarium specimens (I, IAGB, IASI) collected prior to 1997, we found that all of them were wrongly determined, and that they belong actually to *L. densiflorum* (Sîrbu, 2006). *L. virginicum* has been found with certainty in Moldavia since 1997, being collected in the railway station of Socola-Iaşi (herbarium IASI - 17957, leg. Sîrbu and Oprea, 21.VII.1997) (Oprea *et al.*, 1997). However, the current data indicate that it is quite widespread in Moldavia, mainly in railway stations. Since *L. virginicum* is missing in Republic of Moldova (Negru, 2007), and it was discovered only recently in Ukraine (Protopopova *et al.*, 1994), we suppose this species arrived in eastern Romania either from the West (Transylvania), through mountain passes, or, more likely, from the South, along the railway lines linking Bucharest with the major cities of Moldavia.

In Dobrudja, *L. virginicum* was recently reported in a single locality (in Constanța harbour), by Anastasiu *et al.* (2009).

Lepidium oblongum originates in North America (Southern and Central USA) and Central America (Mexico, El Salvador, and Guatemala) (Al-Shehbaz and Gaskin, 2010; DiTomaso and Healy, 2007). There is a single record of this species in Romania (Sîrbu and Oprea, 2011), and according to Akeroyd and Rich (2010), and Drake (2009), it is so far unknown in other regions of Europe. In the railway station of Râmnicu Sărat, the population of *L. oblongum* consists of several dozens of individuals (Sîrbu and Oprea, 2011). Because their seeds have a high capacity of germination (over 80%, in light conditions) (Sîrbu, 2011, unpublished), and the population persisted for three consecutive years in the same place, we can assume that this species is naturalized in Romania, even if it was reported only from a single location, so far. We suppose that it arrived in Râmnicu Sărat railway station with goods or passenger trains, but we do not know currently the ways of its migration from North America to Romania. As it was shown by many authors (*e.g.*, Guo *et al.*, 2012; Mihulka and Pyšek, 2001), there are often positive relationships between the latitudinal distributions occupied by species in their native and exotic ranges. As the primary area of *L. oblongum* extends up to Oregon (on the West coast of the United States) (DiTomaso and Healy, 2007), ca. 44°N, we can suppose that it may have a potential to invade southern regions of Europe, but further studies are required in this regard.

Lag time and invasion curves

The spread of *L. densiflorum* and *L. virginicum* in Romania was initiated, as stated above, at the middle of the last century, about 60, respectively 40 years after registration of the first specimen of these species in the country. This is not surprising since a lag time of over 40-50 years between initial colonization and the onset of rapid population growth and geographic expansion is a common feature of alien plant invasions (Binggeli, 2001; Essl *et al.*, 2009; Pyšek and Prach, 1993). The increase of the invasion slope after 1950 for both species could be related to the intensive railway transport of troops and military equipment, during the Second World War, as well as the environmental disturbances caused by that war.

According to our results, the slopes of invasion curves, which give global estimations of the invasion rates (Delisle *et al.*, 2003) do not differ significantly for *L. densiflorum* and *L. virginicum* in Romania. However, periods of invasibility (*i.e.* sharp increase in slope of the invasion curve) differ between the two species in the second half of the last century. The active spread of *L. virginicum* has been recorded since 1950, with an important advance against *L. densiflorum* (1967). Moreover, evidence of naturalized populations occurred earlier in *L. virginicum* (45 years after first registration in Romania) as against *L. densiflorum* (84 years after first registration in Romania).

Since the invasion curve has an evident increasing trend in the last decade, for both *L. densiflorum* and *L. virginicum*, we can assume that the two species are still far from reaching the saturation phase of invasion in Romania – *i.e.* when the rate of invasion slows down (Essl *et al.*, 2009), and that they will continue to spread in the future.

Habitat preferences and the role of traffic lines in plant spreading

Within their natural range, *L. densiflorum*, *L. virginicum* and *L. oblongum* grow in various disturbed natural and anthropogenic habitats, such as: prairies, pastures, waste grounds, fields, roadsides, railways (all three species) (Al-Shehbaz and Gaskin, 2010; DiTomaso and Healy, 2007), flood plains (*L. densiflorum* and *L. oblongum*), sagebrush flats, gravelly hillsides, rock crevices, seashores, sandy places, open mesa (*L. densiflorum*), llanos, calcareous sand, alluvial terraces (*L. oblongum*) (Al-Shehbaz and Gaskin, 2010) etc. As for the species *L. sativum*, in its native area, it prefers not too warm sites, with rich, open soils, well supplied with water, although it is fairly drought tolerant (Bush 1939; Jansen 2004).

In Central Europe, according to Markgraf (1963), *L.*

densiflorum and *L. virginicum* are usually found on nutrient-rich soils, in railway stations, between the tracks, along roads, on ruderal places, on sands and gravels in harbours, rarely as weeds in crops, entering in composition of the ruderal communities from the orders *Sisymbrietalia* and *Onopordetalia*. In Central Europe, *L. sativum* occasionally escapes from cultivation, in weed communities on newly disturbed fields, rich in nutrients (*Sisymbrion*) (Markgraf, 1963). In Romania, the *Lepidium* species taken into account invade a more limited range of habitats than in their natural areas, but (except *L. oblongum*) comparable with other regions in Europe.

It is well known that the repeated transport of propagules by trains or cars can facilitate the spread and naturalization of alien species along traffic lines (Hulme, 2009; Kowarik, 2003; Von der Lippe and Kowarik 2007). It is probably the case of species studied by us, except *L. sativum*. In more than 86%, and respectively 74% of the total number of records in Romania, *L. virginicum* and respectively *L. densiflorum* grow in ruderal habitats associated with traffic infrastructure, and railway stations have the largest share in the total number of records. The predominance of these species in habitats associated to railways suggests that they immigrated to and spread throughout Romania mainly via the railway transport.

Consequently, it was expected that the spread of both species be influenced by extension of railway network, by intensity of goods and passenger traffic, as well as by disturbance status of habitats associated with railways. But the length of railway network in operation did not change significantly in Romania, in the last 60 years. Between 1950 and 1995, it increased only by 4.8% (from 10853 to 11376 km), and between 1995 and 2010 it even decreased by 5.2%, reaching today below the level of the year 1950 (INSR, 1950-2011). In addition, surprisingly, for the period 1950-2010, we did not found significant positive relationships between the number of records of the species *L. virginicum* and *L. densiflorum* and the rail traffic of goods or passengers. Therefore, we can assume that the spread of both species was determined rather by the favourability of disturbed habitats associated with railways (especially ruderal habitats from train stations). This made possible the germination of seeds accidentally arrived with trains from long distances, and, subsequently, the growth of plants and seed production for several generations. Naturalized plants in new locations have further spread, via anemochorous, epizoochorous (over short distances) or anthropochorous seed dispersal (over variables distances), in ruderal habitats closely associated with railways and roads, as well as in other anthropogenic or natural habitats, on lands situated near transport routes. This agrees with the pattern of stratified dispersal, consisting of long-distance movements, and neighbourhood diffusion (Essl *et al.*, 2009; Kowarik, 2003; Ming and Albrecht, 2004).

Overall, according to our data, it seems that *L. densiflorum* has a more pronounced tendency of diversifying the invaded habitats than *L. virginicum*. Only *L. densiflorum* has been reported so far in Romania as a segetal weed in cereal crops (*Flora Olteniae Exsiccata*, 951, leg. Păun, 21.VI.1976; Vițălariu *et al.*, 1977), or in vineyards, on sandy soils (Negrean, 1980). According to Vițălariu *et al.* (1977), in wheat crops, this species can achieve a significant level of weed infestation (300-400 individuals per m^2). After 1980, however, further cases of crop infestation by this species have not been registered. In natural habitats, only *L. densiflorum* and *L. virginicum* (especially the first one) were found, on sandy lands and river beds, probably in conditions of ruderalization, but reports in this respect were only occasional.

The reliability of the data set

According to the data from literature (Chauvel *et al.*, 2006; Crawford and Hoagland, 2009; Delisle *et al.*, 2003; Pyšek and Prach, 1993; Pyšek and Hulme, 2005), to successfully reconstitute the introduction and invasion of a species on a large geographical scale through floristic data, a sufficient and constant intensity of floristic investigations within that area is necessary. This condition is broadly achieved in Romania, because of the strong and long tradition of floristic research in this country, dating back to the second decade of the 19th century. However, the reconstitution of plant invasions could be affected by the fluctuation in time of the interest shown by botanists to collect herbarium specimens, which may sharply decrease at one time, because a certain degree of saturation can be attained (Chauvel *et al.*, 2006). Our data proved, for instance, that between 1950 and 1980, there was an increasing trend to collect herbarium specimens of *L. densiflorum* and *L. virginicum*, but no other specimen was collected over the next 15 years even though new records continued to appear in the published literature, all that time. Therefore, by using data from multiple sources (herbaria, published literature, field studies), we tried to reduce bias in assessing the patterns of invasion. On the other hand, we have no reasons to believe that botanists have tended to collect or record *L. densiflorum* more than *L. virginicum*, or vice versa. In fact they are very similar morphologically, as already shown, and usually grow in similar habitats. So, the recording effort could be considered similar for the two species over the time. However, such reconstitution remains unavoidably influenced by the magnitude of floristic investigations. Therefore, we suppose that *L. densiflorum* and *L. virginicum* are still under-recorded in Romania, and their genuine distribution could be larger than it is suggested by the currently available data used in this paper.

Conclusions

Four alien species of *Lepidium* were registered till now in the flora of Romania, of which one is casual (*L. sativum*), one naturalized (*L. oblongum*), and two are invasive (*L. densiflorum* and *L. virginicum*). The invasion of the last two species started after a lag time of about 60, respectively 40 years. Although both species were first reported in the northwestern Romania, they subsequently followed different paths of migration in the country. However, the slopes of the invasion curves do not differ significantly between the two species. The predominance of these invasive species in habitats associated to railways suggests that they immigrated to and spread throughout Romania mainly via the railway transport. Naturalized populations along railways have further dispersed to neighbouring suitable habitats, especially ruderal ones, following the pattern of the stratified dispersal. Although invasive, neither *L. densiflorum* nor *L. virginicum* proved to be harmful to crops or natural habitats, so far.

Acknowledgements

This work was partially supported by CNCSIS-UEFISCDI Romania, project number PN II-IDEI_1227. The authors thank the curators of the cited herbaria, the anonymous reviewers for their very helpful comments on the previous version of the manuscript, as well as Mr. Z. Barina and Mr. F. Verloove for providing some important references and herbarium specimens.

References

Akeroyd JR, Rich TCG (2010). *Lepidium* L., p. 398-402. In: Tutin TG, Burges NA, Chater AO, Edmondson JR, Heywood VH, Moore DM, Valentine DH, Walters SM, Webb DA (eds.). Flora Europaea, Vol. 1: *Psilotaceae* to *Platanaceae*, 2nd ed., 1st. paperback printing, Cambridge University Press, Cambridge.

Al-Shehbaz IA (2010). A synopsis of the South American *Lepidium* (*Brassicaceae*). Darwiniana 48:141-167.

Al-Shehbaz IA, Gaskin JF (2010). *Lepidium* L., p. 570-595. In: Boufford D, Freeman C, Gandhi K, Hill M, Kiger R, Poole J, Schmidt H, Shultz L, Strother J, Zarucchi J (Eds.). Flora of North America, Vol. 7, *Magnoliophyta*: *Salicaceae* to *Brassicaceae*, Oxford University Press Inc., New York.

Anastasiu P, Negrean G, Făgăraş M, Samoilă C, Cogălniceanu D (2009). Constanţa harbour (Romania) as a major gateway and reservoir for alien plant species. Acta Horti Bot Bucurest 36:41-60.

Baumgarten JCG (1816). Enumeratio stirpium Magno Transsilvaniae Principatui. Vol. 2, Libraria Camesinae, Vindobonae, 390 p.

Binggeli P (2001). Time-lags between introduction, establishment and rapid spread of introduced environmental weeds. Proc 3rd Internat Weed Sci Congr; 2000 June 6-11; Foz do Iguassu, Brazil, Manuscript number 8-14 p, CD-ROM. Available from: International Weed Science Society, Oxford, MS, USA.

Bush NA (1939). *Lepidium* L., p. 501-524. In: Bush NA (Ed.). The Flora of USSR (in Russian), Vol. 8, Nauka, Moskva-Leningrad.

Chauvel B, Dessaint F, Cardinal-Legrand C, Bretagnolle F (2006). The historical spread of *Ambrosia artemisiifolia* L. in France from herbarium records. J Biogeogr 33:665-673.

Ciocârlan V (2009). Illustrated flora of Romania. Pteridophyta et Spermatophyta (in Romanian). 3rd ed. Edit. Ceres, Bucharest, 1141 p.

Crawford PHC, Hoagland BW (2009). Can herbarium records be used to map alien species expansion over the past 100 years? J Biogeogr 36:651-661.

Csontos P, Vitalos M, Barina Z, Kiss L (2010). Early distribution and spread of *Ambrosia artemisiifolia* in Central and Eastern Europe. Bot Helv 120:75-78.

Degen Á (1902). *Lepidium virginicum* L. Magyarors-Zágbau (*Lepidium virginicum* L. in Ungarn). Mag Bot Lap 1:349-350.

Delisle F, Lavoie C, Jean M, Lachance D (2003). Reconstructing the spread of invasive plants: taking into account biases associated with herbarium specimens. J Biogeogr 30:1033-1042.

DiTomaso JM, Healy EA (2007). Weeds of California and other western states. Vol. 1: *Aizoaceae-Fabaceae*. Agriculture and Natural Resources, University of California, 834 p.

Drake JA (Ed.) (2009). DAISIE handbook of alien species in Europe. Springer Science + Business Media BV, 399 p.

Essl F, Dullinger S, Kleinbauer I (2009). Changes in the spatio-temporal patterns and habitat preferences of *Ambrosia artemisiifolia* during its invasion of Austria. Preslia 81:119-133.

Fedorov AA (1979). Flora of the European USSR (in Russian), Vol. 4. Nauka, Leningrad, 356 p.

Goga ID (1960). Euphorbia *maculata* L. and *Lepidium virginicum* L. and their spread in Banat (in Romanian). Comun Bot /1957-1959/:337-339.

Guo Q, Sax DF, Qian H, Early R (2012). Latitudinal shifts of introduced species: possible causes and implications. Biol Invasions 14:547-556.

Heuffel J (1858). Enumeratio plantarum in Banatu Temesiensi sponte crescentium et frequentius cultarum. Typis Caroli Ueberreuter, Vindobonae, 204 p.

Holmgren PK, Holmgren NH, Barnett LC (1990). Index herbariorum, Part I: the herbaria of the world [Regnum Veg. vol. 120], New York Botanical Garden Press, New York.

Hulme PE (2009). Trade, transport and trouble: managing invasive species pathways in an era of globalization. J Appl Ecol 46:10-18.

INSR (Institutul National de Statistica al României) (1950-2011). Statistical yearbook of Romania (in Romanian). Edit. INSR, Bucharest.

Ivković O (1978). *Lepidium virginicum* L. 1753 (Capparidales, Brassicaceae), a new species in the flora of Serbia (in Serbian). Biosistematika Beograd 4(1):75-79.

Jansen PCM (2004). *Lepidium sativum* L., In: Grubben GJH, Denton OA (Eds.). PROTA 2: Vegetables/Légumes. [CD-Rom]. PROTA, Wageningen, Netherlands.

Karácsonyi C (1995). Flora and vegetation of Satu Mare county (in Romanian). Edit. Muz. Sătmărean, Satu Mare, 196 p.

Kowarik I (2003). Human agency in biological invasions: secondary releases foster naturalisation and population expansion of alien plant species. Biol Invasions 5:281-300.

Lambdon PW, Pyšek P, Basnou C, Hejda M, Arianoutsou M, Essl F, Jaroší V, Pergl J, Winter M, Anastasiu P, Andriopoulos P, Bazos I, Brundu G, Celesti-Grapow L, Chassot P, Delipetrou P, Josefsson M, Kark S, Klotz S, Kokkoris Y, Kühn I,

Marchante H, Perglová I, Pino J, Vilà M, Ziko A, Roy D, Hulme PE (2008). Alien flora of Europe: species diversity, temporal trends, geographical patterns and research needs. Preslia 80:101-149.

Lavoie C, Jodoin Y, Goursaud de Merlis G (2007). How did common ragweed (*Ambrosia artemisiifolia* L.) spread in Quebec? A historical analysis using herbarium specimens. J Biogeogr 34(10):1751-1761.

Lehrer AZ, Lehrer MM (1990). Mapping of fauna and flora of Romania (areographic coordinates) (in Romanian). Edit. Ceres, Bucharest, 295 p.

Markgraf F (1963). *Lepidium* L., p. 401-417. In: Hegi G., Illustrierte Flora von Mitteleuropa. Band 4(1/6), 2 Aufl., Carl Hanser Verlag, München.

McDonald JH (2008). Handbook of biological statistics. Baltimore: Sparky House Publishing, 287 p.

Mihulka S, Pyšek P (2001). Invasion history of *Oenothera* congeners in Europe: a comparative study of spreading rates in the last 200 years. J Biogeogr 28:597-609.

Ming P, Albrecht J (2004). Integrated framework for the simulation of biological invasions in a heterogeneous landscape. Transactions in GIS 8(3):309-334.

Negrean G (1980). *Lepidium densiflorum* and *Lepidium neglectum* in Romania (in Romanian). Stud Comun Muz Satu Mare 4:435-439.

Negrean G (1987). Additions to the flora of Romania (in Romanian). Stud Comun Muz Satu Mare 7-8:447-459.

Negru A (2007). Identification book of plants from the flora of the Republic of Moldova (in Romanian). Edit. Universul, Chişinău, 391 p.

Nyárády EI (1955). *Lepidium* L., p. 418-431. In: Săvulescu T (Ed.), Flora of the Romanian People's Republic (in Romanian), Vol. 3, Edit. Academiei R. P. Române, Bucharest.

Oprea A, Sîrbu C, Paraschiv NL (1997). Flora and vegetation from the Socola-Iaşi railway station, I. Flora (in Romanian). Lucr Şti USAMV Iaşi, Agricult 40:91-97.

Panţu ZC (1912). Contributions to the flora of Bucharest and surroundings (in Romanian), IV. Analele Acad Române, Mem Secţ Şti, Ser II 34:435-598.

Pénzes A (1941). On the recent spread of some species of *Eragrostis* and *Lepidium* (in Hungarian). Bot Közlem 38(3-4):179-180.

Prodan I (1939). Flora for identification and description of plants that grow in Romania (in Romanian), Vol. 1, part. 1, 2nd ed. Tipogr. Cartea Românească, Cluj, 624 p.

Protopopova VV, Mosyakin SL, Shevera MV (1994). *Lepidium virginicum* L. – a new alien species in Ukrainian flora (in Ukrainian). Ukrayins'k Bot Zhurn 51(2-3):225-228.

Protopopova VV, Shevera MV, Mosyakin SL (2006). Deliberate and unintentional introduction of invasive weeds: a case study of the alien flora of Ukraine. Euphytica 148:17-33.

Pyšek P, Hulme P (2005). Spatio-temporal dynamics of plant-invasions: linking pattern to process. Ecoscience 12:302-315.

Pyšek P, Prach K (1993). Plant invasions and the role of riparian habitats: a comparison of four species alien to central Europe. J Biogeogr 20:413-420.

Richardson DM, Pyšek P, Rejmánek M, Barbour MG, Panetta FD, West CJ (2000). Naturalization and invasion of alien plants: concepts and definitions. Diversity Distrib 6:93-107.

Schur JF (1866). Enumeratio plantarum Transsilvaniae. Apud Guilielmum Braumüller, Vindobonae, 984 p.

Simonkai L (1886). Erdély edényes flórájának helyesbített foglalata (Enumeratio florae Transsilvaniae vesculosae critica). Kiadja A Kir. Magyar Természettudományi Társulat, Budapest, 678 p.

Sîrbu C (2006). Floristic and chorological contributions from Moldavia and Muntenia (Romania). Analele Şti Univ "Alexandru Ioan Cuza" Iaşi, Biol veg 52:92-98.

Sîrbu C, Oprea A (2011). Alien plants of the flora of Romania (in Romanian). Edit. "I. I. de la Brad", Iaşi, 733 p.

Soran V (1962). Research on weeds and their association in the Apuseni Mountains (in Romanian). Probl Biol 299-346.

Taiyan Z, Lianli L, Guang Y, Al-Shehbaz IA (2001). *Brassicaceae* (*Cruciferae*), p. 1-193. In: Flora of China, Editorial Committee (Eds.). Flora of China, Vol. 8. Science Press, Beijing and St. Louis, Botanical Garden Press, Missouri.

Terzioğlu S, Anşin R (2001). A chorological study on the taxa naturalized in the Eastern Black Sea Region. Turk J Agric For 25:305-309.

Thellung A (1906). Die Gattung *Lepidium* (L.) R. Br. Eine monographische Studie. Neue Denkschr Schweiz Naturf Ges 41:1-340.

Vinogradova JK, Majorov SR, Chorun LV (2010). The black data book of the flora of Middle Russia (in Russian). GEOS, Moskva, 512 p.

Viţălariu G, Zanoschi V, Turenschi E (1977). Phyto-chorological contributions for Moldavia and Dobrudja (in Romanian). Analele Şti Univ "Alexandru Ioan Cuza" Iaşi, Biol 23:25-27.

Von der Lippe M, Kowarik I (2007). Long-distance dispersal of plants by vehicles as a driver of plant invasions. Conservation Biol 21(4):986-996.

Williamson M, Pyšek P, Jarošyk V, Prach K (2005). On the rates and patterns of spread of alien plants in the Czech Republic, Britain, and Ireland. Ecoscience 12(3):345-354.

Zar JH (2010). Biostatistical analysis. 5th ed. Prentice Hall Inc, New Jersey, 944 pp.

22

Effects of Arbuscular Mycorrhiza Fungi on Growth Characteristics of *Dactylis glomerata* L. under Drought Stress Conditions

Apostolos P. KYRIAZOPOULOS[1]*, Michail ORFANOUDAKIS[1], Eleni M. ABRAHAM[2], Zoi M. PARISSI[2], Nikoleta SERAFIDOU[1]

[1]*Democritus University of Thrace, Department of Forestry and Management of the Environment and Natural Resources, 193 Pantazidou str., 68200 Orestiada, Greece; apkyriaz@fmenr.duth.gr (*corresponding author)*

[2]*Aristotle University of Thessaloniki, Department of Forestry and Natural Environment, Laboratory of Range Science (236), 54124 Thessaloniki, Greece*

Abstract

Limited information is available regarding the selection of effective mycorrhizae and the exploitation of their beneficial effects on the enhancement of the forage production of *Dactylis glomerata* under the predicted warmer and drier conditions in the Mediterranean region. The objective of this study was to test the effects of *Glomus intraradices, Glomus mosseae* and their mix inoculation on growth characteristics and dry matter production of *Dactylis glomerata* in relation to full and limited irrigation. The experiment was conducted in Orestiada, Northeastern Greece. Limited irrigation significantly decreased yield and yield components of *Dactylis glomerata* over the full irrigation. Drought stressed plants had significantly higher root dry weight as a response for better survival under water deficit conditions. The Arbuscular mycorrhizal fungi (AMF) inoculated plants had significant higher shoot dry weight, tiller height and number of leaves in comparison to the non-inoculated plants. On the contrary, under drought stress conditions all AMF plants had lower root dry weights than control plants. Among the studied mycorrhizae species, *Glomus intraradices* performed better than *Glomus mosseae* and their mixture as it increased S/R ratio, tiller height and number of leaves.

Keywords: forage, *Glomus intraradices, Glomus mosseae,* Mediterranean region, orchardgrass

Introduction

Water availability is one of the most ominous abiotic factors limiting the productivity of forage crops in the Mediterranean region. It is predicted that due to climatic change, precipitation will significantly reduce in the Mediterranean region (Bates *et al.*, 2008). The predicted warmer and drier climate conditions will considerably affect the grassland ecosystems' goods and services (Cheddadi *et al.*, 2001). Consequently, significant reductions in productivity are expected and, it is therefore important to adopt appropriate management techniques in order to maintain forage production under the more severe drought conditions.

In the warm and dry Mediterranean environments, arbuscular mycorrhizal (AM) symbiosis could play an important role in alleviating the effects of drought on crop yield. Arbuscular mycorrhizal fungi (AMF) symbiosis could significantly improve plant resistance to drought via increased dehydration-avoidance and increased tolerance (Ruiz-Lozano and Azcón, 2000). However, its efficacy should be linked to the plant AMF functional compatibility. The root colonization by the mycorrhiza increases active absorptive surface area and stimulates water uptake even in water stress conditions. The AMF symbiosis could increase the drought tolerance via the increased soil water movement to the plant roots (Ruiz-Lozano, 2003), and the ability of the mycorrhizal plants to mediate the osmotic stress. These osmotic regulations are related to the sugar level adjustments in the plant shoots (Wu and Xia, 2006). AMF plants postpone declines in water potential during drought stress (Porcel and Ruiz-Lozano, 2004). The ability of the fungus to enhance the water uptake and transport it in the plant, could explain these effects. What is more, the effects of AMF to the drought - plant relation also extend to the soil environment by promoting soil aggregation and thus improving the overall soil water conditions to the rhizosphere (Rillig *et al.*, 2002).

Dactylis glomerata L. is a widespread perennial grass species, distributed throughout Eurasia including the Mediterranean basin (Lumaret, 1988). It is considered as one of the most important forage species due to its high productivity and nutritive value. Although its populations and varieties thrive in a multitude of environments and many of them have been reported as well adapted to drought stress conditions (García and Lindner, 1998; Stewart and Ellison, 2010), little information is available regarding the effects of AMF colonization on its growth characteristics.

The beneficial effect of mycorrhizal colonization on many field crops has been well documented. However, limited information is available concerning the selection of effective mycorrhizae and exploitation of their beneficial effects on enhancing the forage production of *Dactylis glomerata* in the Mediterranean region. The understanding of growth characteristics and dry matter production as affected by AMF colonization is important for the sustainable utilization of *D. glomerata* swards under the predicted future warmer and drier conditions. The general objective of this study was to test the effects of various AMF on growth characteristics and dry matter production of *Dactylis glomerata* in relation to different water availability.

Materials and methods

The experiment was conducted at the farm of the Democritus University of Thrace, in Orestiada, Northeastern Greece (41° 33' N, 26° 31' E), at an altitude of 33 m above sea level. The climate of the area is characterized as sub-Mediterranean. The mean annual precipitation is 515 mm, and the mean annual temperature is 14.2 °C.

Seeds of the perennial grass species *Dactylis glomerata* L. collected from grasslands in the surrounding area during the summer of 2011, were sown into 40 plastic pots (10 lt volume) full of surface soil (0-20 cm) from the Democritus University farm mixed with sand, in March 2012. The soil is silty clay with pH 7.5 (1:2 water), organic matter 20.5 g kg^{-1}, electrical conductivity 0.04 S m^{-1}, total N 1.11 g kg^{-1}, N–NO_3 9 mg kg^{-1}, P (Olsen) 13.2 mg kg^{-1}, and available K 178 mg kg^{-1} (Koutroubas *et al.*, 2012). The total seeding density was seven seeds per pot. Two weeks after the emergence of the seeds, some seedlings were removed from every pot in order to acquire a planting density of two seedlings per pot. The pots were then placed in an open area and subjected to a full irrigation treatment for a month prior to the onset of the experiment.

Two different irrigation treatments were applied once the experiment commenced: (1) up to full capacity and (2) 33% of full capacity. AMF application treatments were defined as: (1) control (no addition of AMF); (2) *Glomus intraradices* Schenck and Smith inoculation; (3) *Glomus mosseae* (Nicol and Gerd) Gerdemann and Trappe (Gm) inoculation; (4) mix of *Glomus intraradices* and *Glomus mosseae* inoculation. In the AMF inoculated pots, 8 ml of 10% w/v water-inoculum solution (25 spores g^{-1}) was applied 3 cm below the depth of seeding in each pot before sowing. The pots were arranged in a completely randomized block (block in irrigation levels) with five replications. All pots were randomized within each irrigation treatment biweekly. Hand weeding was done when required.

The number of tillers per plant, the length of tillers from the base to the tip, the number of leaves per plant and the length of the leaf lamina between the ligule of the youngest fully expanded to the tip of the emerging leaf were measured twice a week, during the growing season (May-June). Subsequently, Leaf Elongation Rate was calculated: Leaf Elongation Rate (LER)=(L2-L1)/t , where L1 is the leaf length at the beginning and L2 is the leaf length at the end of a time (t), (Lemaire and Agnusdei, 2000). The plants from all the pots were removed at late June 2012 and oven-dried at 60 °C for 48 h. Shoot and root samples were weighed in order to determine shoot dry weight (gr $plant^{-1}$) and root dry weight (gr $plant^{-1}$) and to calculate the S/R ratio.

Root samples were cleared and stained and mycorrhizal colonization was evaluated according to Trouvelot *et al.* (1986). It expressed as percentage of mycorrhization (M%) and percentage of arbuscules in the root system (A%).

Data for the shoot dry weight, root dry weight, the S/R ratio, the leaf growth rate, mycorrhizal colonization and the percentage of arbuscules in the root system were analyzed as completely randomized block design with two irrigation treatments and four AMF colonization treatments (Quinn and Keough, 2002) with five replications. Data for the number of tillers, the length of tillers and the number of leaves per plant were analyzed as completely randomized block design in time with two irrigation treatments and four AMF colonization treatments (Quinn and Keough, 2002). General linear models procedure (SPSS 18 for Windows) was used for ANOVA. The LSD at the 0.05 probability level was used to detect the differences among means (Steel and Torrie, 1980).

Results and discussion

Mycorrhizal colonization did not vary with water stress

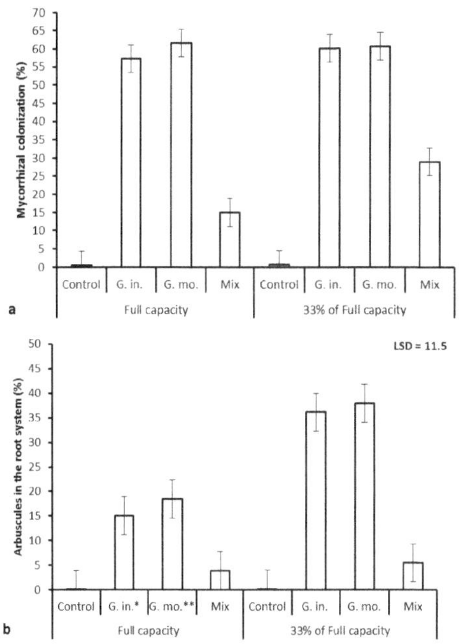

Fig. 1. (a) Percentage of mycorrhizal colonization (Mean ± S.E.) and (b) percentage of arbuscules (Mean ± S.E.) in *Dactylis glomerata* plants grown under well-watered conditions (up to full capacity) and drought stressed (33% of full capacity)

* *Glomus intraradices*

** *Glomus mosseae*

treatment (Fig. 1a). Similar results are common in studies involved potted plants (Auge, 2001). AMF development could be influenced not only by soil water content but also by differences in host and soil factors (Bethlenfalvay *et al.*, 1988). Generally, the colonization of both the studied mycorrhizae species was satisfactory and superior to that of their mixture. Both water stress and mycorrhizae application affected significantly the percentage of arbuscules in the root system (Fig. 1b).

Under limited irrigation the percentage of arbuscules in the root system increased, while reduced arbuscules (%) were found in plants inoculated with the mix of mycorrhizae species.

Water stress and mycorrhizae application affected significantly shoot dry weight, root dry weight, S/R ratio and the number of leaves per plant of *Dactylis glomerata* (Tab. 1). Additionally, significant differences for leaf growth rate and number of tillers per plant were recorded only between the irrigation treatments, while for tiller height only among the tested AMF colonization treatments. The dates of measurements affected significantly the number of leafs and the number of tillers, but did not affect the tillers' height. Significant interaction was observed between the irrigation and the mycorrhizae treatments only for root dry weight, S/R ratio, the number of tillers and tillers' height (Tab. 1), indicating that the water stress effect was not consistent in the mycorrhizae. Moreover, the interaction of irrigation treatment and dates of measurements was significant for the number of leaves and the number of tillers.

The shoot dry weight (across mycorrhizae treatments) and the S/R ratio of limited watered plants reduced

Tab. 1. Statistical significance of F ratios from the analysis of variance for dry weighs of shoot and root, on S/R ratio and on growth parameters

Source of variation	df	A. b[1]	U.b[2]	A/U[3]	LGR[4]	No Leafs	No Tillers	Tiller Height
Water (W)	1	p≤0.05	p≤0.05	p≤0.05	p≤0.05	p≤0.05	p≤0.05	NS
Mycorrhiza (M)	3	p≤0.05	p≤0.05	p≤0.05	NS	p≤0.05	NS	p≤0.05
Date (D)	5					p≤0.05	p≤ 0.05	NS
WXM	3	NS	p≤0.05	p≤0.05	NS	NS	p≤0.05	p≤0.05
WXD	5					p≤0.05	p≤0.05	NS
MXD	15					NS	NS	NS
WXMXD	15					NS	NS	NS

F *Test* at p≤0.05
[1] Aboveground biomass, [2] Underground biomass, [3] Aboveground/Underground, [4] Leaf Growth Rate

Tab. 2. Effects of watering (across mycorrhiza application treatments) on dry weighs of shoot and root, on S/R ratio and on growth parameters (Mean ± S.E.) of *Dactylis glomerata*

Treatment	Watered	Drought stressed	$LSD_{0.05}$
Shoot dry weight (gr)	1.22±0.03 [a]	0.91±0.03 [b]	0.09
Root dry weight (gr)	2.68±0.24 [b]	4.56±0.25 [a]	0.71
S/R	0.65±0.03 [a]	0.24±0.03 [b]	0.09
Number of tillers	6.96±0.16 [a]	6.31±0.16 [b]	0.44
Tiller height (cm)	17.41±0.16 [a]	17.44±0.16 [a]	NS
Number of leaves	6.44±0.15 [a]	5.95±0.15 [b]	0,42
Leaf growth rate (cm/d)	0.18±0.01 [a]	0.14±0.01 [b]	0,03

Means within each row followed by the same letter are not significantly different (LSD Test, p≤0.05).

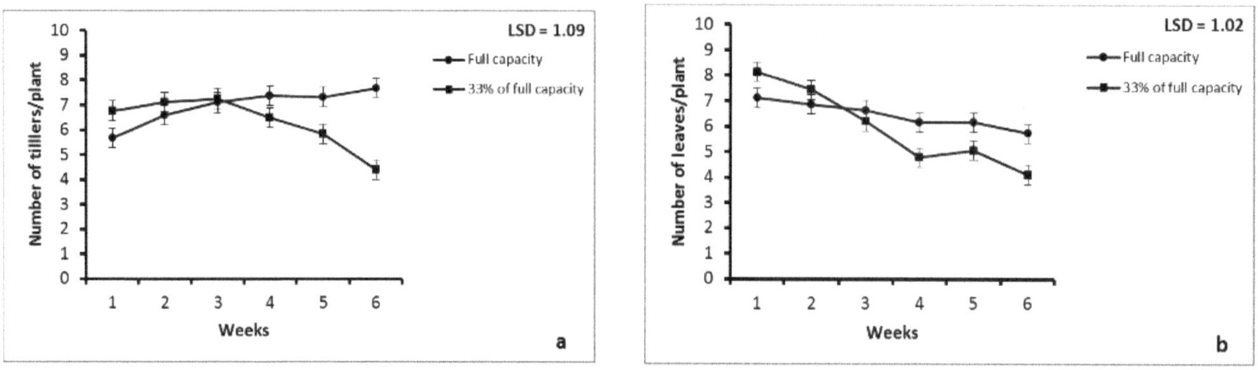

Fig. 2. Number of tillers (a) and number of leaves (b) of *Dactylis glomerata* for the two irrigation treatments (across AMF application treatments) during the experimental period

approximately 25% and 63% respectively, while the root dry weight increased almost 70% (Tab. 2). Similarly, the number of leaves, number of tillers and leaf growth rate were significantly lower for limited watered plants compared to well watered ones, while the tiller height did not significantly differ between them (Tab. 2). The number of tillers and number of leaves under limited irrigation reduced significantly after the fourth and third week respectively (Fig. 2).

The results of the present study suggest that limited irrigation decreases significantly yield and yield components in *Dactylis glomerata*. It has to be noted that the decreased leaf growth rate recorded at plants grown under drought stress conditions has been described as a protection mechanism of grasses grown under drought conditions (Volaire *et al.*, 2009; Wang and Bughrara, 2008). Drought stressed plants had significantly higher root dry weight. This result is associated with the fact that many grass species growing in drought conditions tend to increase their roots in order to increase their active absorptive surface area for water uptake. This response is a water restriction tolerance mechanism for the better survival of grasses under water deficit conditions (Lopez *et al.*, 2013). These results confirm the fact that soil moisture constitutes a critical factor for plant growth and herbage production in the Mediterranean region (Etienne, 1996).

The AMF inoculated plants (across the irrigation treatments) had significant higher shoot dry weight compared to the non-inoculated plants, while there were no significant differences among the mycorrhizae application treatments (Tab. 3). Similarly, the AMF inoculated plants had higher tiller height and number of leaves in comparison to the non-inoculated plants, while the number of tillers and leaf growth rate did not significant differ between them (Tab. 3). Among the mycorrhizae treatments, the plants with *Glomus intraradices* had the highest tiller height and number of leaves although their tiller height did not differ significantly from these of *Glomus mosseae* and their number of leaves from these of their mix. On the contrary, the root dry weight of both *Glomus intraradices* and *Glomus mosseae* inoculated plants was the lowest although this of *Glomus mosseae* inoculated plants did not differ significantly from root dry weight of the mix and the non-inoculated plants. Additionally, the *Glomus intraradices* inoculated plants had the highest S/R ratio, while the mix and non-inoculated plants the lowest one. The opposite trend of the irrigation treatment main effect was observed for plants of the mix mycorrhizae species which had significantly higher root dry weight, lower number of tillers and tillers' height under well watered compared to limited watered conditions (Tab. 4).

Arbuscular mycorrhizal fungi (AMF) inoculation seems to improve *Dactylis glomerata* yield, due to the increase of shoot dry weight, S/R ratio, tiller height and number of

Tab. 3. Effects of mycorrhiza colonization (across watering treatments) on dry weighs of shoot and root, on S/R ratio and on growth parameters (Mean ± S.E.) of *Dactylis glomerata*

Treatment	Control	*Glomus intraradices*	*Glomus mosseae*	Mixed AM	$LSD_{0.05}$
Shoot dry weight (gr)	0.94±0.05 ^{b}	1.14±0.05 ^{a}	1.11±0.05 ^{a}	1.09±0.05 ^{a}	0.14
Root dry weight (gr)	4.33±0.35 ^{a}	2.64±0.35 ^{b}	3.39±0.35 ^{ab}	4.14±0.37 ^{a}	1.00
S/R	0.38±0.05 ^{bc}	0.64±0.05 ^{a}	0.47±0.05 ^{b}	0.30±0.05 ^{c}	0.13
Number of tillers	6.36±0.23 ^{a}	6.92±0.24 ^{a}	6.61±0.23 ^{a}	6.65±0.24 ^{a}	NS
Tiller height (cm)	16.59±0.22 ^{c}	18.26±0.23 ^{a}	17.68±0.22 ^{ab}	17.17±0.23 ^{bc}	0.62
Number of leaves	5.49±0.21 ^{c}	6.81±0.23 ^{a}	6.06±0.21 ^{bc}	6.44±0.23 ^{ab}	0.59
Leaf growth rate (cm/d)	0.14±0.01 ^{a}	0.18±0.01 ^{a}	0.16±0.01 ^{a}	0.16±0.01 ^{a}	NS

Means within each row followed by the same letter are not significantly different (LSD Test, $p \leq 0.05$).

Tab. 4. Effects of watering and mycorrhiza colonization on dry weighs of shoot and root, on S/R ratio and on growth parameters (Mean ± S.E.) of *Dactylis glomerata*

Treatments	Shoot dry weight (gr)	Root dry weight (gr)	S/R	Number of tillers	Tiller height (cm)	Number of leaves	Leaf growth rate (cm/d)
Watered							
Control	1.12±0.07	1.93±0.49 ^{de}	0.64±0.07 ^{b}	7.71±0.32 ^{a}	17.39±0.32 ^{bc}	5.85±0.30	0.16±0.02
Glomus intraradices	1.23±0.07	1.26±0.49 ^{e}	1.01±0.07 ^{a}	7.78±0.35 ^{a}	18.37±0.35 ^{a}	6.94±0.34	0.20±0.02
Glomus mosseae	1.35±0.07	1.94±0.49 ^{de}	0.73±0.07 ^{b}	6.96±0.32 ^{ab}	17.34±0.32 ^{bc}	6.68±0.30	0.18±0.02
Mixed AM	1.19±0.07	5.61±0.49 ^{ab}	0.22±0.07 ^{cd}	5.99±0.32 ^{c}	16.63±0.32 ^{cd}	6.30±0.30	0.17±0.02
Drought stressed							
Control	0.76±0.07	6.74±0.49 ^{a}	0.11±0.07 ^{d}	5.62±0.32 ^{c}	15.78±0.32 ^{d}	5.12±0.30	0.12±0.02
Glomus intraradices	1.05±0.07	4.02±0.49 ^{cd}	0.27±0.07 ^{cd}	6.05±0.32 ^{c}	18.16±0.32 ^{a}	6.67±0.30	0.16±0.02
Glomus mosseae	0.86±0.07	4.82±0.49 ^{bc}	0.22±0.07 ^{cd}	6.26±0.32 ^{bc}	18.03±0.32 ^{ab}	5.44±0.30	0.15±0.02
Mixed AM	0.98±0.07	2.66±0.55 ^{d}	0.38±0.07 ^{c}	7.31±0.35 ^{a}	17.80±0.35 ^{ab}	6.57±0.34	0.15±0.02
$LSD_{0.05}$	NS	1.41	0.19	0.89	0.88	NS	NS

Means within each column followed by the same letter are not significantly different (LSD Test, $p \leq 0.05$).

leaves under up to full and limited irrigation. This could be associated with the higher accumulation of phosphorus, nitrogen and potassium from the soil (Erman *et al.*, 2011; Jacobsen, 1994; Pearson and Jacobsen, 1993).

In contrast, under drought stress conditions all AMF plants had lower root dry weights than control plants. This result is in agreement with Di and Allen (1991) for *Agropyron* cultivars and Cavagnaro *et al.* (2008) for *Solanum lycopersicum* who also found that AMF colonization of plants decreased root biomass. It seems that the AMF inoculated plants cope with water deficit by other mechanisms and not by increasing the root biomass at the expense of shoot biomass. For instance, the AMF inoculated plants of *Agropyron* cultivars performed better under water stress due to their higher water use efficiency compared to non-inoculated ones (Di and Allen, 1991). According to Ruiz-Lozano *et al.* (1995), there is a differentiated effect on plant adaptation to drought among the AMF species. The seven fungal species of genus *Glomus* they studied had the particular ability to alter the physiological parameters that enhance the adaptation of *Lactuca sativa* L. cv. Romana to low soil water content. Additionally, the low root dry weight can be associated with the AMF symbiosis, which improved significantly the resistance of the inoculated plants to drought by reducing the need of an increase to the roots.

Among the studied mycorrhizae species, *Glomus intraradices* performed better than *Glomus mosseae* and their mixture as it increased S/R ratio, tiller height and number of leaves. In contrast to our results, Moradi and Salimi (2013) reported that *Glomus mosseae* was more effective for the growth parameters of *Poa* sp. This confirms that AMF inoculation has positive effects on various plants, but there are variations in response of the varieties, cultivars or species to various AMF species (Long *et al.*, 2010; Naher *et al.*, 2013; Tufenkci *et al.*, 2012; Wang *et al.*, 2008,). The mix of these mycorrhiza species although it increased the growth parameters of *Dactylis glomerata* in comparison to the control, it was inferior to the other AMF application treatments. Moreover, the mix reduces drastically the root system of the plants under limited irrigation, which could detrimentally affect the plants growth under prolonged drought and the sward persistence. This is probably has to do with the interaction between the two mycorrhiza species that reduces their positive effect to plants. The variation upon the growth response is not unusual, with mycorrhizal growth response varying from positive to neutral or negative. These responses are influenced by the symbiotic partners, the genotypes, the developmental stages and the environmental conditions (Facelli *et al.*, 2010; Klironomos, 2003; Smith and Smith, 2011). In particular, the soil conditions could determine how well the non mycorrhizal plants grow in relation to mycorrhizal plants. Negative response also correlates with the higher symbiotic cost in relation to the photosynthetic supply (Jacobsen, 1999). However, the negative growth response as reported here, the reduced number of tillers, may be related to a hormonal response to AMF. Strigolactones, a newly described plant hormone are connected with the number of the tillers in plants (Foo and Reid, 2013). AMF and strigolactones interactions could result to a reduced number of tillers. When plant roots were inoculated by two AMF species, increased the levels of the stigolactones in the shoot leading to tiller reduction.

Conclusions

In conclusion, this study provided evidence that AMF have the potential to be beneficial on growth characteristics and dry matter production of *Dactylis glomerata* grown under different water availability conditions. The beneficial activity of AMF is most likely to be favored under drought stress. Therefore, further studies on *Dactylis glomerata*–AMF symbiosis in the field are needed in order this symbiosis to be tested under plant competition and grazing conditions. The increased number of leaves found in the inoculated plants is an indicator of ameliorated nutritive value (Parissi *et al.*, 2007) but the effects of AMF inoculation on the forage quality of *Dactylis glomerata* also need to be tested.

References

Augé RM (2001). Water relations, drought and vesicular-arbuscular mycorrhizal symbiosis. Mycorrhiza 11:3-42.

Bethlenfalvay GJ, Brown MS, Ames RN, Thomas RS (1988). Effects of drought on host and endophyte development in mycorrhizal soybeans in relation to water use and phosphate uptake. Physiol Plant 72:565-571.

Bates BC, Kundzewicz ZW, Wu S, Palutikof JP (2008). Climate Change and Water. Technical Paper, IPCC Secretariat, Geneva, p. 210.

Cavagnaro TR, Langley AJ, Jackson LE, Smukler SM, Koch GW (2008). Growth, nutrition, and soil respiration of a mycorrhiza-defective tomato mutant and its mycorrhizal wild type progenitor. Funct Plant Biol 35:228-235.

Cheddadi R, Guiot J, Jolly D (2001). The Mediterranean vegetation: What if the atmospheric CO_2 increased? Landscape Ecol 16:667-675.

Di JJ, Allen EB (1991). Physiological responses of six wheatgrass cultivars to mycorrhizae. J Range Manage 44:336-341.

Etienne M (ed.) (1996). Research on temperate and tropical silvopastoral systems: a review, pp: 5-19. In: Western European Silvopastoral Systems. INRA Editions, Paris, France.

Erman M, Demirb S, Ocakc E, Tüfenkci S, Oguz F, Akköprü A (2011). Effects of Rhizobium, arbuscular mycorrhiza and whey applications on some properties in chickpea (*Cicer arietinum* L.) under irrigated and rainfed conditions 1 - Yield, yield components, nodulation and AMF colonization. Field Crop Res 122:14-24.

Facelli E, Smith SE, Facelli JM, Christophersen HM, Smith FA (2010). Underground friends or enemies: model plants help to unravel direct and indirect effects of arbuscular mycorrhizal fungi on plant competition. New Phytol 185:1050-1061.

Foo E, Reid JB (2013). Strigolactones: New Physiological Roles for an Ancient Signal. Plant Growth Regul 32:429-442.

Garcia A. Lindner R (1998). *Dactylis glomerata* genetic resources:

Allozyme frequencies and performance of two subspecies on an acid sandy loam with summer drought. Euphytica 120: 255-264.

Jacobsen I (1994). Research approaches to study the functioning of vesicular arbuscular mycorrhizas in the field. Plant Soil 159:141-142.

Jakobsen I (1999). Transport of phosphorus and carbon in arbuscular mycorrhizas, p. 309-332. In: Varma A, Hock B (Eds.). Mycorrhiza: Structure, Function, Molecular Biology and Biotechnology, Springer, Berlin, Germany.

Klironomos JN (2003). Variation in plant response to native and exotic arbuscular mycorrhizal fungi. Ecology 84:2292-2301.

Koutroubas SD, Fotiadis S, Damalas CA (2012). Biomass and nitrogen accumulation and translocation in spelt (*Triticum spelta*) grown in a Mediterranean area. Field Crop Res 127:1-8.

Lemaire G, Agnusdei M (2000). Leaf tissue turnover and efficiency of herbage utilization, pp. 265-287. In: Lemaire G, Hodgson J, de Moraes A, Nabinger C, de F Carvalho PC (Eds.). Grassland ecophysiology and grazing ecology. CAB International, Wallingford, UK.

Long LK, Yao Q, Huang YH, Yang RH, Guo J, Zhu HH (2010). Effects of arbuscular mycorrhizal fungi on zinnia and the different colonization between *Gigaspora* and *Glomus*. World J Microbiol Biotechnol 26:1527-1531.

López IF, Kemp PD, Dörner J, Descalzi CA, Balocchi OA, García S (2013). Competitive strategies and growth of neighbouring *Bromus valdivianus* Phil. and *Lolium perenne* L. plants under water restriction. J Agron Crop Sci 199:449-459.

Lumaret R (1988). Cytology, genetics and evolution in the genus *Dactylis*. Crit Rev Plant Sci 7:55-91.

Moradi S, Salimi S (2013). Effects of arbuscular mycorrhizal fungi on root morphological properties of *Poa* in drought stress conditions. Int J Agri Crop Sci 5:591-595.

Naher UA, Othman R, Panhwar QA (2013). Beneficial effects of mycorrhizal association for crop production in the tropics - a review. Int J Agr Biol 15:1021-1028.

Parissi ZM, Abraham EM, Karameri E, Kexagia E, Karatassiou M (2007). Comparable study of the nutritive value in Greek populations of *Dactylis glomerata* L. growing in contrasting environments in association to maturity. Grassland Sci Eur 12:371-374.

Pearson JN, Jakobsen I (1993). Symbiotic exchange of carbon and phosphorus between cucumber and three arbuscular mycorrhizal fungi. New Phytol 124:481-488.

Porcel R, Barea JM, Ruiz-Lozano JM (2003). Antioxidant activities in mycorrhizal soybean plants under drought stress and their possible relationship to the process of nodule senescence. New Phytol 157:135-143.

Quinn GP, Keough MJ (2002). Experimental design and data analysis for biologists. Cambridge University Press, Cambridge, UK.

Rillig MC, Treseder KK, Allen MF (2002). Mycorrhizal fungi and global change, pp. 135–160. In: van der Heijden MGA, Sanders IR (Eds.). Mycorrhizal Ecology, Ecological Studies Series 157. Springer, Berlin, Germany.

Ruiz-Lozano JM (2003). Arbuscular mycorrhizal symbiosis and alliviation of osmotic stress. New perspectives for molecular studies. Mycorrhiza, 13: 309-317.

Ruiz-Lozano JM, Azcon R (2000). Symbiotic efficiency and infectivity of an autochthonous arbuscular mycorrhizal *Glomus* sp from saline soils and *Glomus deserticola* under salinity. Mycorrhiza 10:137-143.

Ruiz-Lozano JM, Azcón R, Gómez M (1995). Effects of arbuscular-mycorrhizal *Glomus* species on drought tolerance: physiological and nutritional plant responses. Appl Environ Microbiol 61:456-460.

Smith SE, Smith FA (2011). Roles of arbuscular mycorrhizas in plant nutrition and growth: new paradigms from cellular to ecosystem scales. Annu Rev Plant Biol 62:227-250.

Steel RGD, Torrie JH (1980). Principles and procedures of statistics. 2nd edn. McGraw-Hill, New York, USA.

Stewart AV, Ellison NW (2010). *Dactylis*, pp 73-87. In: Kole C (Ed.). Wild crop relatives: genomic and breeding resources, millets and grasses. Springer-Verlag, Berlin, Germany.

Trouvelot A, Kough JL, Gianinazzi-Pearson V (1986). Mesure du taux de mycorhization VA d'un système radiculaire. Recherche de méthodes d'estimation ayant une signification fonctionnelle, pp 217-221. In: Gianinazzi-Pearson V, Gianinazzi S (Eds.). Physiological and genetical aspects of Mycorrhizae. INRA Press, Paris, France.

Tufenkci S, Demir S, Şensoy S, Ünsal S, Demirer E, Erdinc C, Bicer S, Ekincial A, Kincialp A (2012). The effects of arbuscular mycorrhizal fungi on the seedling growth of four hybrid cucumber (*Cucumis sativus* L.) cultivars. Turk J Agric For 36:317-327.

Volaire F, Norton M, Lelievre F (2009). Summer drought survival strategies and sustainability of perennial temperate forage grasses in Mediterranean areas. Crop Sci 49:2386-2392.

Wang JP, Bughrara SS (2008). Morpho-physiological responses of several fescue grasses to drought stress. HortScience 43:776-783.

Wang C, Li X, Zhou J, Wang G, Dong Y (2008). Effects of arbuscular mycorrhizal fungi on growth and yield of cucumber plants. Commun Soil Sci Plant Anal 39:499-509.

Wu QS, Xia RX (2006). Arbuscular mycorrhizal fungi influence growth, osmotic adjustment and photosynthesis of citrus under well-watered and water stress conditions. J Plant Physiol 163:417-425.

23

The Effect of Zinc on Yield, Yield Components and Micronutrient Concentrations in the Seeds of Safflower Genotypes (*Carthamus tinctorius* L.)

Zehra AYTAC[1*], Nurdilek GULMEZOGLU[2], Zeynep SIREL[1], Inci TOLAY[3], Ayfer ALKAN TORUN[4]

[1]*Eskisehir Osmangazi University, Faculty of Agriculture, Department of Field Crops, Eskisehir, Turkey; zehrak@ogu.edu.tr (*corresponding author)*
[2]*Eskisehir Osmangazi University, Faculty of Agriculture, Eskisehir, Turkey*
[3]*Akdeniz University, Faculty of Agriculture, Department of Soil Science and Plant Nutrition, Antalya, 07058, Turkey*
[4]*Cukurova University, Faculty of Agriculture, Department of Soil Science and Plant Nutrition, Adana, 01030, Turkey*

Abstract

The objective of this study was to determine the effect of zinc (Zn) on yield, yield components and Zn, manganese (Mn), copper (Cu) and iron (Fe) concentrations in seeds of four safflower (*Carthamus tinctorius* L.) genotypes. The research was performed, in a randomized complete block design with three replications. Analysis of variance showed highly significant differences for all the examined traits except harvest index in both years of study. Zinc application significantly affected the plant height, seed weight per plant and 1000-seed weight in the first year and the plant height, seed weight per plant, biological yield per plant, 1000-seed weight and harvest index of the genotypes during the second year. The results of the study showed that Zn uptake differed substantially among the safflower genotypes tested. The Zn, Cu and Mn concentrations in the seeds were increased by Zn application.

Keywords: biological yield, micronutrients, safflower, seed quality, seed yield, zinc deficiency

Introduction

Zinc (Zn) deficiency is one of the most widespread nutritional problems, affecting nearly one-third of the world population (WHO, 2002). In addition, it is known that Zn deficiency not only produces yield losses in plants (Cakmak *et al.*, 1999) but also causes a low accumulation of Zn in the edible parts and seeds of plants (Welch and Graham, 2004). Low Zn intake in the diet results in Zn deficiency in human populations. Serious nutritional problems have been reported, particularly in developed and developing countries, resulting from the consumption of foods low in Zn (Hotz and Brown, 2004; Gibson *et al.*, 2007; Harris *et al.*, 2008; Khoshgoftarmanesh *et al.*, 2010; Nriagu, 2010).

Breeding Zn-resistant plant species and varieties and applying Zn-containing fertilizers to correct Zn deficiencies are two approaches to the improvement of Zn concentration in seeds. To obtain genotypes resistant to Zn deficiency through plant breeding requires a long period of time (Graham and Rengel, 1993; Cakmak *et al.*, 1996; Bouis, 2003; Pfeiffer and McClafferty, 2007; Wissuwa *et al.*, 2008). However, increases in yield and in the level of Zn in seeds are possible and relatively easy with Zn fertilization (Phattarakul *et al.*, 2012; Zou, 2012).

The effects of Zn fertilization on the growth and yield of many plants (Yilmaz *et al.*, 1997; Grewal and Williams, 2000; Khoshgoftarmanesh *et al.*, 2004; Phattarakul *et al.*, 2012; Zou, 2012) have been studied, but few studies have investigated the response of safflower (*Carthamus tinctorius* L.) to Zn fertilization.

Safflower, a multipurposed crop, has been grown for centuries in various regions of the world in view of the orange-red dye obtained from its petals, its many medicinal properties, its feed value and especially its high-quality oil, which is rich in polyunsaturated fatty acids. Interest in this crop has increased in recent years, particularly due to its production in areas with limited rainfall, the preference of consumers for healthy oil, the medicinal uses of the flowers and the extraction of edible dyes from the flowers (Singh, 2007). Safflower is an important oil crop that is cultivated mainly for its seed, which is used to obtain edible oil and as birdseed, and may potentially be produced under low-input conditions (Gyulai, 1996; Peterson, 1996; Elfadl, 2009).

In this paper, the effects of Zn application on the soil, on yield and yield components and the accumulation of Zn, Mn, Cu and Fe in seed under low-Zn field conditions with different safflower genotypes were studied over two years.

Materials and methods

This research was conducted during two years, 2009 and 2010, at the Faculty of Agriculture of Eskisehir Osmangazi University, Eskisehir (39° 48′ N; 30° 31′ E; 789 m elevation). The field experiments included four released safflower varieties ('Yenice', 'Dincer' and 'Remzibey' in Turkey and 'UC-1' in the U.S.) and one high-yielding line

(V-50/63 from Iran). The botanical characteristics of the genotypes were as follows: 'Dincer', non-spiny capitulum and orange petals; 'Yenice', non-spiny capitulum and red petals; 'Remzibey', spiny capitulum; 'UC-1', spiny capitulum and yellow petals; and 'V-50/63', non-spiny capitulum and red petals.

Soil samples were air-dried, passed through a 2 mm sieve and analyzed for pH, $CaCO_3$, organic matter and texture using standard procedures (Rowell, 1996). Plant-available concentrations of Zn, Fe, Cu, and Mn in the soils were determined according to the methods of Lindsay and Norwell (1978) by extraction with DTPA (Diethylenetriamine Pentaacetic Acid). The soil was alkaline, silty loamy and calcareous with a low content of organic matter and had similar amounts of DTPA-extractable Zn, Fe, Mn and Cu (Tab. 1). The amount of DTPA-extractable Zn was less (0.4 mg kg^{-1}) than the amount (0.7 mg kg-1) needed for optimum plant growth (Sillanpaa, 1990).

The long-term (1991-2010) total rainfall, average temperature and relative humidity at the study location were 160.1 mm, 15.4 °C and 58.0%, respectively (Tab. 2). The rainfall, average temperature and relative humidity were higher in 2010 than in 2009. The long-term average temperature was approximately equal to the average temperature during the 2009 growing season.

Tab. 1. Physical and chemical properties of the soils in experimental site

Character	Units	Eskisehir
pH (H_2O)		7.6
Organic matter	%	1.04
$CaCO_3$	%	6.0
Texture		Silty loam
DTPA extractable Fe	mg kg^{-1}	3.4
DTPA extractable Zn	mg kg^{-1}	0.4
DTPA extractable Cu	mg kg^{-1}	0.3
DTPA extractable Mn	mg kg^{-1}	3.5

The experiment followed a randomized complete block design with three replications. Each plot (5.4 m^2) consisted of six rows. The plots were sown at the end of March in 2009 and in the first week of April in 2010, using a seeding rate of 40 kg ha^{-1} in rows with 30-cm spacing, on a well-prepared seed bed. At planting time, phosphorus (P) and nitrogen (N) were applied at a standard rate of 50 kg ha^{-1} P_2O_5 as diammonium phosphate (18-46-0) and 80 kg ha^{-1} N as ammonium nitrate (33-0-0). Plants were irrigated once during emergence and thinned at the rosette stage. Weeds were controlled by hand weeding. Trials were established under rainfed conditions without (-Zn control) and with soil Zn (+Zn: 23 kg Zn ha^{-1} as $ZnSO_4.7H_2O$) applications. Zinc sulfate ($ZnSO_4.7H_2O$) is a common source of Zn fertilizer because it is highly soluble in water and exists in various forms (Hawkins, 1973; Mordvedt and Gilkes, 1993; Yilmaz *et al.,* 1997; Brennan, 2002; Harris *et al.*, 2008).

Yield components were recorded from 10 randomly selected plants in both years. The biological yield per plant, plant height, number of capitula per plant, seed yield per plant, 1000-seed weight, harvest index and seed yield were evaluated for both years. Seed yield in t ha^{-1} was obtained by harvesting the four central rows of the plot by hand. Approximately three months after harvest, as suggested by the International Seed Testing Association (ISTA, 2008), the 1000-seed weight was calculated by counting eight replicates of 100 seeds from each plot and expressing the weight in grams.

At maturity, approximately 10 g of each seed sample were collected from the plots of each genotype. Sub-samples of seed from each plot were dried for 24 h at 105 °C. After the samples were grounded, they were passed through a nylon sieve (0.5 mm), and 0.2 g of each sample was placed in Pyrex reactors in a microwave digestion unit. HNO_3 and H_2O_2, in a proportion of 3:2, were placed in the reactors. The samples were mineralized at 200 °C for half an hour. The samples were then filtered and diluted to a volume of 100 ml with ultra-pure distilled water. The elements were measured in Inductively Coupled Plasma-Atomic Emission Spectrometer (ICP-AES, Jobin, Yvon-Paris). The data presented are the averages of at least three independent experiments, each consisting of three replications. The quality of the analytical process was also controlled through the use of NCS DC73350 certified reference material (poplar leaves). The analysis of these standard reference materials showed good accuracy, with the recovery rates of the metals ranging between 90.3% and 108.9%. The recoveries were compared with the American Organization of Analytical Chemists criteria (AOAC, 1993).

All data were subjected to an analysis of variance (ANOVA) for each character. Mean differences were tested with a least significant difference (LSD) test at a P<0.05 and P<0.01 significance levels. The MSTAT-C computer program (MSTAT, 1984) was used for the analysis.

Tab. 2. Monthly temperature, precipitation and relative humidity, Eskisehir (2009 and 2010)

Months	Temperature (°C)			Precipitation (mm)			Relative humidity (%)		
	2009	2010	Long term*	2009	2010	Long term*	2009	2010	Long term*
March	4.6	6.7	4.9	39.8	27.7	29.6	60.5	59.3	64.2
April	10.0	10.2	9.7	26.0	41.2	44.3	55.7	61.2	62.3
May	14.8	16.4	14.9	28.9	5.7	39,4	50.7	55.3	59.3
June	20.4	19.4	19.2	7.9	46.6	24,4	41.0	59.8	55.0
July	22.2	23.3	22.0	11.4	14.3	13,4	42.9	59.7	51.9
August	21.0	25.3	22.0	2.0	1.5	9,0	42.2	52.0	53.0
Mean	15.5	16.8	15.4				48.8	57.8	58.0
Total				116	137	160.1			

*Long-term average, 1991-2010

Results and discussion

Effect of zinc fertilizer on safflower yield components

The results of the ANOVA showed significant effects ($P<0.05$ or $P<0.01$) of Zn application on plant height, seed weight per plant and 1000-seed weight in both years. Additionally, the biological yield and harvest index were also significant in the second year (Tab. 3). Genotypic differences were significant ($P<0.05$ or $P<0.01$) for all the yield components except the harvest index in both years.

Zinc application produced an increase in the mean of all yield components (Tab. 5 and Tab. 6). Zinc application significantly affected plant height, seed weight per plant and 1000-seed weight in the first year and plant height, seed weight per plant, biological yield per plant, 1000-seed weight and the harvest index of safflower genotypes in the second year. Zinc application significantly affected the biological yield per plant and harvest index of safflower genotypes in the second year but not in the first year. The reason for this difference may be that the years differed climatically. The precipitation rates and temperatures were notably higher in the second year than in the first year. The precipitation rates and the temperature between March and June during vegetative growth were markedly higher in the second year than in the first year. Sufficient vegetative growth increased the Zn uptake of the plant (Tab. 2). Among the genotypes, the highest average biological yield per plant and the highest harvest index were obtained from 'Yenice' (29.17 g and 36.38%, respectively) in the first year. In the second year, the highest average biological yield per plant was obtained in 'Dincer' (20.40 g) and the highest harvest index in 'Remzibey' (31%).

The mean plant height with applied Zn was 11% higher in the first year than the corresponding value without Zn, and 4% higher in the second year. Among the genotypes, the highest average plant height was obtained from 'V-50/63' (85.94 cm) in the first year and from 'Yenice' (93.26 cm) in the second year. Zn deficiency is known to be characterized by short internodes or stunted plants and reduced plant growth (Mengel and Kirkby, 2001). Movahhedy-Dehnavy *et al.* (2009) observed significant differences in plant height in safflower treated with foliar-applied Zn. The plants treated with Zn were 2-4 cm taller than those without Zn.

Tab. 3. Analysis of variance for effects of genotype and Zn application on yield components of safflower in two years (mean square)

Source	df	Plant Height	Number of Capitula	Seed Weight per Plant	Biological Yield per Plant	Thousand Seed Weight	Harvest Index
2009							
Genotypes (G)	4	944.26**	14.82**	33.46**	174.36*	0.54**	149.80ns
Zn Appl. (A)	1	417.76**	0.73ns	4.09*	3.32ns	0.42**	63.45ns
G x A	4	117.72ns	1.39ns	2.80*	13.22ns	0.72**	35.60ns
2010							
Genotypes (G)	4	890.41**	1.70*	3.44*	30.18**	1.17**	35.68ns
Zn Appl. (A)	1	78.70*	0.59ns	21.96**	54.02**	0.29**	196.96*
G x A	4	49.10*	1.36ns	3.26*	12.85ns	3.00**	77.18ns

* Significant at 5% level; ** significant at 1% level; ns: not significant

Tab. 4. Analysis of variance for effects of genotype and Zn Application on seed yield, Zn, Cu, Mn and Fe concentrations of seeds of safflower in two years (mean square)

Source	df	Seed yield	Zn	Cu	Mn	Fe
2009						
Genotypes (G)	4	447699.15**	114.73**	61.28**	19.67**	613.15*
Zn Appl. (A)	1	4543.62ns	178.60**	214.40**	634.80**	0.43ns
G x A	4	136442.48**	12.15**	60.44**	11.19**	120.03ns
2010						
Genotypes (G)	4	370513.76**	23.36**	1.12**	9.52**	265.30*
Zn Appl. (A)	1	169831.72ns	125.99**	187.50**	144.32**	62.43ns
G x A	4	20658.40ns	5.04**	0.27**	4.90**	100.55ns

* Significant at 5% level; ** significant at 1% level; ns: not significant.

The 1000-seed weight and seed weight per plant have the greatest influence on the seed yield of safflower (Mahasi *et al.*, 2006; Arslan, 2007). The mean 1000-seed weight of varieties grown with applied Zn was 8% higher in the first year and 5% in the second year than that for genotypes grown without Zn. Among the genotypes, the highest 1000-seed weight was obtained in 'Dincer' (4.56 g) in the first year and in 'Remzibey' (4.41 g) in the second year. The highest seed weight per plant was obtained in 'Yenice' (10.01 g) and in 'Dincer' (6.31 g) in the first and second year, respectively. Previous studies have shown that the 1000-seed weight is the yield component most affected by Zn fertilization (Kinaci and Kinaci, 2005; Mirzapour and Khoshgoftar, 2006), whereas other yield components were not significantly affected by Zn (Kamali *et al.*, 2009).

The genotype-Zn interaction was significant for seed weight per plant and 1000-seed weight ($P<0.01$) in both years and for plant height in the second year. The highest value of the genotype-Zn interaction for seed weight per plant was found in 'Yenice' in 2009 and in 'Dincer' in 2010. 'Dincer' and 'Remzibey' showed the highest value of the genotype-Zn interaction for 1000-seed weight in 2009 and in 2010, respectively. The highest value of the genotype-Zn interaction for plant height was found in 'Yenice' and in 'V50/63' in 2010. These results indicated that the effects of Zn applications depended on the safflower genotypes. Genotypic differences related to Zn deficiency have been discussed by several researchers (Rengel, 2001; Baligar *et al.*, 2001), and the reason that certain species are able to use soil Zn more efficiently than others is not completely clear (Mengel and Kirkby, 2001).

Seed yield and Zn, Fe, Mn and Cu concentrations in safflower

The seed yield of safflower was not affected significantly by Zn application in none of the years (Tab. 4), although Zn application increased the seed yield (Tab. 7). Seed yield was higher in 2010 than in 2009. The effect of the Zn-genotype interaction on seed yield was significant.

Both with and without Zn application, mean seed yield was higher in the second year than in the first year. Without Zn application in the first year, the mean seed yield was highest in 'Yenice' and lowest in 'UC-1'. Without Zn application in the second year, the mean seed yield was highest in 'Dincer' and lowest in 'V50/63'. Zn application produced a much greater increase in seed yield in the first year (mean 22%) than in the second year (mean 13%). The effect of Zn application on seed yield varied among the genotypes tested in both years. In the first year, Zn application produced the highest increase in seed yield in 'V50/63' (38%), whereas the smallest increase was obtained in 'Remzibey' (9%). Positive effects of Zn application on the seed yield of safflower genotypes have been demonstrated by Movahhedy-Dehnavy *et al.* (2009).

An ANOVA showed significant effects ($P<0.01$) of Zn application in both years on the Zn, Mn and Cu concentrations of safflower seeds. The genotype-Zn interaction was significant for the Zn, Mn and Cu concentrations ($P<0.01$) in both years. Only the genotype

Tab. 5. Mean yield components of safflower genotypes under –Zn and +Zn conditions in two years

Genotypes	Plant height (cm)			Number of capitula			Seed weight per plant (g)		
2009									
	-Zn	*+Zn*	*Mean*	*-Zn*	*+Zn*	*Mean*	*-Zn*	*+Zn*	*Mean*
V50/63	77.82	94.06	85.94	5.07	4.73	4.90	5.55 d	5.78 cd	5.66
Yenice	72.69	88.94	63.83	8.76	9.43	9.09	8.73 b	11.29 a	10.01
Dincer	66.94	64.97	80.81	6.27	5.23	5.75	6.76 c	5.58 d	6.17
Remzibey	64.30	63.36	65.95	5.80	7.03	6.41	3.76 e	4.63 de	4.19
UC-1	51.82	59.55	55.68	5.93	6.97	6.45	3.71 e	4.92 ed	4.31
Mean (Zn)	66.71B	74.18A		6.37	6.68		7.01B	7.55A	
2010									
	-Zn	*+Zn*	*Mean*	*-Zn*	*+Zn*	*Mean*	*-Zn*	*+Zn*	*Mean*
V50/63	82.20 b	92.86 a	87.53	3.98	5.53	4.75	3.53 c	5.85 b	4.69
Yenice	91.20 a	95.32 a	93.26	6.03	5.27	5.65	4.78 b	4.72 b	4.75
Dincer	75.55 bc	72.68 cd	74.11	5.70	6.53	6.11	4.37 b	8.25 a	6.31
Remzibey	68.21 de	74.57 cd	71.39	5.41	4.90	5.15	4.36 b	5.60 b	4.98
UC-1	64.50 e	62.43 e	63.46	5.60	5.90	5.75	3.76 c	4.93 bc	4.34
Mean (Zn)	76.33B	79.57A		5.34	5.63		4.16B	5.87A	

Note: Values are means of three independent replicates. For each trait, means followed by different letters are significantly different from each other for "genotype x Zn interaction" (LSD test, $P < 0.05$ and $P < 0.01$).

Tab. 6. Mean yield components of safflower genotypes under –Zn and +Zn conditions in two years

Genotypes	Biological yield per plant (g)			Thousand seed weight (g)			Harvest Index (%)		
2009									
	-Zn	*+Zn*	*Mean*	*-Zn*	*+Zn*	*Mean*	*-Zn*	*+Zn*	*Mean*
V50/63	17.84	17.63	17.73	3.59 c	3.90 bc	3.74	31.31	33.36	32.33
Yenice	28.05	30.29	29.17	4.20 b	4.34 b	4.27	32.15	40.62	36.38
Dincer	19.56	15.94	17.75	4.19 b	4.94 a	4.56	39.24	34.97	37.10
Remzibey	17.02	17.59	17.30	4.16 b	3.94bc	4.05	22.28	28.51	25.39
UC-1	13.88	18.23	16.05	4.06 b	4.28 b	4.17	27.67	29.74	28.70
Mean (Zn)	19.27B	19.94A		4.04B	4.28A		30.53B	33.44A	30.13
2010									
V50/63	12.83	20.24	16.53	3.20 f	3.27 f	3.23	27.85	29.18	28.51
Yenice	18.56	19.72	19.14	3.99 cd	4.17 bc	4.08	25.80	23.84	24.82
Dincer	19.55	21.26	20.40	3.70 e	3.73 e	3.71	22.41	38.63	30.52
Remzibey	16.08	15.89	15.98	4.19 b	4.64 a	4.41	26.99	35.01	31.00
UC-1	13.42	16.77	15.09	3.81 ed	4.06 bc	3.93	28.02	30.04	29.03
Mean (Zn)	16.09	18.78		3.78B	3.97A		26.21	31.34	28.77

Note: Values are means of three independent replicates. For each trait, means followed by different letters are significantly different (LSD test, $P < 0.05$ and $P < 0.01$).

had a significant effect on Fe concentration in both years (Tab. 4).

In both years, the Zn concentration in the seed was increased by Zn application. In the first and second year, the mean Zn concentration in the seeds was 36 mg kg^{-1} and 38 mg kg^{-1}, respectively, without Zn application, whereas the mean Zn concentration was 41 mg kg^{-1} and 43 mg kg^{-1}, respectively, with Zn application (Tab. 7). Movahhedy-Dehnavy *et al.* (2009) found that Zn application changed the Zn concentration of safflower seeds. Our results were similar to theirs. 'Remzibey' and 'Dincer' varieties showed the highest increase in seeds Zn concentration as a result of Zn application. These results showed that these genotypes can be used for human and animal nutrition. Zinc deficiency in humans and animals has been a serious nutritional problem in Turkey (Cakmak *et al.,* 1999) and in the world (Alloway, 2008). Safflower varieties that show a greater increase in seed Zn in response to Zn application may aid in the solution of this problem. Wissuwa *et al.* (2008) found significant genotypic differences in the concentrations of Zn in grain.

The Cu and Mn concentrations in the seed in all the genotypes increased with Zn application in both years. The Cu, Mn and Fe concentrations in the seeds were 3.96, 5.15 and 50.32 mg kg^{-1}, respectively, without Zn application and 9.31, 14.35 and 50.57 mg kg^{-1}, respectively, with Zn in the first year. In the second year, the Cu, Mn and Fe concentrations in the seeds were 4.75, 4.47 and 61.71 mg kg^{-1}, respectively, without Zn application and 9.75, 8.86 and 58.83 mg kg^{-1}, respectively, with Zn. Copper and Mn were affected by Zn application, whereas the Fe concentration in the seed decreased or did not change markedly. Kaya and Higgs (2002) reported that Zn application might prevent the uptake and transfer of Fe in shoots and fruits.

Tab. 7. Seed yield, Zn, Cu, Mn and Fe concentrations of seeds

Genotypes	Seed yield (kg/ha)			Zn (mg/kg)		Cu (mg/kg)		Mn (mg/kg)		Fe (mg/kg)	
2009											
	-Zn	+Zn	%	-Zn	+Zn	-Zn	+Zn	-Zn	+Zn	-Zn	+Zn
V50/63	1238.2 bc	1706.0 a	38	33.40 h	34.37 g	2.75 d	3.30 d	5.53 ef	16.80 a	36.93	50.00
Yenice	1254.6 bc	1412.6 bc	13	36.90 f	42.93 c	5.80 bc	6.10 b	3.35 f	10.53 cd	50.10	47.77
Dincer	921.1 ef	1249.5 bc	36	40.83 d	43.97 b	4.70 bcd	15.75 a	7.95 ed	15.10 ba	55.17	50.40
Remzibey	953.6 def	1038.3 dce	9	38.47 e	46.73 a	3.60 cd	5.03 bc	5.90 ef	13.33 bc	38.09	43.13
UC-1	738.7 f	828.2 ef	12	30.90 ı	36.90 f	2.95 d	16.35 a	3.03 f	16.00 ab	71.33	61.53
Mean (Zn)	1021.0	1247.0	22	36.00 B	41.00 A	3.96 B	9.31A	5.15B	14.35A	50.32	50.57
2010											
V50/63	999.0	1248.5	25	38.05 g	39.23 e	4.60 de	9.50 b	4.75 cd	9.85 ab	53.33	62.64
Yenice	1138.3	1425.4	25	39.13 e	42.87 c	4.23 e	9.70 ab	4.70 cd	9.40 ab	77.90	65.43
Dincer	1666.8	1756.3	5	40.37 d	45.87 a	4.90 cd	9.83 ab	5.60 c	8.63 b	59.60	59.83
Remzibey	1554.8	1679.7	0	38.33 f	44.10 b	4.23 e	9.57 b	4.00 cd	10.85 a	56.87	52.00
UC-1	1578.6	1580.0	8	36.23 h	40.53 d	5.77 c	10.13 a	3.30 d	5.55 c	60.87	54.23
Mean (Zn)	1388.0	1538.0	13	38.00B	43.00 A	4.75B	9.75A	4.47B	8.86A	61.71	58.83

Note: Values are means of three independent replicates. For each trait, means followed by different letters are significantly different (LSD test, $P < 0.05$ and $P < 0.01$).

Conclusions

The results of the study showed that the yield and yield components of safflower genotypes differed substantially in response to Zn application. It can be concluded that safflower genotype 'V50/63' showed a higher yield with Zn application and that the yield of 'Remzibey' was not strongly affected by Zn deficiency. However, further detailed studies should be conducted to understand the reasons underlying the differential Zn efficiencies of safflower genotypes. 'Remzibey' might be used as a genetic source for obtaining Zn-efficient safflower genotype in breeding programs. The line 'V50/63' was found to be very sensitive to Zn deficiency.

The seed yield and the Zn concentration in the seeds, in the studied safflower genotypes, responded well to Zn application. Therefore, Zn application should be considered by safflower growers in Zn-deficient areas.

References

Alloway BJ (2008). Zinc in soils and crop nutrition. 2nd Ed.,IZA and IFA Brussels, Belgium and Paris, France, 135 p.

Arslan B (2007). The path analysis of yield and its components in safflower (*Carthamus tinctorius* L.). Journal of Biological Sciences 7:668-672.

AOAC (1993). Association of official analytical chemists. Peer Verified Methods Program, Manual of Policies and Procedures, AOAC International, Arlington, VA.

Baligar VC, Fageria NK, He, ZL (2001). Nutrient use efficiency in plants. Communications in Soil Science and Plant Analysis 32:921-950.

Bouis HE (2003). Micronutrient fortification of plants through plant breeding: Can it improve nutrition in man at low cost?. Proc Nutr Soc 62:403-411.

Brennan RF, Holland MDA (2002). Relative effectiveness of soil-applied zinc for four crop species. Australian Journal of Experimental Agriculture 42:985-993.

Cakmak I, Yilmaz A, Ekiz H, Torun B, Erenoglu B, Braun HJ (1996). Zinc deficiency as a critical nutritional problem in wheat production in Central Anatolia. Plant and Soil 180: 165-172.

Cakmak I, Kalayci M, Ekiz H, Braun HJ, Yilmaz A (1999). Zinc deficiency as an actual problem in plant and human nutrition in Turkey: a NATO-Science for Stability Project. Field Crops Research 60:175-188.

Elfadl E, Reinbrecht C, Frick C, Claupein W (2009). Optimization of nitrogen rate and seed density for safflower (*Carthamus tinctorius* L.) production under low-input farming conditions in temperate climate. Field Crops Res 114 (1):2-13.

Gibson RS, Manger MS, Krittaphol W, Pongcharoen T, Gowachirapant S, Bailey KB, Winichagoon P (2007). Does zinc deficiency play a role in stunting among primary school children in NE Thailand? Brazilian Journal Nutrition 97:167-175.

Graham RD, Rengel Z (1993). Genotypic variation in zinc uptake and utilization by plants. Zinc in Soils and Plants. In: Robson AD (Ed.). Kluwer Academic Publishers, Dordrecht, 107-118p.

Grewal HS, Williams R (2000). Zinc nutrition affects alfalfa responses to water stress and excessive moisture. Plant and Soil 23:949-962.

Gyulai J (1996). Market outlook for safflower. In: Mündel HH, Braun J, Daniels C (Eds.). Proceedings of North American Safflower Conference, Great Falls, Montana, January 17-18, Let Bridge, AB, Canada, 15p.

Harris D, Rashid A, Mira G, Arif M, Yunas M (2008). "On-farm" seed priming with zinc in chickpea and wheat in Pakistan. Plant and Soil 306:310.

Hawkins, GW, Martens DC, McCart GD (1973). Response of corn to plowed down and disked-in Zn as zinc sulfate. Communications in Soil Science and Plant Analysis 4: 407-412.

Hotz C, Brown KH (2004). Assessment of the risk of zinc deficiency in populations and options for its control. Food Nutr Bull 25:94-204.

ISTA (2008). International rules for seed testing. Seed Science and Technology 24, Supplement, Zürich, Switzerland.

Kamali N, Mohammadi-Nejad G, Tohidi-Nejad E, Rezaied S (2009). Study of different zinc rates effects on yield and yield components of safflower genotypes in Jiroft region, Iran. Plant Ecophysiology 1:25-29.

Kaya C, Higgs D (2002). Response of tomato (*Lycopersicum esculentum* L.) cultivars to foliar application of zinc when grown in sand culture at low zinc. Scientia Horticulturae 93: 53-64.

Kinaci G, Kinaci E (2005). Effect of zinc application on quality traits of barley in semi arid zones of Turkey. Plant Soil Environment 51:328-334.

Khoshgoftarmanesh AH, Shariatmadari H, Kalbasi M, Karimian N, Khajehpour MR (2004). Zinc efficiency of wheat cultivars grown on a saline calcareous soil. Journal of Plant Nutrition 27:1953-1962.

Khoshgoftarmanesh AH, Schulin R, Chaney RL, Daneshbakhsh B, Afyuni M (2010). Micronutrient-efficient genotypes for crop yield and nutritional quality in sustainable agriculture: A review. Agronomy for Sustainable Development 30:83-107.

Lindsay WL, Norwell WA (1978). Development of a DTPA soil test for Zn, Fe, Mn and Cd. Journal of Soil Science Society of America 42:421-428.

Mahasi, MJ, Pathak RS, Wachira FN, Riungu TC, Kinyua MG, Kamundia JW (2006). Correlation and path coefficient analysis in exotic safflower (*Carthamus tinctorius* L.) genotypes tested in the arid and semi arid lands of Kenya. Asian J Plant Sci 5:1035-1038.

Mengel K, Kirkby EA (2001). Principles of plant nutrition. 5th Ed. (411p). Dordrecht: Kluwer Academic Publishers, 849p.

Mirzapour MH, Khoshgoftar AH (2006). Zinc application effects on yield and seed oil content of sunflower grown on a saline calcareous soil. Journal of Plant Nutrition 29: 1719-1727.

Mordvedt JJ, Gilkes RJ (1993). Zinc fertilizers. In: Robson AD (Eds.). Zinc in Soils and Plants, Dordrecht, Netherlands: Kluwer Academic Publishers, 33-44p.

Movahhedy-Dehnavy M, Modarres-Sanavy SAM, Mokhtassi-Bidgoli A (2009). Foliar application of zinc and manganese improves seed yield and quality of safflower (*Carthamus tinctorius* L.) grown under water deficit stress. Ind Crops Prod 30:82-92.

MSTAT (1984). MSTAT-C Microcomputer statistical program. Michigan State University. East Lansing, MI, USA.

Nriagu J (2010). Zinc deficiency in human health. Encyclopedia of Environmental Health, 789-800.

Pfeiffer WH, McClafferty B (2007). Biofortification: Breeding micronutrient-dense crops. In: Kang MS, Priyadarshan PM (Eds.). Breeding Major Food Staples. Blackwell Science, New York, 61-91p.

Peterson R (1996). Birdseed market Outlook. In: (Mündel, HH, Braun J, Daniels C (Eds.). Proceedings of North American Safflower Conference, Great Falls, Montana, 17-18 January. Lethbridge, AB, Canada, 15p.

Phattarakul N, Rerkasem B, Li LJ, Wu LH, Zou CQ, Ram H, Sohu VS, Kang BS, Surek H, Kalayci M, Yazici A, Zhang FS, Cakmak I (2012). Biofortification of rice grain with zinc through zinc fertilization in different countries. Plant and Soil 361:131-141.

Rengel Z (2001). Genotypic differences in micronutrient use efficiency in crops. Communications in Soil Science and Plant Analysis 32:1163-1186.

Rowell DR (1996). Soil science: methods and applications. Harlow, Longman.

Sillanpaa M (1990). Micronutrient assessment at the country level: An international study. The Government of Finland (FINNDA). Food and Agriculture Organizations of the

United Nations. Rome, Italy.

Singh RJ (2007). Genetic resources. In: Chromosome Engineering and Crop Improvement. CRC Press Inc. Boca Raton, Florida, USA, 168p.

World Health Organization (2002). The World Health Report 2002: Reducing Risks, Promoting Healthy Life. Geneva, Switzerland: World Health Organization.

Welch RM, Graham RD (2004). Breeding crops for enhanced micronutrient content. Plant Soil 245:205-214.

Wissuwa M, Ismail AM, Graham RD (2008). Rice grain zinc concentrations as affected by genotype, native soil-zinc availability, and zinc fertilization. Plant and Soil 306:37-48.

Yilmaz A, Ekiz H, Torun B, Gultekin I, Karanlik S, Bagci SA, Cakmak I (1997). Effect of different zinc application methods on grain yield and zinc concentration in wheat cultivars grown on zinc-deficient calcareous soils. Journal of Plant Nutrition 20:461-471.

Zou CQ, Zhang YQ, Rashid A, Ram H, Savasli E, Arisoy RZ, Ortiz-Monasterio I, Simunji S, Wang ZH, Sohu V, Hassan M, Kaya Y, Onder O, Lungu O, Yaqub MM, Joshi AK, Zelenskiy Y, Zhang FS, Cakmak I (2012): Biofortification of wheat with zinc through zinc fertilization in seven countries. Plant and Soil 361:119-130.

The Simultaneous Effect of Water Supply and Genotype on Yield Quantity, Antioxidants Content and Composition of Processing Tomatoes

Lajos HELYES[1], Andrea LUGASI[2], Hussein G. DAOOD[3], Zoltán PÉK[1]*

[1]*Szent István University, Horticultural Institute, Páter K.út 1., H-2103 Gödöllő, Hungary; Helyes.Lajos@mkk.szie.hu, Pek.Zoltan@mkk.szie.hu (*corresponding author)*

[2]*Budapest Business School, College of Commerce, Catering and Tourism, Department of Catering; Alkotmány u. 9-11., Budapest, H-1054, Hungary; lugasi_andrea@citromail.hu*

[3]*Szent István University, Regional Knowledge Center, Páter K. u. 1., Gödöllő, H-2100, Hungary; Daood.Hussein@fh.szie.hu*

Abstract

An open field experiment was carried out two years, in order to study the simultaneous effect of water supply and examined varieties on the yield parameters (marketable yield, °Brix and °Brix yield) and main antioxidant components (carotenoids, mainly lycopene, phenolic compounds and tocopherols) in two different type of processing tomatoes. It was conducted in the Experimental Farms of the Institute of Horticulture at Szent István University, Gödöllő, Hungary. 'Brixsol' F_1 (normal average fruit weight) and 'Strombolino' F_1 (cherry type) varieties were used. Climatic conditions were significantly different in the two seasons. Increasing the water supply increased fruit yield but reduced antioxidants of individual fruits, by inducing more and larger fruits, and by its dilution effects. We have measured significantly higher carotenoids and polyphenols content by processing cherry type hybrid 'Strombolino' F_1. In general, we can state that irrigation decreased (most cases significantly) carotenoids, polyphenol and tocopherol concentrations in tomato fruits.

Keywords: carotenoid, polyphenol, processing tomato, tocophenol

Introduction

Phytonutrients (also referred to as phytochemicals) are certain natural components of plants. Vegetables (tomato, sweet pepper, eggplant, broccoli, carrot, onion, etc.) are rich sources of phytonutrients. WHO places low fruit and vegetable intake sixth on its list of 20 risk factors for mortality worldwide (WHO, 1990). The importance of tomato cultivation has been growing rapidly in the last decades in the world. Beside the fresh market tomatoes, the processing tomato has a great importance in the food industry, basically because of its health promoting features (Clinton *et al.*, 1996). Bioactive composition of tomatoes is complex and very difficult to assess. Levels of plant metabolites are strongly affected by genetic and environmental factors as well as transportation and storage conditions (Hanson *et al.*, 2004). The growth factors such as light, temperature, humidity, type of soil, application of fertilizers, damage caused by microorganisms and insects, stress induced by UV radiation, heavy metals, and pesticides all alter metabolite composition of plants (Orcutt and Nilsen, 2000; Helyes and Lugasi, 2006; Helyes *et al.*, 2006; Pék *et al.*, 2010).

There are several reports on the effect of irrigation on tomato yield, dry matter and fruit soluble solids content. Pernice *et al.* (2010) found, that tomatoes produced an average yield of almost 100 t ha^{-1}. Yield was greatly influenced by the water regime; when comparing no irrigation to the reduced irrigation condition a considerable increase was observed in average yield (+45.7%). Similarly Liu *et al.* (2011) reported that drip irrigation compared to nonirrigation, increased tomato fruit size by 32%. Others (Patanè and Cosentino, 2010) also reported this marked influence of irrigation on tomato fruit size. Liu *et al.* (2011) found that irrigation also increased marketable fruit yield by 127% and total fruit yield by 66%, while it decreased soluble solids content by 19% which was also reported by Favati *et al.* (2009). In summary the greatest effect of increasing soil water deficit is the rise in soluble solids, total solids and fruit firmness, and a decrease in fruit size and yield (Patanè and Cosentino, 2010).

More than 700 carotenoids have been isolated from different plant sources. Lycopene is the most important carotenoid in tomato fruit (Rao, 2006). Chlorophyll breaks down and carotenoids, mostly lycopene, accumulate during ripening (Biacs and Daood, 2000). García-Valverde *et al.* (2013) reported from 83.2 to 97.6 mg kg^{-1} fw. total lycopene and from 3.0 to 6.1 mg kg^{-1} fw. β-carotene in the fruits of five processing tomato cultivars. Liu *et al.* (2011) reported that lycopene content decreased by 8% in irrigated compared to nonirrigated conditions. While Riggi *et al.* (2008) found that in the well watered treatment higher amounts of lycopene were measured, regardless of the ripening stage. They also found that β-carotene content was positively affected by water stress. Particularly in the first 2 ripening stages β-carotene/lycopene ratio was also influenced by water stress and suggested that, the carotenoid biosynthetic pathway is more 'β-carotene accumulation' oriented under soil water deficit conditions. Favati *et al.* (2009) also found that compared to well irrigated the lycopene and β-carotene concentration was higher in less irrigated tomatoes.

The term phenolic antioxidant refers to both simple phenolic acids and flavonoids. Tomato fruits are also rich in polyphenols that amount to the largest part of the antioxidant capacity of the soluble phase (Inert *et al.*, 2009).

In the case of tomatoes, several studies point to high levels of vitamin E in the processed product (Abushita *et al.*, 2000; Seybold *et al.*, 2004). Tomatoes at the ripening stage contain α-tocopherol and γ-tocopherol at the average concentration of 3.5 and 1.2 g/g, respectively (Abushita *et al.*, 1997).

The aim of the present study was to evaluate the influence of genetic and environmental factors (temperature and water supply) on yield quantity and quality, especially on the content of carotenoids (lycopene and its isomers, β-carotene), phenolic compounds (flavonoids and phenolic acids and total phenolics) and tocopherols in a cherry tomato, compared with a normal fruit weight variety.

Materials and methods

Plant material

This experiment was conducted at the Experimental Farm of the Institute of Horticultural, Szent István University, in Gödöllő, Hungary. The experimental field is on brown forest soil, with mechanical composition of sand, sandy-clay and the subsoil water is below 5 m, therefore it cannot influence the water turnover. 'Brixsol' F_1 (normal average fruit weight) 'Strombolino' F_1 (cherry type) seeds were sown on the 2nd of April in 2010 and 7th of April in 2011 in greenhouse and transplanted on the 14th of May 2010 and 12th of May in 2011. The experimental design was randomized block, number of replications were four for each treatment. Tomato plants were arranged in double (twin) rows with a distance of 1.2 and 0.4 m between the rows and 0.3 m between the plants. Crop density was 4.2 plant/m^2. Regularly irrigated and control (with no irrigation) were the two different water supply treatments applied in the present work. The amount of irrigation supply was calculated from weather forecasting data of the Hungarian Meteorological Service (2011). The amount of daily irrigation demand was estimated from expected daily potential evapotranspiration according to a previous study (Helyes and Varga, 1994).

In drip irrigation the water was given in accordance with air temperature (daily irrigation water (mm) = average daily temperature×0.2). National Meteorological Institute forecasts were used to calculate the probable air temperature. It is important to note that precipitation of 2010 was very extreme, because the highest rainfall in the last hundred years was measured that year in Hungary. Water amounts of 558 and 410 mm in 2010 and 524 and 159 mm in 2011 were usable for plants in irrigation treatment and unirrigated control respectively during the vegetation period (Fig. 1). Fruits were measured at harvesting on the 6th of August in 2010 and 26th of July and 26th of August in 2011.

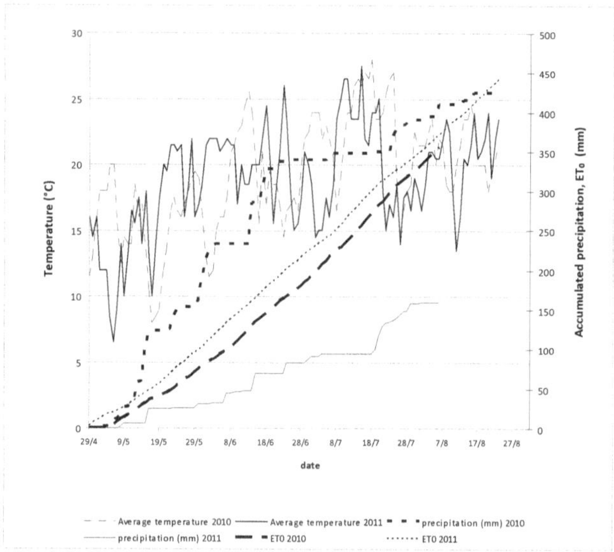

Fig. 1. Meteorological data during tomato vegetation period in 2010 and 2011

Measurement of environmental parameters

Changes of the environmental factors were monitored via measuring three parameters. During the experiment we defined the air temperature (°C), relative humidity (RH %) and incoming light intensity (μmol $m^{-2}s^{-1}$). Temperature, relative humidity and photosynthetically active radiation (not shown) were measured six times per hour using a SKYE DataHog micrometeorological instrument, placed at two meters height (Skye Instruments Ltd, Llandrindrod Wells, UK).

Analysis of quality components

Extraction of carotenoids, tocopherols. Lipids and fat-soluble pigments and tocopherols from raw tomato were extracted according to a previously described procedure with slight modification (Abushita *et al.*, 1997). Five-gram samples from tomato fruits were taken in triplicate (at least) and disintegrated in a crucible mortar in the presence of quartz sand. The water was removed by adding 25 mL of methanol with further disintegration of the aggregating bulk. The mixture was then transferred quantitatively to a 100 mL conical flask, and 70 mL of a 6:1 dichloroethane-methanol solution was added. The mixture was shaken for 15 min by a mechanical shaker. A few drops of distilled water were added when the dichloroethane phase was not clearly separated from the polar phase (water + methanol). A separatory funnel separated the two phases, and the lower layer containing lipids in dichloroethane was dried over anhydrous sodium sulfate. Finally, the organic solvent was evaporated under vacuum by rotary evaporator (IKA® RV

10, Sigma-Aldrich Ltd., Budapest, Hungary) at not higher than 40 °C. The residues were either redissolved in the HPLC eluent for analysis of carotenoids or saponified for analysis of tocopherols.

HPLC equipments and conditions. To analyse carotenoid and flavonoid type compounds, a Waters Alliance liquid chromatographic instrument consisting of a Model 2696 Separation Module (Gradient pump, autosampler and column heater) and a Model 2695 photodiode array detector was used. Operation and data processing were performed by Empower software.

The separation of carotenoids was performed on Cross-linked-C18 (240 _ 4.6 mm i.d.) 3µm ISIS column using gradients elution starting with 20% water in acetone and ending with 5% water in acetone with a flow rate of 0.7 mL/min according to Daood *et al.* (2013). The column effluents were detected at their maximum absorption wavelength lycopene, β-carotene and zeaxanthin were identified and quantified using standard materials purchased from Sigma-Aldrich Ltd. (Budapest, Hungary). Identification of other compounds was based on their retention time, spectral characteristics and mass spectrum data array). They were quantified as either lycopene- or β-carotene-equivalent according to their spectral characteristics (Fig. 2).

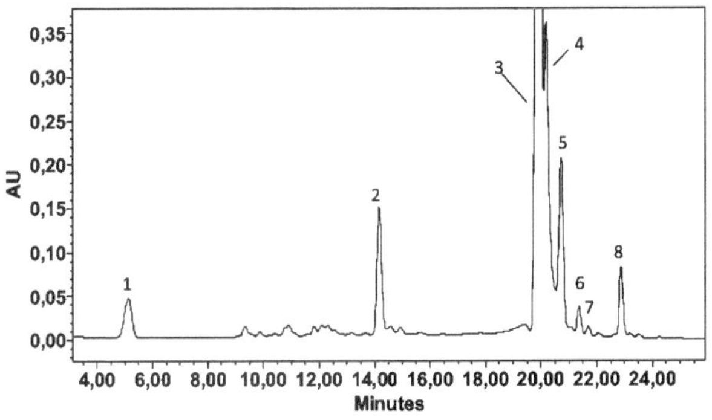

Fig. 2. HPLC profile of tomato carotenoids separated on cross-linked C18 column with gradient elution of water in acetone (for more details see text). Peak identification as: 1: zeaxanthin, 2: lycoxanthin, 3: lycopene, 4 and 5: cis-lycopene, 6: unidentified, 7 gamma-caotene, 8: beta-carotene

For tocopherol analysis, a combination of a Beckman 114M isocratic pump, a model RF-535 Shimadzu fluorometric detector, and a Waters-740 Data Module integrator was used. The separation was performed on Lichosorb 5µm (250 _ 4.6 mm i.d.) with a mobile phase consisting of 99.5:0.5 n-hexane-ethanol. The fluorometric detector was set at 295 and 320 nm as the excitation and emission wavelength, respectively (Abushita *et al.*, 1997) (Fig. 3).

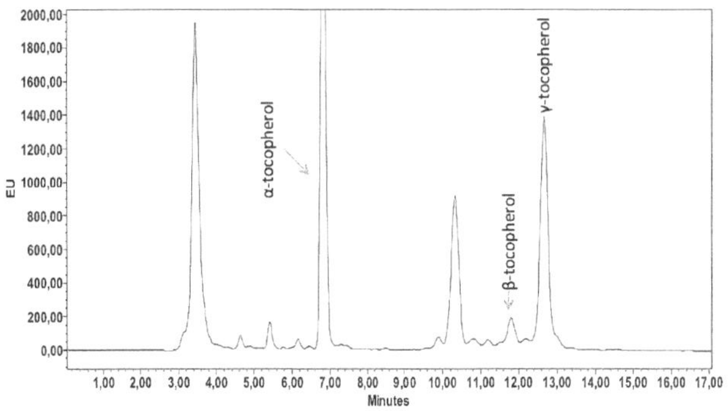

Fig. 3. Separation of tocopherol analogues from tomato by normal-phase HPLC using n-hexane-ethanol, 99.5:0.5, as the mobile phase and fluorescence detection (excitation 295 nm, emission 320 nm)

The phenolic compounds were determined by RP-HPLC. Sample preparation and extraction: Tomato sample was homogenised by a warring blender, 10 g was weighed into an Erlenmeyer flask, 10 ml distilled water was added to the sample and placed in ultrasonic bath for 30 sec., 15 ml of 2% acetic acid in methanol was added to the mixture. Mechanical shaker shook the mixture for 15 minutes. The flask was kept at 4 °C for overnight. Next day, the filtrate was cleaned further by passing through 0.45 µm HPLC syringe filter before injection onto HPLC column for analysis of phenolic compounds. The analytical column was EC NUCLEODUR Sphinx RP (Symetron Ltd. Budapest, Hungary), 3µm, 150 x 4.6. The mobile phase was [B] 1% formic acid in water, [C] acetonitril and a flow rate of 0.6 ml/min. Injected volume was 20 µl. Phenols having absorption maxima around 280 nm were quantified as gallic acid- equivalent, while flavonoids were quantified as rutin-equivalent at 320 nm and 355 nm. Peaks of phenol components were identified by comparing their spectral characteristics and retention with those of available standards (apigenin, caffeic acid glucoside, chlorogenic acid, sinapic acid glucoside, and ferulic acid; Sigma-Aldrich Ltd. Budapest, Hungary) (Pék *et al.*, 2013) (Fig. 4).

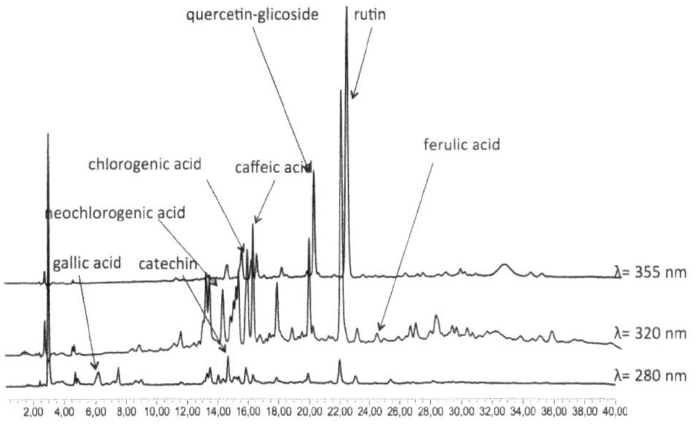

Fig. 4. HPLC profile of tomato phenols separated EC NUCLEODUR Sphinx RP with gradient elution of water in formic acid and acetonitril

Statistical analysis

The data were analysed by two-factor analysis of variance (ANOVA) with repetitions and the means separated using the Student's test at p=0.05.

Results and discussion

The effect of irrigation on tomato yield parameters depends on the actual weather conditions, basically the air temperature and precipitation (quantity and distribution).

Fig. 1 shows daily average temperature and precipitations during the vegetation period in these two years. It is also important to note that precipitation of 2010 was very extreme, because the highest rainfall in the last hundred years was measured that year in Hungary.

Average daily precipitation was almost three times more (4.1 mm) in 2010, than in 2011 (1.4 mm) during the crop season.

The two cultivars differ basically in average fruit weight, so normal processing type 'Brixsol' F1 produced higher yield than cherry type 'Strombolino' F1, but it was significant only in 2011. The irrigated plants gave significantly higher yield quantity in both years, while unirrigated plants showed yield loss particularly in 2011 (Tab. 1), which is in agreement with previous studies of processing tomatoes (Barbagallo *et al.*, 2013; Helyes *et al.*, 2012).

The effect of irrigation on processing tomato is complex; through the number of flowers and the percentage of fruit set it first increases the number of fruits per plant, and then enlarges the size of fruits (Wudiri and Henderson, 1985).

Better water supply caused lower °Brix, than control. The soluble solid content of fruits was often very high without irrigation. In spite of this, the level of Brix yield per hectare remarkably increased as a result of significantly higher yield quantity (Tab. 2). Greatest effect of increasing soil water deficit was the rise in fruit firmness, soluble solids and a decrease in fruit size and yield, which is in agreement with Patanè and Cosentino (2010).

In this study we evaluated total carotenoid, polyphenol and tocopherol contents and their composition depending on tested cultivars and water supplement in 2010 and 2011. During the two-year experiment, we found that the total carotenoid concentration of the tomato fruits ranged from 52.5 to 109.2 mg kg^{-1}. This is nearly a twofold difference. Cherry type tomatoes contain significantly higher lycopene concentration than large fruited cultivars (Helyes and Lugasi, 2006; Kuti and Konuru, 2005). StC gave the highest total carotenoid content (109.2 mg kg^{-1}) in July 2011, because of the cooler weather conditions (Dumas *et al.*, 2003) (Tab. 3). During the ten days preceding the 26th of July harvest, the average maximum temperature was 25.2 °C, the average minimum temperature was 14.5 °C and every day was a little rainy and cloudy. On the contrary, the ten days preceding the 26th of August harvest was much warmer. The average daily minimum temperature was 14.2 °C, the average maximum temperature was 29.3 °C and the maximum temperature reached or was above 30 °C five times. It is important to note that the average *cis*-lycopene ratio was significantly higher in case of 'Strombolino' F_1 in 2011, particularly in the second harvest (Tab. 3). Since irrigation also decreased lycopene concentration, thus to

Tab. 2. Soluble solid content (°Brix) and Brix yield (t ha^{-1}) of tomatoes under different water supply in 2010 and 2011 (n=4; ±SD)

Cultivars and treatments	2010		2011	
	°Brix	Brix yield	°Brix	Brix yield
BI	4.58 ± 0.12^{a}	1.80 ± 0.42^{a}	5.33 ± 0.17^{a}	5.44 ± 0.55^{c}
BC	5.03 ± 0.21^{a}	1.35 ± 0.16^{a}	7.25 ± 0.65^{bc}	3.54 ± 0.84^{b}
STI	5.44 ± 0.08^{b}	1.52 ± 0.15^{a}	6.63 ± 0.19^{b}	3.59 ± 0.62^{b}
STC	5.98 ± 0.12^{c}	1.44 ± 0.34^{a}	7.48 ± 0.49^{c}	1.97 ± 0.48^{a}

BI: 'Brixsol' irrigated, BC: 'Brixsol' control, StI: 'Strombolino' irrigated, StC: 'Strombolino' control.
Values are reported as the mean ± standard deviation. Means in a column with different letters are significantly different at p=0.05.

Tab. 1. Quantitative yield parameters (t ha^{-1}) of tomatoes under different water supply in 2010 and 2011 (n=4; ±SD)

Cultivars and treatments	2010			2011		
	Red yield	Green yield	Marketable yield	Red yield	Green yield	Marketable yield
BI	39.2 ± 8.1^{b}	10.2 ± 2.2^{b}	49.4 ± 5.1^{b}	102.1 ± 7.0^{c}	18.7 ± 3.4^{c}	120.8 ± 5.5^{c}
BC	26.8 ± 2.1^{a}	4.2 ± 0.6^{a}	31.0 ± 1.6^{a}	48.8 ± 7.2^{a}	14.4 ± 1.7^{b}	63.2 ± 3.8^{b}
StI	27.9 ± 7.1^{ab}	14.5 ± 3.9^{b}	42.4 ± 5.5^{b}	54.1 ± 7.8^{b}	5.6 ± 3.7^{a}	59.7 ± 5.7^{b}
StC	24.0 ± 1.6^{a}	6.0 ± 2.1^{a}	30.0 ± 1.8^{a}	26.4 ± 4.7^{a}	7.1 ± 2.8^{a}	33.5 ± 3.7^{a}

BI: 'Brixsol' irrigated, BC: 'Brixsol' control, StI: 'Strombolino' irrigated, StC: 'Strombolino' control.

Tab. 3. Carotenoid components (mg kg^{-1} FW) of tomatoes under different water supply in 2010 and 2011 (n=4; ±SD)

Harvesting date	Cultivars and treatments	Zeaxanthin	Lycoxanthin	Lycopene	9Z+13Z Lycopene	β-carotene	γ-carotene	Phytoene
06.8.2010	BI	0.33 ± 0.06^{c}	0.83 ± 0.09^{ab}	43.6 ± 3.12^{a}	3.40 ± 0.98^{a}	2.85 ± 0.25^{bc}	0.19 ± 0.03^{a}	1.27 ± 0.19^{a}
	BC	0.35 ± 0.04^{c}	0.52 ± 0.04^{a}	41.7 ± 4.10^{a}	6.81 ± 1.38^{b}	2.47 ± 0.34^{b}	0.23 ± 0.03^{ab}	1.72 ± 0.08^{a}
	StI	0.47 ± 0.07^{c}	1.05 ± 0.07^{ab}	48.3 ± 4.78^{a}	4.55 ± 0.74^{a}	3.63 ± 0.50^{c}	0.32 ± 0.04^{c}	2.02 ± 0.21^{b}
	StC	0.57 ± 0.01^{c}	1.49 ± 0.06^{b}	62.6 ± 2.77^{b}	2.98 ± 1.21^{a}	3.27 ± 0.17^{c}	0.28 ± 0.02^{c}	1.69 ± 0.13^{a}
26.7.2011	StI	0.28 ± 0.10^{b}	0.86 ± 0.20^{a}	73.5 ± 15.8^{bc}	8.30 ± 1.66^{bc}	2.77 ± 0.23^{b}	0.24 ± 0.01^{b}	1.61 ± 0.19^{a}
	StC	0.18 ± 0.03^{a}	1.55 ± 0.23^{b}	95.8 ± 10.2^{c}	7.51 ± 1.93^{b}	2.40 ± 0.31^{b}	0.21 ± 0.03^{a}	1.51 ± 0.07^{a}
26.8.2011	BI	0.21 ± 0.02^{a}	0.69 ± 0.15^{a}	39.3 ± 11.7^{a}	9.95 ± 0.57^{c}	1.16 ± 0.14^{a}	0.31 ± 0.01^{c}	1.95 ± 0.21^{ab}
	BC	0.37 ± 0.14^{bc}	1.32 ± 0.35^{b}	48.9 ± 6.4^{a}	15.11 ± 1.01^{e}	1.12 ± 0.11^{a}	0.29 ± 0.05^{c}	2.33 ± 0.50^{abc}
	StI	0.24 ± 0.01^{ab}	0.78 ± 0.28^{a}	57.3 ± 19.4^{ab}	13.60 ± 0.32^{d}	3.45 ± 0.26^{c}	0.25 ± 0.04^{ab}	2.76 ± 0.32^{c}
	StC	0.41 ± 0.07^{c}	1.53 ± 0.42^{b}	72.1 ± 24.5^{abc}	19.57 ± 1.85^{f}	3.70 ± 0.79^{c}	0.48 ± 0.11^{d}	2.55 ± 0.57^{bc}

BI: 'Brixsol' irrigated, BC: 'Brixsol' control, StI: 'Strombolino' irrigated, StC: 'Strombolino' control.
Values are reported as the mean ± standard deviation. Data in the same column bearing the same superscript letter are not significant at p=0.05.

improve fruit yield and lycopene concentration of tomatoes, Sánchez-Rodríguez *et al.* (2012) suggest moderate water stress. Irrigation probably indirectly affected lycopene concentration by inducing more and larger fruits, and thus had a dilution effect on ingredients. By the higher lycopene production per unit area the higher yield could account for the concentration loss of individual fruits (Helyes *et al.*, 2012).

The all-*trans* form of lycopene of tomato fruits was higher, while *cis*-isomers of lycopene were lower in all of the cultivars and harvests. The average *cis*-lycopene concentration was significantly higher in 2011 than in 2010, and it was nearly two-fold higher in case of 'Strombolino' at the second harvest date in 2011. It seems probable that the warmer temperatures and a better light supply activated the biosynthesis of *cis*-isomers of lycopene in 2011, which is in agreement with Kuti and Konuru (2005).

Tab. 4 shows the effect of water supply and examined varieties on the concentration of polyphenol groups (flavonoids and phenolic acids). We have measured significantly lower flavonoids and phenolic acids concentration in irrigated samples, which is contradictory to the findings of Sánchez-Rodríguez *et al.* (2012), in tomato grown in a growth chamber near optimal environmental conditions. Phenolic acids results showed the same pattern as flavonoids and ranged between 30.7-154.1 mg kg^{-1}. Irrigated samples gave lower values, in agreement with Sánchez-Rodríguez *et al.* (2012), who reported lower phenolic acids concentration of cherry tomato fruits under moderate water stress conditions.

In the examined tomato samples we could identify nine components of polyphenols. These are the following: neochlorogenic acid, chlorogenic acid, caffeic acid, ferulic acid, gallic acid, quercetin, rutin, catechin, naringin (Tab. 5.). In our experiment the average rutin content ranged from 41.7 to 98.8 mg kg^{-1}. This is more than a double difference, and in all samples, the rutin concentration showed significant decrease with irrigation. The second largest concentration was found to be of chlorogenic acid. We could not measured detectable amount of naringin in 'Brixsol' F1. Riga *et al.* (2008) found strong negative correlation between cumulative temperature during the 45 days before harvest and total phenolic concentration. Also lower phenolics content of fresh tomato fruit was reported at higher temperatures (Incerti *et al.*, 2009), and our results could confirm this temperature effect. Probably seasonal

Tab. 4. Concentration of polyphenols (mg kg^{-1} FW) of tomatoes under different water supply in 2010 and 2011 (n=4, ±SD)

Harvesting date	Cultivars and treatments	Flavonoids (mg kg^{-1})	Phenolic acids (mg kg^{-1})
06.8.2010	BI	25.7±2.5 a	30.7±3.6 a
	BC	69.9±8.7 cd	39.9±4.9 b
	StI	81.5±16.2 d	65.9±8.2 c
	StC	106.9±5.3 e	78.3±3.7d
26.7.2011	StI	69.0±6.4 d	109.0±15.5 e
	StC	58.6±4.3 c	154.1±19.4 f
26.8.2011	BI	22.6±3.8 a	31.7±4.0 a
	BC	37.7±6.9 b	42.2±5.6 b
	StI	81.0±11.0 d	101.1±12.7 e
	StC	120.1±20.4 e	112.5±20.2 e

BI: 'Brixsol' irrigated, BC: 'Brixsol' control, StI: 'Strombolino' irrigated, StC: 'Strombolino' control.
Values are reported as the mean ± standard deviation. Data in the same column bearing the same superscript letter are not significant at p=0.05.

Tab. 5. Concentration of main polyphenols (mg kg^{-1} FW) of cherry tomato under different water supply (n=4, ±SD)

Harvesting date	Cultivars and treatments	Quercetin-glucoside	Rutin	Catechin	Naringin	Neocloro-genic acid	Clorogenic acid	Caffeic acid	Ferulic acid	Gallic acid
06.8. 2010	BI	$7.2±0.3^{a}$	$14.3±1.6^{a}$	$4.3±0.5^{c}$	nd	$4.8±0.6^{a}$	$12.4±1.2^{b}$	$7.7±0.2^{b}$	$2.9±0.4^{b}$	$2.9±1.1^{c}$
	BC	$15.3±1.2^{cd}$	$45.8±5.9^{cd}$	$8.8±1.6^{d}$	nd	$9.0±0.9^{b}$	$10.7±1.0^{b}$	$8.7±1.0^{b}$	$3.7±1.0^{b}$	$7.7±0.9^{f}$
	StI	$13.6±2.2^{bc}$	$46.6±10.0^{cd}$	$6.0±2.1^{cd}$	$15.3±1.6^{c}$	$11.9±1.4^{c}$	$26.6±3.2^{c}$	$19.7±2.5^{e}$	$3.1±0.3^{b}$	$4.7±0.7^{de}$
	StC	$12.8±0.5^{b}$	$68.1±2.5^{e}$	$10.4±1.4^{d}$	$15.5±0.8^{c}$	$20.5±0.8^{e}$	$30.2±0.9^{d}$	$19.4±1.1^{e}$	$3.6±0.3^{b}$	$4.5±0.4^{d}$
26.7. 2011	StI	$13.2±0.9^{bc}$	$41.7±1.9^{c}$	$3.3±0.4^{b}$	$0.3±0.03^{a}$	$13.3±1.8^{cd}$	$45.8±6.9^{e}$	$6.3±0.6^{a}$	$5.8±0.6^{c}$	$3.0±0.4^{c}$
	StC	$13.4±1.8^{bc}$	$52.2±4.9^{d}$	$3.0±0.4^{b}$	$0.4±0.05^{b}$	$12.0±1.4^{c}$	$84.3±12.4^{f}$	$8.0±0.7^{b}$	$7.1±0.5^{d}$	$1.2±0.1^{a}$
26.8. 2011	BI	$7.8±1.8^{a}$	$11.7±1.2^{a}$	$3.2±0.8^{bc}$	nd	$4.8±0.6^{a}$	$10.9±1.5^{b}$	$5.0±0.8^{a}$	$0.3±0.1^{a}$	$1.5±0.1^{b}$
	BC	$13.1±3.4^{bc}$	$19.1±2.4^{b}$	$5.6±1.10^{c}$	nd	$9.2±1.10^{b}$	$7.4±1.02^{a}$	$6.2±0.8^{a}$	$2.0±0.6^{b}$	$1.4±0.2^{ab}$
	StI	$13.2±1.2^{bc}$	$65.5±9.2^{de}$	$2.0±0.3^{a}$	$0.4±0.06^{b}$	$15.0±1.4^{d}$	$22.6±2.9^{c}$	$11.4±1.8^{d}$	$9.5±1.8^{e}$	$6.5±0.7^{ef}$
	StC	$18.1±3.4^{d}$	$98.8±16.2^{f}$	$2.7±0.4^{ab}$	$0.4±0.05^{b}$	$28.9±4.2^{f}$	$31.5±3.0^{d}$	$13.1±2.9^{d}$	$12.6±4.5^{e}$	$6.5±1.2^{def}$

BI: 'Brixsol' irrigated, BC: 'Brixsol' control, StI: 'Strombolino' irrigated, StC: 'Strombolino' control, nd: not detected.
Values are reported as the mean ± standard deviation. Data in the same column bearing the same superscript letter are not significant at p=0.05.

Tab. 6. Effect of season and water supply on concentration of tocopherol in tomato fruits ($\mu g\ g^{-1}$ FW) (n=4) (±SD)

Harvesting date	Cultivars and treatments	α-tocopherol	β-tocopherol	γ-tocopherol	γ-tocotrienol
06. 8. 2010	BI	7.07 ± 0.24^{d}	0.71 ± 0.01^{c}	2.04 ± 0.05^{b}	0.19 ± 0.01^{a}
	BC	4.85 ± 0.30^{b}	0.53 ± 0.07^{ab}	2.02 ± 0.03^{b}	0.20 ± 0.01^{a}
	StI	6.19 ± 0.49^{c}	0.65 ± 0.05^{bc}	2.25 ± 0.09^{c}	0.26 ± 0.01^{b}
	StC	7.14 ± 0.22^{d}	0.57 ± 0.04^{b}	2.26 ± 0.01^{c}	0.24 ± 0.01^{b}
26. 7. 2011	StI	6.85 ± 0.68^{cd}	0.41 ± 0.11^{a}	2.87 ± 0.59^{c}	0.63 ± 0.07^{d}
	StC	7.69 ± 0.72^{d}	0.36 ± 0.12^{a}	3.43 ± 0.66^{cd}	0.45 ± 0.11^{c}
26. 8. 2011	BI	3.93 ± 0.65^{a}	0.88 ± 0.10^{d}	1.21 ± 0.28^{a}	0.33 ± 0.07^{bc}
	BC	6.78 ± 1.37^{cd}	0.91 ± 0.04^{e}	1.55±0.24 a	0.26 ± 0.04^{b}
	StI	4.29 ± 0.60^{a}	0.86 ± 0.15^{de}	3.35 ± 0.31^{d}	0.39 ± 0.07^{c}
	StC	9.49 ± 1.67^{e}	0.83 ± 0.14^{cde}	3.77 ± 1.05^{cd}	0.56 ± 0.04^{cd}

BI: 'Brixsol' irrigated,BC: 'Brixsol' control, StI: 'Strombolino' irrigated, StC: 'Strombolino' control.
Values are reported as the mean ± standard deviation. Data in the same column bearing the same superscript letter are not significant at p=0.05.

variation of phenolic compounds is also closely correlated with the temperature regime during the fruit development period of tomato.

We have also measured tocopherol content and composition. Four components of tocopherols were identified in the examined tomato samples. There were determined α-tocopherol, γ-tocopherol, β-tocopherol and γ-tocotrienol were determined in decreasing percentage respectively α-tocopherol represented 48-70% of total tocopherols and γ-tocopherol had the second largest concentration (between 22-38%) depending on treatment, while the remaining two were lower than 10% (Tab. 6). Control plants gave higher average α-tocopherol concentration (except BI sample, in the first harvest in 2010), which is in agreement with Hwang *et al.* (2012).

Conclusions

The climatic conditions, particularly the precipitation was significantly different in the two investigated seasons, because 2010 was the rainiest year, but during 2011 there was drought. Increasing the water supply increased fruit yield but reduced °Brix significantly. On the other hand it is very important to mention that Brix yield increased significantly. Many studies have demonstrated that the ingredient content is fundamentally determined by the genetic nature of varieties, but it does not exclude the fact that environmental factors also strongly affect it. Irrigation decreased ingredients contents of individual fruits, by inducing more and larger fruits, and by its dilution effects, but the higher yield could compensate for the concentration loss of individual fruits by higher ingredients production per unit area. We have measured significantly higher carotenoids and polyphenols content in processing cherry type hybrid 'Strombolino' F_1. Cherry type shows significantly higher carotenoid contents, mainly lycopene, than large fruited cultivars. α-tocopherol is the most abundant tocopherol in tomato fruits. In general, we can state that irrigation decreased (in most cases significantly) carotenoids, polyphenol and tocopherol concentration in tomato fruits.

Acknowledgements

This study was funded in part by Research Centre of Excellence (17586-4/2013/TUDPOL) Szent István University and KTIA_AIK_12-1-2012-0012 project.

References

Abushita AA, Daood HG, Biacs PA (2000). Change in carotenoids and antioxidant vitamins in tomato as a function of varietal and technological factors. J Agric Food Chem 48:2075-2081.

Abushita AA, Hebshi EA, Daood HG, Biacs PA (1997). Determination of antioxidant vitamins in tomatoes. Food Chem 60:207-212.

Barbagallo RN, Silvestro ID, Patanè C (2013). Yield, physicochemical traits, antioxidant pattern, polyphenol oxidase activity and total visual quality of field-grown processing tomato cv. Brigade as affected by water stress in Mediterranean climate. J Sci Food Agr 93:1449-1457.

Biacs PA, Daood HG (2000). Lipoxygenase-catalysed degradation of carotenoids from tomato in the presence of antioxidant vitamins. Biochem Soc Trans 28:839-845.

Clinton SK, Emenhiser C, Schwartz SJ (1996). Cis-trans lycopene isomers, carotenoids, and retinol in the human prostate. Cancer Epidemiol Biomark Prev 5:35-51.

Daood HG, Bencze G, Palotás G, Pék Z, Helyes L (2013). HPLC analysis of carotenoids from tomato using cross-linked C-18 column and MS detection. J Chromatogr Sci doi:10.1093/chromsci/bmt139.

Dumas Y, Dadomo M, Di Lucca G, Grolier P (2003). Effects of environmental factors and agricultural techniques on antioxidantcontent of tomatoes. J Sci Food Agric 83:369-382.

Favati F, Lovelli S, Galgano F, Miccolis V, Di Tommaso T, Candido V (2009). Processing tomato quality as affected by irrigation scheduling. Sci Hort 122:562-571.

García-Valverde V, Navarro-González I, García-Alonso J, Periago MJ (2013). Antioxidant Bioactive Compounds in Selected Industrial Processing and Fresh Consumption Tomato Cultivars. Food Bioprocess Tech 6: 391-402.

Hanson PM, Yang RY, Wu J, Chen JT, Ledesma D, Tsou SCS, Lee TC (2004). Variation for antioxidant activity and antioxidants in tomato. J Amer Soc Hort Sci 129:704-711.

Helyes L, Lugasi A (2006): Formation of certain compounds having technological and nutritional importance in tomato fruits during maturation. Acta Alim 35:183-193.

Helyes L, Lugasi A, Pék Z (2006). Tomato fruit quality and content depend on stage of maturity. Hort Sci 41:1400-1401.

Helyes L, Lugasi A, Pék Z (2012). Effect of irrigation on processing tomato yield and antioxidant components. Turk J Agr For 36:702-709.

Helyes L, Pék Z, Lugasi A (2008). Function of the variety technological traits and growing conditions on fruit components of tomato (*Lycopersicon lycopersicum* L Karsten), Acta Alim 37:427-436.

Helyes L, Varga G (1994). Irrigation demand of tomato according to the results of three decades. Acta Hort 376:323-328.

Hungarian Meteorological Service (2011). Temperature forecast of Gödöllő, 15 Aug. 2011.

Hwang ES, Stacewicz-Sapuntzakis M, Bowen PE (2012). Effects of Heat Treatment on the Carotenoid and Tocopherol Composition of Tomato. J Food Sci 77:C1109-C1114.

Incerti A, Navari-Izzo F, Pardossi A, Izzoa R (2009). Seasonal variations in polyphenols and lipoic acid in fruits of tomato irrigated with sea water. J Sci Food Agr 89:1326-1331.

Kuti JO, Konuru HB (2005). Effects of genotype and cultivation environment on lycopene content in red-ripe tomatoes. J Sci Food Agr 85:2021-2026.

Liu K, Zhang TQ, Tan CS, Astatkie T (2011). Responses of fruit yield and quality of processing tomato to drip-irrigation and fertilizers phosphorus and potassium. Agronomy J 103:1339-1345.

Orcutt DM, Nilsen ET (2000). The physiology of plants under stress soil and biotic factors. New York, JohnWiley and Sons Inc 696 p.

Patanè C, Cosentino SL (2010). Effects of soil water deficit on yield and quality of processing tomato under a Mediterranean climate. Agr Water Management 97:131-138.

Pék Z, Daood H, Nagyné MG, Neményi A, Helyes L, (2013). Effect of environmental conditions and water status on the bioactive compounds of broccoli, Cent Eur J Biol 8:777-787.

Pék Z, Helyes L, Lugasi A (2010). Color changes and antioxidant content of vine and post-harvest ripened tomato fruits. HortScience 45:466-468.

Pernice R, Parisi M, Giordano I, Pentangelo A, Graziani G, Gallo M, Fogliano V, Ritieni A (2010). Antioxidants profile of small tomato fruits: Effect of irrigation and industrial process. Sci Hort 126:156-163.

Rao VA (2006). Tomatoes, Lycopene and Human Health: Preventing Chronic Diseases. Caledonian Science Press, Inverness 238 p.

Riga P, Anza M, Garbisu C (2008). Tomato quality is more dependent on temperature than on photosynthetically active radiation. J Sci Food Agr 88:158-166.

Riggi E, Patanè C, Ruberto G (2008). Content of carotenoids at different ripening stages in processing tomato in relation to soil water availability. Australian J Agric Res 59:348-353.

Sánchez-Rodríguez E, Ruiz JM, Ferreres F, Moreno DA (2012). Phenolic profiles of cherry tomatoes as influenced by hydric stress and rootstock technique. Food Chem 134:775-782.

Seybold C, Frohlich K, Bitsch R, Otto K, Böhm V (2004). Changes in contents of carotenoids and vitamin E during tomato processing. J Agric Food Chem 52:7005-7010.

World Health Organization (1990). Diet, nutrition, and the prevention of chronic diseases. Report of a WHO Study Group. WHO Technical Report Series. No. 916. p 148.

Wudiri BB, Henderson DW (1985). Effects of water stress on flowering and fruit set in processing-tomatoes. Sci Hort 27:189-198.

Comparative Analysis of Mineral Elements and Essential Amino Acids Compositions in *Juglans sigillata* and *J. regia* Walnuts Kernels

Meizhi ZHAI*, Zhenyuan WANG, Dan WANG, Jing XU, Guanzhao SHI

*Northwest Agriculture & Forestry University, Walnut Research Center, College of Forestry, Yangling, Shaanxi, China; plum-zhai@163.com (*corresponding author)*

Abstract

Walnut kernel is famous for its high nutritional and economic values. The kernel is usually considered to be a good source of minerals and essential amino acids. In this paper, mineral elements (calcium, magnesium, iron, manganese, copper, and zinc) and essential amino acids (phenylalanine, valine, threonine, isoleucine, leucine, methionine, and lysine) compositions of kernels from 11 kinds of walnuts (*Juglans sigillata*) and 17 kinds of walnuts (*Juglans regia*) originated from China were determined by ICP-MS and HPLC, respectively. The order of nutritive mineral elements depending on their content (mg/100g) of samples was Mg> Ca> Zn> Mn> Fe> Cu in *J. regia*, while the order in *J. sigillata* was Mg> Ca> Mn> Fe > Zn > Cu. For essential amino acids, the order depending on the content (mg/g) of the essential amino acids in *J. regia* samples was leucine> isoleucine> valine> phenylalanine> lysine> threonine> methionine, while the order in *J. sigillata* was leucine> isoleucine> lysine> phenylalanine> valine> threonine> methionine. The kernels of walnuts (*J. regia* and *J. regia*) are good sources of health foods and dietary supplements. 'Y029' in *Juglans sigillata* and 'XJ004' in *Juglans regia* provided the best profiles of mineral elements and essential amino acids in comparison to others.

Keywords: essential amino acid, HPLC, ICP-MS system, mineral element, walnut kernel

Introduction

Walnuts (*Juglans regia* and *Juglans sigillata*) are widely distributed in China. There is a various of walnuts differing in varieties, productivity, physical and chemical features of their kernel. Some of them have been evaluated as promising and may serve as germplasm sources for breeding (Warmund *et al.,* 2009; Martínez *et al.,* 2010; Qin *et al.,* 2012; Albuquerque *et al.,* 2013). As we all know, the kernel of walnut as food has remarkable nutrients such as quinones, oils, tannins, essential fatty acids, albumin, mucilage, mineral matter, amino acids, and so on. Thus, walnut is also considered to be a good source of minerals and essential amino acids. In generally, significant quantities of magnesium (Mg), calcium (Ca), phosphorus (P), potassium (K), and iron (Fe) are present in the kernels. Researchers are conducted to study on the contents of nutrition elements in cultivated walnuts and their impact on human health. It was found that an adequate intake of walnut kernel could be beneficial to lowering blood cholesterol and improving arterial function (Sabate *et al.,* 1993; Savage, 2001; Nash and Westpfal, 2005; Patel, 2005). Moreover, the amino acids regarded as essential nutrients for humans are phenylalanine, valine, *threonine, tryptophan, isoleucine, methionine, leucine, lysine*, and *histidine* (Young, 1994). Importantly, cysteine (or sulphur-containing amino acids), tyrosine (or aromatic amino acids), and arginine are essential for infants and growing children (Imura and Okada, 1998; Joint, 2007). The essential amino acids play a key role in the development of body since they are not synthesized by people. While the amino acids including *arginine, cysteine, glycine, glutamine, histidine, proline, serine,* and *tyrosine* are considered conditionally essential (Reeds, 2000; Fürst and Stehle, 2004). For example, tyrosine is essential for the patients with the disease of phenylketonuria (PKU). In the present study, Wu *et al.* (2007) reported that there were 16 kinds of amino acids in the 4 kinds of walnut kernels besides of glutamic acid. The amino acids included 7 kinds of essential amino acids, in which the content of leucine was the highest. In comparison, the content of amino acids from Pao-walnut kernels was significantly ($p<0.01$) lower than those from other groups. In addition, amino acids were also determined in other varieties of walnut kernels (Yanbin, 2002; Qingzhu *et al., 2003*) Alternatively, mineral elements have been identified as valuable compositions in kernels of walnuts. The previous research found that there were 5 main kinds of mineral elements (K, P, Mg, Ca, and Mn) in kernels of walnut cultivars 'Franquette' and 'Hartley' originated from France and California, respectively, in which the contents of K, P, and Mg were significant differences according to the factors like the varieties of walnuts, growth conditions, and climate (Lavedrine *et al., 2000*). The mineral elements in 5 kinds of walnut genotypes ('Sebin-Type-I', 'Körcegöz', 'Karabodur', 'Tozam', and 'Güvenli') cultivated in Turkey were also determined by Caglarirmak (2003). It was found that the mean contents of mineral elements were 316.0, 270.0, 85.0, 90.0, 2.0, 2.5, 2.9, and 1.0 mg/100g for P, K, Ca, Mg, Zn,

Mn, Cu, and Fe, respectively. Chemical analysis revealed the composition to be 13.77% protein, 62.84% fat, 18.67% total carbohydrates, 1.81% ash, and 2.98% moisture (Caglarirmak, 2003). Recently, selenium (Se) was also found in 15 kinds of walnut kernels (*J. regia*) grown in Erzincan and the contents of Se were ranged between 7.25 and 57.67 ng g^{-1} (Ozrenk *et al.,* 2012).

Furthermore, by the analysis of chemical compositions in walnut kernels from New Zealand, USA ('Tehama' and 'Vina'), and Europe ('Esterhazy', 'G120', 'G139'), the results showed that the contents of proteins were between 13.6% and 18.1%, and the contents of amino acids were similar among cultivars (Savage, 2001). In addition to proteins, amino acids, and minerals, *g*-tocopherol and fiber were also found in kernels of walnuts (Sze-Tao and Sathe, 2000; Ozrenk *et al.,* 2012). In this paper, the aim of our study was to determine the contents of mineral elements and essential amino acids in kernels from 11 kinds of walnuts *J. sigillata* and 17 kinds of walnuts *J. regia* collected in China.

Materials and methods

Materials

The kernels from 11 kinds of walnuts *J. sigillata* ('Y021', 'Y026', 'Y029', 'Y041', 'Y042', 'Y075', 'Y082', 'Y096', 'Y098', 'Y099', and 'Y101') and 17 kinds of walnuts *J. regia* ('XJ004', 'XJ005', 'XJ008', 'XJ012', 'XJ017', 'XJ018', 'XJ019', 'XJ020', 'XJ029', 'SX05', 'SX13', 'SS004', 'SAK002', 'SHZ002', 'X1', 'X9', and 'SS002') were obtained in Shaanxi (lat. 34°12'~34°20'N, long. 108°~108°7'E) and Yunnan (lat. 27°08'~27°15'N, long. 101°31'~102°11'E) provinces in China, respectively. The climate is characterized by the annual mean rainfall of 635.1 mm and mean temperature of 12.9 °C for Shaanxi province, and 1 300 mm and 16 °C for Yunnan province. Kernels were kept in glass jars until analyses at refrigerator. In all stages of experiments, dry and mature kernels were used. Azotic acid (65%, Fluka), oxygenated water (33%), ethanol, triethylamine, phenyl isothiocyanate (PITC, 99.9%), and ultrapure water were used as received.

Methods

The digestion microwave system (Milestone Inc., USA) was used for the mineralization of solid sample. 0.5 g samples, 6 mL 65% nitric acid, and 2 ml 33% hydrogen peroxide were introduced in Teflon recipients. Then, the oven temperature was increased to 180 °C at 5 °C min^{-1} and held for 20 min. After cooling to room temperature, the liquid samples were transferred into the marked glass balloons and the total volume of 50 ml was obtained with ultrapure water, and ultrahigh-purity commercial acids were used to prepare all reagents, standards and walnut kernel samples. After digestion treatment, samples were filtrated through whatman No 42 (Özcan *et al.*, 2010). The filtrates collected for the mineral elements determination were analyzed by ICP-MS (Perkin-Elmer Elan 9000). The ICP torch was fassele type and ceramic alumina torch injector. Nebulizer was tip cross flow and the flow of gas is 0.93 L min^{-1}. Additionally, the parameters of the spray chamber were sweeps/reading 20 s, reading/ replicate 2, and number of replicates 5.

The compositions of amino acids were determined by amino acid analyzer with a Pico-Tag column. Accurately weighed sample (0.1 g) was hydrolyzed with 600 μL HCl (6 M) at 110 °C for 18 h under nitrogen. After hydrolysis, the hydrolysate was dried, mixed with ethanol / triethylamine / water solution (2:2:1, v/v/v) and dried again. Then the dried sample was derivatized with an ethanol/triethylamine/water /PITC (7:1:1:1, v/v/v/v) solution at 25 °C for 20 min under nitrogen and then dried. Finally, the sample was dissolved in 50 μL of 5 mM sodium acetate buffer (pH 7.6, 40 °C) containing 6% (v/v) acetonitrile and analyzed by HPLC. Norleucine was used as an internal standard to calculate percent recovery of amino acids, while tryptophan was determined by the method of colorimetric (No. 3)(Spies and Chambers, 1948). The contents of amino acids were reported as grams of amino acid per 100 g of protein.

Statistical analyses

All the data were analyzed using the SPSS 16.0 statistical software for analysis of variance using ANOVA and Duncan's least significant difference (LSD at $p < 0.05$) for statistical significance. This research was performed by 3 duplicates with a replicate, and data were reported as the mean ± standard error of the mean.

Results and discussion

Mineral elements

As known, the kernels of walnut have a remarkable amount of mineral nutrition that is beneficial to body health among common foodstuffs. Actually, the kernels contain many kinds of minerals, in which Mg and Ca were the major minerals. Tab. 1 shows the results of all the mineral compositions in kernels from 11 kinds of walnuts *J. sigillata* and 17 kinds of walnuts *J. regia*. Ca, Mg, Fe, Mn, Zn, and Cu were determined by atomic absorption spectrometry. From the Tab. 1, Mg and Ca showed the higher contents than others in the determined samples. The order depending on the contents of elements (mg/100g) in *J. regia* samples was Mg> Ca> Zn> Mn> Fe> Cu, while the order in *J. sigillata* was Mg> Ca> Mn> Fe> Zn> Cu. Recent studies showed that osteoporosis, hypertension, coronary heart disease, congestive heart failure, arrhythmia, diabetes mellitus, asthma, migraine headaches, and pre-eclampsia may be associated with the deficiency of Mg (*Touyz 2004*). In Tab. 1, the contents of Mg varied between 401.47 ('X1') and 749.27 mg/100g ('SS002') in kernels of *J. regia*, 538.40 ('Y075') and 771.93 mg/100g ('Y096') in kernels of *J. sigillata*. The mean values of Mg were 579.18 mg/100g in *J. regia* and 647.57 mg/100g in *J. sigillata*. The ratio of the mean contents of Mg in *J. sigillata* to *J. regia* was 1.11.

It was followed by Ca, which is an essential nutrient and a vital electrolyte for human body (Bryant *et al., 1999*). In addition, it is well known that Ca plays an important role in the development of bone and teeth. The determination showed that the contents of Ca were changed between 155.10 ('Y096') and 482.03 mg/100g ('Y101'), and the mean value was 288.42 mg/100g in *J. sigillata*. The contents of Ca varied between 253.90 ('XJ004') to 504.07 mg/100g ('SS004') in *J. regia*, and the recorded mean value was 352.18 mg/100g. The ratio of the mean contents of Ca in *J. regia* to *J. sigillata* was 1.22.

Tab. 1. The contents of mineral elements (mg/100g, dry weight) in kernels from 11 kinds of walnuts (*J. sigillata*) and 17 kinds of walnuts (*J. regia*)

No.	Samples	Mineral Elements					
		Ca	Mg	Fe	Mn	Zn	Cu
	J. sigillata						
1	Y021	266.63 ± 0.18 c	651.33 ± 0.48 f	11.29 ± 0.02 d	13.38 ± 0.03 e	8.85 ± 0.03 b	6.17 ± 0.02 de
2	Y026	268.23 ± 0.32 d	698.87 ± 0.73 i	10.88 ± 0.16 c	12.55 ± 0.04 e	7.66 ± 0.03 a	3.85 ± 0.04 b
3	Y029	282.57 ± 0.45 e	653.13 ± 0.63 fg	10.34 ± 0.06 b	6.84 ± 0.04 b	12.37 ± 0.04 g	5.59 ± 0.05 cd
4	Y041	285.27 ± 0.80 g	625.43 ± 0.38 c	9.44 ± 0.06 a	5.47 ± 0.10 a	10.43 ± 0.17 d	3.65 ± 0.07 b
5	Y042	285.37 ± 0.54 g	653.50 ± 1.02 g	9.71 ± 0.06 a	18.43 ±0.05 f	10.93 ± 0.03 e	5.71 ± 0.06 cd
6	Y075	284.50 ± 0.46 fg	538.40 ± 0.40 a	11.56 ± 0.04 d	24.51 ± 0.07 g	13.63 ± 0.05 h	4.87 ± 0.03 c
7	Y082	283.23 ± 0.44 ef	579.20 ± 0.47 b	13.14 ± 0.09 f	10.29 ± 0.21 d	11.56 ± 0.21 f	5.37 ± 0.07 cd
8	Y096	155.10 ± 0.45 a	771.93 ± 0.46 j	10.67 ± 0.19 c	9.35 ± 0.12 c	10.31 ± 0.07 d	4.96 ± 0.06 c
9	Y098	261.10 ± 0.46 b	640.30 ± 0.32 e	13.72 ± 0.13 g	24.50 ± 0.26 g	10.37 ± 0.10 d	5.72 ± 0.06 cd
10	Y099	318.60 ± 0.45 h	631.20 ±0.43 d	12.55 ± 0.07 e	19.28 ± 0.16 f	11.39 ± 0.10 f	6.91 ± 1.05 e
11	Y101	482.03 ± 0.55 i	680.03 ± 1.19 h	12.33 ± 0.11 e	13.14 ± 0.07 e	9.95 ± 0.06 c	1.46 ± 0.05 a
	J. regia						
1	XJ004	253.90 ± 1.80 a	483.30 ± 0.44 c	10.45 ± 0.08 k	4.46 ± 0.03 c	10.38 ± 0.03 e	3.23 ± 0.04 d
2	XJ005	295.27 ± 0.49 b	614.53 ± 0.30 kl	9.84 ± 0.06 j	5.48 ± 0.04 f	9.64 ± 0.08 d	3.35 ± 0.02 de
3	XJ008	423.37 ± 0.40 k	616.73 ± 0.50 l	8.83 ± 0.03 g	9.01 ± 0.07 l	12.69 ± 0.05 h	5.30 ± 0.06 k
4	XJ012	336.17 ± 0.64 h	557.70 ± 0.44 f	9.58 ± 0.10 i	7 .34 ± 0.07 i	12.55 ± 0.11 h	5.10 ± 0.06 j
5	XJ017	310.57 ± 0.35 e	607.37 ± 0.43 j	8.45 ± 0.07 f	5.12 ± 0.06 e	11.56 ± 0.14 f	4.42 ± 0.04 g
6	XJ018	328.33 ± 0.35 f	567.30 ± 0.26 g	7.86 ± 0.05 e	4.86 ± 0.03 d	13.32 ± 0.05 j	3.43 ± 0.06 e
7	XJ019	341.97 ± 1.35 i	581.27 ± 0.52 h	9.26 ± 0.03 h	8.37 ± 0.08 k	13.13 ± 0.08 ij	4.64 ± 0.03 h
8	XJ020	299.53 ± 0.38 c	642.50 ± 0.96 m	7.26 ± 0.06 d	3.93 ± 0.01 b	12.65 ± 0.05 h	4.86 ± 0.03 i
9	XJ029	295.37 ± 0.64 b	594.13 ± 0.52 i	8.37 ± 0.03 f	5 .88 ± 0.07 g	13.11 ± 0.06 i	5.21 ± 0.05 jk
10	SX05	331.43 ± 0.28 m	525.53 ± 0.58 e	9.97 ± 0.07 j	15.48 ± 0.14 n	12.07 ± 0.07 g	4.15 ± 0.03 f
11	SX13	451.40 ± 0.72 g	739.00 ± 0.56 n	7.05 ± 0.03 c	17.12 ± 0.02 o	12.11 ± 0.05 g	3.42 ± 0.05 e
12	SS004	504.07 ± 0.43 n	566.80 ± 1.82 g	5.85 ± 0.04 b	7.56 ± 0.04 j	12.09 ± 0.01 g	1.37 ± 0.03 a
13	SAK002	326.23 ± 1.88 f	613.07 ± 1.41 k	4.44 ± 0.06 a	2.73 ± 0.06 a	7.93 ± 0.06 c	3.48 ± 0.05 e
14	SHZ002	302.73 ± 0.81 d	479.97 ± 1.24 b	7.34 ± 0.09 d	10.53 ± 0.05 m	7.12 ± 0.05 b	2.11 ± 0.05 b
15	X1	448.40 ± 1.01 l	401.47 ± 0.38 a	5.74 ± 0.03 b	15.55 ± 0.05 n	6.84 ± 0.04 a	2.17 ± 0.03 b
16	X9	405.00 ± 0.43 j	506.20 ± 0.43 d	9.84 ± 0.04 j	9.13 ± 0.04 l	12.69 ± 0.06 h	4.11 ± 0.05 f
17	SS002	333.27 ± 0.64 g	749.27 ± 0.75 o	8.72 ± 0.05 g	6.26 ±0.03h	9.61 ± 0.06 d	2.95 ±0.03c

The same letter indicated that the values were not significantly different at $P < 0.05$, according to Duncan's multiple range test.

Zn is one of the most abundant nutritionally essential elements in the human body. It is found 85% of the whole Zn in muscle and bone, 11% in the skin and the liver, and the remaining in all the other tissues (Tapiero and Tew, 2003). In this paper, Zn was limited between 6.84 ('X1') and 13.32 mg/100g ('XJ018') in *J. regia*, 7.66 ('Y026') and 13.60 mg/100g ('Y075') in *J. sigillata* (Tab. 1). The mean values of Zn were 11.15 (*J. regia*) and 10.68 mg/100g (*J. sigillata*). As the second most abundant transition metal ion after iron in the human body, Zn is believed to be an essential factor in many biological processes such as brain function and pathology, gene transcription, immune function, and mammalian reproduction (Vallee and Falchuk, 1993; Berg and Shi, 1996; Xu *et al.*, 2010).

The contents of Mn varied between 2.73 mg/100g in 'SAK002' and 17.12 mg/100g in 'SX13' from *J. regia*, 5.47 mg/100g in 'Y075' and 24.51 mg/100g in 'Y075' from *J. sigillata*. In addition, the mean contents of Mn were 8.17 mg/100g in *J. regia* and 14.34 mg/100g in *J. sigillata*. The ratio of the mean contents of Mn in *J. sigillata* to *J. regia* was 1.76. In human body, Mn is an important glandular regulator and involved in the metabolism of glucydes, lypides, and protides. At the same time, Mn plays an important physiological role in the nervous system and cardiac function (Ding *et al., 2014*).

It is well known that Fe is a key kind of element for health and normal performance of organism. In this paper, the contents of Fe were between 4.44 mg/100g in 'SAK002' and 10.45 mg/100g in 'XJ004' from *J. regia*, 9.44 mg/100g in 'Y041' and 13.72 mg/100g in 'Y098' from *J. sigillata*. The mean values of Fe were 8.17 mg/100g in *J. regia* and 11.42 mg/100g in *J. sigillata*. The ratio of the mean contents of Fe in *J. sigillata* to *J. regia* was 1.40.

Cu is a kind of dynamic element and associated with anti-infectious, antiviral, anti-inflammatory. In addition, Cu is an essential trace element in living organisms, functioning as a cofactor for many enzymes (Tapiero *et al.*, 2003; Fraga, 2005). The kernels of walnuts are rich sources of Cu, and the variation of Cu was limited between 1.37 ('SS004') and 5.30 mg/100g ('XJ008') in *J. regia*, 1.46 ('Y099') and 6.91 mg/100g ('Y099') in *J. sigillata*. The mean contents of Cu were 3.72 mg/100g in *J. regia* and 4.93 mg/100g in *J. sigillata*. The ratio of the mean contents of Cu in *J. sigillata* to *J. regia* was 1.33.

As a result, the order depending on the contents of elements in *J. regia* samples was Mg> Ca> Zn> Mn> Fe> Cu, while the order in *J. sigillata* was Mg> Ca> Mn> Fe> Zn> Cu, which was similar with the order of mineral contents (Mg> Ca> Mn> Zn> Fe> Cu) in walnut kernel (*J. regia* L.) from USDA National Nutrient Database (Tab. 2)(*USDA, 2013*).

Compositions of essential amino acids

The compositions of essential amino acids are summarized in Tab. 3. The essential amino acids including phenylalanine, valine, threonine, isoleucine, leucine, methionine, and lysine were detected by HPLC. As can be seen in Tab. 3, the highest content of essential amino acid was leucine in both *J. sigillata* and *J. regia*, while the lowest one was methionine. The order depending on the contents of the essential amino acids (mg/g) in *J. regia* kernels was leucine> isoleucine> valine> phenylalanine> lysine> threonine> methionine, while the order in *J. sigillata* was leucine> isoleucine> lysine> phenylalanine> valine> threonine> methionine.

The mean content of threonine in *J. sigillata* was 16.24 mg/g. In comparison, the highest one was 18.32 mg/g in kernels of 'Y075', which made up to 10.26% of total threonine in *J. sigillata*. While the highest content of threonine in kernels of 'XJ004' was 17.72 mg/g, which reached 7.51% of total threonine in *J. regia* and the mean value of threonine in *J. regia* was 13.87 mg/g. From the analyzed results, the ratio of total contents of threonine in *J. sigillata* to *J. regia* was 1.17.

Tab. 2. Mineral contents of walnut (*J. regia*) kernel (mg/100g, dry weight) from USDA National Nutrient Database and walnuts (*J. regia* and *J. sigillata*)

Minerals	USDA				Walnuts (*J. regia* and *J. sigillata*)			
	Units	Value per 100 grams	No. of Data Points	Std. Error	Units	Value per 100 grams	No. of Data Points	Std. Error
Ca	mg	98	7	3.009	mg	327.13	28	1.572
Fe	mg	2.91	7	0.086	mg	9.45	28	0.424
Mg	mg	158	7	1.204	mg	606.05	28	1.885
Zn	mg	3.09	7	0.083	mg	10.96	28	0.368
Cu	mg	1.586	7	0.026	mg	4.20	28	0.265
Mn	mg	3.414	7	0.122	mg	10.59	28	1.138

Tab. 3. Compositions of essential amino acids (mg/g, dry weight) in kernels from 11 kinds of walnuts (*J. sigillata*) and 17 kinds of walnuts (*J. regia*)

NO.	Samples	Essential amino acid compositions						
		Threonine	Valane	Methionine	Isoleucine	Leucine	Phenylalanine	Lysine
	J. sigillata							
1	Y021	17.65±0.0021 i	4.45±0.0012 a	7.73±0.0012 i	29.08±0.0018 d	38.17±0.0019 e	20.24±0.0018 j	22.93±0.0015 g
2	Y026	13.62±0.0019 a	16.76±0.0033 b	5.47±0.0018 a	25.17±0.0017 a	32.54±0.0023 a	14.55±0.0023 a	18.34±0.0021 cd
3	Y029	17.68±0.0015 i	23.47±0.0021 k	6.16±0.0023 c	33.18±0.0032 k	41.64±0.0018 i	20.69±0.0015 k	19.72±0.0012 ef
4	Y041	16.33±0.0018 f	20.14±0.0018 d	6.73±0.0012 f	31.62±0.0012 g	38.90±0.0094 f	19.18±0.0023 f	18.77±0.0026 de
5	Y042	16.78±0.0018 g	22.39±0.0012 i	6.66±0.0018 e	32.73±0.0023 i	42.04±0.0018 j	20.07±0.0038 i	22.12±0.0015 g
6	Y075	18.32±0.0026 j	23.41±0.0023 j	7.05±0.0019 h	32.85±0.0026 j	40.69±0.0021 h	19.94±0.0020 h	19.10±0.0018 def
7	Y082	14.36±0.0015 b	20.79±0.0012 e	6.53±0.0023 d	30.92±0.0023 e	38.03±0.0019 d	18.16±0.0029 e	15.45±0.0027 b
8	Y096	15.34±0.0018 d	21.25±0.0023 f	5.93±0.0019 b	28.34±0.0018 b	35.35±0.0023 b	17.93±0.0012 c	13.43±0.0993 a
9	Y098	17.42±0.0009 h	21.68±0.0015 h	6.65±0.0023 e	31.96±0.0029 h	40.35±0.0015 g	19.74±0.0020 g	18.77±0.0023 de
10	Y099	15.97±0.0020 e	21.34±0.0021 g	6.82±0.0026 g	31.02±0.0026 f	38.98±0.0015 f	18.08±0.0017 d	19.79±0.0015 f
11	Y101	15.14±0.0018 c	19.14±0.0015 c	6.62±0.0020 e	28.92±0.0026 c	36.67±0.0020 c	17.73±0.0015 b	17.80±0.0028 c
	J. regia							
1	XJ004	17.72± 0.0032 o	23.93± 0.0035 p	7.47±0.0021 l	33.53±0.0012 o	42.93± 0.0026 o	22.26± 0.0024 p	22.33± 0.0021 l
2	XJ005	12.98±0.0020 d	18.11±0.0026 m	5.33±0.0019 a	26.22±0.0023 c	32.98±0.0029 d	15.74±0.0017 c	20.49±0.0029 h
3	XJ008	13.19±0.0015 e	16.29±0.0018 g	6.49±0.0029 g	29.29±0.0012 k	35.73±0.0019 g	17.34±0.0015 i	19.69±0.0015 e
4	XJ012	11.12±0.0045 a	13.54±0.0018 c	7.12±0.0026 k	22.61±0.0012 a	31.02±0.0029 a	14.94±0.0027 a	16.91±0.0026 b
5	XJ017	13.75±0.0020 g	17.57±0.0012 k	6.23±0.0023 e	29.76±0.0026 l	36.01±0.0023 i	16.98±0.0015 f	19.97±0.0032 f
6	XJ018	13.88±0.0017 h	17.25±0.0010 i	6.81±0.0015 j	28.19±0.0017 g	36.28±0.0032 j	17.47±0.0020 j	20.10±0.0012 g
7	XJ019	13.27±0.0026 f	15.73±0.0012 e	5.63±0.0015 b	27.98±0.0009 f	34.75±0.0009 e	17.54±0.0015 k	19.48±0.0015 d
8	XJ020	11.97±0.0009 c	15.98±0.0025 f	5.34±0.0018 a	24.54±0.0018 b	32.23±0.0015 b	15.88±0.0017 d	20.71±0.0029 i
9	XJ029	12.99±0.0026 d	16.95±0.0012 h	6.74±0.0018 i	28.51±0.0020 i	35.74±0.0024 g	17.04±0.0015 g	21.32±0.0024 j
10	SX05	14.10±0.0015 i	17.35±0.0012 j	6.24±0.0015 e	29.93±0.0022 m	37.14±0.0018 k	18.54±0.0025 m	18.88±0.0024 c
11	SX13	14.44±0.0025 j	17.56±0.0009 k	6.04±0.0015 d	29.23±0.0027 j	37.74±0.0012 l	18.62±0.0023 n	20.03±0.0015 fg
12	SS004	13.13±0.0019 e	14.62±0.0020 d	5.73±0.0019 c	26.31±0.0015 d	32.65±0.0022 c	16.04±0.0015 e	16.78±0.0021 a
13	SAK002	15.68±0.0015 m	18.34±0.0015 o	6.60±0.0033 h	27.19±0.0012 e	35.81±0.0023 h	17.25±0.0023 h	22.22±0.0026 k
14	SHZ002	15.76±0.0009 n	18.26±0.0018 n	6.49±0.0015 g	28.29±0.0015 h	35.61±0.0026 f	18.22±0.0033 l	22.23±0.0021 k
15	X1	11.30±0.0021 b	2.74±0.0018 a	5.58±0.0018 b	44.46±0.0024 p	41.96±0.0020 n	15.48±0.0013 b	27.47± 0.0026 m
16	X9	15.50±0.0018 l	5.54±0.0015 b	6.63±0.0032 h	47.53±0.0012 q	50.19±0.0441 p	18.85±0.0023 o	29.43±0.0023 n
17	SS002	15.09±0.0023 k	17.92±0.0035 l	6.34±0.0018 f	30.43±0.0009 n	38.26±0.0026m	18.25±0.0022 l	21.30±0.0018 j

The same letter indicated that the values were not significantly different at $P < 0.05$, according to Duncan's multiple range test.

The following one is valine, which can accelerate the wound healing, treatment of liver failure, raise blood sugar levels, and increase the content of *growth hormone*. In Tab. 3, the highest content of valine was 23.47 mg/g for 'Y029', which reached 10.93% of the total aline in the kernels of 'Y029' from *J. sigillata*, and the mean value of valine was 19.53 mg/g for *J. sigillata*. At the same time, the highest content (23.93 mg/g, 'XJ004') of valine in *J. regia* was very similar to that in *J. sigillata*. While the highest content was just up to 8.94% of the total content of valine and the mean value was 15.75 mg/g for *J. regia*. The ratio of the mean values of valine in *J. sigillata* and *J. regia* was 1.24.

Moreover, Tab. 3 shows that the lowest content of essential amino acids was methionine. The mean value of methionine was just 6.58 mg/g in *J. sigillata*, while the highest content of methionine was 7.73 mg/g in 'Y021'. Meanwhile, the highest content of methionine was 7.47 mg/g ('XJ004') in *J. regia* and the mean value was 6.28 mg/g, which was very similar to the mean content of methionine in *J. sigillata*. Cysteine and methionine are 2 kinds of sulfur-containing proteinogenic amino acids. Loss of methionine has been linked to senile greying of hair and leads to the accumulation of hydrogen peroxide in hair follicles, which reduces the effectiveness of tyrosinase to loss color of hair (Wood *et al., 2009*)

What's more, isoleucine was the second highest content of essential amino acids in both *J. sigillata* and *J. regia*. The highest values of isoleucine were 33.18 mg/g in 'Y029' from *J. sigillata* and 47.53 mg/g in 'X9' from *J. regia*. The mean contents of isoleucine were 30.53 mg/g in *J. sigillata* and 30.24 mg/g in *J. regia*. A research showed that methionine supplement in the diet specifically increased mitochondrial ROS production and mitochondrial DNA oxidative damage in rat liver mitochondria offering a plausible mechanism for its hepatotoxicity (Gomez *et al., 2009*) In all the determined essential amino acids, the mean content of leucine was the highest, corresponding the research by Wu *et al.* (2007).

The highest content of leucine in *J. sigillata* was 42.04 mg/g ('Y042'), while in *J. regia* that was 50.19 mg/g ('X9'). In addition, the mean values of leucine were 38.49 mg/g (*J. sigillata*) and 36.88 mg/g (*J. regia*). As a dietary supplement, leucine has been found to slow the degradation of muscle tissue by increasing the synthesis of muscle proteins in aged rats (Combaret *et al., 2005*). However, the supplementation of leucine with long-term didn't increase muscle mass or strength in healthy elderly men (Verhoeven *et al., 2009*). Until now, the dietary supplemental of leucine has not be associated the prime reason for muscular growth or optimal maintenance for the entire population.

Furthermore, phenylalanine is an essential amino acid and required to assure normal anabolism (Pimentel *et al.,* 2014). Phenylalanine as a nutritional supplement is used in the manufacture of food and drink products for its famous functions of analgesic and antidepressant. In this study, the content of phenylalanine in 'Y029' (*J. sigillata*) was the highest (20.69 mg/g) and the mean value of phenylalanine was 18.76 mg/g. While the highest content of phenylalanine in *J. regia* was 22.26 mg/g in 'XJ004' and the mean level of phenylalanine was 17.44 mg/g. In addition, phenylalanine hydroxylase (PAH) is responsible for converting dietary phenylalanine into tyrosine in the liver. However, when PAH is lacked or changed lowly, people will be dangerous for phenylketonuria (PKU). Therefore, In order to reach safe concentrations of phenylalanine in blood, PKU patients need to follow a Phe-restricted diet (Giovannini *et al.,* 2007; Rocha *et al.,* 2012).

Finally, as shown in Tab. 3, the highest content of lysine in *J. sigillata* was 22.93 mg/g ('Y021'), while the highest one in *J. regia* was 29.43 mg/g ('X9'). Lysine is a precursor compound of arginine and histidine. In view of lysine, the high content of lysine negatively influenced the uptake of arginine in vivo (Wu and Meininger, 2002). Lysine residues were essential for both receptor-dependent proinflammatory and receptor- independent cytolytic activities (Cheung et al., 2014). In comparison, the results for the composition of essential amino acid in *J. sigillata* and *J. regia* from China were consistent with several published reports including those for Spanish almonds (Calixto *et al.,* 1981), New Zealand hazelnuts (Savage and McNeil, 1998), Turkish Tombul hazelnut (Alasalvar *et al.,* 2003), Chilean hazelnut (Villarroel *et al., 1987*), almond, pecan, pine nut, and pistachio (*Ruggeri et al.,* 1998). What's more, it was favorable with the data of amino acid composition for edible nuts from USDA (Tab. 4)(USDA, 2013). It can be seen that the ratio of the contents of essential amino acids in both *J. sigillata* and *J. regia* to the data from USDA were more than two-fold, even the highest ratio of isoleucine nearly reached five-fold. Therefore, the kernels of *J. sigillata* and *J. regia* were considered to be good sources of essential amino acids. In addition, researchers reported that lysine was the first limiting essential amino acid in total walnut proteins (Ruggeri *et al.,* 1998; Sze-Tao and Sathe, 2000) Compared to the recommended essential amino acid pattern from WHO/FAO for an adult (Joint, 2007), all the determined kernels contain adequate amounts of all the essential amino acids.

Tab. 4. Essential amino acids contents of walnut kernel (mg/g, dry weight) from USDA National Nutrient Database and walnuts (*J. regia* and *J. sigillata*)

Essential amino acids	USDA	*J. sigillata*	*J. regia*	Ratios	
				J. sigillata/USDA	*J. sigillata*/USDA
Threonine	5.96	16.24	13.87	2.72	2.33
Isoleucine	6.25	30.53	30.24	4.88	4.84
Leucine	11.70	38.49	36.88	3.29	3.15
Lysine	4.24	18.75	21.14	4.42	4.99
Methionine	2.36	6.58	6.28	27.9	2.66
Phenylalanine	7.11	18.76	17.44	2.64	2.45
Valine	7.53	19.53	15.75	2.59	2.09

Conclusions

The kernels of walnuts contained relatively high amounts of Mg, Ca, Zn, Mn, Fe, and Cu, in which Mg and Ca were the major mineral elements. The contents of mineral elements were different between *J. regia* and *J. sigillata*. The order of nutritive mineral elements depending on their content (mg/100g) of samples was Mg> Ca> Zn> Mn> Fe> Cu in *J. regia*, while the order in *J. sigillata* was Mg> Ca> Mn> Fe> Zn> Cu. In comparison, leucine was the highest content of essential amino acids in all the investigated samples. The order depending on the content (mg/g) of the essential amino acids in *J. regia* samples was leucine> isoleucine> valine> phenylalanine> lysine> threonine> methionine, while the order in *J. sigillata* was leucine> isoleucine> lysine> phenylalanine> valine> threonine> methionine. Moreover, 'Y029' in *Juglans sigillata* and 'XJ004' in *Juglans regia* provided the best profiles of mineral elements and essential amino acids in comparison to others. Importantly, the kernel of walnut is good source of health food and dietary supplement for mineral elements and essential amino acids.

Acknowledgement

Authors are grateful to the financial support from the State Forestry Administration (201004027).

References

Alasalvar C, Shahidi F, Liyanapathirana CM, Ohshima T (2003). Turkish tombul hazelnut (*Corylus avellana* L.). 1. Compositional characteristics. J Agric Food Chem 51(13):3790-3796.

Albuquerque TG, Costa HS, Sanches-Silva A, Santos M, Trichopoulou A, D'Antuono F, Alexieva I, Boyko N, Costea C, Fedosova K (2013). Traditional foods from the Black Sea region as a potential source of minerals. J Sci Food Agric 93(14):3535-3544.

Bryant RJ, Cadogan J, Weaver CM (1999). The new dietary reference intakes for calcium: implications for osteoporosis. J Am Coll Nutr 18(Sup5):406S-412S.

Caglarirmak N (2003). Biochemical and physical properties of some walnut genotypes (*Juglans regia* L.). Food/Nahrung, 47(1):28-32.

Calixto FS, Bauza M, Martinez de Toda F, Argamenteria A (1981). Amino acids, sugars, and inorganic elements in the sweet almond (*Prunus amygdalus*). J Agric Food Chem 29(3):509-511.

Cheung GY, Kretschmer D, Queck SY, Joo H-S, Wang R, Duong AC, Nguyen TH, Bach T-HL, Porter AR, DeLeo FR (2014). Insight into structure-function relationship in phenol-soluble modulins using an alanine screen of the phenol-soluble modulin (PSM) α3 peptide. FASEB J 28(1):153-161.

Combaret L, Dardevet D, Rieu I, Pouch M-N, Béchet D, Taillandier D, Grizard J, Attaix D (2005). A leucine-supplemented diet restores the defective postprandial inhibition of proteasome-dependent proteolysis in aged rat skeletal muscle. J Physiol 569(2):489-499.

Ding Y, Luo Y, Fu J (2014). Effects of Mn (II) on peroxynitrite nitrifying fibrinogen. Bio-med Mater Eng 24(1):901-907.

Fraga CG (2005). Relevance, essentiality and toxicity of trace elements in human health. Mol Aspects Med 26(4):235-244.

Fürst P, Stehle P (2004). What are the essential elements needed for the determination of amino acid requirements in humans? J Nutr 134(6):1558S-1565S.

Giovannini M, Verduci E, Salvatici E, Fiori L, Riva E (2007). Phenylketonuria: dietary and therapeutic challenges. J Inherit Metab Dis 30(2):145-152.

Gomez J, Caro P, Sanchez I, Naudi A, Jove M, Portero-Otin M, Lopez-Torres M, Pamplona R, Barja G (2009). Effect of methionine dietary supplementation on mitochondrial oxygen radical generation and oxidative DNA damage in rat liver and heart. J Bioenerg Biomembr 41(3):309-321.

Hao Y, Wang K, Wang S, Zhuang Y (2002). Analysis of the protein and fat among pricocious varieties of walnut. Food Science 23(10):123-125.

Imura K, Okada A (1998). Amino acid metabolism in pediatric patients. Nutrition 14(1):143-148.

Joint W (2007). Protein and amino acid requirements in human nutrition. World Health Organization Technical Report Series 935:1-265.

Lavedrine F, Ravel A, Villet A, Ducros V, Alary J (2000). Mineral composition of two walnut cultivars originating in France and California. Food Chem 68(3):347-351.

Martínez ML, Labuckas DO (2010). Lamarque AL and Maestri DM, Walnut (*Juglans regia* L.): genetic resources, chemistry, by-products. J Sci Food Agric 90(12):1959-1967.

Nash SD, Westpfal M (2005). Cardiovascular benefits of nuts. Am J Cardiol 95(8):963-965.

Özcan MM, İman C, Arslan D (2010). Physico-chemical properties, fatty acid and mineral content of some walnuts (*Juglans regia* L.) types. Agricultural Sciences 1(2):62-67.

Ozrenk K, Javidipour I, Yarilgac T, Balta F, Gündogdu M (2012). Fatty acids, tocopherols, selenium and total carotene of pistachios (*P. vera* L.) from Diyarbakir (Southeastern Turkey) and walnuts (*J. regia* L.) from Erzincan (Eastern Turkey). Food Sci Technol Int 18(1):55-62.

Patel G (2005). Essential fats in walnuts are good for the heart and diabetes. J Am Diet Assoc 105(7):1096-1097.

Pimentel FB, Alves RC, Costa AS, Fernandes TJ, Torres D, Almeida MF, Oliveira M (2014). Nutritional composition of low protein and phenylalanine-restricted dishes prepared for phenylketonuric patients. LWT-Food Sci Technol 57(1):283-289.

Qin Z, Guo X, Lin Y, Chen J, Liao X, Hu X, Wu J (2012). Effects of high hydrostatic pressure on physicochemical and functional properties of walnut (*Juglans regia* L.) protein isolate. J Sci Food Agric 93(5):1105-1111.

Zhang Q, Ding X, Cheng Z, Kan J, Zhao G (2003). The

development of research on walnut protein. Journal of Cereals and Oils 5:21-23.

Reeds PJ (2000). Dispensable and indispensable amino acids for humans. J Nutr 130(7):1835S-1840S.

Rocha JC, Spronsen FJV, Almeida MF, Soares G, Quelhas D, Ramos E, Guimarães JT, Borges N (2012). Dietary treatment in phenylketonuria does not lead to increased risk of obesity or metabolic syndrome. Mol Genet Metab 107(4):659-663.

Ruggeri S, Cappelloni M, Gambelli L, Nicoli S, Carnovale E (1998). Chemical composition and nutritive value of nuts grown in Italy. Ital J Food Sci 10(3):243-252.

Sabate J, Fraser GE, Burke K, Knutsen SF, Bennett H, Lindsted KD (1993). Effects of walnuts on serum lipid levels and blood pressure in normal men. New Engl J Med 328(9):603-607.

Savage G (2001). Chemical composition of walnuts (*Juglans regia* L.) grown in New Zealand. Plant Food Hum Nutr 56(1):75-82.

Savage G, McNeil D (1998). Chemical composition of hazelnuts (*Corylus avellana* L.) grown in New Zealand. Int J Food Sci Nutr 49(3):199-203.

Spies JR, Chambers DC (1948). Chemical determination of tryptophan. Anal Chem 20(1):30-39.

Sze-Tao KWC, Sathe SK (2000). Walnuts (*Juglans regia* L): proximate composition, protein solubility, protein amino acid composition and protein in vitro digestibility. J Sci Food Agric 80(9):1393-1401.

Tapiero H, Tew KD (2003). Trace elements in human physiology and pathology: zinc and metallothioneins. Biomed Pharmacothe 57(9):399-411.

Tapiero H, Townsend D, Tew K (2003). Trace elements in human physiology and pathology. Copper. Biomed Pharmacother 57(9):386-398.

Touyz RM (2004). Magnesium in clinical medicine. http://www.ncbi.nlm.nih.gov/pubmed/14977544.

U.S. Department of Agriculture Research Service (2013). [Online]. USDA National Nutrient Database for Standard Reference, Release 26. Nutrient Data Laboratory Home. Available: ttp://www.ars.usda.gov/Services/docs.htm?docid=8964.

Vallee BL, Falchuk KH (1993). The biochemical basis of zinc physiology. Physiol Rev 73(1):79-118.

Verhoeven S, Vanschoonbeek K, Verdijk LB, Koopman R, Wodzig WK, Dendale P, van Loon LJ (2009). Long-term leucine supplementation does not increase muscle mass or strength in healthy elderly men. Am J Clin Nutr 89(5):1468-1475.

Villarroel M, Biolley E, Schneeberger R, Ballester D, Ramirez S (1987). Amino acid composition of Chilean hazel nuts. Food Chem 25(2):155-158.

Warmund MR, Elmore J, Drake M, Yates MD (2009). Descriptive analysis of kernels of selected black and Persian walnut cultivars. J Sci Food Agric 89(1):117-121.

Wood J, Decker H, Hartmann H, Chavan B, Rokos H, Spencer J, Hasse S, Thornton M, Shalbaf M, Paus R (2009). Senile hair graying: H_2O_2-mediated oxidative stress affects human hair color by blunting methionine sulfoxide repair. FASEB J 23(7):2065-2075.

Wu G, Meininger CJ (2002). Regulation of nitric oxide synthesis by dietary factors. Annu Rev Nutr 22(1):61-86.

Wu K, Xiao Q, Tang L, Pang Y, Wei L, Zhang L (2007). Differences analysis on the contents of crude fat and amino acids in kernel of walnut. Nonwood Forest Research 25(2):15-18.

Xie X, Smart TG (1991). A physiological role for endogenous zinc in rat hippocampal synaptic neurotransmission. Nature 349:521-524.

Xu Z, Yoon J, Spring DR (2010). Fluorescent chemosensors for Zn^{2+}. Chem Soc Rev 39(6):1996-2006.

Young VR (1994). Adult amino acid requirements: the case for a major revision in current recommendations. J Nutr 124(Sup8):1517S.

Sensory Quality of 'Cherry' Tomatoes in Relation to 1-MCP Treatment and Storage Duration

Marek GAJEWSKI*, Katarzyna MAZUR, Jadwiga RADZANOWSKA, Katarzyna KOWALCZYK, Monika MARCINKOWSKA, Klaudyna RYL, Karolina KALOTA

*Warsaw University of Life Sciences – SGGW, Department of Vegetable and Medicinal Plants, Nowoursynowska 166, 02-787 Warszawa, Poland; marek_gajewski@sggw.pl (*corresponding author)*

Abstract

1-MCP (1-methylcyclopropene), the ethylene receptors blocker, is used in horticultural practice for prolonging 'shelf-life' of several species of fruits and vegetables. The objective of this study was to determine the effect of 1-MCP treatment on sensory characteristics of 'cherry' type tomatoes (*Solanum lycopersicum* L. var. *cerasiforme*) after 4-week storage. In the experiment the fruits were harvested at pink and light-red fruit stages (3rd and 5th stage, according to USDA classification), and stored for 3 and 4 weeks at 12 °C, then moved to 20 °C for 2 days to obtain marketable maturity. 1-MCP was applied after harvest in concentration of 1 ppm, for 12 hours at 18 °C. The plants belonging to the 'Dasher F_1' cultivar were grown in a greenhouse in rockwool slabs and fruits were harvested in June. Sensory analysis was performed in expert panel with Quantitative Descriptive Analysis (QDA), using 12 sensory descriptors concerning smell, taste, flavour and texture of the fruits. Also semi-consumer test of liking was carried out. As the control freshly harvested red fruits at optimal consumer's maturity were used. The 1-MCP treated fruits harvested in the light-red stage were of better sensory quality than the untreated ones, similar to freshly harvested, but in the case of the pink fruits, the quality was rated lower. It can be concluded that 1-MCP treatment was effective in delaying ripening of the fruits by better keeping their firmness and colour, however in the case of fruits harvested in earlier stage, overall sensory quality and consumer acceptance was scored lower than for the freshly harvested ones, mainly due to more intensive sour taste and lower sweetness.

Keywords: tomato fruit, *Solanum lycopersicum*, sensory attributes, quantitative descriptive analysis, consumer test

Introduction

Tomato, one of the most popular vegetables worldwide, is consumed for its sensory value, attractive appearance and nutritional value. Recently, 'cherry' type tomatoes, with small fruits became popular in Europe. Tomato is a climacteric fruit and its ripening process can be accelerated by ethylene. There are reports pointing out potential benefits of 1-MCP (1-methylcyclopropene) in postharvest of fruits and vegetables resulting in delaying their ripening or senescence processes. 1-MCP was introduced to horticultural practice in 1999 in the USA. For vegetables and fruits it is used as the formulation known as SmartFresh^{TM}. 1-MCP is safe for human, quickly diffuses from the plant tissue after the treatment (Blankenship and Dole, 2003; Watkins, 2006). It is believed that mechanism of 1-MCP action lays in blocking ethylene receptors in plant cells. It results in delaying fruits and vegetables ripening or senescence and therefore extends their storage ability (Fan and Mattheis, 2000; Wills and Ku, 2002; Watkins, 2006). The effect depends on 1-MCP concentration and time of the treatment. It was reported that 1-MCP reduced production of ethylene, cell respiration, restrained fruit colour changes, softening, taste and aroma degradation as well as diseases of fruits (Moretti *et al.*, 2002; Wills and Ku, 2002; Mostofi *et al.*, 2003). There are reports that 1-MCP treatment markedly affects also ripening of tomato fruits by inhibiting ethylene production (Opiyo and Ying, 2005; Beno-Moualem *et al.*, 2004; Tassoni *et al.*, 2006; Watkins, 2006; Tadesse *et al.*, 2012), significantly reduces respiration rates in the fruits (Guillén *et al.*, 2007), delays change of fruits colour and reduces softening and decay of stored fruits (Guillén *et al.*, 2006). Between 1-MCP treated and untreated ripe tomato fruits no differences in titratable acidity, soluble solids, fructose and glucose contents were reported (Opiyo and Ying, 2005; Ergun *et al.*, 2006; Guillén *et al.*, 2006). Some authors pointed out possibilities of use of the treatment for prolonging shelf life of tomatoes. According to Mir *et al.* (2004) one treatment with 1-MCP prolonged ripening of both breaker and pink tomato fruits for 4 to 6 days and delayed loss of firmness of the fruits. Vijay *et al.* (2010) reported that the treatment is effective even for tomato fruits stored at high temperatures (25-30 °C). According to Huber (2008) 1-MCP treatment of tomatoes should be done when the fruits start to ripe, what results in their better quality. However, literature data on the effectiveness of 1-MCP treatment of tomato fruits in different ripening stages are insufficient, especially as regards 'cherry' type tomatoes.

Tomato fruits are chilling sensitive, so the optimum temperature for storage of tomato fruits harvested at the mature green or breaker stage of maturity is 12.5-13 °C (Roberts *et al.*, 2002; Getinet *et al.*, 2008; Suslow and Cantwell, 2009). In the experiments of Gajewski (2000), tomatoes harvested at the breaker stage of maturity kept well their quality for 4 weeks at the temperature of 12-13 °C. Due to low storage ability of tomato fruits, various methods of prolonging storage period for the fruits are investigated, including controlled atmosphere storage and heat treatment (Sayed *et al.*, 2004; Toor and Savage, 2006).

Sensory characteristics of tomato fruit is a very important component of its quality and decides on a high degree on consumer's acceptance of the fruit (Azodanlou *et al.*, 2003). For the unification of sensory methods ISO standards were approved (Anonymous, 1996, 1999). There are some reports on relationship between sensory quality and physical or chemical characteristics of different fruits (Gajewski and Arasimowicz, 2004), but just a few reports concerning sensory characteristics of 'cherry' tomatoes. However, some authors indicate that negative influence of 1-MCP treatment on sensory characteristics of tomatoes is possible (Cliff *et al.*, 2009). It is not clear if it affects consumer acceptance of the fruit. Tomatoes of 'cherry' type are a perishable crop, so method of prolonging their shelf life would be suitable for practice.

The objective of this study was to determine the effect of 1-MCP treatment on sensory quality traits of 'cherry' tomatoes (*Solanum lycopersicum* L. var. *cerasiforme*). The results of the study could explain in what degree this method of postharvest treatment positively or adversely affects sensory quality of the fruits and their consumer acceptance.

Materials and methods

The work was carried out at Warsaw University of Life Sciences. The cultivar used in the study was 'Dasher F_1'. This cultivar is a standard 'cherry' type one and is characterized by small (15-17 g weight) red fruits, which are slightly elongated in shape. The fruits were obtained from tomato plants grown in the experimental greenhouse with controlled climate conditions (temperature and humidity). Standard coconut slabs were used as the growing medium in the greenhouse. Tomato seeds were sown at the beginning of January, then the seedlings were transplanted to the final place of cultivation in February, at the density of 2.7 plants per m^2. Drop irrigation system was applied in the greenhouse for watering and fertilizing the plants. Nutrients concentrations ($mg{\cdot}dm^{-3}$) in the feeding solution were kept on optimal level for tomato plants, and were as follows: N-NO_3 – 210, P – 60, K – 340, Mg – 50, Ca – 200, Fe – 2.0, Mn – 0.6, B – 0.3, Cu – 0.15, Zn – 0.3, Mo – 0.05.

The fruits for the experiment were harvested at pink fruit and light-red fruit stages (3rd and 5th maturity stages, according to USDA classification), selected in respect of uniformity, and stored for 3 and 4 weeks at experimental cold store, at the temperature of 12 °C and 85% RH, than moved for 2 days to 20 °C to obtain marketable maturity. 1-MCP was applied in the form of SmartFresh™ (Agrofresh Inc.) formulation directly after harvest, in special air-tight steel chambers (capacity of 1 m^3) in concentration of 1 $\mu L{\cdot}L^{-1}$, for 12 hours at 18 °C. The conditions of the treatment were chosen according to previous own experiments and literature data (Wills and Ku, 2002; Watkins, 2006). As the control in sensory evaluation of stored fruits harvested in both stages of maturity freshly harvested fruits, at optimal consumer's maturity (i.e. red, ripe fruits) were used.

Sensory analysis was carried out in sensory laboratory of the Department, equipped according to ISO standard (Anonymous, 1999). The established expert panel, consisting of 12 persons, previously selected and trained according to ISO standard (Anonymous, 1996), carried out the evaluation using Quantitative Descriptive Analysis (QDA) method. At the first part of the QDA procedure 'brainstorming' sessions were run to determine sensory attributes for the tomatoes. Panelists received samples of 'cherry' tomatoes, varying in sensory properties, and generated a set of 12 descriptors (Tab. 1) for odour, appearance, texture and flavour/taste of the fruits. Also, overall quality impression was taken into account as a separate descriptor.

Tab. 1. Quality descriptors used in the quantitative descriptive analysis of 'cherry' tomatoes

No.	Descriptors	Definition	Anchoring points
1	'tomato' odour	characteristic odour of tomato	low intensity – high intensity
2	foreign odour	unusual odour for tomato fruit	low intensity – very intensity
3	flesh colour intensity	red colour intensity for half-cut fruit	pale red – dark red
4	skin thickness	impression during chewing the fruit	thin skin – thick skin
5	flesh firmness	impression during chewing the fruit	soft fruit – firm fruit
6	flesh meliness	impression during chewing the fruit	low – high meliness
7	flesh juiciness	impression during chewing the fruit	not juicy – very juicy
8	'tomato' flavour	characteristic flavour for tomato	low intensity – high intensity
9	sour taste	basic taste	low intensity – high intensity
10	sweet taste	basic taste	low intensity – high intensity
11	bitter taste	basic taste	none – very intensive
12	foreign taste	unusual taste for tomato fruit	none – very intensive
13	Overall quality	general sensory quality impression	low quality – high quality

The main analysis was performed in separate booths, equipped with computers. Every assessor received samples of the tomatoes. The sample replicates were served in coded small plastic containers, covered by lids, in a totally random order. The assessments were marked on the monitors on non-structural lines with anchoring points (described in Tab. 1). Results were converted to numerical values (from 0 to 10 points). The analysis was performed during two independent sessions for each of maturity stage. For coding samples and for initial processing of the data, Analsens software was used. Semi-consumer test of 'cherry' tomatoes liking was also performed alongside with QDA procedure. For this evaluation non-structural scale was also used, with anchoring points: 'I do not like it' – 'I like it very much'.

For the analysis of variance Anova was applied (StatgraphicsPlus 4.1™ software) and Tukey's HSD test was used to determine which values differ significantly at $\alpha \leq 0.05$. Correlation coefficients between scores for overall quality impression or for consumer liking and for several sensory quality descriptors were also calculated. For the calculation all single scores of assessors were taken into account and significance levels were determined. S

Results and discussion

Storage conditions concerning temperature and relative humidity used in the study for the tomato fruits were in accordance with recommendations (Saltweit, 1997) for not fully ripen tomatoes, and enabled slow ripening of the fruits during the storage. The QDA (Quantitative Descriptive Analysis) method used in the study is often applied for detailed description of sensory characteristics of food products. In this method an assumption is done that sensory quality is a complex of many descriptors (attributes), which can be individually estimated by a consumer (Meilgaard *et al.*, 1999). In the analysis 12 sensory quality attributes were used, and, additionally, overall sensory quality impression was also scored. The set of sensory attributes elaborated by the panelists related to all aspects of sensory quality of tomatoes. The attributes are given in Tables 2-5 and devided on those concerning odour, mechanical properties (texture), colour and taste/flavour. Scores for the attributes are given in the tables for both storage periods separately (i.e. for 3 and 4 weeks). It should be noted that control fruits (freshly harvested fruits in red maturity stage) were used as a 'standard' in each of the experts sessions and since the fruits could vary in quality, the scores for control samples used for pink or light-red stored fruits evaluation might vary. The discussion of results obtained in the study is restricted, since there is shortage of data in literature on the influence of 1-MCP on sensory characteristics of 'cherry' tomatoes.

Tab. 2. Scores for sensory attributes concerning odour, colour and texture for 'cherry' tomato fruits after 3-week storage (scale 0-10)

Maturity stage at harvest	Treatment	Sensory attributes						
		'Tomato' odour	Foreign odour	Flesh colour	Skin thickness	Flesh firmness	Flesh mealiness	Flesh juiciness
Light-red	Control fruit	6.83 d	0.04 a	6.18 c	5.52 a	3.79 a	1.19 a	7.69 c
	Untreated	6.33 c	0.00 a	6.79 d	5.75 a	4.81 ab	1.29 a	7.28 bc
	1-MCP treated	6.18 c	0.19 a	5.40 b	5.94 ab	5.05 ab	1.45 a	7.16 b
	Means	6.45 B	0.08 A	6.12 A	5.74 A	4.55 A	1.31 A	7.38 A
Pink	Control fruit	5.69 b	0.25 a	7.11 d	5.07 a	3.29 a	1.36 a	7.41 bc
	Untreated	5.41 a	0.32 a	6.17 c	6.07 ab	4.36 ab	1.60 a	7.02 b
	1-MCP treated	5.48 a	0.05 a	4.21 a	6.64 b	5.78 b	1.18 a	6.17 a
	Means	5.53 A	0.21 A	5.83 A	5.93 A	4.48 A	1.38 A	6.87 A

Note to Tab. 2-6: different letters in each column are used for marking significant differences at $\alpha \leq 0.05$; small letters - between objects, capitals - between means of maturity stages

The results of sensory analysis for attributes concerning odour, colour and mechanical properties of tomato fruits are shown in Tab. 2 and Tab. 3. After 3 weeks of storage the fruits harvested in pink fruit stage were characterized by a lower intensity of typical 'tomato' aroma compared to the fruits harvested in the red stage (Tab. 2).

The influence of 1-MCP treatment on this quality attribute was insignificant for both maturity stages compared to stored untreated fruits. However, freshly harvested red fruits (control) showed a higher intensity of the tomato odour. With extending storage period up to 4 weeks, the differences between treated and untreated fruits remained insignificant (with a tendency to a higher 'tomato' odour intensity for 1-MCP treated fruits), but the difference between stored and freshly harvested control fruits increased (Tab. 3).

Foreign odour (untypical odour of tomato fruit) was detected at a very low level in all fruit samples, independent on maturity stage, but increased a little after 4 weeks of storage. There were found significant differences between experimental objects in the case of flesh colour. For the fruits stored for 3 and 4 weeks the paler red colour was detected in 1-MCP treated samples than in untreated ones or in the control. No significant differences between maturity stages as regards skin thickness attribute were found. The thickest skin was detected in 1-MCP treated pink fruits, both after 3 and 4 weeks of storage. In the case of flesh firmness a tendency to higher firmness for 1-MCP treated stored fruits compared to the untreated ones was found, and for the fruits stored for 4 weeks their firmness was rated significantly lower than for the freshly harvested ones.

This finding is in accordance with reports of Moretti *et al.* (2002), Wills and Ku (2002) and Mostofi *et al.* (2003) pointing out that 1-MCP treatment delays tomato fruit ripening, including their softening. With extending storage period from 3 to 4 weeks flesh firmness showed a tendency to decrease, which indicated that ripening process in 1-MCP treated fruits was still going slowly.

As regards flesh mealiness, no significant differences were found for experimental objects, both after 3 and 4 weeks of storage. This attribute was rated on a very low level, for the fruits stored for 4 weeks, especially. Flesh juiciness was rated as the highest for control of both light-red and pink fruits evaluated after 3 weeks of storage, but after 4 weeks significant differences were detected between untreated pink fruits and other objects (except the control for the pink fruits).

Tab. 3. Scores for sensory attributes concerning odour, colour and texture for 'cherry' tomato fruits after 4-week storage (scale 0-10)

Maturity stage at harvest	Treatment	Sensory attributes						
		'Tomato' odour	Foreign odour	Flesh colour	Skin thickness	Flesh firmness	Flesh mealiness	Flesh juiciness
Light-red	Control fruit	4.95 a	0.43 ab	4.28 a	6.40 a	3.84 c	0.91 a	6.47 b
	Untreated	5.29 ab	0.83 b	5.45 b	6.45 a	2.25 a	0.50 a	6.49 b
	1-MCP treated	5.49 b	0.70 b	4.77 a	6.74 ab	2.87 b	0.35 a	6.55 b
	Means	5.24 A	0.65 A	4.83 A	6.53 A	3.99 A	0.59 A	6.50 A
Pink	Control fruit	5.43 b	0.02 a	5.04 ab	6.24 a	3.29 bc	0.99 a	6.28 ab
	Untreated	4.96 a	1.08 b	5.45 b	7.27 b	2.73 b	1.20 a	5.96 a
	1-MCP treated	5.19 ab	0.88 b	4.29 a	7.41 b	2.98 b	0.60 a	6.59 b
	Means	5.19 A	0.66 A	4.93 A	6.97 A	3.00 A	0.93 A	6.28 A

Results of sensory analysis concerning attributes of flavour/taste are presented in Tab. 4 and Tab. 5. It can be seen that intensity of this characteristic flavour for tomato was rated as the highest for control in evaluations performed both after 3 and 4 weeks of storage, but the differences between objects were not always statistically significant. The intensity of this flavour showed a tendency to decrease with increasing time of storage. Fruits harvested in pink stage of maturity were characterized by a lower tomato flavour intensity after storage than fruits harvested in later, light-red stage. It should be noted that sour taste, when present in a high intensity, is an undesirable trait in tomato fruits and usually also in most other foods. In the study it was found that 1-MCP treated fruits were characterized by slightly higher sour taste than the untreated fruits, both after 3 and 4 weeks of storage. The control fruits showed the lowest intensity of this taste. By contrast, sweet taste intensity was the highest for control fruits. It is not a surprise because these fruits ripened on the plant to obtain full maturity and they could accumulate more assimilates. As regards the effect of 1-MCP treatment, a tendency was found (not significant for 3 weeks of storage) to lower sweet taste intensity for the 1-MCP treated fruits than for the untreated ones. As we previously reported (Mazur *et al.*, 2013), 1-MCP treated 'cherry' tomatoes showed significantly higher titratable acidity and a little lower sugars content after two-week storage, so results of this sensory evaluation probably reflects the phenomenon. There were found no differences in bitter taste intensity between the objects, including the control fruits, which is a positive finding since this taste is an undesirable one in most foods. The same can be said as regards foreign taste, which was detected by the panelists on a very low level in all fruit samples.

Tab. 4. Scores for sensory attributes concerning flavour/taste and overall quality for 'cherry' tomato fruits after 3-week storage (scale 0-10)

Maturity stage at harvest	Treatment	Sensory attributes					
		'Tomato' flavour	Sour taste	Sweet taste	Bitter taste	Foreign taste	Overall quality
Light-red	Control fruit	6.97 b	3.37 a	5.57 c	0.15 a	0.00 a	7.73 c
	Untreated	6.67 ab	3.33 a	5.08 c	0.33 a	0.00 a	7.35 bc
	1-MCP treated	6.22 ab	4.77 b	4.06 b	0.29 a	0.00 a	7.70 c
	Means	6.62 A	3.82 A	4.90 A	0.26 A	0.00 A	7.59 B
Pink	Control fruit	6.90 b	3.15 a	5.39 c	0.12 a	0.00 a	7.05 b
	Untreated	6.23 ab	4.18 ab	3.70 b	0.14 a	0.03 a	6.99 ab
	1-MCP treated	5.92 a	5.44 c	2.87 a	0.08 a	0.07 a	6.40 a
	Means	6.35 A	4.26 A	3.99 A	0.11 A	0.03 A	6.81 A

Tab. 5. Scores for sensory attributes concerning flavour/taste and overall quality for 'cherry' tomato fruits after 4-week storage (scale 0-10)

Maturity stage at harvest	Treatment	Sensory attributes					
		'Tomato' flavour	Sour taste	Sweet taste	Bitter taste	Foreign taste	Overall quality
Light-red	Control fruit	6.38 c	4.49 ab	3.99 cd	0.65 a	0.60 a	7.12 d
	Untreated	5.30 b	3.20 a	3.45 bc	0.24 a	0.71 a	4.75 b
	1-MCP treated	5.25 b	4.58 b	3.12 ab	0.30 a	0.30 a	4.98 b
	Means	5.64 A	4.09 A	3.52 A	0.40 A	0.54 A	5.62 A
Pink	Control fruit	5.72 b	3.27 a	4.36 d	0.33 a	0.01 a	6.05 c
	Untreated	4.23 a	3.94 a	3.28 ab	0.34 a	0.22 a	3.95 a
	1-MCP treated	4.31 a	4.82 b	2.79 a	0.23 a	0.31 a	4.48 ab
	Means	4.76 A	4.01 A	3.48 A	0.30 A	0.18 A	4.83 A

Overall quality descriptor shows how experts perceive, in general, the fruit quality and therefore it summarizes the individual sensory impressions. It can be seen that the scores for overall quality are highly differentiated, for 4-week storage especially (Tab. 4 and Tab. 5). In the case of the fruits stored for 3 weeks, the quality of the fruits harvested in light-red stage was rated on similar level for 1-MCP treated, untreated and control fruits, but in the case of fruits harvested in pink maturity stage, the stored fruits, whether treated or untreated, were rated lower than the control ones. After 4 weeks of storage the differences between stored and control fruits became even bigger, and the 1-MCP treated fruits showed a tendency to higher scores than the untreated ones.

Semi-consumer hedonic test of tomatoes liking gives the information how the fruits are perceived by the expert panel from a consumer's point of view. The performed test was not a strict full consumer test, since the panel was small and not representative for a broad spectrum of consumers, but obtained information, even if it is limited, is quite useful. Hedonic scores for tomatoes liking after 3 weeks of storage were slightly differentiated, and the lowest scores got 1-MCP treated pink fruits (Tab. 6). However, in the case of fruits harvested in light-red stage a tendency to higher scores for the 1-MCP treated ones is evident. When we look at the results after 4-week storage, a positive effect of the treatment is evident, for the pink fruits especially, where the difference between the treated and untreated fruits is statistically significant.

Tab. 6. Scores for consumer liking for 'cherry' tomato fruits after storage (scale 0-10)

Maturity stage at harvest	Treatment	Consumer liking	
		3-week storage	4-week storage
Light-red	Control fruit	7.82 c	6.82 c
	Untreated	6.94 b	4.15 b
	1-MCP treated	7.63 bc	4.49 b
	Mean	7.46 B	5.15 A
Pink	Control fruit	7.13 b	6.73 c
	Untreated	6.60 ab	3.24 a
	1-MCP treated	6.14 a	4.63 b
	Mean	6.62 A	4.87 A

The results described above show that 1-MCP treatment influences some of sensory quality traits, but the effect is differentiated and depends on maturity stage of the fruits at harvest and also on storage duration. What is also important, the treatment does not cause noticeable increase of intensity of the most negatively perceived quality attributes – foreign odour and taste, bitter taste, and foreign taste. Moreover, overall quality impression for the 1-MCP treated fruits after storage period is high enough, compared to the untreated ones, therefore the treated stored fruits can compete with the freshly harvested ripe fruits on food market.

The results obtained in our study do not confirm fully Cliff *et al.* (2009) report. These authors pointed out that 1-MCP treatment could negatively influence sensory characteristics of tomato fruits. However, tomato cultivars used by the authors were different from traditional big fruits.

Tab. 7. Correlation coefficients for sensory quality evaluation of 'cherry' tomatoes

Scores for	Sensory attributes				
	'Tomato' flavour	Sour taste	Sweet taste	Flesh firm-ness	Flesh juici-ness
Overall quality impression	0.79**	0.20	0.20	0.55*	0.60*
Consumer's liking for tomatoes	0.72*	-0.15	0.45°	0.53*	0.48°

Explanation: ° - correlation significant at $p<0.10$, *correlation significant at $p<0.05$, **correlation significant at $p<0.01$

In Tab. 7 correlation coefficients between some sensory attributes and scores for overall quality and for consumer liking are shown. It can be seen that overall quality impression is significantly correlated with characteristic tomato taste intensity, as well as with flesh firmness and flesh juiciness. However, the correlation between overall quality and sour taste or with sweet taste was weak. The same sensory attributes were important in hedonic evaluation of liking in semi-consumer test.

Conclusions

The study showed that 1-MCP treatment of 'cherry' tomatoes affects some sensory quality descriptors of the stored fruits compared with the untreated ones. The influence of the treatment is mostly positive and reflects increased storability of the treated fruits. It can be also concluded that 1-MCP treatment of 'cherry' tomatoes does not negatively influences overall sensory quality impression for the fruits stored for 3 weeks, compared with freshly harvested ripe fruits, and compared with the stored untreated fruits results in a higher overall quality. However, when the storage period is prolonged up to 4 weeks, quality of stored fruits, even 1-MCP treated ones, shows significant degradation in respect of most sensory attributes in comparison with the stored ones. Also, consumer liking for the stored fruits is rated lower in that case, mainly due to decrease of flavour characteristic to tomatoes.

Acknowledgements:

The study was supported by NCN, grant N N310 728640.

References

Anonymous (1996). Sensory analysis. Experts. PN-ISO 8586-2.

Anonymous (1999). Sensory analysis. Methodology. PN-ISO 6564.

Azodanlou R, Darbellayb C, Luisierc J-L, Villettazc J-C, Amadoa R (2003). Development of a model for quality assessment of tomatoes and apricots. Lebensm-Wiss Technol 36:223-233.

Beno-Moualem D, Gusev L, Dvir O, Pesis E, Meir S, Lichter A (2004). The effects of ethylene, methyl jasmonate and 1-MCP on abscission of cherry tomatoes from the bunch and expression of endo-1,4-glucanases. Plant Sci 167:499-507.

Blankenship SM, Dole JM (2003). 1-Methylcyclopropene: a review. Postharvest Biol Tec 28:1-25.

Cliff M, Lok S, Changwen Lu, Toivonen PMA (2009). Effect of 1-methylcyclopropene on the sensory, visual, and analytical quality of greenhouse tomatoes. Postharvest Biol Tec 53:11-15.

Ergun M, Sargent SA, Huber DJ (2006). Postharvest quality of grape tomatoes treated with 1-methylcyclopropene at advanced ripeness stages. HortScience 41(1):183-187.

Fan X, Mattheis JP (2000). Yellowing of broccoli in storage is reduced by 1-methylcyclopropene. HortScience 35:885-887.

Gajewski M (2000). Storability of tomato (*Lycopersicon esculentum* Mill.) glasshouse cultivars from field growing with supports. Annales Universitas Mariae Curie-Skłodowska, Lublin, Supl. 8:311-317 [in Polish with English summary].

Gajewski M, Arasimowicz D (2004). Sensory quality of eggplant fruits (*Solanum melongena* L.) as affected by cultivar and maturity stage. Polish Journal of Food and Nutrition Sciences 13/54 (3):249-254.

Getinet H, Seyoum T, Woldetsadik K (2008). The effect of cultivar, maturity stage and storage environment on quality of tomatoes. J Food Eng 87:467-478.

Guillén F, Castillo S, Zapata PJ , Martinez-Romero D, Serrano M, Valero D (2007). Efficacy of 1-MCP treatment in tomato fruit. 1. Duration and concentration of 1-MCP treatment to gain an effective delay of postharvest ripening. Postharvest Biol Tec 43:23-27.

Guillén F, Castillo S, Zapata PJ, Martínez-Romero D, Valero D, Serrano M (2006). Efficacy of 1-MCP treatment in tomato fruit. Effect of cultivar and ripening stage at harvest. Postharvest Biol Tec 42:235-242.

Huber DJ (2008). Suppression of ethylene responses through application of 1-methylcyclopropene: a powerful tool for elucidation ripening and senescence mechanisms in climacteric and nonclimacteric fruits and vegetables. HortScience 43:106-111.

Mazur KZ, Gajewski M, Czechowska E, Decyk J, Metera A M, Marcinkowska MM, Wtulich JA, Przybył JL (2013). Effect of 1-MCP Treatment on Quality of Stored 'Cherry' Tomatoes. CAMA IX International Controlled and Modified Atmosphere Research Conference, Trani, Italy, 3-7 June 2013:52.

Meilgaard M, Civille GV, Carr BT (1999). Sensory Evaluation Techniques. 3rd ed., CRC Press, Boca Raton, London.

Mir N, Canoles M, Beadury R (2004). Inhibiting Tomato Ripening with 1-Methylcyclopropene. J Amer Soc Hort Sci 129(1):112-120.

Moretti CL, Araújo AL., Marouelli WA, Silva WC. (2002). 1-Methylcyclopropene delays tomato fruit ripening. Hortic Bras 20(4):659-663.

Mostofi Y, Toivonen PMA, Lessani H, Babalar M, Lu CW (2003). Effects of 1-methylcyclopropene on ripening of greenhouse tomatoes at three storage temperatures. Postharvest Biol Tec 27:285-292.

Opiyo AM, Ying TJ (2005). The effects of 1-methylcyclopropene treatment on the shelf life and quality of cherry tomato (*Lycopersicon esculentum* var. *cerasiforme*) fruit. Int J Food Sci Tech 40:665-673.

Roberts PK, Sargent SA, Fox AJ (2002). Effect of storage temperature on ripening and postharvest quality of grape and mini-pear tomatoes. Proc Fla State Hort Soc 115:80-84.

Sayed A, Nakano K, Maezawa S (2004). Combined effect of heat treatment and modified atmosphere packaging on the color development of cherry tomato. Postharvest Biol Tec 34:113-116.

Suslow TV, Cantwell M (2009). Tomato. Recommendation for maintaining postharvest quality. www.postharvest.ucdavis/producefacts/veg/tomato.

Tadesse TN, Farneti B, Woltering E (2012). Effect of Ethylene and 1-Methylcyclopropene on Color and Firmness of Red and Breaker Stage Tomato stored at Different Temperatures. American Journal of Food Technology 7(9):542-551.

Tassoni A, Watkins CB, Davies PJ (2006). The effect of 1-MCP treatment on ripening in tomato fruit: changes in ethylene, polyamines and associated gene expression. J Exp Bot 57:3313-3325.

Toor RK, Savage GP (2006). Changes in major antioxidant components of tomatoes during post-harvest storage. Food Chem 99:724-727.

Vijay P, Rakesh P, Girish CS (2010) Ripening of tomato (*Solanum lycopersicum* L.). Part I: 1-methylcyclopropene mediated delay at higher storage temperature. J Food Sci Technol 47(5):519-526.

Watkins CB (2006). The use of 1-methylcyclopropene (1-MCP) on fruits and vegetables. Biotechnology Advances 24:389-409.

Wills RBH, Ku VVV (2002). Use of 1-MCP to extend the time to ripen of green tomatoes and postharvest life of ripe tomatoes. Postharvest Biol Tec 26:85-90.

Determination of *Cyclotrichium niveum* Essential Oil and Its Components at Different Altitudes

Memet INAN[1*], Ahmet Zafer TEL[2]

[1]*Adiyaman University, Kahta Vocational School, Medicinal and Aromatic Plants, 02400, Adiyaman, Turkey; minan@adiyaman.edu.tr (*corresponding author)*
[2]*Adiyaman University, The Faculty of Science and Letters, Department of Biology, 02400 Adiyaman, Turkey; aztel@adiyaman.edu.tr*

Abstract

Cyclotrichium niveum (Boiss.) Manden. & Scheng. is a perennial species of Lamiaceae family of high importance due to essential oils compounds. Essential oil rates and essential oil components were determined in plants of this species collected from different altitudes. In order to have specific data, the plants were harvested from their natural growing area; the samples were picked up during full flowering period, from three different altitudes (890 m, 1239 m and 1605 m) in order to enhance that essential oil rates are directly influenced by this aspect. The analysis of essential oil extracted included 42 components and pulegone was the main component among these. Whereas the lowest pulegone rate was determined as 59.9% at 890 m altitude, the highest value (68.12%) was determined at 1605 m. It was also determined that as the altitude increased, the rate of this particular component in essential oil also increased, while the rate of other important components (such as isomenthone, limonene, 1,8-cineole, isopulegone) decreased.

Keywords: *Cyclotrichium niveum,* elevation, pulegone, secondary metabolites

Introduction

It is known that several internal and extrinsic factors are efficient on the secondary metabolites in medicinal and aromatic plants (Chalchat *et al.*, 1997; Hudaip *et al.*, 2002; Kofidis and Bosabilidis, 2008; Panou-Filotheou *et al.*, 2001; Takaloo *et al.*, 2012). Essential oils included among the secondary metabolites group produced by the plants have been intensely used in food industry due to their aromatic and protective properties (Smith-Palmer *et al.*, 2001). Structures of the secretory hairs and the amount of secreted substances existing on the surface of leaves of plants growing at different altitudes can be different. It has been reported by some researchers that however the number of epidermal tissues on the surface of leaf and the density of hair on leaf surface, the amount of secreted essential oil increases with the altitude (Bosabalidis, 2002; Kofidis and Bosabilidis, 2008). Similarly, the rate of essential oil and essential oil components are affected by different altitudes. It has also been determined in several studies that have been carried out with different plants, that as the altitude increases, the rate of essential oil decreases (Delazar *et al.*, 2011; Gulbaba *et al.*, 2002; Haider *et al.*, 2009), which is in agreement with our conclusions as well.

Cyclotrichium niveum (Bois.) Mandan. & Scheng. a member of the family Lamiaceae is a perennial species endemic for Turkey, which have a significant release of essential oil (Başer *et al.*, 1994; Dirmenci *et al.*, 2010; Sanon *et al.* 2007; Tel, 2009). Essential oils are complex compounds and their chemical composition and concentration of elements show great changes. Essential oils, terpenoids, phenylpropanoids and benzenoids, existing in the plants as the secondary metabolites, can be present as fatty acid derivatives and amino acid derivatives (Lodhia *et al.*, 2009).

In dried leaves of *C. niveum,* essential oil exists at a rate of 5.6% (Baytop, 1984). Depending upon the picking period of the plant and the used sample organ, this amount of essential oil can vary between 1.5-2.1% (Alim *et al.*, 2009; Başer, 2002; Cetinus *et al.*, 2007; Goze *et al.*, 2010). The main components of this essential oil are pulegone and isomenthone (Başer, 2002). It has been determined that in the essential oil samples taken out from different organs of plants, the rate of pulegone varies between 32.5-56.4% and the rate of isomenthone varies between 34.2-35.4% (Başer *et al.*, 1994). It has been reported that 76.84% pulegone and 6.65% isomenthone exists in the essential oil extracted from the upper section of plants (Cetinus *et al.*, 2007). Alim *et al.* (2009) reported that they determined 26 leading components in *C. niveum* essential oil from plants picked in full blossoming period; the rate of pulegone was determined as 50.46%. the rate of isomenthone was determined as 34.53%, while limonene (% 2.11), 1,8-cineole (% 1.91), Y-element (% 1.53), gama-pentene (% 1.46) were the subsequent.

Studies related to essential oil components existing in different organs of *C. niveum* species have been carried out and reported in speciality literature. This study aimed to determine the effect of different altitudes upon the rate of essential oil and essential oil components in *C. niveum* plants growing naturally.

Materials and methods

Plant Material

C. niveum plants were collected from three different altitudes upon the road of Adıyaman/ Kahta/ Değirmenbaşı - Malatya (West of Mount Nemrut). The dates on which plants were picked up, the coordination and pick-up altitudes are presented in Tab. 1. The determinations of plants collected in full-flowering period was conducted by Ahmet Zafer Tel (Adıyaman University, Faculty of Arts and Science, Department of Biology) and their herbariums were recorded in the aforementioned faculty (Voucher No: Tel 8881). Because blossoming was earlier at lower altitudes, the dates for collection of the plants differ. Whereas flowering in low altitudes was at the beginning of July, it occurred towards the end of July as the altitude increased. After the plants were picked up, they were naturally dried in a shady, breezy area, and stalk-leaf separation was performed.

Tab. 1. Collecting dates, coordination and altitudes of sample picking locations for *Cyclotrichium niveum* (Bois.) Mandan & Scheng plants

Location	Collected Date	Coordination	Altitude (m)
Location 1 (L_1)	04.07.2013	37°57'44" North 38°40'20" East	890
Location 2 (L_2)	12.07.2013	38°01'08" North 38°43'51" East	1239
Location 3 (L_3)	25.07.2013	38°00'30" North 38°45'36" East	1605

Essential Oil Analysis

The leaves separated from stalks were boiled in Clevenger type steam distillation device for 3 hours, with 3 replications from the samples taken out of each location. The averages determined with 3 replications were found and compounded oils were put into bottles for GC/MS analysis and stored at +4 °C temperature.

Gas Chromatography Mass Spectrometry (GC/MS) Analysis

The analysis of essential oils of *C. niveum* was carried out by using Thermo Scientific Focus Gas Chromatograph equipped with MS, auto sampler and TR-5MS (0.25 mm x 60 m i.d, film thickness 0.25). The carrier gas was helium (99.9%) at a flow rate of 1 mL/min; ionization energy was 70 eV. Mass range m/z 50-650 amu. Data acquisition was scan mode. MS transfer line temperature was 250 °C, MS ionization source temperature was 220 °C, the injection port temperature was 220 °C. The samples were injected with 250 split ratio. The injection volume was 1 μL. Oven temperature was programmed in the range of 50 to 220 °C at 3 °C /min. The structure of each compound was identified by comparison with their mass spectrum (FFNSC13 and Wiley9 library). The data were handled using Xcalibur software program.

Soil and Temperature

Soil samples from the experimental locations contained low organic matter (0.32%), salt content of 0.017%, pH of 7.90% and the texture was clay (Celik, 2012). According to the long-term meteorological record, the average temperature values in July were 26.8 °C in the research area.

Results and discussion

Essential Oil

The rate of essential oil can be affected by several factors. Some studies have revealed that the altitude negatively affects the rate of essential oil and yield (Delazar *et al.*, 2011; Gulbaba *et al.*, 2002; Haider *et al.*, 2009). In this study that we carried out on *C. niveum*, we determined that as the altitude increased, the rate of essential oil obtained from plants decreased. In accordance to this, the rate of essential oil in plants picked up from L_1 (890 m), L_2 (1239 m) and L_3 (1605 m) locations were 5.5%, 4.8% and 4.1%, respectively (Tab. 2). These determined data were higher than the results reported by other researchers (Alim *et al.*, 2009; Başer, 2002; Cetinus *et al.*, 2007; Goze *et al.*, 2010). This difference might probably be resulted from the section of plant, climate, geographical position, soil structure, etc.

Chemical Composition of the Essential Oil

The components of essential oil obtained from *C. niveum* plants are presented in Tab. 2, while the changes of significant components, at different altitudes of the 3 locations, are shown in Fig. 1.

As a result of GC/MS analysis, specifically 42 compounds were determined as leading in *C. niveum* essential oil. Total rate of components in essential oil extracted from samples collected at the lowest altitude was 95.12%, 96.87% for the second location, and 96.13% for the highest altitude location. Pulegone, isomenthone, 1,8-cineole and limonene components included almost 86% of the total essential oil structure. According to the average obtained for the three locations, pulegone was the main component, having almost 65% of the essential oil composition.

Puleone amount increased with the altitude. In accordince to this, whereas the lowest pulegone value was obtained from L_1 point as 59.9%, the rate of pulegone in plants picked up from L_2 and L_3 was 65.23% and 68.12%, respectively. These values we determined for pulegone component are higher than the values of some researchers (Alim *et al.*, 2009; Başer *et al.*, 1994), Cetinus *et al.* (2007); they have reported lower values than ours (50.46%). This situation can be caused by different locations and altitudes.

As the altitude increased, the rates of some components in essential oil, such as limonene, 1,8- cineole, isomenthone, isopulegone, exo-isocamphonone, were determined to decrease, in contrast to pulegone. A similar case was also observed for some other constituents, while components such as verbenone, citronellol, ocimenone, spathulenol, selinene, and naphthalenone were determined to increase in rate, once with the altitude.

Tab. 2. Essential oil rates and components obtained from *C. niveum* plants collected from different altitudes (%)

No	Retention Time	Components (%)	Location L_1	L_2	L_3
		Essential Oil (%)	5.5	4.8	4.1
1	9.43	2-hexenal	0.05	0.04	0.05
2	10.92	Cyclopentane	-	0.05	0.05
3	11.6	α- pinene	0.86	0.67	0.59
4	11.77	1-octene,3-ethyl	-	0,04	0.04
5	12.79	Sabinene	0.44	0.21	0.24
6	13.05	β- pinene	1.72	1.01	1.06
7	13.77	Psedulimonene	0.06	0.05	0.05
8	14.44	α- terpinene	0.04	-	-
9	14.54	limonene	3.26	2.57	2.17
10	14.72	1,8- cineole	4.49	1.71	1,31
11	15.87	trans-sabinene	0.06	-	-
12	16.27	Terpinolene	0.06	0.05	0.05
13	16.56	Linalool	0.18	0.12	0.22
14	16.96	Myrtanal	0.05	0.04	0.04
15	17.97	iso-carveol	0.04	-	-
16	18.33	Cyclobutene	0.23	0.12	0.16
17	18.63	Menthone	0.13	0.16	0.13
18	18.95	isomenthone	10.58	7.23	6.53
19	19.13	carvenone	0.77	0.19	0.05
20	19.26	isopulegone	2.03	1.59	1.64
21	19.84	terpineol	0.17	0.09	0.11
22	19.94	carvone	0.15	0.09	-
23	20.25	verbenone	-	0.21	0.31
24	20.41	citronellol	-	0.04	0.05
25	20.48	cuminaldehyde	0.1	0.25	0.05
26	21.32	pulegone	59.9	65.23	68.12
27	22.35	thymol	0.18	0.06	0.08
28	22.54	pamabel	0.07	0.04	0.06
29	22.96	azido-2-phenylethene	0.09	0.07	-
30	23.22	trans-carvyl acetate	0.04	0.22	0.04
31	24.08	ocimenone	0.76	0.85	0.88
32	24.23	eugenol	0.07	0.28	-
33	26.02	β- elemene	0.12	0.09	0.08
34	26.65	citronellic acid	0.21	0.45	0.14
35	27.26	exo-isocamphonone	11.03	10.02	5.47
36	27.69	germacrene	0.52	0.58	0.44
37	28.07	bicyclogermacrene	1.21	1.49	1.23
38	28.29	eugenyl acetate	0.04	0.43	-
39	30.18	spathulenol	0.17	0.19	0.22
40	31.61	napthy-1-ol	0.04	0.05	0.05
41	34.09	selinene	0.04	0.07	0.11
42	36.42	naphthalenone	0.22	0.36	0.61
Total			95,12	96,87	96,13

The results we determined related to isomenthone component, varying from 6.53 for L_3 to 10.58 for L_1 ,had lower values than the results reported in previous studies (Alim *et al.*, 2009; Başer *et al.*, 1994).

Some of the components could not be determined in essential oils of the plants collected from the lowest altitude (cyclopentane, verbenone, citronellol) nor at the highest altitude (components such as α- terpinene, trans-sabinene, iso-carveol, carvone) (Tab. 2).

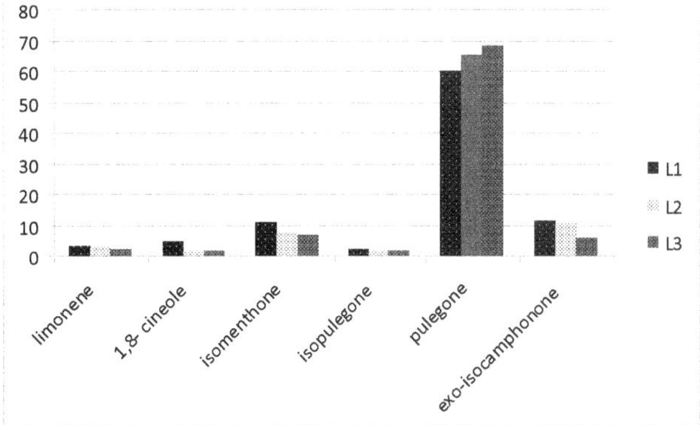

Fig 1. Changes of important components of *C. niveum* essential oil extracted from samples from different altitudes

On the other hand, several components, such as limonene, 1,8-cineole obtained from plants picked up from the lowest altitude (L_1) were determined to be higher in comparison with the data recorder for higher altitudes.

There has been no literature related data for exo-isocamphonone component, in studies carried out on essential oil extracted from plants. Whereas in the current study, this component was determined with 11.03% at the lowest altitude, while its rate at the highest altitude decreased to 5.47%.

These changes in essential oil composition are probably caused by photoperiod and air temperature at different altitudes.

Conclusion

All important components determined at high rate in *C. niveum* essential oil were negatively or positively affected by altitude. Even more, the data determined that as the altitude increased, the essential oil in *C. niveum* plants decreased proportionally. The highest rate of pulegone, as one of the most important components of the essential oil, was determined at high altitudes, but some other important components, such as isomenthone, 1,8-cineole, limonene, isopulegone decreased in rate as the altitude increased.

Acknowledgement

The authors would like to thank Prof. Dr. Elif Sultan Giray for her contribution to this work.

References

Alim A, Goze I, Cetin A, Atas AD, Vural N, Donmez E (2009). Antimicrobial Activity of the Essential Oil of *Cyclotrichium niveum* (Boiss.) Manden. Et Scheng. Afr J Microbiol Res 3(8):422-425.

Başer KHC, Sarikardasoglu S, Tümen G (1994). The Essential Oil of *Cyclotrichium niveum* (Boiss.) Manden. Et Scheng. J Essen Oil Res 6:9-12.

Başer KHC (2002). Aromatic biodiversity among the flowering plant taxa of Turkey. Pure Appl Chem 74(4):527-545.

Baytop T (1984). Therapy With Medicinal Plants in Turkey (Past and Present). Publication of Istanbul University, 480 p .

Bosabalidis AM (2002). Structural features of *Origanum* sp. In S.E. Kintzios (ed.), Oregano. The Genera *Origanum* and *Lippia*. Taylor and Francis Press, London, 64 p.

Chalchat JC, Garry RP, Michet A (1997). Variation of the chemical composition of essential oil of *Mentha piperita* L. during the growing time. J Essent Oil Res 9:463-465.

Celik, A (2012). Determination of the potential raw materials of the non-arable areas of the Adıyaman povience for use in birck-ceramic production. Cukurova University, Institute of Natural and Applied Sciences, Adana, Turkey, phD Diss. Abstr.128-147.

Cetinus ŞA, Goze İ, Saraç B, Vural N (2007). Scavenging Effect and Antispasmodic Activity of The Essential Oil of *Cyclotrichium niveum.* Fitoterapia 78:129-133.

Delazar A, Bahmani M, Shoar HH, Raisi AT, Asnaashari S, Nahar L, Sarker SD (2011). Effect of Altitude, Temperature and Soil on Essential Oil Production in *Thymus fedtschenkoi* Flowers in Osko and Surrounding areas in Iran. J Essen Oil Bearing Plants 14(1):23-29.

Dirmenci T, Dündar E, Deniz G, Arabacı T, Martin E, Jamzad Z (2010). Morphological, karyological and phylogenetic evaluation of *Cyclotrichium*: a piece in the tribe *Mentheae* puzzle. Turk J Bot 34:159-170.

Goze I, Cetin A, Goze A, (2010). Investigation of Effects of Essential Oils of *Origanum minutiflorum* O. Schwarz PH Davis and *Cyclotrichium niveum* (Labiatae) Plants on Angiogenesis in Shell-Less chick Embryo Culture. African J Biotech 9(14):2156-2160.

Gulbaba GA, Özkurt N, Kurkcuoglu M, Başer KHC (2002). Identification of Rosemary (*Rosmarinus officinalis* L.) Populations in Mersin and Adana Province and Leaf Essential Oil Yields and Their Compositions. Eastern Mediterranean Forestry Research Institute Turkey 1300-7912.

Haider F, Kumar N, Banerjee S, Naqvi AA, Bagchi GD (2009). Effect of Altitude on the Essential Oil Constituents of *Artemisia roxburghiana* Besser var. *purpurascens* (Jacq.) Hook. J Essen Oil Res 21(4):303-304.

Hudaib M, Speroni E, Di Pietra AM, Cavrini V (2002). GC/MS evaluation of thyme (*Thymus vulgaris* L.) oil composition and variations during the vegetative cycle. J Pharm Biomed Anal 29:691-700.

Kofidis G, Bosabilidis AM (2008). Effects of altitude and season on glandular hairs and leaf structural traits of *Nepeta nuda* L. Bot Stud 49:363-372.

Lodhia MH, Bhatt KR, Thaker VS (2009). Antibacterial activity of essential oils from palmarosa, evening primrose, lavender and tuberose. Indian J Pharm Sci 71(2):134-136.

Panou-Filotheou H, Bosabalidis AM, Karataglis S (2001). Effects of copper toxicity on leaves of oregano (*Origanum vulgare* subsp. *hirtum*). Ann Bot 88:207-214.

Sanon B, Azaz AD, Dirmenci T (2007). Identified Saprophytic Microfungi on the *Cyclotrichium* (Boiss.) Manden.& Scheng. species distributed in Turkey. J App Bio Sci 1(3):57-59.

Smith-Palmer A, Stewart J, Fyfe L (2001). The potential application of plant essential oils as natural food preservatives in soft cheese. Food Microbiol 18:463-470.

Takaloo SG, Hassani A, Hassanpouraghdam MB, Meshkatalsadat MH, Pirzad A, Heidari M (2012). Essential Oil Content and Composition of *Thymus migricus* Klokov & Desj-Shost. Affected by Plant Growth Stage and Wild Habitat Altitude. Romanian Biotechnological Letters 17(1):6982-6988.

Tel AZ (2009). Contributions to the flora of Nemrut Mountain (Adıyaman/Turkey). Biological Diversity and Conservation 2(1):36-60.

Seasonal Variations in Total Antioxidant Capacity and Total Phenolics Content of Leaves of *Phyllostachys* Taxa Using Different Extraction Methods

András NEMÉNYI[1]*, Éva STEFANOVITSNÉ-BÁNYAI[2], Szonja Szimóna BURJÁN[1], Zoltán PÉK[1], Attila HEGEDŰS[3], Csaba GYURICZA[4], Lajos HELYES[1]

[1] *Szent István University, Faculty of Agricultural and Environmental Sciences, Institute of Horticulture, Páter K. u. 1., Gödöllő, 2100 Hungary; nemenyi.andras@mkk.szie.hu (*corresponding author)*

[2] *Corvinus University of Budapest, Faculty of Food Science, Department of Applied Chemistry, Villányi út 64-70., Budapest, 1118 Hungary*

[3] *Corvinus University of Budapest, Faculty of Horticultural Science, Department of Genetics and Plant Breeding, Villányi út 64-70., Budapest, 1118 Hungary*

[4] *Szent István University, Faculty of Agricultural and Environmental Sciences, Institute of Plant Production, Páter K. u. 1., Gödöllő, 2100 Hungary*

Abstract

Changes in total phenolic content (TP) and total antioxidant capacity (AC) during the vegetation period (April-November) were analysed in *Phyllostachys aureosulcata* f. *aureocaulis (PAA), P. flexuosa (PF), P. humilis (PH), P. sulphurea* var. *sulphurea (PSS).* Different extraction methods were compared: infusion or decoction with water or aqueous methanol. The highest significant value for TP was measured in the case of infusion at 90 °C for 5 min, which lead to the highest value for AC. During the vegetation period the highest values of TP were measured in April and May in the case of *PAA* (409.5; 314.9 and 258.7; 119.0 μg GA/ml) and *PH* (388.4; 411.6 and 252.9; 253.3 μg GA/ml). There was a clear trend in the changes of TP, with high values in April and November and a peak during August-September. Similar to TP, the highest values of AC were measured in April and May in *PH* (519.7; 566.3 and 513.5; 510.4 μg AA/ml) and *PAA* (534.5; 337.8 and 394.9; 275.4 μg AA/ml). We compared the change of TP in all taxa with stress index values derived from daily maximum and minimum air temperature, cumulative precipitation plus irrigation and cumulative evapotranspiration values. A close correlation $R^2 = 0.32$ (p=0.001) was found between stress index values and the change in TP of all taxa. The correlation is even closer $R^2 = 0.52$ (p=0.001) with *PF, PH* and *PSS.* Our results with *PAA* can be explained by possible inter/intra-specific differences in freeze tolerance and cold-acclimation.

Keywords: bamboo, *Bambusoideae, herba, Poaceae,* tea

Introduction

The bamboo genus *Phyllostachys* originated in the warm temperate regions of SE China and has long been cultivated and has naturalized in neighbouring countries such as Korea and Japan (Ohrnberger, 2002). *Phyllostachys* taxa have significant importance in the agriculture of Asian countries, mainly in forestry and olericulture (Kleinhenz and Midmore, 2001), but are also used in traditional Korean herbal medicine (Kim *et al.*, 1995) and as ornamentals (Ohrnberger, 2002). The determination of human health promoting properties and phytochemicals of horticultural crops, including the measurement of antioxidant activity or total phenolics has been one of the targets of horticultural research in the past decades (Kwon *et al.*, 2008; Im *et al.*, 2010). The leaves of bamboo (*Herba Phyllostachys*) are used in traditional Chinese and Korean medical pharmacology (Kim *et al.*, 1995; Chen and Chen, 2004), but also other *herba* from other bamboo taxa are used in Korea (Park and Lim, 2009; Hwang *et al.*, 2007; Park *et al.*, 2007). These *herba* have high antioxidant properties (Hu *et al.*, 2012; You *et al.*, 2010; Park *et al.*, 2007) and contain flavonoids (Kim *et al.*, 2010), phenolic acids (Park *et al.*, 2007), chlorogenic acid derivatives (Kweon *et al.*, 2001), phenolic and flavone glycosides (Li *et al.*, 2008; Park *et al.*, 2007). Leaf extracts have been shown to have anticarcinogenic (Lin *et al.*, 2008; Shin *et al.*, 2003), cardioprotective (Fu *et al.*, 2006) effects, relieve lipotoxicity (Kim *et al.*, 1995; Panee *et al.*, 2008; Ryou *et al.*, 2012), protect against oxidative stress-induced glaucoma (Lee *et al.*, 2010; Kim *et al.*, 2013). Leaf extracts can synergistically be used in combination with other compounds in the treatment of immune/allergic diseases related to mast cells (Kim *et al.*, 2010), treatment of leukaemia (Kim *et al.*, 2007), in the prevention and treatment of diabetic complications (Jung *et al.*, 2007) and autoimmunity (Kim *et al.*, 2007). Bamboo leaves are also used as food additives (Oh and Lim, 2010), and bamboo leaf extracts also exhibit antimicrobial activity during food storage (Park and Lim, 2010). The studies of Jeong *et al.* (2008), In *et al.* (2010) and Kim *et al.* (2012) have evaluated the nutritional components and antioxidant activities of *Sasa borealis*, maengjong-juk (*Phyllostachys pubescens*) and ojuk (*Phyllostachys nigra*) leaf water extracts prepared by traditional tea manufacturing processes. Therefore, the aim of the present study was to investigate the seasonal variations in antioxidant activity and total phenolic content of different *Phyllostachys* bamboo taxa leaves using different extraction methods.

Materials and methods

Plant material and sample preparation

For the comparison of different extraction methods, a sample comprised of 30 leaves that were randomly collected from a similar exposure and position within the canopy of *Phyllostachys flexuosa,* on the 30th of November 2010. The bamboos are growing in the Botanical Garden of Eötvös Lóránd University (BG ELTE), Budapest, Hungary (47°29" N latitude, 19°05'02" E longitude and 108 m altitude). Four repetitive samples were collected.

For the examination of seasonal effect, 30 leaves per each sample from a similar exposure and position within the canopy were randomly collected on the last day of each month, from April to November 2011, from the following *Phyllostachys* taxa: *P. aureosulcata f. aureocaulis, P. flexuosa, P. humilis, P. sulphurea var. sulphurea.* Four repetitive samples were collected each month while winter sampling (from December to March) was neglected to avoid the effect of unpredictable leaf tissue freeze damage.

Measurement of environmental parameters and irrigation

Air temperature (°C) and precipitation (mm) were recorded during the experiment (Fig. 1). Air temperature was measured six times per hour by a meteorological instrument. Plants were irrigated by overhead sprinklers between April 1 and October 10 every 14 days with 40 mm of water. Potential evapotranspiration (ET_0) was calculated from that used for other vegetable crops (Helyes and Varga, 1994; Helyes *et al.*, 2013); the amount of daily water demand was calculated based on the daily average

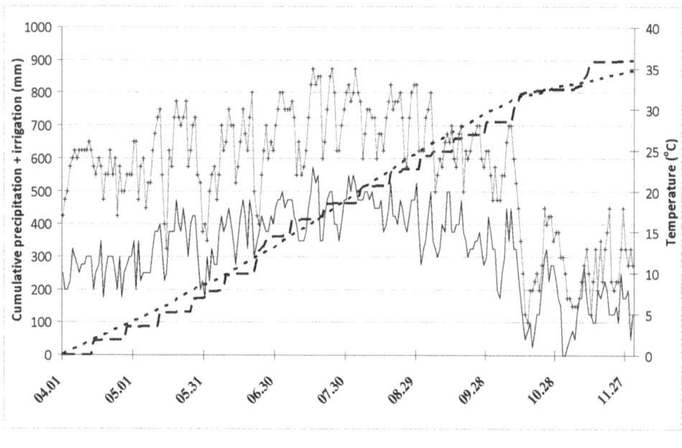

Fig. 1. Change of daily minimum, maximum air temperature (°C), cumulative precipitation plus irrigation (mm) and cumulative potential evapotranspiration (mm) during the vegetation period (April-November). Grey line (+), black line, dotted line, dashed line represents maximum temperature, minimum temperature, cumulative potential evapotranspiration, cumulative precipitation plus irrigation respectively.

temperature (in °C) divided by five and was expressed in millimetres:

$$I_d = \left(\frac{T_{min} + T_{max}}{2}\right) / 5$$

Stress index (SI) was calculated for each sampling date (month) from monthly cumulative values of daily minimum air temperature Tmin <= 5 °C (SI value 5/0) + daily maximum air temperature Tmax <= 5 °C (SI value 5/0) + daily average air temperature Tave <= 5 °C (SI value 5/0) + cumulative precipitation plus irrigation < cumulative potential evapotranspiration (SI value 1/0).

Preparation of *Phyllostachys* leaf extracts

To compare the different extraction methods extracts were prepared from 65 °C air dried leaves of *Phyllostachys flexuosa.* Leaves were cut into small pieces, and 1 g of dried leaves were infused or decocted with 50 ml water (80, 90, or 100 °C; 5, 10, 20 min or 24 h or with aqueous methanol (25 °C; water/ methanol 80/ 20, v/v 72 h). For the investigation of the effect of seasonal variation, 65 °C air dried leaves of the different *Phyllostachys* taxa were also cut into small pieces, and 1 g of dried leaves were infused with 50 ml water (90 °C, 5 min or 24 h) or with aqueous methanol (25 °C; water/ methanol 80/ 20, v/v 72 h). The methanolic and aqueous extracts were stored at room temperature for 24 h. After centrifugation (13,000 rpm, 10 min) the supernatants were stored at -20 °C until the analyses.

Determination of total phenolics content

The total amount of soluble phenols (TP) was determined using Folin-Ciocalteu reagent according to the method of Singleton and Rossi (1970) and determined spectrophotometrically at 760 nm. The content of soluble phenols was calculated from a standard curve obtained by different concentrations of gallic acid (GA) and given in µg GA/ml.

Determination of total antioxidant capacity by FRAP assay

The total antioxidant capacity (AC) related to ascorbic acid was determined spectrophotometrically using the FRAP (Ferric Reducing Antioxidant Power) method according to Benzie and Strain (1996). It is based on the reduction of the Fe^{3+}-TPTZ complex to the ferrous form at low pH. This reduction was monitored by measuring the absorption change at 593 nm. Results were expressed as µg equivalents of ascorbic acid (AA).

Statistical analysis

Results were expressed as the average plus/minus standard deviations. The data were analysed by two-factor analysis of variance (ANOVA) with repetitions and the means separated using the LSD test at p=0.05. Regression analysis was performed using Statistica 9 software.

Results and discussion

Effect of different extraction methods on total phenolic content and antioxidant capacity in *Phyllostachys flexuosa*

Fig. 2 shows that the highest TP values were obtained by aqueous infusion, compared to decoction or methanolic extraction. The highest significant value of 492.3±12.7 μg GA/ml for TP was measured in the case of infusion at 90 °C for 5 min which lead similarly the highest measured value (539.2±10.8 μg AA/ml) in the case of AC, but this latter was not significantly the highest compared to infusion at 80 °C for 5 min (531.4±19.5 μg AA/ml). The second and third highest (14 and 21% lower) TP values were measured in the case of infusion at 90 °C for 10 and 20 min (421.3±8.4 and 391.9±5.6 μg GA/ml) respectively, but the latter was not significantly higher than infusion at 80 °C for 20 min (365.6±25.7 μg GA/ml). TP value of infusion at 80 °C for 20 min (365.6±25.7 μg GA/ml) was not significantly lower than TP value of decoction at 20 min (363.4±6.7 μg GA/ml). In the case of AC there was no significant difference between the second, third and fourth highest measured values (531.3±19.3; 511.9±14.6; and 508.6±11.7 μg AA/ml) of infusion at 80 °C for 5 and 20 min and at 90 °C for 10 min respectively. AC values of infusion at 80 °C for 10 min and at 90 °C for 20 min (472.8±9.2 and 466.4±24.6 μg AA/ml) were significantly lower (by 7-12%) than the above extraction methods but were not significantly different compared to each other. Methanolic extraction for 72 h gave similarly high AC values (382.3±11.6 μg AA/ml) compared to decoction for 20 min (381.7±19.4 μg AA/ml) but these values were not significantly different from each other. AC values of methanolic extraction for 72 h (382.3±11.6 μg AA/ml) and decoction for 20 min (381.7±19.4 μg AA/ml) were 29% lower than infusion at 90 °C for 5 min (539.2±10.8 μg AA/ml). The method of extraction and the used solvent had a significant effect on extraction yield of polyphenols

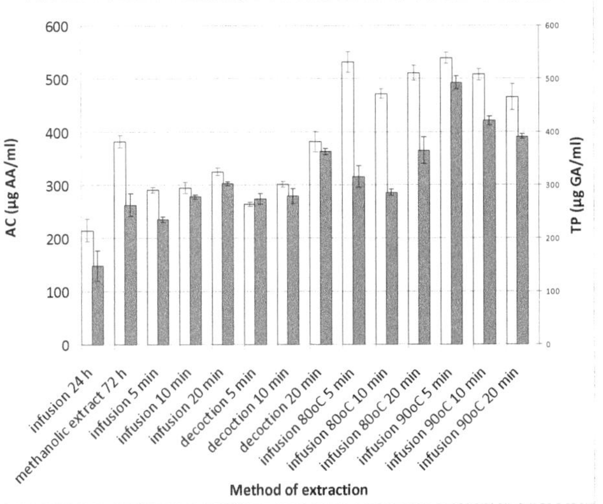

Fig. 2. Total phenolics content (TP) (μg GA/ml) and antioxidant capacity (AC) (μg AA/ml) of leaf extracts of *Phyllostachys flexuosa* using different extraction methods and times (n=4, ±SD). White and grey columns refer to antioxidant capacity and total phenolic content, respectively.

and on antioxidant activity in teas, but results vary depending on which solvent and method produces the highest levels. In *Camellia sinensis* tea, Wang and Helliwell (2001), found aqueous ethanol as the best solvent, Turkmen *et al.* (2006) suggested N,N-dimethylformamide, while Khokhar and Magnusdottír (2002) reported water to be the best solvent of cathecins. Park and Lim (2009) have found that in the leaves of bamboo *Sasa borealis* the total polyphenol contents of the 70% ethanol and the water crude extracts were not significantly different, while in the 70% ethanolic extract the total flavonoid contents were significantly higher than in water crude extracts. Our extraction results with aqueous infusion were similar to those of Perva-Uzunalic *et al.* (2006) who have found the highest extraction efficiency of *Camellia sinensis* tea cathecins with aqueous infusion at 80 °C for 20 min (97%) and at 95 °C for 10 min (90%). Our results were also comparable to those of Jeong *et al.* (2008) with *Sasa borealis* leaf tea, with the specification that they found the highest PT in 80% methanol extract, and also with those of Kim *et al.* (2012), although freeze-drying in their case can account for the much higher values of TP found in *P. nigra* infused with 80 °C water for 10 min. The TP values of aqueous extractions in this study were similar to those of In *et al.* (2010) in the case of maengjong-juk (*Phyllostachys pubescens*) leaf tea.

Changes in total phenolic content and antioxidant capacity throughout the vegetation period in different *Phyllostachys* taxa

Changes in total phenolics content

Tab. 1 shows the results of all three extraction methods (infusion 90 °C, 5 min or 24 h or aqueous methanol 72 h). The highest values were obtained in the case of infusion at 90 °C, for 5 min or 24 h. Only in May, August and September the values were slightly higher (11-14% higher on average) with infusion at 90 °C for 24 h compared to infusion at 90 °C for 5 min. The lowest (33-39% lower on average) values were measured for aqueous methanol for 72 h. During the course of the vegetation period the highest values of TP were measured in April and May in the case of *P. aureosulcata f. aureocaulis* and *P. humilis* with infusion at 90 °C for 24 h or with aqueous methanol for 72 h, and the difference was not significant. Only in the case of infusion at 90 °C for 5 min were the TP values of *P. humilis* significantly higher than *P. aureosulcata f. aureocaulis* during April and May. Throughout the vegetation period the highest TP values were more or less consistently measured in these two taxa. There was no significant change in TP values of *P. flexuosa* with infusion at 90 °C for 5 min or with aqueous methanol for 72 h between April and August, while values of infusion at 90 °C for 24 h showed significant slight decline (by 13%) in this species until August. Changes in TP values of *P. flexuosa, P. humilis* and *P. sulphurea var. sulphurea* from September to November were significantly different and showed the same tendency with the three extraction methods. *P. flexuosa* had the lowest values during most months, with the significantly lowest value in October with aqueous methanolic extraction for 72 h, but the difference in other months compared especially to *P. sulphurea var. sulphurea* was not always significant. If we compared the changes in all taxa shown in Tab. 1, there was a clear trend in the changes of TP during the vegetation period, with high values in April and November, which is consistent with the findings of Zhang *et al.* (2002) who have found the highest total flavonoid level in *Phyllostachys nigra* leaves between November and April and of Ni *et al.* (2012) who have reported that total phenolics and total flavonoids in leaves of bamboo *Sasa argenteostriatus* were the highest from November to March. In addition, the same tendency was measured in leaves of bamboos *Indocalamus herklotsii, I. decorus* and *I. latifolius* by Su *et al.* (2011) and in *Phyllostachys heterocycla, Pleioblastus amaru, Dendrocalamus oldhami* and *Acidosasa edulis* by Lü *et al.* (2011). TP values declined during May, June and July and the change was significant for most extraction methods, months and taxa except for *P. flexuosa*, as described above. This decline is also reflected in the report of Zhang *et al.* (2002) and Ni *et al.* (2012). Our results clearly show a peak in August and September which has not been reported in the literature before, with an abrupt drop and low values in October. *P. aureosulcata f. aureocaulis* was the only taxon which had exactly the opposite tendency in October compared to the other taxa, with the highest significantly TP values (292.6±9.2; 386.0±13.2; 207.9±2.8 μg GA/ml) with all three extractions (infusion 90 °C, 5 min or 24 h or aqueous methanol 72 h. This indicated a putative effect of the environmental factors on total phenolic contents of bamboo leaves, which required further analyses.

Tab. 1. Changes in total phenolics (TP) content (μg GA/ml) and antioxidant (AC) capacity (μg AA/ml) using different extraction methods (M-methanol; A-aqueous infusion; 90- infusion at 90 °C for 5 min) *Phyllostachys* taxa between April and November. PAA-*Phyllostachys aureosulcata f. aureocaulis*, PF-*P. flexuosa*, PH-*P. humilis*,PSS-*P. sulphurea* var. *sulphurea*

Month	Taxon	TPM	ACM	TPA	ACA	TP90	AC90
April	PAA	258.7±8.0	534.5±9.2	409.5±12.0	334.9±4.0	266.1±12,1	337.8±12.8
	PF	183.9±4.8	334.8±3.2	252.9±2.0	261.3±2.5	257.9±4.4	339.2±7.2
	PH	252.9±8.0	519.7±8.4	388.4±8.1	339.5±4.1	391.3±4.3	513.5±13.2
	PSS	178.1±12.1	302.3±14.0	259.1±20.4	223.8±10.8	244.6±2.0	329.2±25.2
May	PAA	119.0±8.0	394.9±11.2	314.9±10.5	225.3±11.6	214.0±22.7	275.4±13.6
	PF	171.1±16.2	328.7±6.8	279.8±8.8	220.7±17.6	232.2±9.2	302.0±20.4
	PH	253.3±4.0	566.3±26.4	411.6±2.1	339.5±2.0	395.0±14.7	510.4±15.2
	PSS	154.1±8.7	280.6±4.0	241.7±8.0	191.5±6.7	195.0±8.8	259.9±7.6
June	PAA	112.7±4.1	399.5±8.3	278.1±1.2	193.6±7.2	240.1±16.4	355.5±4.8
	PF	148.8±28.0	339.4±4.0	262.4±4.7	157.5±23.6	240.1±18.2	346.4±19.2
	PH	205.4±4.3	442.8±30.0	334.7±24	244.1±18.4	295.0±28.0	408.5±22.7
	PSS	109.5±14.0	175.1±19.2	178.1±17.6	69.5±6.8	190.5±22.1	264.5±18.0
July	PAA	144.6±2.5	344.6±3.6	251.2±4.0	143.1±2.0	226.4±21.6	320.9±21.2
	PF	132.2±6.1	260.9±4.0	212.4±4.9	135.7±6.4	225.2±11.2	324.6±16.0
	PH	139.3±16.2	291.9±9.6	238.8±8.1	124.1±15.2	191.3±15.2	266.8±7.2
	PSS	90.5±6.0	119.0±11.2	126.4±8.0	31.9±6.8	145.5±22.6	209.5±14.1
August	PAA	178.5±14.3	383.0±4.0	290.1±8.2	187.7±8.	184.3±12.3	331.8±1.2
	PF	153.3±18.4	329.9±6.8	255.8±12.3	151.6±8.0	268.6±20.0	373.5±20.0
	PH	181.0±8.0	430.8±12.4	372.7±28.7	190.9±6.4	177.3±19.6	394.2±3.2
	PSS	190.1±2.7	272.0±2.0	256.2±18.0	163.2±6.8	272.3±3.2	446.3±15.6
September	PAA	153.7±4.8	365.8±7.6	283.9±6.4	131.2±4.8	202.5±16.0	376.7±8.8
	PF	185.1±4.8	386±16.0	311.2±8.1	175.0±7.2	254.1±4.1	406.7±18.4
	PH	196.7±24.1	437.6±2.8	335.5±14.7	194.2±22.4	268.2±28.4	378.1±7.2
	PSS	174.8±2.0	320.7±12.4	281.0±5.2	141.0±1.6	294.6±24.2	400.2±11.2
October	PAA	207.9±2.8	438.0±16.8	386.0±13.2	208.7±1.2	292.6±9.2	350.1±7.2
	PF	56.6±16.2	142.8±15.6	93.8±11.2	22.5±15.2	81.4±4.0	163.2±7.2
	PH	155.8±3.6	191.8±8.0	212.4±14.4	133.3±6.0	249.2±23.2	367.0±20.4
	PSS	73.6±8.0	139.4±19.2	126.9±18.0	33.4±4.8	103.7±2.8	149.4±11.2
November	PAA	180.2±16.8	314.9±4.8	270.7±17.6	190.6±18.0	235.1±15.6	329.8±2.0
	PF	252.1±16.7	303.1±16.0	255.8±14.0	232.0±27.6	245.9±6.0	363.2±27.6
	PH	190.1±4.0	413.3±5.6	311.6±15.2	185.9±2.0	318.6±8.1	420.2±22.4
	PSS	154.1±6.8	249.6±8.4	205.0±3.2	134.8±2.4	232.2±14.4	298.3±21.6
Month		*	***	*	***	NS	NS
Genotype		*	***	***	***	*	*
Month^{x}Genotype		*	***	**	***	*	*

NS - Non-significant,* significant at P=0.05,** significant at P=0.01 ,*** significant at P=0.001.

Changes in total antioxidant capacity

Results for AC in the case of all three extraction methods (infusion 90 °C, 5 min or 24 h or aqueous methanol 72 h) are shown in Tab. 1. The highest AC values were obtained in the case of infusion at 90 °C, for 5 min. Somewhat lower (8-17% lower on average) values were measured with aqueous methanol for 72 h. Only in April and May, values were similar or slightly higher (12% higher on average) with aqueous methanolic extraction for 72 h compared to infusion at 90 °C for 5 min. The lowest (31-59% lower on average) values were measured for infusion at 90 °C for 24 h. Similar to the results described for TP during the examined period the highest values of AC were measured in April and May also in *P. humilis* and *P. aureosulcata f. aureocaulis* but this time in the case of aqueous methanolic extraction for 72 h and infusion at 90 °C for 5 min, but the difference was not always significant. Only in the case of infusion at 90 °C for 5 min were the AC values of *P. humilis* significantly higher than those of *P. aureosulcata f. aureocaulis* during both months of April and May. During the course of the examined period similar to the results of TP the highest AC values were more or less consistently measured in these two taxa. As the results of TP, there was no significant change in AC values of *P. flexuosa* with aqueous methanol for 72 h or with infusion at 90 °C for 5 min or between April and August, while values of infusion at 90 °C for 24 h showed a significant decline (by 50%) in this species until August. From September to November changes in AC values of *P. flexuosa, P. humilis* and *P. sulphurea var. sulphurea* were significantly different and showed the same tendency with the three extraction methods, as in the case of TP. *P. flexuosa* and *P. sulphurea f. sulphurea* had the lowest values during most months, with the lowest value for *P. flexuosa* in October with infusion at 90 °C for 24 h, but the difference was not significant compared to *P. sulphurea var. sulphurea.* When changes in all taxa were compared (see Tab. 1), there was a clear trend in the changes of AC during the vegetation period, with high values in April and November, which is similar to the findings of Zhang *et al.* (2002) who have found the highest antioxidant activity in *Phyllostachys nigra* leaves, between autumn and spring, with the strongest measured activity in February. The results of Ni *et al.* (2012) also showed that the highest antioxidant activity in bamboo *Sasa argenteostriatus* leaves occurred in December and remained generally high from October to April. AC values declined during May, June and July and the change was significant for most months, extraction methods and all taxa except for *P. flexuosa*, as described above. This decline is also reflected in the report of Zhang *et al.* (2002) and Ni *et al.* (2012). Our results clearly showed a peak in August and September which has not been reported in the literature before, then a steep decline and low values in October. *P. aureosulcata f. aureocaulis* was the only taxon which had exactly the opposite tendency in October compared to the other taxa, with significantly AC higher values with infusion at 90 °C for 24 h or aqueous methanol for 72 h. Since both Hu *et al.* (2012) and Ni *et al.* (2012) have found a positive correlation between bamboo total phenolics and antioxidant activity our results were also in agreement with those of Su *et al.* (2011) and Lü *et al.* (2011).

Correlation between environmental parameters and total phenolic content

A late summer peak in TP values has not been reported in the literature before and since *Phyllostachys* is neither native nor naturalized in Hungary, we examined the possible environmental cause of this phenomenon. Environmental parameters including daily maximum and minimum temperature (°C), cumulative precipitation plus irrigation (CPI) and cumulative potential evapotranspiration (CE) values, for the examined April to November period, are shown on Fig. 1. The temperature range was more or less the same till early May, and remained with larger variations till the end of May. From the beginning of June both daily minimum (were above +10 °C) and maximum (were above +15 °C) values started to rise and levelled off between mid-July to mid-August. From mid-August temperatures started to decline and the lowest daily minimum and maximum temperatures (at or below +5 °C) were recorded from mid-October. CPI values were similar or lower than the CE values till late June. From late June CPI values were higher than the CE values until the end of July. From early August CPI values were lower than CE values until mid-October. From mid-October, CPI values were similar or higher than CE. The above changes in CPI compared to CE can be interpreted, as there was probably water surplus during the period between late June and end of July and between mid-October and end of November. Abiotic stress factors are known to induce phenylpropanoid metabolism (Dixon and Paiva, 1995) including temperature (Janda *et al.*, 1999; Pál *et al.*, 2013) and water stress (De Abreu and Mazzafera, 2005). To find correlation between TP and water and/or temperature stress, changes of TP in the case of all taxa and extraction methods expressed as % of maximum value were compared, with stress index values derived from daily maximum and minimum temperature, CPI and CE values for the examined April to November period. A close correlation $R^2 = 0.32$ (p=0.001) was found between stress index values and the change in TP shown on Fig. 3. By comparing the change of TP in *P. flexuosa, P. humilis* and *P. suphurea var. sulphurea,* with stress index values the correlation is even closer $R^2 = 0.52$ (p=0.001). Our results were in agreement with the findings of Oh *et al.* (2009) who have found that low temperatures caused an increase in total phenolics content in lettuce. The same was reported for *Hypericum perforatum* (Zobayed *et al.*, 2005). The effect of altitudinal variation also caused an increase in total phenolic and flavanoid contents and antioxidant activity in bamboo *Indocalamus latifolius* (Ni *et al.*, 2013). Our findings are especially in conjunction with those of De Abreu and Mazzafera (2005) who have found that water or low temperature stress caused an increase of phenolic contents in *Hypericum brasiliense*. During the cold-acclimation of plants an accumulation of secondary metabolites, including TP was reported by Cansev *et al.* (2012) and Pennycooke *et al.* (2005). Our interesting results with *P. aureosulcata f. aureocaulis* in which TP values immediately rose after exposure to low temperatures in October, while the other taxa produced high values only later in November, could be explained by the different freeze tolerance of these species, *P. aureosulcata* being the most freeze tolerant of the four taxa (Ohrnberger, 2002), and also by possible inter/intra-specific differences in responsiveness to low temperature exposure during cold-acclimation as reported for other plants (Li *et al.*, 2002; Li *et al.*, 2005; Guárdia *et al.*, 2013). Previous available studies on bamboos have not attempted to find reasons for seasonal changes in TP, only defined the period of the year optimal for harvest of bamboo leaves rich in TP. Our findings suggested that climatically applicable exposure to moderate water stress and low temperatures could enhance the phytochemical content of bamboo leaves intended for harvest as *herba* or food additive. But further specific experiments involving temperature and or water stress with regards to phytochemical content of bamboo leaves need to be carried out to underline this.

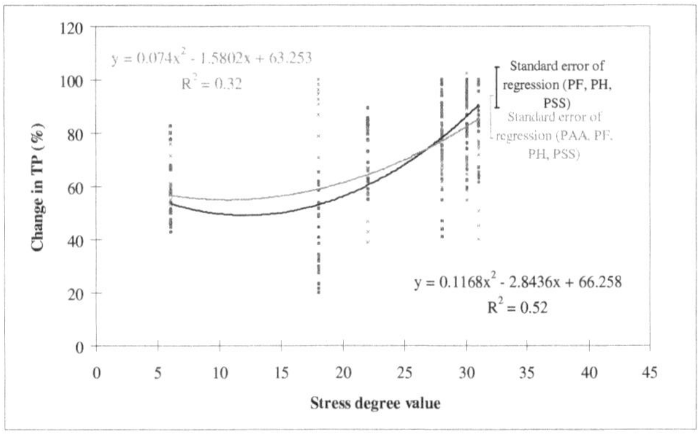

Fig. 3. Correlation between stress index and change in TP content (%) expressed as average of all four taxa PAA-*Phyllostachys aureosulcata f. aureocaulis*, PF-*P. flexuosa*, PH-*P. humilis*, PSS-*P. sulphurea var. sulphurea* or three taxa except *Phyllostachys aureosulcata f. aureocaulis*. (p=0.001). Black and grey dots and lines refer to (PF, PH, PSS) and (PAA, PF, PH, PSS) respectively.

Conclusions

The highest significant TP and AC values were obtained by aqueous infusion at 90 °C for 5 min. During the vegetation period the highest values of TP and AC were measured in April and May in the case of *PAA* and *PH*. The lowest values during most months were measured in *PF* and *PSS*. Changes of TP and AC during the vegetation period gave high values in April and November in all taxa. A peak in TP and AC detected in August-September has not been reported in the literature before. Compared to the other taxa *PAA* produced exactly the opposite tendency in October, with significantly the highest TP and AC values. A close correlation was found between stress index values and the change in TP showing that water or low temperature stress caused an increase of phenolic contents. The results of *PAA* could possibly be explained by inter/intra-specific differences during cold-acclimation in the responsiveness to low temperature exposure.

Acknowledgements

This study was funded by TÁMOP-4.2.1. B-11/2/KMR 2011-0003 and Research Centre of Excellence 17586-4/2013/TUDPOL Szent István University and KTIA_AIK_12-1-2012-0012 project and OTKA K84290 project. Attila Hegedűs acknowledges a János Bolyai Scholarship, Hungarian Academy of Sciences.

References

Benzie IFF, Strain JJ (1996). The ferric reducing ability of plasma (FRAP) as a measure of "antioxidant power": the FRAP assay. Anal Biochem 239(1):70-76.

Cansev A, Gulen H, Celik G, Eris A (2012). Alterations in total phenolic content and antioxidant capacity in response to low temperatures in olive (*Olea europea* L. 'Gemlik'). Plant Arch 12(1):489-494.

Chen JK, Chen TT (2004). Chinese Medical Herbology and Pharmacology. 1st Ed. Art of Medicine Press, City of Industry, 1267 p.

De Abreu IN, Mazzafera P (2005). Effect of water and temperature stress on the content of active constituents of *Hypericum brasiliense* Choisy. Plant Physiol Biochem 43(3):241-248.

Dixon RA, Paiva NL (1995). Stress-induced phenylpropanoid metabolism. The Plant Cell 7(7):1085-1097.

Fu XC, Wang MW, Li SP, Wang HL (2006). Anti-apoptic effect and the mechanism of orientin on ischaemic/reperfused myocardium. J Asian Nat Prod Res 8(3):265-272.

Guárdia M, Savé R, Diaz R, Vilanova A, Aletá N (2013). Genotype and environment: two factors related to autumn cold hardiness on Persian walnut (*Juglans regia* L.). Ann Forest Sci 70(8):791-800.

Helyes L, Neményi A, Pék Z, Berki M, Daood HG (2013). Effect of variety and water supply on phytochemical (phenolics and carotenoids) content and composition of processing tomato. Acta Hort 971:93-98.

Helyes L, Varga Gy (1994). Irrigation demand of tomato according to the results of three decades. Acta Hort 376:323-328.

Hu C, Xu D, Chen H, Yuan K (2012). Contents of the total flavonoids and the total phenols and antioxidant activities in the leaf from different species of *Phyllostachys*. Adv Mat Res 343-344:1103-1108.

Hwang JH, Choi SY, Ko HC, Jang MG, Jin YJ, Kang SI, Park JG, Chung WS, Kim SJ (2007). Anti-inflammatory effect of the hot water extract from *Sasa quelpaertensis* leaves. Food Sci Biotechnol 16(5):728-733.

Im JS, Lee EH, Lee JN, Kim KD, Kim HY, Kim MJ (2010). Sulforaphane and total phenolics contents and antioxidant activity of radish according to genotype and cultivation location with different altitudes. Korean J Hortic Sci 28(3):335-342.

In MJ, Park MK, Kim S, Chae HJ, Chae MW, Sone J, Ji H-S, Han KS, Kim DC (2010). Composition analysis and antioxidative activity of maengjong-juk (*Phyllostachys pubescens*) leaves tea. J Appl Biol Chem 53(2):116-119.

Janda T, Szalai G, Tari I, Páldi E (1999). Hydroponic treatment with salicylic acid decreases the effects of chilling injury in maize (*Zea mays* L.) plants. Planta 208(2):175-180.

Jeong CH, Choi SG, Heo HJ (2008). Analysis of nutritional components and evaluation of functional activities of *Sasa borealis* leaf tea. Korean J Food Sci Technol 40(5):586-592.

Jung SH, Lee JM, Lee HJ, Kim CY, Lee EH, Um BH (2007). Aldose reductase and advanced glycation endproducts inhibitory effect of *Phyllostachys nigra*. Biol Pharm Bull 30(8):1569-1572.

Khokhar S, Magnusdottír SGM (2002). Total phenol, catechin, and caffeine contents of teas commonly consumed in the United Kingdom. J Agr Food Chem 50(3):565-570.

Kim DS, Son EJ, Kim M, Heo YM, Nam JB, Ro JY, Woo S-S 2010. Antiallergic herbal composition from *Scutellaria baicalensis* and *Phyllostachys edulis*. Planta Med 76(7):678-682.

Kim KA, Kang KD, Jo H, Lee HJ, Kim CY, Um BH, Jung SH (2013). Compound isolated from the leaves of *Phyllostachys nigra* protects NMDA-induced retinal degeneration. J Food Biochem 37(4):465-474.

Kim NJ, Lee SJ, Kwon CH, Hong ND (1995). Antilipoperoxidant effects of leaves of *Phyllostachys bambusoides* S. et Z. Korean J Pharmacognosy 26(4):368-376.

Kim SH, Kim TS, Kim SJ, Seong CN, Lee OH, Lee HJ, Yoo JC (2007). Inhibition of interleukin-12 production in mouse macrophages via suppression of nuclear factor-xB binding activity by *Phyllostachys nigra* var. *henonis*. Immunopharm Immunot 29(1):131-139.

Kim SH, Kim TS, Lee HJ, Yoo JC (2007). Enhancement of 1,25-dihydroxyvitamin D3- and all-trans retinoic acid-induced differentiation of human leukemia HL-60 cells by *Phyllostachys nigra* var. *henonis*. Immunopharm Immunot 29(1):119-129.

Kim SM, Jeon JS, Kang SW, Kim WR, Lee KD, Um BH. (2012). Composition analysis and antioxidant activity of ojuk (*Phyllostachys nigra* Munro) leaf tea and shoot tea. J Appl Biol Chem 55(2):95-101.

Kleinhenz V. Midmore D (2001). Aspects of bamboo agronomy. Adv Agron 74:99-146.

Kweon MH, Hwang HJ, Sung HC (2001). Identification and antioxidant activity of novel chlorogenic acid derivatives from bamboo (*Phyllostachys edulis*). J Agr Food Chem 49(10):4646-4655.

Kwon YD, Ko EY, Hong SJ, Park SW (2008). Comparison of sulforaphane and antioxidant contents according to different parts and maturity of broccoli. Korean J Hortic Sci 26(3):344-349.

Lee HJ, Kim K-A, Kang KD, Lee EH, Kim CY, Um BH, Jung SH (2010). The compound isolated from the leaves of *Phyllostachys nigra* protects oxidative stress-induced retinal ganglion cells death. Food Chem Toxicol 48(6):1721-1727.

Li C, Puhakainen T, Welling A, Viherá-Aarnio A, Ernstsen A, Junttila O, Heino P, Palva ET (2002). Cold acclimation in silver birch (*Betula pendula*). Development of freezing tolerance in different tissues and climatic ecotypes. Physiol Plantarum 116(4):478-488.

Li C, Wu N, Liu S (2005). Development of freezing tolerance in different altitudinal ecotypes of *Salix paraplesia*. Biol Plantarum 49(1):65-71.

Li T, Park MH, Kim MJ, Ryu BM, Kim MJ, Moon GS (2008). Tachioside, an antioxidative phenolic glycoside from bamboo species. Food Sci Biotechnol 17(6):1376-1378.

Lin Y, Collier AC, Liu W, Berry MJ, Panee J (2008). The inhibitory effect of bamboo extract on the development of 7, 12-dimethylbenz[a]anthracene (DMBA)-induced breast cancer. Phytother Res 22(11):1440-1445.

Lü ZL, Ren ML, Quyang YL, Lin X, Zhang BL (2011). Seasonal variations of flavonoid contents in leaves of four of bamboo species. Beijing Linye Daxue Xuebao/Journal of Beijing Forestry University 33(4):81-84.

Ni Q, Wang Z, Xu G, Gao Q, Yang D, Morimatsu F, Zhang Y (2013). Altitudinal variation of antioxidant components and capability in *Indocalamus latifolius* (Keng) McClure leaf. J Nutr Sci Vitaminol 59(4):336-342.

Ni Q, Xu G, Wang Z, Gao Q, Wang S, Zhang Y (2012). Seasonal variations of the antioxidant composition in ground bamboo *Sasa argenteastriatus* leaves. Int J Mol Sci 13(2):2249-2262.

Oh H-K, Lim H-S (2010). Quality characteristics of the hamburger patties with bamboo (*Sasa borealis*) leaf extract with/without cooked rice. Korean J Food Sci An 30(5):833-841.

Oh M-M, Carey EE, Rajashekar CB (2009). Environmental stresses induce health-promoting phytochemicals in lettuce. Plant Physiol Bioch 47(7):578-583.

Ohrnberger D (2002). Bamboos of the world: Annotated nomenclature and literature of the species and higher and lower taxa. 1st Ed. Elsevier Science B.V., Amsterdam, 596 p.

Pál M, Gondor OK, Janda T (2013). Role of salicylic acid in acclimation to low temperature. Acta Agron Hung 61(2):161-172.

Panee J, Liu W, Lin Y, Gilman C, Berry MJ (2008). A novel function of bamboo extract in relieving lipotoxicity. Phytother Res 22(5):675-680.

Park HS, Ju HL, Hyun JK, Hyun JC, Lee IS (2007). Antioxidant flavone glycosides from the leaves of *Sasa borealis*. Arch Pharm Res 30(2):161-166.

Park YO, Lim HS (2009). Antioxidant activities of bamboo (*Sasa borealis*) leaf extract according to extraction solvent. J Korean Soc Food Sci Nutr 38(12):1640-1648.

Park YO, Lim HS (2010). Antimicrobial activity of bamboo (*Sasa borealis*) leaves fraction extracts against food poisoning bacteria. J Korean Soc Food Sci Nutr 39(12):1745-1752.

Pennycooke JC, Cox S, Stushnoff C (2005). Relationship of cold acclimation, total phenolic content and antioxidant capacity with chilling tolerance in petunia (*Petunia x hybrida*). Environ Exp Bot 53(2):225-232.

Perva-Uzunalic A, Skerget M, Knez Z, Weinreich B, Otto F, Grüner S (2006). Extraction of active ingredients from green tea (*Camellia sinensis*): Extraction efficiency of major cathecins and caffeine. Food Chem 96(4):597-605.

Ryou SH, Kang MS, Kim KI, Kang YH, Kang JS (2012). Effects of green tea or *Sasa quelpartensis* bamboo leaves on plasma and liver lipids, erythrocyte Na efflux, and platelet aggregation in ovariectomized rats. Nutr Res Pract 6(2):106-112.

Shin DH, Kim JC, Kim SR, Oh H, Park IC, Oh KS, Chung HJ, Jang JS, Kim SH (2003). The effect of *Phyllostachys nigra var henonis* Strapf. leaf extract in the gamma-irradiated mice. Korean J Vet Res 43(1):49-55.

Singleton VL, Rossi JA (1965). Colometry of total phenolics with phosphomolibdic-phosphotungstic acid reagents. Am J Enol Viticult 161(3):144-158.

Su CH, Liu GH, Wang FS, Ding YL, Xue JH (2011). Seasonal changes of *Indocalamus* leaf active ingredients contents. Chinese J Appl Ecol 22(9):2471-2476.

Turkmen N, Sari F, Velioglu YS (2006). Effects of extraction solvents on concentration and antioxidant activity of black and black mate tea polyphenols determined by ferrous tartrate and Folin-Ciocalteu methods. Food Chem 99(4):835-841.

Wang H, Helliwell K (2001). Determination of flavonols in green and black tea leaves and green tea infusions by high-performance liquid chromatography. Food Res Int 34(2-3):223-227.

You Y, Kim K, Yoon HG, Lee J, Chun J, Shin DH, Park J, Jun W (2010). Chronic effect of ferulic acid from *Pseudosasa japonica* leaves on enhancing exercise activity in mice. Phytother Res 24(10):1508-1513.

Zhang Y, Wu XQ, Yu ZY (2002). Studies on seasonal variation of flavonoids and lactones in bamboo leaves. Linchan Huaxue Yu Gongye/Chemistry and Industry of Forest Products 22(2):65-69.

Zobayed SMA, Afreen F, Kozai T (2005). Temperature stress can alter the photosynthetic efficiency and secondary metabolite concentration in St. John's wort. Plant Physiol Bioch 43(10-11):977-984.

Influence of Ecological Conditions on Seeds Traits and Essential Oil Contents in Anise (*Pimpinella anisum* L.)

Milica G. ACIMOVIC[1*], Jasna KORAC[2], Goran JACIMOVIC[2], Snezana OLJACA[1], Lana DJUKANOVIC[3], Vesna VUGA-JANJATOV[4]

[1]*University of Belgrade, Faculty Agriculture, Nemanjina 6, 11080 Zemun, Serbia; acimovicbabicmilica@gmail.com (*corresponding author)*
[2]*University of Novi Sad, Faculty of Agriculture, Trg Dositeja Obradovica 8, 21000 Novi Sad, Serbia*
[3]*Institute for Plant Protection and Environment, Teodora Drajzera 9, 11000 Belgrade, Serbia*
[4]*Agricultural High School Sabac, Vojvode Putnika 56, 15000 Sabac, Serbia*

Abstract

Anise (*Pimpinella anisum* L.) is an annual plant of the Apiaceae family, widely cultivated for the seed and essential oil. Under field condition, anise is cultivated in a vast number of countries including Serbia. Field experiments were carried out during two growing seasons, at three localities, in order to determine the effect of different soil and climatic conditions on the quality of anise seed principal traits (thousand seed weight, germination energy and total germination) as well as on the content and quality of its essential oil. During the experiment it was found that the value of aniseed principal traits was significantly lower in hotter and drier year in comparison to the year with moderate conditions, because of more favourable weather conditions for seed development during this year. A significantly higher concentration of essential oil was also accumulated in the moderate year in comparison to the dry and hot one. This can be attributed to a longer period of fruit formation and synthesis of essential oils and better climatic conditions. It can be concluded that drought caused a significant decrease in thousand seed weight, germination energy and total germination as well as essential oil content in anise. Contrary to this, the content of *trans*-anethole was significantly higher in the dry year. It can be assumed that under stress conditions the amount of *trans*-anethole in the essential oil increases, because in such conditions the plants produce more secondary metabolites, substances that prevent oxidation processes in the cells.

Keywords: essential oil content, germination, *Pimpinella anisum,* seed weight, *trans*-anethole

Introduction

Anise (*Pimpinella anisum* L.) is an annual plant of the Apiaceae family, widely cultivated for fruit and essential oil. This plant, originating from the Mediterranean region, is cultivated under field conditions in a vast number of countries, especially in Southern Europe and Southeast Asia, but also in the United States, China and Chile (Embong *et al.*, 1977). Since anise favors warm climatic conditions throughout the growing season, it is particularly cultivated in subtropical regions (Ullah *et al.*, 2013).

Anise is an important raw natural material mostly used in medicine, pharmaceutics, perfumery and cosmetic industries. It is also used as functional food; it is added to confectionary products (honey cookies, candies) and alcoholic beverages (liqueurs and sweet flavoured wines). Anise fruits, also known as aniseed, contain essential oil (*Anisi aetheroleum*). Recent research found that this essential oil has an antimicrobial and antifungal properties (Kubo and Himejima, 1991; Kosalec *et al.*, 2005; Ozcan and Chalchat 2006; Yazdani *et al.*, 2009) and antioxidant effect on the human health (Gülcin *et al.*, 2003; Tepe *et al.*, 2006; Rajeshwari *et al.*, 2011). The primary constituent of anise essential oil is *trans*-anethole, a volatile phenylpropanoid, which comprises 80 to 90% of the oil (Tabanca *et al.*, 2006; Orav *et al.*, 2008). As a predominant component, it gives off the characteristic odour and has sweet aromatic flavour (Koeduka *et al.*, 2009).

The quality of anise is determined mainly on the basis of the essential oil content and its composition. Both of these parameters are significantly affected by environmental factors, i.e. soil type and weather conditions during the year (temperature, precipitation, etc.) especially during the development of anise fruit (stages of plant maturity) and by the applied agronomic practices (Zehtab-Salmasi *et al.*, 2001; Omidbaigi *et al.*, 2003; Tuncturk and Yildirim, 2006; Özel, 2009; Jevdjović *et al.*, 2012). Also, the concentration of essential oil may significantly vary among anise fruits of different origin (Tabanca *et al.*, 2006; Orav *et al.*, 2008).

The aim of this study was to determine the effect of different localities, with different soil and microclimatic conditions during two growing seasons, on the quality of anise seed principal traits (thousand seed weight, germination energy and total germination), on its essential oil content, as well as on the content of *trans*-anethole, the main constituent of the essential oil.

Materials and Methods

Experiment locations

Field experiments were carried out during two years (Y), at three localities (L) in the Vojvodina Province, Republic of Serbia: L1 – Mošorin (45°18' N, 20°09' E, 111 m above sea level), L2 – Veliki Radinci (45°02' N, 19°40' E, 110 m above sea level) and location L3 – Ostojićevo (45°54' N, 20°09' E, 88 m above sea level). All sites are located in Vojvodina region (northern Serbia), which is mostly a flat area located in the southern part of the Pannonian lowland.

Soil conditions

In Vojvodina region, the dominant soil type is calcareous chernozem on loess terrace. Individual soil samples were taken from the top soil layer (0-30 cm) and analyzed according to standardized methods adopted in Serbia and EU countries for the purpose of determining agrochemical properties of soil types on each locality. At all three locations, all soil had a neutral reaction to the soil solution and had moderately humus content (2.2-2.7%). The soil at localities L1 and L2 were highly calcareous chernozem, whereas the content of $CaCO_3$ at L3 was significantly lower. The content of available phosphorus and potassium was the highest on locality L1 (81.6 mg $100g^{-1}$ of soil), while at L2 and L3 the content of these nutrients was moderate to optimal (results of soil analyses are not shown in this paper).

Climate conditions

Climate in Vojvodina region is moderate continental with some tendencies towards continental. The region is located in a semi-arid area where the variations as amount of precipitation, air temperature and other important climatic elements among years are substantial. Data for basic meteorological elements per month - precipitation (P, mm) and average air temperatures per month (T, °C) for both analyzed years and for all three localities is shown in Tab. 1. Weather conditions in the analysed years indicate that, in the period of active growing season (IV-IX), both years were hotter in relation to long-term average values (LTA) and were moderately to severely dry.

Tab. 1. Precipitation and temperature conditions at analyzed locations

	Year	Month												GS[1]	WP[2]	Yearly (1.X-30.IX)
		I	II	III	IV	V	VI	VII	VIII	IX	X	XI	XII			
								L1 – Mošorin								
T (°C)	2011	0	0	6	13	17	21	22	23	20	11	3	4	19	-	12
	2012	2	-5	8	13	17	23	25	24	21	14	10	3	20	-	13
	LTA[3]	0	2	6	12	17	20	22	21	17	12	6	1	18	-	11
P (mm)	2011	25	37	26	23	63	37	62	2	25	35	2	49	212	266	478
	2012	43	67	4	83	51	31	48	4	14	48	36	55	230	200	430
	LTA	39	34	38	48	60	87	68	58	47	47	50	50	360	255	617
								L2 - Veliki Radinci								
T (°C)	2011	0	0	7	13	17	21	22	24	21	11	4	4	20	-	12
	2012	2	-4	8	13	17	22	25	24	20	14	10	1	20	-	12
	LTA	1	2	7	11	17	20	21	21	17	11	6	1	18	-	11
P (mm)	2011	31	31	15	20	46	69	93	0	19	25	5	46	247	246	493
	2012	40	52	4	85	72	31	40	0	14	47	31	41	242	172	414
	LTA	37	36	38	52	56	86	65	63	53	54	54	42	375	261	636
								L3 – Ostojićevo								
T (°C)	2011	0	0	7	13	17	21	22	24	21	11	3	4	20	-	12
	2012	1	-5	8	13	17	23	25	24	20	14	9	0	20	-	12
	LTA	0	1	6	11	17	20	22	21	17	11	5	1	18	-	11
P (mm)	2011	16	18	36	10	47	31	101	0	30	37	0	38	219	235	454
	2012	37	57	4	60	52	24	55	7	21	68	27	46	219	173	392
	LTA	31	31	33	48	53	74	59	54	49	43	45	41	337	224	561

[1]GS – Growing Season (IV-IX); [2]WP – Winter Precipitation (X-III), [3]LTA – Long-Term Averages (1965-2012)

Conducting the experiment

The anise seed used in this study was a widely grown early ripening cultivar, named N-210. In both years, at all three localities, anise was sown by hand at the end of March and during the first decade of April (depending on the year; Tab. 2), which is optimal time for sowing anise in Serbian climate. It was done directly into the soil with seedling depth of 2-3 cm, in continuous rows, 35 cm apart. One experimental plot contained five rows, but only the middle row from each plot was harvested to eliminate the marginal effect.

After germination of the plants in the field, they were thinned by hand in order to get 70 plants per meter. The plots were kept weed-free by hand weeding and hoeing, without the application of herbicides. The plants were harvested manually in August (2011) and at the end of July (2012) at the stage of their full maturity.

The seed materials were analyzed 3 months after harvesting in order to determinate the following: thousand seed weight (TSW), germination energy (GE), total germination (TG), as well as quantity and quality of anise essential oil (essential oil content – EOC and *trans*-anethole content – TAC).

Thousand seed weight was determined and the germination test was carried out in Seed testing laboratory of the Agricultural Extension Service in Sremska Mitrovica. Eight samples from every locality, containing 100 seeds each, were taken and weighed. Their average value was multiplied by 10 to calculate the thousand seed weight. Standard germination was carried out in a plastic box (10x15x6 cm) containing moist plated paper. There were six replications from each locality. A replicate consisted of a single box with 100 seeds. The plastic boxes were incubated at alternating temperatures of 20/30 °C with 8/16 h light/darkness regime and illumination was provided by white fluorescent tubes.

The seedlings were evaluated 7 and 21 days after sowing. The first count test was performed with a standard germination test. The percentage of normal seedlings was recorded 7 days after sowing, according to the International rules for seed testing (ISTA).

Essential oil content was determined at the Faculty of Chemistry, Belgrade, by hydrodistillation using Clevenger-type apparatus. The seed from each locality (10 g, in six replications) was crushed in an electric mill. The oil quality was assessed by applying combined gas chromatography-mass spectrometry. GC-MS analysis was performed by using an Agilent 6890 gas chromatograph coupled with an Agilent 5973 Network mass selective detector (MSD) in positive ion electron impact (EI) mode. The separation was achieved by using Agilent 19091S-433 HP-5MS fused silica capillary column, 30 m x 0.25 mm i.d. and 0.25 μm film thickness. The GC oven temperature was programmed from 60 °C to 285 °C at a rate of 4.3 °C/min. Helium was used as the carrier gas; inlet pressure was 25 kPa; linear velocity was 1ml/min at 210 °C. Injector temperature: 250 °C; injection mode: splitless. MS scan conditions: source temperature, 200 °C; interface temperature, 250 °C; energy, 70 eV; mass scan range, 40-350 amu. Identification of components was done on the basis of retention index and by comparison with reference spectra (Wiley and NIST databases).

Tab. 2. Sowing and harvest dates and phonological observations - developmental stages of anise and their duration (days) during two years at three analyzed localities

Developmental Stages / Year	2011						2012					
	L1	days	L2	days	L3	days	L1	days	L2	days	L3	days
Sowing date	26.03.	-	01.04.	-	06.04.	-	01.04.	-	04.04.	-	14.04.	-
Germination	11.04.	16	26.04.	25	12.05.	36	19.04.	18	22.04.	18	28.04.	14
Stem elongation	17.05.	36	28.05.	32	06.06.	25	05.06.	47	08.06.	47	10.06.	43
Flowering	18.06.	32	29.06.	30	09.07.	33	18.06.	13	20.06.	12	01.07.	21
Fruit formation	10.07.	22	18.07.	19	02.08.	24	01.07.	13	13.07.	23	16.07.	15
Maturation (harvest)	06.08.	27	16.08.	29	29.08.	27	28.07.	27	31.07.	18	04.08.	19
Total length of growing season	-	133	-	135	-	145	-	118	-	118	-	112

Statistical analyses

The experiments were conducted in six replications at each locality. The results were processed as randomized complete block design combined over years by analysis of variance (ANOVA, GenStat Release v.9.1., Rothamsted Experimental Station, trial version), whereby in the sums of squares the year appears as an additional component. The assumed mathematical model of the experiment in a completely randomized design was:

$$X_{ijk}=\mu+\alpha_i+\beta_j+(\alpha\beta)_{ij}+\varepsilon_{ijk},$$

where α_i and β_j represent effects of localities and years, $(\alpha\beta)_{ij}$ is their interaction, and ε_{ijk} is error, i.e. random variable with distribution $N(0,\sigma^2)$. The significance of the influence of localities, conditions of the years, as well as their interaction were tested using the F-test of ANOVA. Duncan's Multiple Range Test at 5% probability level (p=0.05) was used to test the significance of differences between mean values of analysed localities and the years (Statistica v.9. software; StatSoft, Inc., 2007). Correlation analysis was performed by using GenStat Release v.9.1. software according to Pearson's methods to determine the relationship among the analysed traits.

Results and discussion

The F-test used for the analysis of trait variance (Tab. 3) indicates that the thousand seed weight, germination energy and total germination were affected by weather conditions – with statistically significant effect ($p<.001$), localities (L) and their interaction (Y×L) during the years (Y). In the case of thousand seed weight, this interaction was significant only for probability level $p<0.05$. Annual conditions and locations had a significant effect on essential oil content, while the effect of their interaction was not significant. The effect of weather conditions during the years was also highly significant for the *trans*-anethole content. Although in this case the effect of localities was absent, the effect of YxL interaction was statistically significant. However, based on the percentage of individual sources of variation in the total sum of squares, it can be concluded that the localities had a dominant influence (48.6%) on the overall variability of thousand seed weight in the experiment, whereas the annual conditions had a dominant influence on germination energy, total germination and essential oil content (37.6; 44.5 and 36.7%, respectively). The strongest influence of the

Tab. 3. Selected indicators of ANOVA for weight of 1000 seeds (WS), germination energy (GE), total germination (TE), quantity and quality of aniseed essential oil (EO and TA)

Source of variation	d.f.	WS (g)		GE (%)		TG (%)		EO (%)		TA (% in EO)	
		s.s. (%)	F-pr.	s.s. (%)	F-pr.	s.s. (%)	F-pr.	s.s. (%)	F-pr.	s.s. (%)	F-pr.
Year (Y)	1	37,0	<,001**	37,6	<,001**	44,5	<,001**	36,7	<,001**	74,1	<,001**
Locality (L)	2	48,6	<,001**	26,4	<,001**	27,8	<,001**	27,2	<,001**	1,0	0,542^{ns}
Y x L	2	3,7	0,039*	24,9	<,001**	17,6	<,001**	2,5	0,333^{ns}	6,2	0,034*
Repetition	5	0,4	0,966^{ns}	2,3	0,294^{ns}	2,7	0,120^{ns}	10,4	0,133^{ns}	1,4	0,873^{ns}
Y x Repetition	5	0,7	0,913^{ns}	2,0	0,359^{ns}	2,1	0,216^{ns}	1,7	0,894^{ns}	1,9	0,768^{ns}
Residual	20	9,6		6,9		5,3		21,4		15,4	
Total	35	100		100		100		100		100	

d.f. – degrees of freedom; s.s. – sum of square (as % of total s.s.); F-pr. – probability of the F-test of ANOVA
** Statistically significant at the level $p<0.01$; * significant at the level $p<0.05$; ^{ns}non significant

years was evident on the variability of *trans*-anethole content (74.1%), while the impact of the localities and YxL interaction was significantly lower.

Thousand seed weight is an important yield parameter which has direct impact on the crop's final yield and quality of the seed. By observing the average thousand seed weight value for both analyzed years, it can be concluded that there were significant differences between localities, i.e. the highest values were obtained in locality L3 and they were significantly higher than those in L1 and L2 (Fig. 1). On average, the thousand seed weight values obtained in all three localities in 2012 were significantly lower than the values from the previous year. This can be explained by the fact that the temperatures in June 2012, and especially in July 2012 (critical periods for the developmental stages of anise: flowering and seed formation; shown in Tab. 2) were much higher (for 1-3 °C) than in the previous year. In addition, the total length of growing season in 2011 was longer (133-145 days) than in 2012 (112-118 days), which resulted in a longer period of seed formation, also shown in Tab. 2.

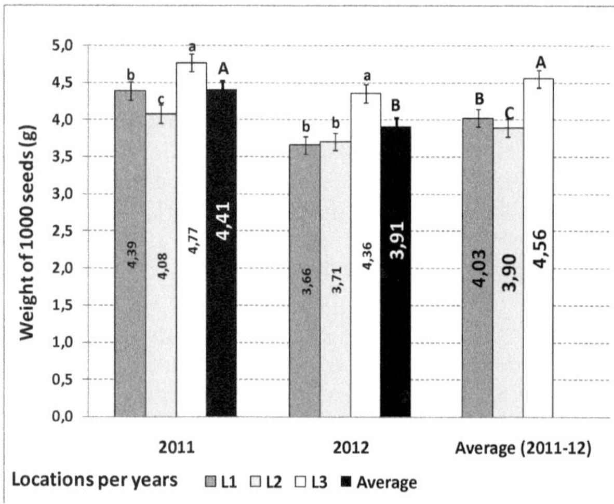

Fig. 1. The effects of years and localities on weight of 1000 seeds.

Bars represent differences between individual means which were deemed to be significant at $p<0.05$. Values followed by different small letter (a-c) on the bars indicate differences between localities in each year at the same significance level. Values followed by capital superscripts (A-C) in the same way relate to the average values per years (2011 and 2012) and localities (in average for both years)

More severe drought in 2012 caused a decrease in the 1000 seed weight. The average weight of 1000 seeds in 2012 was 3.91 g, which was lower by 11.3% than in 2011. According to Ipek *et al.* (2004); Jevdjović and Maletić, (2006); Tuncturk and Yildirim (2006) and Darzi *et al.* (2012), the mass of 1000 aniseed ranges from 2.08-5.46 g. These values are in accordance with our experiment, where the mass of 1000 seeds ranged from 3.66 to 4.77 g depending on the climatic conditions during the year and localities. Diederichsen (1996) reported that in the years with unfavourable weather conditions, coriander plants had a minimum weight of 1000 seed, which was confirmed by our results. The differences in the seed weight can be attributed to temporary variability of weather conditions (Hendrix and Trapp, 1992). The authors also reported that larger parsnip seed has greater germination than smaller seed, while Hasanah and Setyaningsih (1993) reported that seed size has not affected the germination of coriander.

Germination energy throughout the experiment had an average of 79.76% (Tab. 4) and was statistically significantly higher in the first year of the study (85.64%) than in the second (73.88%). As previously mentioned, during the first investigated year, climatic conditions were more favourable for seed formation, i.e. the higher thousand seed weight was obtained, which probably resulted in higher values of germination energy and total germination.

On average, there were statistically significant differences between localities in both years, i.e. the highest germination energy was obtained in L3 (86.27%), followed by L2 and the lowest germination energy was obtained in L1 (74.36%). Generally, in our research, anise had rather high germination energy (78.46-91.04% in 2011 and 62.13-81.50% in 2012) and total germination (87.38-93.71% in 2011 and 68.42-84.67% in 2012). Jevdjović and Maletić, (2006), found that germination energy ranged from 42.50 to 73.00% and total germination from 45.25 to 83.00% and that these values greatly depended on fertilization rate and year conditions. Similarly to germination energy, our results indicate that total germination was higher in the first investigated year, when it was up to 90.26%, in relation to 79.13% in year 2012. On average, there were significant differences between localities in both years, i.e. the highest total germination values were obtained in L3 and the lowest in L1. Other comparison and further comments are very similar to the relations obtained for germination energy.

A significantly higher concentration of essential oil was synthesized from plants in 2011 (3.93%) compared to the second investigated year (3.52%) (Fig. 2a). This might be due to suitable environmental conditions during essential oil accumulation. Higher essential oil content in 2011 can be attributed to better climatic conditions, to a longer period of seed formation and synthesis of essential oils (27-29 days in 2011 in relation to 18-27 days in 2012). The period of seed formation and filling in 2011 lasted longer and was prolonged by two months - July and August, when the temperatures in July were at the LTA level, and in August they were even higher (+2-3 °C in relation to LTA). Precipitation in July was at the LTA level at locality L1, which is significantly higher than the LTA values at L2 and L3 (+28 and +42 mm, respectively), whereas in August there was no rainfall. It can be concluded that the first part of this critical period provided favourable conditions for seed formation and filling, while the second, warmer and drier period favoured the synthesis of essential oil and fruit ripening.

Tab. 4. Germination energy and total germination (%) of aniseed depending on years and localities

Locality / Year	Germination energy (%)			Total germination (%)		
	2011	2012	Average	2011	2012	Average
L1	86,58^{b}	62,13^{d}	74,36^{C}	88,88^{b}	68,42^{d}	78,68^{C}
L2	78,46^{c}	78,00^{c}	78,23^{B}	87,38^{b}	84,67^{c}	86,03^{B}
L3	91,04^{a}	81,50^{c}	86,27^{A}	93,71^{a}	84,29^{c}	89,00^{A}
Average	85,64^{A}	73,88^{B}	79,76	90,26^{A}	79,13^{B}	84,69

Values followed by small superscript letter (a-d) across the years indicate differences between localities in each year and between years at the significance level p<0.05. Values followed by capital superscripts (A-C) in the same way relate to the averaged values per years (2011, 2012) and localities (average for both years).

Unlike the previous year, in 2012, the period of seed formation and filling was shorter and occurred only in July, with significantly higher temperatures in relation to LTA (+3-4 °C) and lower amounts of rainfall (from -4 mm in the locality L3 to -25 mm in L2; comparing to the LTA values). Thus, drought in 2012 caused lower content of essential oil in anise. Higher temperatures and precipitation deficiency during this year led to a decrease of 10.4% in essential oil content compared to 2011.

There were significant differences regarding the essential oil concentration among localities. In both years, the highest essential oil content was obtained in the locality L2 (4.08 and 3.79%, respectively) and the lowest in L3. Higher amount of precipitation in the period of aniseed filling might have influenced lower concentration of essential oil in anise (3.78 and 3.23%) in locality L3, in comparison to other localities. The fact that the locality has a significant role in the essential oil formation was indicated by Dražić *et al.*, (2007). In their research, the content of essential oil of anise grown in five localities in Serbia ranged from 0.5-1.2%. According to Orav *et al.* (2008), essential oil content from anise seed originating from different European countries ranged from 1 to 5%, which is more consistent with our results. Essential oil content of anise obtained from experiments carried out in Iran ranged from 2.21 to 2.75% (Zehtab-Salmasi *et al.*, 2001) and in investigations carried out in Turkey from 2.66 to 2.74% (Tuncturk and Yildrim, 2006). However, a higher level of anise essential oil, 3.5 to 5.5%, was observed in another experiment carried out in Iran (Omidbaigi *et al.*, 2003).

Contrary to the content of essential oil, the content of *trans*-anethole in essential oil was significantly higher in 2012 (Fig. 2b) in comparison to 2011. In both years, the differences between localities were minor when compared to the other investigated traits. This observation is in accordance with the F-test of ANOVA analysis (Tab. 3), where it was found that the year had an important influence on the *trans*-anethole content, while the effect of the locality was absent. Also, on average, there were no significant differences between localities in both years.

Throughout our experiment, the average content of *trans*-anethole was 94.78%, which is similar to an earlier study conducted by Orav *et al.* (2008). The content of *trans*-anethole was highly influenced by the year, while the effect of the locality was absent. In contrast to the content of the essential oil, the content of *trans*-anethole in the essential oil was significantly higher in the dry 2012 (96.35%) in comparison with 2011 (93.20%). On average, there were no significant differences between localities in both years.

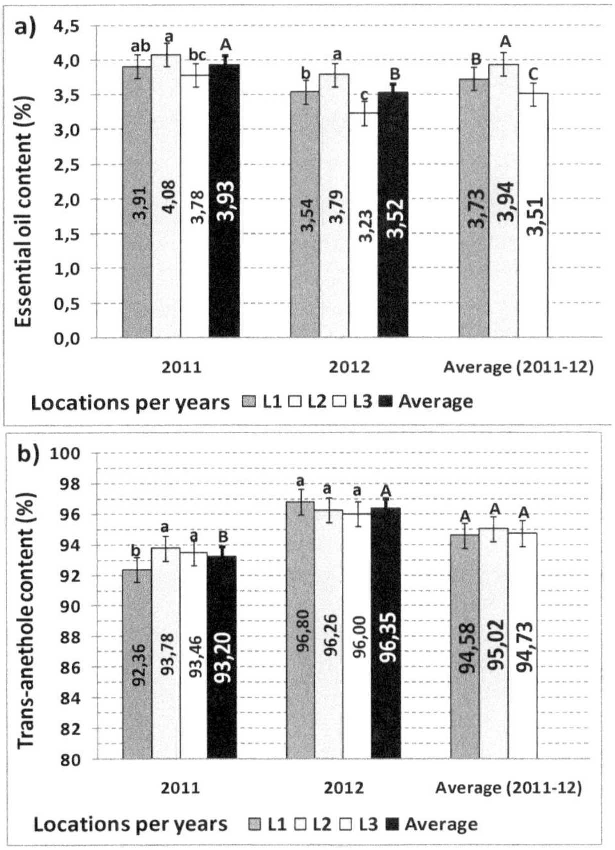

Fig. 2. The effects of years and localities on Essential oil content (a) and Trans-anethole content in EO (b)
Bars represent differences between individual means which were deemed to be significant at p<0.05. Values followed by different small letters (a-c) on the bars indicate differences between localities in each year at the p<0.05. Values followed by capital superscripts (A-C) in the same way relate to the average values per years (2011 and 2012) and localities (in average for both years)

It can be assumed that under stress conditions the amount of *trans*-anethole in the essential oil increases, because in such conditions the plants produce more secondary metabolites, the substances that prevent oxidation processes in the cells.

Correlation analysis (Tab. 5) shows that germination energy was in strong positive correlation with thousand seed weight (r=0.799*) and in negative correlation with *trans*-anethole content (r=-0.641*). Total germination was in the higher positive correlation with germination energy (r=0.952*) than with thousand seed weight (r=0.694*) and it was in moderate correlation with essential oil content (r=0.369*) as well as in strong negative correlation with *trans*-anethole content (r=-0.648). Also, *trans*-anethole content was in moderate negative correlations with thousand seed weight and the essential oil content (r=-0.604* and -0.543*, respectively).

Tab. 5. Simple correlation coefficients (r) between investigated parameters

Parameters	Weight of 1000 seeds (g)	Essential oil content (%)	*Trans*-anethole content (%)	Germination energy (%)
Essential oil content (%)	0.052	-	-	-
Trans-anethole content (%)	-0.604*	-0.543*	-	-
Germination energy (%)	0.799*	0.234	-0.641*	-
Total germination (%)	0.694*	0.369*	-0.648*	0.952*

Zehtab-Salmasi *et al.* (2001) show that the percentage of essential oil significantly decreases due to delayed sowing, while water deficit resulted in the increase of the essential oil content. A recent research confirms that the amount of the essential oil obtained from different species produced under drought conditions was either maintained or enhanced, depending on the extent of stress (Zehtab-Salmasi *et al.*, 2001). Water deficit during stem elongation and umbel formation reduces anise growth, grain yield and oil production (Zehtab-Salmasi *et al.*, 2001). In a two-year experiment in Pančevo, Serbia, Jevdjović *et al.* (2012) analyzed the effect of NPK fertilization on seed yield, quality and content of anise essential oil.

The results showed that, in comparison to the control plot, significantly higher mean values were obtained for all examined traits in the fields where NPK fertilizer was applied. Anise oil content was higher in the dry 2011 compared to the extremely wet 2010, while the overall seed germination rate of anise did not significantly differ between the years.

The correlation analysis shows that total germination is in positive correlation with germination energy, thousand seed weight and essential oil content, but in strong negative correlation with *trans*-anethole content. Also, *trans*-anethole was in moderate negative correlation with thousand seed weight and essential oil content. Kiralan *et al.* (2009) found that there is a correlation between essential oil yield and fruit size in coriander. Small fruits have a higher percentage of essential oil than large fruits.

Conclusions

Drought causes a significant decrease in thousand seed weight, germination energy and total germination as well as content of essential oil in anise. Contrary to this, *trans*-anethole content was significantly higher in the dry year. It can be assumed that under stress conditions the amount of *trans*-anethole in the essential oil increases, because in such conditions the plants produce more secondary metabolites, the substances that prevent oxidation processes in the cells.

Acknowledgements

This work was supported by the Ministry of Science and Technological Development, Republic of Serbia (Project TR-31051).

References

Darzi MT, Seyedhadi MH, Rejali F (2012). Effect of the application of vermicompost and phosphate solubilizing bacterium on the morphological traits and seed yield of anise (*Pimpinella anisum* L.). J Med Plants Res 6:215-219.

Diederichsen A (1996). Coriander (*Coriandrum sativum* L.). Promoting the conservation and use of underutilized and neglected crops. (3 Eds.) Institute of plant genetics and Crop Plant research, Gatersleben/International plant genetic resources institute, Rome, Italy.

Dražić S, Živanović T, Prodanović S (2007). Stability of productive traits of genotypes of cultivated medicinal plants of the family *Apiaceae*. Biotechnol Biotec Eq 21:100-106.

Embong MB, Hadziyev D, Molnar S (1977). Essential oils from species grown in Alberta. Anise oil (*Pimpinella anisum*). Can J Plant Sci 57:681-688.

Gülcin I, Oktay M, Kirecci E, Küfrevioglu ÖI (2003). Screening of antioxidant and antimicrobial activities of anise *Pimpinella anisum* L. seed extracts. Food Chem 83:371-382.

Hasanah M, Setyaningsih P (1993). Studies on coriander (*Coriandnun sadvum* Linn.) phenology and the effect of umbel position and seed size on seed viability. Journal of Spice and Medicinal Crops 11:42-49.

Hendrix SD, Trapp EJ (1992). Population demography of *Pastinaca sativa* (Apiaceae): Effect of seed mass on emergence, survival and recruitment. Am J Bot 79:365-375.

Ipek A, Demirayak S, Gurbuz B (2004). A study on the adaptation of some anise (*Pimpinella anisum* L.) population to Ankara conditions. Tarim Bilim Derg 10:202-205.

Jevdjović R, Maletić R (2006). Effects of application of certain types of fertilizers on anise seed yield and quality. Journal of Agricultural Sciences, Belgrade 51:117-122.

Jevdjović R, Todorović G, Marković T, Kostić M, Sivčev I, Stanković S (2012). Effect of fertilization on yield, seed quality and content of essential oil of anise (*Pimpinela anisum* L.) and dill (*Anethum graveolens* L.). Proc 7th CMAPSEEC 248-254.

Kiralan M, Calkoglu E, Ipek A, Bayrak A, Gurbuz B (2009). Fatty acid and volatile oil composition of different coriander (*Coriandrum sativum*) registrated varieties cultivated in Turkey. Chem Nat Compd 45:100-102.

Koeduka T, Baiga TJ, Noel JP, Pichersky E (2009). Biosynthesis of *t*-anethole in anise: characterization of t-anol/isoeugenol synthase and an O-methyltransferase specific for a C7-C8 propenyl side chain. Plant Physiol 149:384-394.

Kosalec I, Pepeljnjak S, Kustrak D (2005). Antifungal activity of fluid extract and essential oil from anise fruits (*Pimpinella anisum* L., Apiaceae). Acta Pharmaceut 55:377-385.

Kubo I, Himejima M (1991). Anethole, a synergist of polygodial against filamentous microorganisms. J Agric Food Chem 39:2290-2292.

Omidbaigi R, Hadjiakhoondi A, Saharkhiz M (2003). Changes in content and chemical composition of *Pimpinella anisum* L. oil at various harvest time. J Essent Oil Bear Pl 6(1):46-50.

Orav A, Raal A, Arak E (2008). Essential oil composition of *Pimpinella anisum* L. fruits from various European countries. Nat Prod Res 22:227-232.

Ozcan MM, Chalchat JC (2006). Chemical composition and antifungal effect of anise (*Pimpinella anisum* L.) fruit oil at ripening stage. Ann Microbiol 56:353-358.

Özel A (2009). Anise (*Pimpinella anisum* L.): changes in yields and component composition on harvesting at different stages of plant maturity. Exp Agric 45:117-126.

Rajeshwari CU, Abirami M, Andallu B (2011). In vitro and in vivo antioxidant potential of anise (*Pimpinella anisum*). Asian J Exp Biol Sci 2:80-89.

Tabanca N, Demirci B, Kirimer N, Baser KHC, Bedir E, Khan IA, Wedge DE (2006). Gas chromatographic-mass spectrometric analysis of essential oil from *Pimpinella* species gathered from Central and Northern turkey. J Chromatogr A 1117(2):194-205.

Tepe B, Akpulat AH, Sokmen M, Daferera D, Yumrutas O, Aydin E, Polissiou M, Sokmen M (2006). Screening of the antioxidtaive and antimicrobial properties of the essential oil of *Pimpinella anisum* and *Pimpinella flabellifolia* from Turkey. Food Chem 97:719-724.

Tuncturk M, Yildirim B (2006). Effect of seed rates on yield and yield components of anise (*Pimpinella anisum* L.). Indian J Agr Sci 76:679-681.

Ullah H, Mahmood A, Ijaz M, Tadesse B, Honermeier B (2013). Evaluation of anise (*Pimpinella anisum* L.) accessions with regard to morphological characteristics, fruit yield, oil contents and composition. J Med Plants Res 7(29):2177-2186.

Yazdani D, Rezazadeh S, Amin G, Zainal Abidin MA, Shahnazi S, Jamalifar H (2009). Antifungal activity of dried extract of anise (*Pimpinella anisum* L.) and star anise (*Illicum verum* Hook.f.) against dermatophyte and saprophyte fungi. Journal of Medicinal Plants 8:24-29.

Zehtab-Salmasi S, Javanshir A, Omidbaigi R, Alyari H, Ghassemi-Golezani K (2001). Effects of water supply and sowing date on performance and essential oil production of anise (*Pimpinella anisum* L.). Acta Agronomica Hungarica 49:75-81.

Investigation of Correlation between Traits and Path Analysis of Confectionary Sunflower Genotypes

Mehmet SINCIK*, A. Tanju GOKSOY

*Uludag University, Faculty of Agriculture, Department of Field Crops, 16059, Bursa, Turkey; sincik@uludag.edu.tr (*corresponding author); agoksoy@uludag.edu.tr*

Abstract

This study examined the relationships between various traits associated with seeds and crude protein yields of confectionary sunflower, as well as the direct and indirect effects of these traits on crude protein yield. Eight open-pollinated confectionary sunflower populations and two open-pollinated confectionary sunflower cultivars used as control varieties were evaluated during two growing seasons. Certain agronomical and technological traits such as plant height, head diameter, seed yield, 1000-seed weight, kernel ratio, crude protein ratio and crude protein yield were measured. The results for the individual years and the combined data for both years revealed that there were significant positive correlations between the seed yield and the plant height, head diameter, 1000-seed weight and crude protein yield. The crude protein yield was also positively and significantly correlated with the 1000-seed weight and the seed yield, but the correlations with the plant height, head diameter, kernel ratio and crude protein ratio were not significant. The path coefficients from path analysis indicated that the seed yield had the maximum positive direct effect on the crude protein yield, followed by the crude protein ratio. The greatest positive indirect effects on the crude protein yield were attributed to the 1000-seed weight, plant height and head diameter through their effects on seed yield. The magnitudes of the direct effects of the seed yield and the crude protein ratio on the crude protein yield were 79.2% and 56.9%, respectively. Selection based on seed yields, 1000-seed weights and crude protein ratios could be used to improve crude protein yield in confectionary sunflower breeding programs.

Keywords: character association, crude protein yield, *Helianthus annuus* L., seed yield

Introduction

Two types of sunflower (*Helianthus annuus* L.), oilseed and confectionary sunflower, are commonly used for cultivation (Lu and Hoeft, 2009). Oilseed sunflower is the most important oilseed crop in Turkey. Recently, confectionary sunflower production has increased, mainly in the Middle and East Anatolia Region of Turkey. Thus, confectionary sunflower breeding has become more prominent over the last decade. The seeds of confectionary and oilseed sunflowers are distinguished by differences in their hullability, shell color, seed weight, morphology and kernel to pericarp weight ratio, as well as the crude protein contents (Hladni *et al.*, 2011). Confectionary sunflower produces large seeds with low oil contents, used in baking and snack applications (Lu and Hoeft, 2009). Although the favored seed color in Turkey is white with grey stripes, consumers from Balkan countries such as Serbia, Bulgaria, Moldova and Romania prefer black seeds. The confectionary seeds preferred by Turkish individuals are usually bigger and longer, and their 1000-seed weights are higher (Kaya, 2004).

Determining the associations between traits of confectionary sunflower cultivars is an important aspect of breeding programs that aim to improve the yield and quality of confectionary sunflowers characteristics. An understanding of the relationships between yield and the factors that affect yield is a prerequisite for designing an effective breeding program (Velkov, 1980). Yield is a complex characteristic and is a function of several component characteristics and their interactions with the environment (Chikkadevaiah *et al.*, 2002). Hence, the selection of characteristics that have a high heritability and a positive association with yield is an important requirement for improving yields through breeding programs. Many previous studies have reported that seed yields are positively correlated with the numbers of seeds per head, 1000-seed weight, head diameters and plant heights of confectionary sunflowers (Singh *et al.*, 1985; Lal *et al.*, 1997; Teklewold *et al.*, 2000; Shankar *et al.*, 2006; Kholghi *et al.*, 2011). Hladni *et al.* (2009) suggested that the two most important criteria for introducing new confectionary hybrids into production are a high number of seeds per head and a high crude protein yield. The first criterion is important for farmers, while the latter is important for consumers. Hladni *et al.* (2011) reported a very strong positive correlation between crude protein yields and the crude protein contents of seeds, seed yields and 1000-seed weight.

Path analysis is useful for partitioning a correlation coefficient into its direct and indirect effects (Dewey and Lu, 1959). A number of studies have shown that the number of seeds per head can have a highly positive direct effect on seed yield (Habib *et al.,* 2006; Shankar *et al.,* 2006; Kholghi *et al.,* 2011). However, Hladni *et al.* (2011) reported that seed yield, the crude protein content of seeds and the 1000-seed weight had a large influence on the crude protein yield of confectionary sunflowers.

The aim of this study was to investigate the relationships among various characteristics of confectionary sunflowers and their seed and crude protein yields and to determine the direct and indirect effects of the examined characteristics on crude protein yields.

Materials and methods

Eight open pollinated confectionary sunflower populations (CSP-1, CSP-2, CSP-3, CSP-4, CSP-5, CSP-6, CSP-7 and CSP-8) and two open pollinated confectionary sunflower cultivars ('Çiğdem 1' and 'İnegöl Alası') used as control varieties were sown in the experimental area of the Department of Field Crops, Faculty of Agriculture, Uludağ University, Bursa (latitude 40° 22' N, longitude 29° 12' E, altitude 155 m), Turkey, in 2009 and 2010. Bursa is located in the Southern Marmara region and has a Mediterranean-type climate. The average annual rainfall is 713 mm, and the mean monthly temperature is 14.4 °C. The soil of this site is classified as Vertisol typic habloxrert. Soil analyses revealed that the soil is a slightly alkaline (pH = 7.3) clay loam, rich in potassium (933 kg ha^{-1}), with a moderate amount of phosphorus (70 kg ha^{-1}) and an organic matter content of 1.2%.

Field experiments were conducted in a randomized, complete block design with three replications. The row-to-row distance was 70 cm, and the plant-to-plant distance was 30 cm. Each 14 m^2 plot consisted of four 5.0 m long rows. Sixty kilograms per hectare of both nitrogen (N) and phosphorus (P) were applied as a combined fertilizer (20-20-0) prior to sowing. Another 60 kg N ha^{-1} were added when the plants reached their reproductive stage. Before emergence, Linuron was sprayed at a rate of 0.20 mL m^{-2} for weed control. Plants were irrigated three times at different growth periods: heading, flowering and milking stages. Hand hoeing was performed after each irrigation event. Plots were harvested by hand during the period of physiological maturation and then threshed to determine the seed yield and other characteristics.

The plant height (cm), head diameter (cm), seed yield (kg ha^{-1}), 1000-seed weight (g), kernel ratio (%), crude protein ratio (%) and crude protein yield (kg ha^{-1}) were recorded. Plant height and head diameter were measured on ten plants selected randomly from each of the plots in both years. The crude protein ratio was determined by using the Kjeldahl method. An analysis of variance was performed for each trait for each individual year and for the combined data from both years. The simple correlation coefficients among all of the studied traits were calculated according to the method given by Singh and Chaudhary (1979). Path coefficients were analyzed as described by Dewey and Lu (1959) to determine the direct and indirect effects of different agronomical and technological traits on the crude protein yields.

Results

The mean values and statistical differences of the observed traits among the ten confectionary sunflower genotypes are presented in Tab. 1. There was a significant effect of the year on the differences observed in the examined traits among the genotypes. Therefore, the two experimental years were evaluated separately. In addition, significant differences were found among the genotypes for every trait in both experimental years. The mean values of the examined genotypes revealed that the plant heights (110.3-207.0 cm), head diameters (11.6-15.3 cm), seed yields (2552.0-3410.9 kg ha^{-1}), 1000-seed weights (82.0-108.0 g), kernel ratios (54.0-62.0%), crude protein ratios (18.8-23.4%) and crude protein yields (567.4-727.6 kg ha^{-1}, respectively, in 2009. However, in 2010, the ranges for the same traits were 122.7-216.3 cm, 13.8-18.5 cm, 2717.9-3713.3 kg ha^{-1}, 104.0-125.7 g, 50.0-56.0%, 18.3-23.1% and 547.4-756.1 kg ha^{-1}, respectively. The results indicated that the genotypes, CSP-3, CSP-4 and CSP-5 were the most promising lines for a majority of the studied traits (Tab. 1).

The correlation coefficients between the seed yields and the plant heights ($r=0.768$; $p<0.01$), head diameters ($r=0.733$; $p<0.05$), 1000-seed weights ($r=0.831$; $p<0.01$)

Tab. 1. Average values of examined traits of confectionary sunflower genotypes

Genotypes	PH (cm)	HD (cm)	SY (Kg ha^{-1})	TSW (g)	KR (%)	CPR (%)	CPY (Kg ha^{-1})
			2009				
CSP-1	195.0 b-d	14.6 ab	2910.1 cd	90.0 cd	58.0 ab	22.6 ab	658.1 ab
CSP-2	207.0 a	15.0 ab	2993.0 bc	93.0 cd	54.0 b	19.3 cd	567.4 c
CSP-3	168.0 e	13.0 ab	3103.6 b	96.0 bc	62.0 a	23.4 a	727.6 a
CSP-4	196.0 bc	12.9 ab	3440.3 a	106.0 a	57.0 ab	20.5 b-d	706.3 ab
CSP-5	192.0 cd	14.3 ab	3380.7 a	108.0 a	56.0 ab	18.8 d	635.0 bc
CSP-6	192.7 cd	15.3 a	3420.4 a	104.0 a	54.0 b	20.7 b-d	707.3 ab
CSP-7	110.3 f	13.3 ab	2552.0e	82.0 e	57.0 ab	22.4 ab	570.5 c
CSP-8	201.3 ab	14.0 ab	3410.9 a	102.0 ab	56.0 ab	19.3 cd	657.2 ab
Cigdem	166.3 e	11.6 b	2783.5 d	86.0 de	56.3 ab	23.1 a	641.4 bc
İnegol	187.0 d	15.6 a	3360.2 a	103.0 ab	54.0 b	21.3 a-c	718.9 a
LSD (0.05)	*8.15*	*3.49*	*138.9*	*7.67*	*6.62*	*2.22*	*74.3*
			2010				
CSP-1	175.3 d	13.9 d	3010.4 d	109.3 cd	55.0 ab	19.4 de	582.0 d
CSP-2	194.0 c	14.3 d	2853.7 e	106.3 d	54.7 a-c	19.2 d-f	547.4 d
CSP-3	213.7 a	16.4 bc	3713.3 a	125.7 a	50.3 e	18.3 f	681.6 bc
CSP-4	204.7 b	16.3 bc	3310.5 c	104.0 d	53.3 bc	23.1 a	756.1 a
CSP-5	212.3 a	15.9 c	3637.0 ab	118.3 b	55.3 a	19.5 c-e	708.4 a-c

CSP-6	213.3 a	17.4 ab	3360.5 c	106.0 d	56.0 a	21.4 b	718.3 ab
CSP-7	122.7 e	13.8 d	2717.9 e	115.0 bc	51.3 de	20.4 bc	554.1 d
CSP-8	194.3 c	18.5 a	3547.2 b	104.3 d	50.0 e	18.9 d-f	660.8 bc
Cigdem	202.0 b	15.7 c	3583.6 ab	119.0 b	53.0 cd	18.4 ef	662.7 bc
İnegol	216.3 a	16.7 bc	3613.8 ab	119.3 b	55.0 ab	19.6 cd	654.5 c
LSD (0.05)	*5.74*	*1.42*	*155.1*	*5.86*	*1.89*	*1.02*	*64.1*
			Two-year average				
CSP-1	185.2 d	14.2 b	2960.4 e	99.7 cd	56.5	20.9 ab	620.3 e
CSP-2	200.5 ab	14.6 ab	2923.1 e	99.5 cd	54.3	19.2 cd	557.5 f
CSP-3	190.8 c	14.7 ab	3408.0 bc	110.8 a	56.2	20.8 ab	704.2 a-c
CSP-4	200.3 ab	14.6 ab	3375.2 c	105.0 b	55.2	21.8 a	731.7 a
CSP-5	202.2 ab	15.1 ab	3508.7 a	113.2 a	55.7	19.1 d	671.4 b-d
CSP-6	203.0 a	16.4 a	3390.1 bc	105.2 b	55.0	21.0 ab	713.0 ab
CSP-7	116.5 e	13.5 b	2633.5 f	98.5 d	54.2	21.4 ab	562.5 f
CSP-8	197.8 b	16.3 a	3478.2 ab	103.2 bc	53.0	19.0 d	659.3 c-e
Cigdem	184.2 d	13.7 b	3183.9 d	102.5 b-d	54.7	20.8 ab	651.6 de
İnegol	201.7 ab	16.2 a	3487.0 ab	111.2 a	54.5	20.4 bc	686.1 a-d
LSD (0.05)	*6.81*	*2.58*	*142.3*	*6.60*	*ns*	*1.67*	*67.1*

Note: Different letters between genotypes denote significant differences (LSD test, P<0.05) PH: Plant height; HD: Head diameter; SY: Seed yield; TSW: Thousand seed weight, KR: Kernel ratio, CPR: Crude protein ratio, CPY: Crude protein yield.

and crude protein yields ($r=0.852$; $p<0.01$) were positive and significant for the two-year combined data. In contrast, seeds yields were negatively and non-significantly associated with the crude protein ratios ($r=-0.264$; ns) and kernel ratios ($r=-0.053$, ns). Relatively similar results were also obtained for the individual years (Tab. 2).

The correlation analyses of the combined two years of data revealed that crude protein yields were only significant and positively correlated with the 1000-seed weights ($r = 0.699$; $p<0.05$) and seed yields ($r = 0.852$; $p<0.01$), but not significantly correlated with the plant heights, head diameters, kernel ratios and crude protein ratios. The correlation results for the individual years in terms of the crude protein yield were also similar to the two-year combined data. However, the significant correlation between the crude protein yields and 1000-seed weights observed in the combined data was not significant for neither individual year. The kernel ratio and the crude protein ratio had insignificant correlative associations with all of the other observed traits (Tab. 2). In addition, plant height was positively and significantly associated with head diameter ($r=0.633$; $p<0.01$).

The results of the path analysis revealed that the seed yield and the crude protein ratio had the greatest direct positive effects on the crude protein yield in both experimental years and in the combined data (Tab. 3). For the two-year combined data, the direct effect of seed yield on crude protein yield was 79.2%, with a positive

Tab. 2. Correlation coefficient of the analyzed traits with crude protein yield

Traits	HD	SY	TSW	KR	CPP	CPY
			PH			
2009	0.451^{ns}	0.716*	0.661*	-0.323^{ns}	-0.586^{ns}	0.312^{ns}
2010	0.633*	0.789**	0.108^{ns}	0.318^{ns}	-0.057^{ns}	0.684*
2-yr average	0.633*	0.768**	0.511^{ns}	-0.172^{ns}	-0.400^{ns}	0.581^{ns}
			HD			
2009		0.415^{ns}	0.413^{ns}	-0.550^{ns}	-0.480^{ns}	0.068^{ns}
2010		0.747*	-0.124^{ns}	-0.196^{ns}	0.058^{ns}	0.708*
2-yr average		0.733*	0.505^{ns}	-0.270^{ns}	-0.388^{ns}	0.520^{ns}
			SY			
2009			0.980**	-0.257^{ns}	-0.612**	0.636*
2010			0.444^{ns}	-0.101^{ns}	-0.266^{ns}	0.730*
2-yr average			0.831**	-0.053^{ns}	-0.264^{ns}	0.852**
			TSW			
2009				-0.243^{ns}	-0.650*	0.573^{ns}
2010				-0.177^{ns}	-0.534^{ns}	0.006^{ns}
2-yr average				-0.284^{ns}	-0.202^{ns}	0.699*
			KR			
2009					0.600^{ns}	0.260^{ns}
2010					0.295^{ns}	0.086^{ns}
2-yr average					0.318^{ns}	0.270^{ns}
			CPR			
2009						0.221^{ns}
2010						0.426^{ns}
2-yr average						0.264^{ns}

Note: *: significant at p=0.05 probability level; **: significant at p=0.01 probability level; ns: not significant; PH: Plant height; HD: Head diameter; SY: Seed yield; TSW: Thousand seed weight, KR: Kernel ratio, CPR: Crude protein ratio, CPY: Crude protein yield.

correlation coefficient (r=0.852; p<0.01). The crude protein ratio had the second greatest direct effect on crude protein yields. According to the two-year combined data, the direct effect of crude protein ratio on crude protein yield was 56.9%, but the positive correlation was not significant (r=0.264; ns). In general, the direct effect of the 1000-seed weight was negative and of a moderate level. The percentages of this negative direct effect were 18.4% and 11.7% for the 2010 and combined data, respectively. The direct effects of the other traits on crude protein yields were smaller in magnitude and inconsistent between years (Tab. 3). The highest positive indirect effects on crude protein yield were obtained for the 1000-seed weight (+0.910), plant height (+0.842) and head diameter (+0.803) through their impacts on seed yield. The 1000-seed weight, plant height and head diameter had indirect effects (through seed yield) on crude protein yield of 75.5%, 72.0% and 70.9%, respectively. The analysis of the two-year combined data also revealed positive indirect effects of similar magnitudes. In addition, the kernel ratio exhibited a positive indirect effect on crude protein yield of 40.9-63.7% via its impact on crude protein ratios in the both experimental years and in the combined data. The indirect effects of the other characteristics on crude protein yield were smaller in magnitude (Tab. 3).

Discussion

There was a high degree of variability among the confectionary sunflower genotypes in terms of all of the measured traits, in both experimental years (Tab. 1). In general, the mean values of most of the studied

Tab. 3. Direct and indirect effects of agronomic traits on crude protein yield

Traits	Direct effect	%	Indirect effect											
			PH		HD		SY		TSW		KR		CPR	
			PC	%	PC	%	PC	%	PC	%	PC	%	PC	%
							2009							
PH	-0.002	0.11			0.016	1.09	0.865	58.61	0.013	0.88	-0.001	0.05	-0.579	39.25
HD	-0.036	3.51	-0.001	0.07			0.500	49.04	0.008	0.79	-0.001	0.13	-0.474	46.44
SY	1.207	65.35	-0.001	0.06	0.015	0.80			0.019	1.04	-0.001	0.03	-0.604	32.69
TSW	0.019	1.05	-0.001	0.06	0.015	0.79	1.182	63.55			-0.001	0.03	-0.642	34.50
KR	0.002	0.03	0.001	0.06	-0.020	2.12	-0.311	33.37	-0.005	0.51			0.593	63.67
CPR	0.987	56.15	0.001	0.05	-0.017	0.98	-0.738	42.00	-0.013	0.73	0.002	0.08		
							2010							
PH	-0.011	0.90			-0.160	13.40	0.939	78.68	-0.023	1.89	-0.023	1.95	-0.038	3.18
HD	-0.252	20.58	-0.007	0.55			0.888	72.42	0.026	2.11	0.014	1.17	0.039	3.16
SY	1.190	71.49	-0.008	0.51	-0.188	11.33			-0.093	5.57	0.007	0.44	-0.177	10.64
TSW	-0.209	18.35	-0.001	0.10	0.031	2.75	0.528	46.39			0.013	1.13	-0.356	31.26
KR	-0.073	15.24	-0.003	0.71	0.049	10.32	-0.120	25.08	0.037	7.69			0.196	40.95
CPR	0.666	58.91	0.001	0.05	-0.015	1.30	-0.316	27.96	0.112	9.86	-0.022	1.91		
							Two year average							
PH	-0.026	2.26			0.014	1.24	0.842	72.03	-0.072	6.14	0.018	1.58	-0.196	16.75
HD	0.023	2.02	-0.017	1.48			0.803	70.91	-0.071	6.26	-0.029	2.55	-0.190	16.76
SY	1.096	79.17	-0.020	1.47	0.017	1.21			-0.116	8.42	0.006	0.41	-0.129	9.31
TSW	-0.140	11.65	-0.013	1.12	0.012	0.96	0.910	75.53			0.030	2.52	-0.099	8.21
KR	0.107	28.88	-0.004	1.22	-0.006	1.66	0.058	15.61	-0.039	10.72			0.155	41.89
CPR	0.489	56.87	0.011	1.23	-0.009	1.03	-0.289	33.59	0.028	3.30	0.034	3.97		

Note: PH: Plant height; HD: Head diameter; SY: Seed yield; TSW: Thousand seed weight, KR: Kernel ratio, CPR: Crude protein ratio, PC: Path coefficient.

characteristics of the genotypes were higher in the second year than those recorded in the first year. These higher mean values were likely attributed to the impact of the greater amount of spring precipitation that occurred in the second year compared to the first year. The genotypes exhibited considerable differences between years in most of the observed traits because of the significant genotype x environment interactions.

A successful breeding program for confectionary sunflower requires knowledge of a well-defined association between agronomical and technological characteristics. In this study, the correlations between seed yield and the plant height, head diameter, 1000-seed weight and crude protein yield were positive and highly significant. Previous studies have similarly reported that the association between seed yield and the 1000-seed weight is positive and significant (Shankar *et al.*, 2006; Hladni *et al.*, 2010; Hladni *et al.*, 2011; Kholghi *et al.*, 2011). In contrast to these results, Shrinivasa (1982) and Tariq *et al.* (1992) reported a negative correlation between seeds yield and the 1000-seed weight. On the other hand, our findings are in agreement with those of Marinkovic (1992), Tahir *et al.* (2002) and Vidhyavathi *et al.* (2005), who reported that the association between seed yield and the head diameter is positive and significant. However, our results did not support the conclusion of Manivannan *et al.* (2007), who reported that the correlation between seed yield and the head diameter was negative and non-significant. This discordance among results is most likely associated with differences in the genetic material and environmental conditions used in these studies. Hladni *et al.* (2011) reported that the two most important criteria for introducing new confectionary hybrids into production are a high seed yield and a high crude protein yield. The same authors noted that there is a very strong positive correlation between seed yield and crude protein yield. The earlier findings of Hladni *et al.* (2011) conform to our results. In this study, the correlations between all of the other traits measured were generally not significant and were also inconsistent between years. This lack of significant correlations for these traits found in our study is not consistent with results of various previous studies that reported positive and significant correlations between seed crude protein content and seed yield (Hladni *et al.*, 2010 and 2011), seed crude protein content and kernel ratio (Hladni *et al.,* 2011), 1000-seed weight and head diameter (Marinkovic, 1992; Kholghi *et al.*, 2011), and seed crude protein content and crude protein yield (Hladni *et al.*, 2011).

As demonstrated by the path analysis results, seed yield had the greatest direct positive effect on crude protein yields, and crude protein ratio the second greatest direct positive effect. These results correspond to those of Hladni *et al.* (2011), who reported that the seed yield, seed crude protein content and 1000-seed weight all had a large influence on crude protein yield. In addition, Kholghi *et al.* (2011) found that the number of seeds per head, 1000-seed weight and head diameter had positive direct effects on seed yields of confectionary sunflowers.

Conclusion

The results from this study indicated that seed yield was positively and significantly correlated with plant height, head diameter, 1000-seed weight and crude protein yield. Crude protein yield was positively and significantly correlated with 1000-seed weight and seed yield, but not significantly correlated with plant height, head diameter, kernel ratio and crude protein ratio. The path analysis results revealed that seed yield, followed by crude protein ratio, had the greatest direct positive effect on crude protein yield. The greatest positive indirect effects on crude protein yield were exhibited by the 1000-seed weight, plant height and head diameter though their impacts on seed yield. Therefore, breeding programs based on selection for seed yield, 1000-seed weight and crude protein ratio may be used to improve the crude protein yield of confectionary sunflowers.

Acknowledgements

This research work was supported by The Office of Scientific Research Projects of Uludag University (Project No. 2009/17. Project Leader: Assoc. Prof. Dr. Mehmet Sincik).

References

Chikkadevaiah SHL, Nandini R (2002). Correlation and path analysis in sunflower. Helia 25(37):109-118.

Dewey DR, Lu KH (1959). A correlation and path coefficient analysis of components of crested wheat grass seed production. Agron J 51(9):515-518.

Habib H, Mehdi SS, Rashid A, Anjum MA (2006). Genetic association and path analysis for seed yield in sunflower (*Helianthus annuus* L.). Pak J Agri Sci 43:131-135.

Hladni N, Jocic S, Miklic V, Dušanic N, Saftic-Pankovic D, Radeka I, Lecic N (2009). The evaluation of values of new confectionery sunflower hybrids. Proc 50th Oil Industry Conf, Herceg Novi, 57-61p.

Hladni N, Jocic S, Miklic V, Mijic A, Saftic-Pankovic D, Škoric D (2010). Effect of morphological and physiological traits on seed yield and oil content in sunflower. Helia 33(53):101-115.

Hladni N, Jocic S, Miklic V, Safticpankovic D, Kraljevic-Balalic M (2011). Interdependence of yield and yield components of confectionary sunflower hybrids. Genetika 43(3):583-594.

Kaya Y (2004). Confectionery sunflower production in Turkey. Proceeding of 16th International Sunflower Conference. August 29-September 2. Fargo, USA. 817-822.

Kholghi M, Bernousi I, Darvishzadeh R, Pirzad A, Hatami Maleki H (2011). Collection, evaluation and classification of Iranian confectionary sunflower (*Helianthus annuus* L.) populations using multivaraite statistical techniques. Afr J Biotechnol 10:5444-5451.

Lal GS, Bhaderiya VS, Singh AK (1997). Genetic association and path analysis in elite lines of sunflower. Crop Res Hisar 13(3):631-634.

Lu G, Hoeft E (2009). Sunflower, p. 125-168. In: Kole C, Hall TC (Eds.). Compendium of transgenic crop plants, vol 2. Wiley-Blackwell, London.

Manivannan N, Karthika R, Punitha B, Vindhiyavarman P, Muralidharan V (2007). Association pattern among the yield attributes in varieties and hybrids of sunflower (*Helianthus annuus* L.). Madras Agric J 94:84-88.

Marinkovic R, (1992). Path coefficient analysis of some yield components of sunflower. Euphytica 60(3):201-205.

Shankar VG, Ganesh M, Ranganatha ARG, Bhave MHV, (2006). A study on correlation and path analysis of seed yield and yield components in sunflower (*Helianthus annuus* L.) Agric Sci Digest 26(2):87-90.

Shrinivasa MK (1982). Inheritance of fertility restoration and oil content in sunflower. M.Sc (Ag.) Thesis, University of Agricultural Sciences, Bangalore, India.

Singh RK, Chaudhary BD (1979). Biometrical methods in quantitative genetic analysis. Kalyani Publishers, New Delhi, India, 303 p.

Singh JV, Yadva TP, Kharb RPS (1985). Correlation and path coefficient analysis of sunflower. Indian J Agric Sci 95(1):243-246.

Tahir MHN, Imran M, Hussain MK (2002). Evaluation of sunflower (*Helianthus annuus* L.) inbred lines for drought tolerance. Int j Agric Biol 4(3):398-400.

Tariq M, Idress G, Tahir A (1992). Genetic variability and correlation studies in sunflower. Sarhad J Agric 8(6):659-663.

Teklewold A, Jayaramaiah H, Jagadeesh BN (2000). Correlations and path analysis of physiomorphological characters of sunflower (*Helianthus annuus* L.) as related to breeding method. Helia 23(32):105-114.

Velkov VN (1980). Relation between yield and some characters in sunflower. Genetica Selektsiya 13b(5):329-338.

Vidhyavathi R, Mahalakshmi P, Manivannan N, Muralidhran N (2005). Correlation and path analysis in sunflower (*Helianthus annuus* L.). Agric Sci Digest 25(1):6-10.

31

Pharmacological Benefits of Herbal Formulations in the Management of *Psoriasis vulgaris*

Andreea Nicoleta BOCA[1], Alexandru TATARU[2], Anca Dana BUZOIANU[1],

Carlo PINCELLI[3], Carmen SOCACIU[4,5]*

[1]"Iuliu Hatieganu" University of Medicine and Pharmacy, Department of Pharmacology, Toxicology and Clinical Pharmacology, 6 Pasteur Street, Cluj-Napoca, Romania; boca.andreea@umfcluj.ro; abuzoianu@umfcluj.ro

[2]"Iuliu Hatieganu" University of Medicine and Pharmacy, Department of Dermatology, 3-5 Clinicilor Street, Cluj-Napoca, Romania; dr.tataru@yahoo.com

[3]University of Modena and Reggio Emilia, Department of Medicine, Institute of Dermatology, Laboratory of Cutaneous Biology, 71 Via Del Pozzo, Modena, Italy; carlo.pincelli@unimore.it

*[4]University of Agricultural Sciences and Veterinary Medicine, Faculty of Food Science and Technology, 3-5 Mănăştur St., Cluj-Napoca, Romania; carmen.socaciu@usamvcluj.ro (*corresponding author)*

[5]Center for Applied Biotechnology CCD-BIODIATECH, Proplanta Cluj-Napoca, Romania; csocaciu@proplanta.ro

Abstract

Psoriasis is a chronic inflammatory skin disease, with an important impact on the patient's quality of life. Its incidence and prevalence are continuously increasing. The complex ethiopathology of this disorder is only partially known; there is a clear genetic predisposition, which associates a number of environmental triggering factors such as an unbalanced diet and lifestyle. The conventional therapeutic options are not always satisfactory in terms of efficiency and safety, therefore, complementary and alternative medicine approaches are frequently chosen by patients, mostly as self-medication. This review, based on recent literature flow data, outlines the pharmacological benefits of herbal formulations with antipsoriatic activity. It also reveals the molecules responsible for their effects, as well as their interference with the metabolic and immunopathogenic mechanisms of this disease. An important number of plants have been proved to act as antipsoriatic agents, many botanical-based preparations containing key-phytochemical molecules (belonging mainly to phenolics, triterpenoids and phytosterols or unsaturated fatty acids, as mentioned in specific phyto-pharmaceutical databases). Specific mechanisms of action, which can explain their activity (such as lipoxygenase inhibition, antioxidant, anti-inflammatory, anti prostaglandin), were recently described. Only some of these formulations have been actively tested *in vitro* or *in vivo*. Most publications in the field agree on the need for more *in vitro* and *in vivo* studies, especially clinical assessment on patients with *Psoriasis vulgaris*. These would provide more accurate data on the efficacy and safety of such herbal formulations for this disease.

Keywords: activation pathways in psoriasis, anti-inflammatory activity, complementary and alternative medicine, medicinal plants, phytochemicals, *Psoriasis vulgaris*

Background

Psoriasis vulgaris is a chronic inflammatory skin disorder, with a worldwide distribution and an increasing incidence and prevalence. Recent epidemiological studies show its impact on a level of 1.5-4%, with important local variations depending on ethnicity and climate (Jacobson *et al.*, 2011).

Psoriasis vulgaris has a complex ethiopathology, scientifically proven by a genetic predisposition associated with environmental triggering factors, such as unbalanced diet, stress, trauma and infection being the most frequently mentioned ones (Griffiths *et al.*, 2007).

Most studies on the immunopathogenesis of psoriasis point to a primary T cell-mediated hyperproliferation of keratinocytes, leading to the well-known lesions (Nickoloff and Nestle, 2004; Homey and Meller, 2008; Jeon *et al.*, 2013).

Epidermal infiltrate of oligoclonal CD8+ T and dermal presence of CD4+ T cells is actually a constant feature of chronic psoriasis lesions. These cells could be responding to specific antigens present in the HLA-Cw6 binding pocket of keratinocytes, or to bacterial superantigens, which cause an important activation of T-cells. The activation and remission episodes might be due to the changing balance between effector and suppressor CD4+ and CD8+ T cells. Additionally, natural killer T cells, dendrite cells and a number of cytokines (interleukins, interferon) are present in these lesions. The complex and only partially elucidated interaction between T cells, their resulting cytokines and dendrite cells leads to the pathogenic inflammation.

A number of pathogenic models for psoriasis have been discussed and there are still many unknown aspects on the matter. Type 1 pathway of cytokines and inflammation producing gene products have been linked to this disorder. The activated T helper 1(Th 1) and T cytotoxic 1(Tc 1) cells release IFNγ, which through signal transducer and activator of transcription 1 (STAT1) induces a large group of inflammatory genes. Th17, the latest characterized subset of Th cells, are also involved in the immunopathogenesis of psoriasis (Fig. 1).

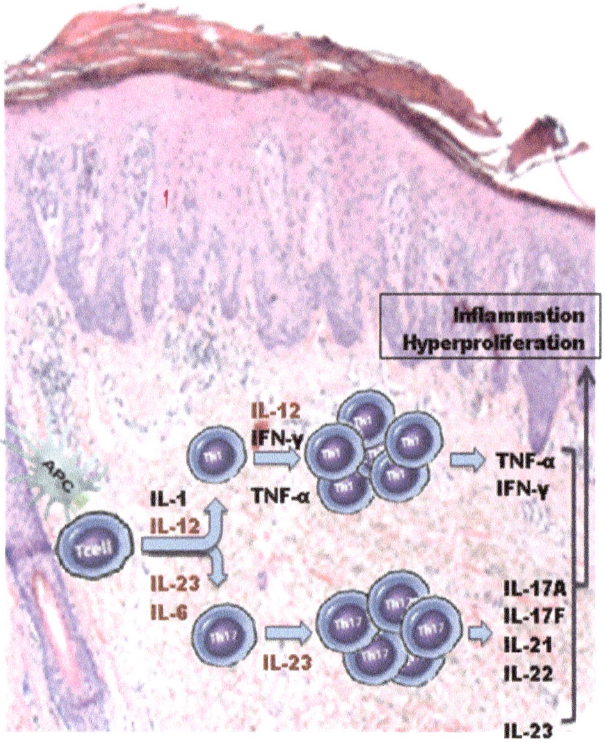

Fig. 1. Th1 and Th17 in the imunopathogenesis of psoriasis

Clinically, psoriasis is characterized mainly by its cutaneous involvement, but it can also affect joints and nails. Its typical signs, the red scaly patches, can cover large areas of the body, being easy to notice and therefore stigmatizing, with a strong impact on the quality of life (Kurd *et al.*, 2010).

These scaling of psoriatic lesions are due to particular histological features. The cellular cycle of keratinocytes is decreased from the normal period of 30 days to 3-5 days, in which they transit from the basal to the cornified layer. These cells present parakeratosis, an altered differentiation pattern, nuclei being present even in the cornified layer. In addition, there is an aberrant expression of antigens associated to the differentiation. The local erythema is caused by the hyper proliferation of the vascular endothelial cells, which are contorted and dilated. Pathognomonic features of psoriasis are the Munro's microabscesses, formed by the focal migration of neutrophils into the epidermis (Schon *et al.*, 2005).

Psoriasis is a chronic recurrent condition, so remission periods and acute episodes alternate in an unpredictable way. While some patients can be in remission for years, others may present rapid progression to severe forms.

The morphology, distribution and severity of the cutaneous manifestations are highly variable, ranging from small, localized papules, to large plaques, covering the whole body, intensely symptomatic.

The treatment is also chronic, having a significant impact on the patient's quality of life. According to the clinical form and severity, patients can use topical products, or require systemic therapy. The classical therapeutic options include immunosuppressive agents such as different corticosteroids, azathioprine, cyclosporine, methotrexate, or one of the biologic agents (Gottlieb, 2005). New insights into the immunologic mechanisms of psoriasis might lead to the development of more targeted therapeutic interventions (Asadullah *et al.*, 2002).

Currently available options have variable efficacy and frequently associate well-known side effects. In addition, not all patients are responsive to them; some cases are actually worsened. Another problem arises from the important costs of these therapies.

All these decrease the adherence to treatment, affect the trust patients have in their physicians and increase the percentage of those starting to use self-medication (Magin *et al.*, 2009). Kivelevitch *et al.* (2012) suggest that up to 30-40% of the population is non-adherent to the various therapeutic options. It is reported that for their skin diseases patients show an increasing preference towards natural approaches, likely due to the perceived decreased risk for adverse effects of such therapies (Magin *et al.*, 2006).

For a proper efficacy assessing of natural compound's safety, both *in vitro* research and *in vivo* preclinical (on animal models) and clinical studies are essential. *In vitro* research can help assess the safety of the compounds and provide important details on their specific cellular and also molecular activity. Development of animal models is of a real value for such studies, especially xenotransplantation models using human skin grafted on to immunodeficient mice. They allow studying the behavior of different cell types in human tissue. Even if no animal model mimics psoriasis completely, many aspects are present in these (Danilenko, 2008), which also helped prove that specific T cell subsets can induce psoriatic lesions in uninvolved skin from psoriasis patients, supporting the immunopathogenic pathways discussed above (Schon, 1999). Clinical studies are the final step, critical for a valid assessment of efficacy and safety.

Complementary and alternative medicine (CAM) promotes plant based formulas approaches the use of natural products, mainly based on individual plants extracts or complex formulas (dietary supplements) combined with diet-based therapies, which demonstrated their efficacy, but many times their benefits are proven only by limited reported experience.

The prevalence of CAM application is stronger in developing countries, but there is an increasing demand also in developed countries. Natural formulas of dietary supplements (known as natural remedies) have an important preventive role and are reported to be used by patients with various dermatologic problems, more frequently than by the general population. It is estimated that up to 50% of dermatologic patients use CAM, alone or in combination with conventional therapies (Fuhrmann *et al.*, 2010; Jensen, 1990). Plant-based preparations are used not only in CAM, but also in conventional dermatology (e.g. use of psoralens, podophyllin, pyrethrins, oatmeal).

Plant-based preparations include individual medicinal herbs or combined formulas, which may enhance their efficacy and reduce side effects (Bensky and Barolet, 1990). According to Mantle *et al.* (2001), approximately one-third of them are useful skin disorders, including psoriasis, compared to only 1-3% of synthetic drugs.

Psoriasis patients frequently use alternative therapies, especially herbal products. Their preference for natural approaches is mostly explained by the perceived decreased potential for adverse effects of CAM therapies (Magin *et al.*, 2006). Thus, for a proper management of their patient's condition, it is important for physicians to be aware of the current evidence concerning plant-based preparations.

Recently (Deng *et al.*, 2013) made a meta-analysis of Chinese plants with antipsoriatic activity. The most commonly used herbs were *Sophora flavescens* root and *Lithospermum erythrorhizon* root, which were proven to have anti-inflammatory, anti-proliferative, anti-angiogenic and tissue repair actions. These actions may explain their benefits of the topical multi-herbal formulations in psoriasis.

Tabs. 1 and 2 summarize data about the main effects of *Aloe vera* and other plants in the management of *Psoriasis vulgaris*, in relation to their composition in bioactive compounds. The efficiency and safety of plant-based preparations is critically evaluated, considering the involvement of natural molecules on the cellular mediated

Tab. 1. Bioactive components from *Aloe vera*, with antipsoriatic activity (from Duke Phytochemical and Ethnobotanical Databases, 2013)

Compound	Activity
Provitamin A (beta-carotene), vitamin C and E	antioxidant
Bradykinase	enzyme with anti-inflamatory activity
Enzymes: aliiase, alkaline phosphatase, amylase, carboxypeptidase, catalase, cellulase, lipase and peroxidase	breakdown of carbohydrates and lipids
Calcium, chromium, copper, zinc selenium, magnesium, potassium	essential minerals for the various enzymatic reactions, some having also antioxidant potential
Polymannan	anti-inflammatory and immunostimulatory properties
Alprogen	glycoprotein with antiallergic properties
C-glucosyl chromone	a novel anti-inflammatory compound
Anthraquinones (aloin and emodin, aloesin)	analgesic, antibacterial and antiviral properties, a moderate photoprotective effect
β-sitosterol, campesterol, cholesterol and lupeol	anti-inflammatory action, antiseptic and analgesic potential
Auxins and gibberellins	plant hormones efficient in wound healing and inflammation
Salicylic acid	anti-inflammatory and antibacterial activity
Lignin	enhances cutaneous absorption of other ingredients, when mixed in topical preparations

Tab. 2. Other plants with antipsoriatic activity and the key-molecules with their associated activity (according to Duke Phytochemical and Ethnobotanical Databases, 2013 and other references)

Plant name (Latin, common)	Key-molecules	Other References
Allium sativum var. *sativum* L. (garlic) *Angelica sinensis* (angelica) *Balanites aegyptiacus* (Brussel-Sprout) *Daucus carota* (carrot) *Glycine max* (soybean) *Morus alba* (white mulberry)	Arachidonic acid Carotenoids Phospholipids	Pazyar and Feily, 2011
Capsicum annuum (bell pepper, cherry pepper) *Capsicum frutescens* L.(hot pepper) *Zingiber officinale* (Ginger)	Capsaicin Capxanthin	
Curcuma longa (turmeric) Zingiber officinale (ginger)	Curcumin	Hatcher *et al.*,2008 Noorafshan & Ashkani-Esfahani, 2013 Jantarat, 2013
Indigo naturalis (indigo)		Lin *et al.*, 2012
Hippophae rhamnoides(seabuckthorn)	PUFAs Phytosterols	Zeb, 2004 Khan *et al.*, 2010; Kumar *et al.*, 2011

immunological mechanisms, in the antioxidant defense and apoptotic effects. The interference of natural molecules with cellular pathways can explain scientifically their beneficial effects at low doses, without secondary effects. These findings are assessed in recent studies and reviews, standing as a proof of the benefic potential of natural phytochemicals in such a chronic skin disease.

Recently, a relevant literature reveals the efficacy of plants and herbal preparations to be used as food supplements or teas or as ingredients of dermocosmetic formulas (May *et al.*, 2012; Kuchekar and Bhise, 2012; Bhuchar *et al.*, 2012; Chandrasekaran *et al.*, 2012) some of them being investigated by clinical trials (Su *et al.*, 2011; Kaur and Kumar, 2012; Al-Khafaji, 2012; Rahman *et al.*, 2012; Li *et al.*, 2012; Lu *et al.*, 2012; Deng *et al.*, 2013, Yu *et al.*, 2013), as well by topical treatments (Bos and Spuls, 2008).

The topical formulations based on conventional excipients are replaced by new biocompatible and biodegradable vehicles (liposomes, microemulsions, solid-lipid nanoparticles for bioactive molecules), in order to improve their efficacy and safety as well to allow a controlled delivery (Katare *et al.*, 2010). Such vehicles are very compatible with lipophilic or hydrophilic plants extracts as well with pure compounds (psoralens, corticosteroids, levulinic acid, theophiline, vitamin D analogies).

Plants with antipsoriatic potential: scientific evidence and preclinical/clinical studies

Aloe vera (*Aloe barbadensis miller*) is a green succulent plant, belonging to the Liliaceae family, widely spread in some of the dry regions of Africa, Asia, Europe and America. "The plant of immortality", as Egyptians called it, has been used since ancient times for its health, beauty, medicinal and skin care properties. Two thousand years ago, Greek scientists considered it the universal panacea; Alexander the Great and Christopher Columbus used it for treating soldier's wounds (Surjushe *et al.*, 2008).

To the anecdotal use, scientific research added new data regarding the use of *Aloe vera* for various medical purposes, especially dermatological conditions including wound healing, herpes simplex, atopic dermatitis, seborrheic dermatitis, acne, diaper dermatitis, lichen planus, aphthous stomatitis, human papilloma virus and frostbite. For these indications, studies like those of Surjushe *et al.* (2008), or Feily and Namazi (2009) report various efficacy and sometimes, conflicting results.

Aloe vera contains around 75 potentially active constituents; in relation to their cutaneous benefits, its most relevant compounds are described in Tab. 1 (Atherton, 1998; Surjushe *et al.*, 2008). Some *in vitro* and *in vivo* studies assessed the efficacy and safety of *Aloe vera* in psoriasis (Morelli *et al.*, 2010; Deng et *al.*, 2013)

Its anti-inflammatory activity is most likely the most relevant one for this pathology. *Aloe* has been reported to inhibit the cyclooxygenase pathway and to reduce prostaglandin E2 production (Hutter *et al.*, 1996).

Clinical studies on the use of *Aloe vera* preparations in plaque psoriasis are few in number; Syed *et al.* (1996) and Choonhakarn *et al.* (2010) agree on its efficacy in topical applications, reporting improvement of PASI and DLQI scores (more than with placebo, but less than with steroids); but there is also conflicting data Paulsen *et al.* (2005). Therefore, although there are some promising data, clinical effectiveness needs more scientific confirmation.

Use of topical *Aloe vera* gel is generally well tolerated, rarely causing allergic reactions, as reported by Mantle *et al.* (2001). It should be avoided in patients with known hypersensitivity to *Aloe vera* or plants of the *Liliaceae* family (e.g. garlic, onions, tulips). Through oral ingestion, it may lower blood sugar levels, so caution is advised in patients with diabetes, hypoglycemia, or glucose intolerance.

Cayenne pepper (*Capsicum frutescens*), due to its main ingredient, **capsaicin**, has important medical benefits in decreasing the intensity of different types of neurological induced pain. Topical capsaicin has modulating effects on substance P, thus it is efficient in a number of neuralgias: post herpes zoster neuralgia, diabetic neuropathy, notalgia paresthetica, reflex sympathetic dystrophy and also for hemodialysis-induced pruritus (Reuter *et al.*, 2010).

In vitro studies Yu CS (2011), as well as *in vivo* data, including a double-blind clinical trial, report efficacy for topical capsaicin in psoriasis, especially the pruritic forms (Ellis *et al.*, 1993; Bernstein *et al.*, 1986; Arnold and van de Kerkhof, 1994). The use of topical capsaicin can be limited by the temporary erythema and burning sensation on the site of application.

Indigo naturalis is a dark-blue powder obtained by crushing, fermenting and adding calcium to the leaves of indigo-producing plants such as: *Baphicacanthus cusia, Strobilanthes formosanus, Polygonum tinctorium* and *Isatis indigotica. Indigo naturalis* has been used for many years by traditional Chinese medicine for the treatment of various dermatoses.

A number of components have been isolated from *indigo naturalis*, the most active ones being indigo, isoindigo, tryptanthrin and indirubin. Experimental data on them report anti-proliferative effects on epidermal keratinocytes, anti-inflammatory activities and effects on regulatory T cells (Plitzko *et al.*, 2009; Lin *et al.*, 2009; Li *et al.*, 2011). These could explain in part the efficacy of *indigo naturalis* in plaque psoriasis, which has been confirmed by patients and discussed by Reuter *et al.* (2010) and Deng *et al.* (2013).

In psoriatic lesions *indigo naturalis* has been shown to upregulate claudin-1 expression and restore tight junction function; synergistic effect was reported for indirubin, inigo and tryptanthrin (Lin *et al.*, 2013). Clinical studies by Lin and his group (2007, 2008, 2011) on topical formulations of *indigo naturalis*, including oils and ointments, report efficacy in plaque psoriasis, nail psoriasis, also recalcitrant forms, in both adult and pediatric patients. Few cases of itching for a couple of days after starting the therapy were the only adverse events indicated.

These results suggest that topical application of *indigo naturalis* may be a safe and effective therapy for psoriasis, but due to the limited number of patients included in the studies and variety of the *indigo naturalis* preparations used, more *in vitro* and *in vivo* research is needed.

Mahonia aquifolium (*Oregon grape*) is a flowering plant belonging to the Berberidaceae family. It is native from North America, being spreading on the west coast from Southeast Alaska to Northern California. *Mahonia* is an evergreen shrub, with leathery, pinnate leaves, often with of spiny leaflets and dense clustered racemes of yellow flowers, sometimes fragrant, followed by black or purple berries.

This plant is used for various purposes: a number of cultivars and hybrids have been developed as decorative plants; the resistant holly-like leaves are sometimes used by florists for greenery; its fruits are used in the traditional diets of Pacific Northwest aboriginal people; the juice is used to make wine; dyes are extracted from the inner bark and roots (yellow dye), as well as from the berries (purple dye).

The use of *Mahonia aquifolium* for medical purposes goes back to Indian tribes, which were using it to treat dyspepsia. Nowadays, it is still of significant medical interest for its analgesic, anti-inflammatory, antioxidant and hepatoprotective effects; these are due principally to its rich content in alkaloids like berberine, palmatine, jatrorrhizine, berbamine and oxyacanthine (Chao *et al.*, 2013).

For plaque psoriasis, meta-analysis including both scientific *in vitro* and *in vivo* relevant data agrees on the efficacy of *Mahonia aquifolium* preparations (Deng *et al.*, 2013; Reuter *et al.*, 2010).

In vitro studies report *Mahonia* alkaloids to reduce the expression of adhesion molecules, proliferation and activation markers (Augustin *et al.*, 1999), inhibition of inhibition of 5-lipoxygenase and antioxidant properties of protoberberine and aporphine alkaloids (Misík *et al.*, 1995; Müller and Ziereis, 1994).

Clinical data reported by Bernstein *et al.* (2006), Gulliver and Donsky (2005) and Wiesenauer and Ludtke (1996) show promising results in psoriasis cases for topical formulations (*Mahonia aquifolium* ointments or creams) compared to placebo or to classic therapeutic options (corticoids, vitamin D derivatives). Preparations of *Mahonia* were well tolerated, with rare side effects including rash, a burning sensation when applied and staining of clothes.

Olive (*Olea europaea*) oil and olive leaves extract have been used since ancient times to moisturize and help rejuvenate damaged skin. Nowadays, many health benefits of olive oil are supported by scientific data and new positive attributes keep being discovered.

Both olive oil and olive leaves extract contain oleic and palmitic acids and a spectrum of polyphenols: flavonoids, lignans and secoiridoid glucosides. These compounds have anti-inflammatory and anti-oxidant effects, benefiting a number of skin conditions (e.g. atopic dermatitis, acne, seborrheic dermatitis, diaper dermatitis), stretch marks and scars (Viola and Viola, 2009).

Olive oil has been used in psoriasis, based on anecdotal evidence, since Hippocrates, when doctors were instructed to rub olive oil on psoriasis lesions. Scientific evidence on the matter is still limited, but there are few studies reporting efficacy of olive oil mixed with other natural ingredients, likely due to their anti-inflammatory and hydrating effects (Al-Waili, 2003; Bjorneboe *et al.*, 1988).

Sea buckthorn (*Hippophae rhamnoides*) is a deciduous shrub belonging to the Elaeagnaceae family, which is cultivated in various parts Europe and Asia for its nutritional and medicinal benefits. Its fruits, but also seeds and other parts, are appreciated for their contents rich in vitamins (A, B1, B9, B12, C, E, K and P), carotenoids (beta carotene, lycopene, and zeaxanthin), flavonoids, phytosterols and unsaturated fatty acids. These components confer *sea buckthorn* therapeutic benefits including cardiovascular protection, antiulcer activity, antitumoral, anti-inflammatory, antioxidant, antibacterial, antiatherogenic and hypoglycemic effects (Patel *et al.*, 2012).

Sea buckthorn extracts have antibacterial, anti-inflammatory, antioxidant and regeneration-promoting properties, being historically used for various skin conditions like dermatoses and wound healing. It is now being tried as a natural remedy for psoriasis, in an effort to control the periodic flare-ups. Scientific studies in this direction show promising results (Ganju *et al.*, 2005, Patel *et al.*, 2012).

Chamomile is the common name given to daisy-like plants of the Asteraceae family (*Matricaria recutita, Chamomilla recutita*). It has a long history of medical use, either in oral administration, as tea, or for topical applications.

Various chemical compounds are present within chamomile, each with different effects. Antiseptic properties are due to bisabolol and its derivatives, and also to chamazulene, farnesene and matricin. Anti-inflammatory effects are principally due to a number of flavonoids: apigenin, quercetin, luteolin and patuletin. Anecdotic use of chamomile extracts includes various inflammatory skin conditions like atopic dermatitis, rosacea, seborrheic dermatitis, mucous membrane inflammations, for and also wound healing (Ross, 2008).

The anti-inflammatory effects of chamomile, and partially also the antiseptic ones, justify possible benefits on psoriatic lesions (Graf, 2000; Aertgeerts *et al.*, 1985). Still, scientific studies on the matter are very limited, so more research in this direction is needed to properly assess its efficacy in psoriasis.

Because chamomile can cause uterine contractions, leading to miscarriage, pregnant women are advised not consume it. There is also risk of allergic reaction, contact dermatitis, or even anaphylaxis. It is still unclear whether individuals with, reported allergies to chamomile were actually exposed to it, or to plants with similar appearance (Facchinetti *et al.*, 2012).

Potential antipsoriatic activities have been reported also for preparations of *Alpinia galangal, Sphaerantus indicus, Annona squamosa, Camptotheca acuminate* nut, *Daucus carota L., Curcuma longa*, Kukui nut, Neem tree and for topical formulae mixing various herbs, but here is limited scientific assessment on their regard (Brown *et al.*, 2005; Deng *et al.*, 2013; Pandey *et al.*, 1994; Saelee *et al.*, 2011).

Conclusions

Psoriasis patients frequently use plant-based preparations. Physicians should be aware of the likely use of these therapeutic options, and the current evidence concerning their efficacy and safety.

Experimental data indicate mainly *Aloe vera*, capsaicin, chamomile, *indigo naturalis*, *Mahonia aquifolium*, but also of *Alpinia galangal, Annona squamosa, Camptotheca acuminate nut, Capsicum frutescens, Curcuma longa,* Kukui nut and Neem tree, as having anti-inflammatory, anti-proliferative and other actions of benefit to psoriatic lesions. The majority of evidence coming from clinical trials, though not numerous, agrees on the beneficial effects of these preparations, and their good safety profile, no serious side effects being reported.

Due to the limited number of *in vitro* data, and small size of most clinical studies, the proper magnitude of efficacy and safety cannot be measured with accuracy, so the current promising results should encourage more scientific research assessing the potential of these preparations.

Future *in vitro* research, followed by clinical assessment of selected plant-based formulations, will supply more scientific evidence in support of their efficacy in psoriasis. In addition, carefully chosen combinations, along with new delivery systems with increased topical absorption and improved stability, might increase the efficacy of these natural remedies.

Increasing the knowledge on their molecular mechanisms of action and scientifically proven benefits will help include plant-based preparations in the daily clinical management of psoriasis.

Acknowledgement

This work was partly supported by the project POSDRU 107/1.5.S/78702 to Andreea Nicoleta Boca.

References

Aertgeerts P, Albring M, Klaschka F, Nasemann T, Patzelt-Wenczler R, Rauhut K, Weigl B (1985). Comparative testing of Kamillosan cream and steroidal (0.25% hydrocortisone, 0.75% fluocortin butyl ester) and non-steroidal (5% bufexamac) dermatologic agents in maintenance therapy of eczematous diseases. Z Hautkr 60:270-277.

Al-Khafaji M (2012). Psoriasis. A case study. Chinesische Medizin 27:121-132.

Al-Waili NS (2003). Topical application of natural honey, beeswax and olive oil mixture for atopic dermatitis or psoriasis: partially controlled, single-blinded study. Complement Ther Med 11(4):226-234.

Arnold WP, van de Kerkhof PC (1994). Topical capsaicin in pruritic psoriasis. J Am Acad Dermatol 31(1):135.

Asadullah K, Volk HD, Sterry W (2002). Novel immunotherapies for psoriasis. Trends Immunol 23:47-53.

Atherton P (1998). *Aloe vera* revisited. Br J Phytother 4:76-83.

Augustin M, Andrees U, Grimme H *et al.* (1999). Effects of *Mahonia aquifolium* ointment on the expression of adhesion, proliferation, and activation markers in the skin of patients with psoriasis. Forsch Komplementarmed 6:619-621.

Bensky D, Barolet R (1990). Chinese herbal medicine: Formulas and Strategies. Seattle: Eastland Press.

Bernstein JE, Parish LC, Rapaport M, Rosenbaum MM, Roenigk HH Jr (1986). Effects of topically applied capsaicin on moderate and severe psoriasis vulgaris. J Am Acad Dermatol 15(3):504-507.

Bernstein S, Donsky H, Gulliver W, Hamilton D, Nobel S, Norman R (2006). Treatment of mild to moderate psoriasis with Reliéva, a *Mahonia aquifolium* extract - a double-blind, placebo-controlled study. Am J Ther 13(2):121-126.

Bhuchar S, Katta R, Wolf J (2012). Complementary and alternative medicine in dermatology: An overview of selected modalities for the practicing dermatologist. Am J Clin Dermatol 13(5):311-317.

Bjorneboe A, Smith AK, Bjorneboe GE, Thune PO, Drevon CA (1988). Effect of dietary supplementation with n-3 fatty acids on clinical manifestations of psoriasis. Br J Dermatol 118(1):77-83.

Bologna JL, Jorizzo JL, Schaffer JV (2012). Dermatology. 3rd ed. Philadelphia: Elsevier Saunders.

Bos JD, Spuls PI (2008). Topical treatments in psoriasis: today and tomorrow. Clin Dermatol 26(5):432-437.

Brown AC, Koett J, Johnson DW *et al.* (2005). Effectiveness of kukui nut oil as a topical treatment for psoriasis. Int J Dermatol 44:684-687.

Chandrasekaran K, Ramachandran T, Vigneswaran C (2012). Effect of medicinal herb extracts treated garments on selected diseases. IJTK 11(3):493-498.

Chao J, Liao JW, Peng WH, Lee MS, Pao LH, Cheng HY (2013). Antioxidant, Analgesic, Anti-Inflammatory, and Hepatoprotective Effects of the Ethanol Extract of *Mahonia oiwakensis.* Stem. Int J Mol Sci 14(2):2928-2945.

Choonhakarn C, Busaracome P, Sripanidkulchai B (2010). A prospective, randomized clinical trial comparing topical *Aloe vera* with 0.1% triamcinolone acetonide in mild to moderate plaque psoriasis. J Eur Acad Dermatol Venereol 24:168-172.

Danilenko DM (2008). Review paper: preclinical models of psoriasis. Vet Pathol 45:563-575.

Deng S, May BH, Zhang AL, Lu C, Xue CC (2013). Plant extracts for the topical management of psoriasis: A systematic review and meta-analysis. Br J Dermatol 168:954-967.

Deng S, May BH, Zhang AL, Lu C, Xue CC (2014). Topical herbal formulae in the management of psoriasis: Systematic review with meta-analysis of clinical studies and investigation of the pharmacological actions of the main herbs. Phytother Res 28(4):480-497.

Deng S, May BH, Zhang AL, Lu C, Xue CC (2013). Topical herbal medicine combined with pharmacotherapy for psoriasis: a systematic review and meta-analysis. Arch Dermatol Res 305(3):179-189.

Duke Phytochemical and Ethnobotanical Databases http://www.ars-grin.gov/duke/.

Ellis CN, Berberian B, Sulica VI, *et al.* (1993). A double-blind evaluation of topical capsaicin in pruritic psoriasis. J Am Acad Dermatol 29:438-442.

Facchinetti F, Pedrielli G, Benoni G, Joppi M, Verlato G, Dante G, Balduzzi S, Cuzzolin L (2012). Herbal supplements in pregnancy: unexpected results from a multicentre study. Hum Reprod 27(11):3161-3167.

Feily A, Namazi MR (2009). *Aloe vera* in dermatology: a brief review. G Ital Dermatol Venereol 144(1):85-91.

Fuhrmann T, Smith N, Tausk F (2010). Use of complementary and alternative medicine among adults with skin disease: updated results from a national survey. J Am Acad Dermatol 63:1000-1005.

Ganju L, Padwad Y, Singh R, *et al.* (2005). Anti-inflammatory activity of seabuckthorn (*Hippophae rhamnoides*) leaves. Int Immunopharmacol 5(12):1675-1684.

Gottlieb AB (2005). Psoriasis: emerging therapeutic strategies. Nat Rev Drug Discov 4:19-34.

Graf J (2000). Herbal anti-inflammatory agents for skin disease. Skin Therapy Lett 5(4):3-5.

Griffiths CE, Barker JN (2007). Pathogenesis and clinical features of psoriasis. Lancet 370:263-271.

Gulliver WP, Donsky HJ (2005). A report on three recent clinical trials using *Mahonia aquifolium* 10% topical cream and a review of the worldwide clinical experience with *Mahonia aquifolium* for the treatment of plaque psoriasis. Am J Ther 12:398-406.

Hatcher H, Planalp R, Cho J, Torti FM, Torti SV (2008). Curcumin: From ancient medicine to current clinical trials. Cell Mol Life Sci 65(11):1631-1652.

Homey B, Meller S (2008). Chemokines and other mediators as therapeutic targets in psoriasis vulgaris. Clin Dermatol 26(5):539-545.

Hutter JA, Salman M, Stavinoha WB, Satsangi N, Williams RF, Streeper RT, Weintraub ST (1996). Antiinflammatory C-glucosyl chromone from *Aloe barbadensis*. J Nat Prod 59(5):541-543.

Jacobson CC, Kumar S, Kimball AB (2011). Latitude and psoriasis prevalence. J Am Acad Dermatol 65:870-873.

Jantarat C (2013). Bioavailability enhancement techniques of herbal medicine: A case example of curcumin. International J Pharmacy and Pharmaceutical Sci 5:493-500.

Jensen P (1990). Use of alternative medicine by patients with atopic dermatitis and psoriasis. Acta Derm Venereol 70:421-424.

Jeon YJ, Kim BH, Kim S, Oh I, Lee S, Shin J, Kim TY (2013). Rhododendrin ameliorates skin inflammation through inhibition of NF-κB, MAPK, and PI3K/Akt signaling. Eur J Pharmacol 714(1-3):7-14.

Katare OP, Raza K, Singh B, Dogra S (2010). Novel drug delivery systems in topical treatment of psoriasis: rigors and vigors. Symposium-Psoriasis 76:612-621.

Kaur A, Kumar S (2012). Plants and plant products with potential antipsoriatic activity - A review. Pharm Biol 50(12):1573-1591.

Khan BA, Akhtar N, Mahmood T (2010). A comprehensive review of a magic plant, *Hippophae rhamnoides*. Pharmacognosy J 2:58-61.

Kivelevitch DN, Tahhan PV, Bourren P, Kogan NN, Gusis SE, Rodríguez EA (2012). Self-medication and adherence to treatment in psoriasis. Int J Dermatol 51(4):416-419.

Kuchekar S, Bhise K (2012). Formulation and development of antipsoriatic herbal gelcream. JSIR 71:279-284.

Kumar R, Phani Kumar G., Chaurasia OP, Singh SB (2011). Phytochemical and pharmacological profile of seabuckthorn oil: A review. Research Journal of Medicinal Plant 5:491-499.

Kurd SK, Troxel AB, Crits-Christoph P, Gelfand JM (2010). The risk of depression, anxiety and suicidality in patients with psoriasis: A population-based cohort study. Arch Dermatol 146(8):891-895.

Li D, Wu YS, Wang C (2011). Pharmacokinetics research on anti-inflammatory effect and analgesic effect of *indigo naturalis*. Chin J Exp Tradit Med Formul 17:137-140.

Li N, Li YQ, Li HY, Guo W, Bai YP (2012). Efficacy of externally applied chinese herbal drugs in treating psoriasis: A systematic review. Chinese Journal of Integrative Medicine 18:222-229.

Lin YK (2011). *Indigo naturalis* oil extract drops in the treatment of moderate to severe nail psoriasis: a small case series. Arch Dermatol 147:627-629.

Lin YK, Chang CJ, Chang YC *et al.* (2008). Clinical assessment of patients with recalcitrant psoriasis in a randomized, observer-blind, vehicle-controlled trial using *indigo naturalis*. Arch Dermatol 144:1457-1464.

Lin YK, Chen HW, Leu YL, Yang YL, Fang Y, Hwang TL (2013). *Indigo naturalis* upregulates claudin-1 expression in human keratinocytes and psoriatic lesions. J Ethnopharmacol 145(2):614-620.

Lin YK, Leu YL, Yang SH, Chen HW, Wang CT, Pang JH (2009). Anti-psoriatic effects of *indigo naturalis* on the proliferation and differentiation of keratinocytes with indirubin as the active component. J Dermatol Sci 54(3):168-174.

Lin YK, See LC, Chang YC, Huang YH, Chen JL, Tsou TC, Leu YL, Shen YM (2011). Treatment of psoriatic nails with *indigo naturalis* oil extract: a non-controlled pilot study. Dermatology 223(3):239-243.

Lin YK, Wong WR, Chang YC *et al.* (2007). The efficacy and safety of topically applied *indigo naturalis* ointment in patients with plaque-type psoriasis. Dermatology 214:155-161.

Lin YK, See L-C, Huang YH, Chang YC, Tsou TC, Leu YL, Shen YM. (2012). Comparison of refined and crude *indigo naturalis* ointment in treating psoriasis: Randomized, observer-blind, controlled, intrapatient trial. Arch Dermatol 148(3):397-400.

Lu CJ, Yu JJ, Deng, JW (2012). Disease-syndrome combination clinical study of psoriasis: Present status, advantages, and prospects Chin J Integr Med 18(3):166-171.

Magin PJ, Adams J, Heading GS, Pond CD (2009). Patients with skin disease and their relationships with their doctors: a qualitative study of patients with acne, psoriasis and eczema. Med J Aust 190(2):62-64.

Magin PJ, Adams J, Heading GS, Pond DC, Smith W (2006). Complementary and alternative medicine therapies in acne, psoriasis, and atopic eczema: results of a qualitative study of patients' experiences and perceptions. J Altern Complement Med 12(5):451-457.

Mantle D, Gok MA, Lennard TW (2001). Adverse and beneficial effects of plant extracts on skin and skin disorders. Adverse Drug React Toxicol Rev 20(2):89-103.

May BH, Zhang AL, Zhou W, Lu, CJ, Deng S, Xue, CCL. (2012). Oral herbal medicines for psoriasis: A review of clinical studies. Chin J Integr Med 18: 172-178.

Misík V, Bezáková L, Máleková L, Kostálová D (1995). Lipoxygenase inhibition and antioxidant properties of protoberberine and aporphine alkaloids isolated from *Mahonia aquifolium*. Planta Med 61:372-373.

Morelli V, Calmet E, Jhingade V (2010). Alternative therapies for common dermatologic disorders, part 2. Prim Care 37:285-296.

Müller K, Ziereis K (1994). The antipsoriatic *Mahonia aquifolium* and its active constituents; I. Pro- and antioxidant properties and inhibition of 5-lipoxygenase. Planta Med 60(5):421-424.

Nickoloff BJ, Nestle FO (2004). Recent insights into the immunopathogenesis of psoriasis provide new therapeutic opportunities. J Clin Invest 113(12):1664-1675.

Noorafshan A, Ashkani-Esfahani S (2013) A review of therapeutic effects of curcumin. Curr Pharm Des 19(11):2032-2046.

Pandey SS, Jha AK, Kaur V (1994). Aqueous extract of neem leaves in treatment of *Psoriasis vulgaris*. Indian J Dermatol Venereol Leprol 60:63-67.

Patel CA, Divakar K, Santani D, Solanki HK, Thakkar JH (2012). Remedial Prospective of *Hippophae rhamnoides* Linn. ISRN Pharmacol 2012:436857.

Paulsen E, Korsholm L, Brandrup F (2005). A double-blind, placebo-controlled study of a commercial *Aloe vera* gel in the treatment of slight to moderate psoriasis vulgaris. J Eur Acad Dermatol Venereol 19:326-331.

Pazyar N, Feily A (2011). Garlic in dermatology. Dermatology Reports 3(1):e4.

Plitzko I, Mohn T, Sedlacek N, Hamburger M (2009). Composition of *indigo naturalis*. Planta Med 75(8):860-863.

Rahman M, Alam K, Ahmad MZ, *et al.* (2012) Classical to current approach for treatment of psoriasis: a review. Endocr Metab Immune Disord Drug Targets 12(3):287-302.

Reuter J, Merfort I, Schempp CM (2010). Botanicals in dermatology: an evidence-based review. Am J Clin Dermatol 11:247-267.

Reuter J, Wölfle U, Weckesser S, Schempp C (2010). Which plant for which skin disease? Part 1: Atopic dermatitis, psoriasis, acne, condyloma and herpes simplex. J Dtsch Dermatol Ges 8(10):788-796.

Roopashree TS, Dang R, Shobha Rani RH, Narendra C (2008). Antibacterial activity of antipsoriatic herbs: *Cassia tora, Momordica charantia* and *Calendula officinalis.* IJARNP 1:20-28.

Ross SM (2008). Chamomile: a spoonful of medicine. Holist Nurs Pract 22(1):56-57.

Saelee C, Thongrakard V, Tencomnao T (2011). Effects of Thai medicinal herb extracts with anti-psoriatic activity on the expression on NF-κB signaling biomarkers in HaCaT keratinocytes. Molecules 16(5):3908-3932.

Schon MP (1999). Animal models of psoriasis-what can we learn from them? J Invest Dermatol 112(4):405-410.

Schon MP, Boehncke WH, Brocker EB (2005). *Psoriasis: Clinical manifestations, pathogenesis and therapeutic perspectives.* Discov Med 5(27):253-258.

Su YS, Chen GS, Lan CCE (2011). Infantile generalized pustular psoriasis: A case report and review of the literature. Dermatologica Sinica 29:22-24.

Surjushe A, Vasani R, Saple DG (2008). *Aloe vera*: a short review. Indian J Dermatol 53(4):163-166.

Syed TA, Ahmad SA, Holt AH (1996). Management of psoriasis with *Aloe vera* extract in a hydrophilic cream: a placebo-controlled, double-blind study. Trop Med Int Health 1:505-509.

Viola P, Viola M (2009). Virgin olive oil as a fundamental nutritional component and skin protector. Clin Dermatol 27:159-165.

Wiesenauer M, Ludtke R (1996). *Mahonia aquifolim* in patients with *Psoriasis vulgaris* - an intraindividual study. Phytomedicine 3:231-235.

Yu CS (2011). Study on HIF-1α gene translation in psoriatic epidermis with the topical treatment of capsaicin ointment. ISRN Pharm 2011:821874.

Yu JJ, Zhang CS, Zhang AL, May B, Xue CC, Lu C (2013) Add-on effect of chinese herbal medicine bath to phototherapy for *Psoriasis vulgaris*: A systematic review. Evid Based Complement Alternat Med 2013:673078.

Zeb A (2004). Important therapeutic uses of sea buckthorn (*Hippophae rhamnoides*): a review. J Biol Sci 4(5):687-693.

Effect of Preharvest Calcium Treatments on Sweet Cherry Fruit Quality

Deniz EROGUL

Ege University, Faculty of Agriculture, Department of Horticulture, 35100, İzmir, Turkey; deniz.erogul@ege.edu.tr

Abstract

In this study, the effects of different foliar calcium compounds on fruit cracking and quality of sweet cherry variety '0900 Ziraat' were investigated. Calcium caseinate, calcium chloride, calcium hydroxide and calcium nitrate were used as foliar sprays. Calcium applications reduced the cracking index 38% to 66% compared to cherries that did not receive foliar treatment. The most efficient applications for decreasing cracking were calcium hydroxide and calcium chloride. Calcium chloride and calcium hydroxide were determined to be the most effective compounds, reducing cracking by 62% and 66%, respectively. Calcium caseinate also decreased fruit cracking, by nearly 50% compared to the control. However, the leaves and fruit were covered with white film. This white film can be easily removed, as postharvest processing of cherries uses water, and normally no residue remains. Preharvest foliar calcium treatments significantly decreased fruit cracking, but cracking may not be completely eliminated. None of the calcium compounds affected removal force, average fruit weight, total soluble solids, or titratable acidy. Another potentially important finding was that foliar treatment with calcium chloride significantly increased firmness of sweet cherry fruit. Calcium chloride application increased the firmness of cherries by an average of 12% compared to other compounds.

Keywords: calcium, fruit cracking, fruit quality, preharvest sprays, sweet cherry

Introduction

Turkey has suitable conditions for growing cherry (*Prunus avium* L.) because of its location in the temperate zone and its geographical advantages which include its climatic advantages a great range providing chilling as well as earliness. Thus, it allows a long harvest period and day length. The total sweet cherry production was 2.225.519 tons in the world and Turkey was the leading worldwide sweet cherry producer country in 2012 with 480.748 tons (FAO, 2014).

Cherry production in the Aegean region starts with the harvest of early season varieties in May and continues until mid-August in the inner parts of the region where mid- and late-season cherry varieties are grown. Fruit cracking occurs in years that receive excess rainfall before the harvest period, and presents a significant problem for cherry growers. These rains cause significant economic loss as a result of the cracking, decay, and decrease in fruit quality.

Cracking in sweet cherries caused by rainfall shortly before harvest is one of the major limits to successful production of sweet cherry in many parts of the world (Sekse, 1995). A number of other factors affect cherry cracking, including fruit firmness and maturity stage, crop load, orchard temperature, irrigation status, duration of rainfall, and other environmental conditions (Rupert *et al.*, 1997; Simon, 2006). Crack-resistant varieties can reduce cracking rate but are not effective in completely eliminating or controlling the problem (Cline and Trought, 2007). Resistant cultivars and application of plant nutrients are used to help prevent cracking.

Calcium is involved in the construction of cell walls and is the major component contributing to the mechanical properties of plant tissues; calcium is also the most widely studied nutritional factor in relation to fruit cracking (Shear, 1975; Huang *et al.*, 2005). It is known that calcium decreases the permeability of cell membranes. Dipping fruits at optimum harvest stage in calcium solutions was reported to decrease water uptake rate and cracking index of fruit by 15% to 30% (Ackley, 1956). Various studies have examined the potential to reduce cracking in cherry fruit by spray application of calcium compounds such as calcium chloride [$CaCl_2$], calcium hydroxide [$Ca(OH)_2$] and calcium nitrate [$Ca(NO_3)_2$] (Callan, 1986; Meheriuk *et al.*, 1991; Yamamoto *et al.*, 1992; Rupert *et al.*, 1997; Marshal and Weaver, 1999). Studies on calcium caseinate are limited, but this compound has been reported to significantly decrease cracking in cherry (Ono *et al.*, 1954).

In this study, to examine the prevention or reduction of cracking in cherry, different calcium sprays were applied before the harvest period. Various forms of calcium were combined to perform trials on *Prunus avium* L. '0900 Ziraat', and the effect of these treatments on cherry fruit cracking and quality were determined.

Materials and methods

Plant material

The experiment was carried out in 2012 on 8 year old

'0900 Ziraat' sweet cherry trees (*Prunus avium* L.) grafted onto '*Prunus mahaleb*' rootstock in Kemalpaşa, İzmir Province, Turkey. Variety '0900 Ziraat' is commonly cultivated in most cherry-growing regions. In the experimental orchard, 'Stella' variety was used as a pollinizer and drip irrigation was provided. Standard methods were used for pruning, cultivating, nutrition, and pest management.

Pre-harvest treatments

Four different calcium compounds were used as preharvest foliar sprays and were applied 30, 20, and 10 days before harvest. The treatments included the following: control (water with surfactant), 0.5% calcium nitrate [$Ca(NO_3)_2$], 0.5% calcium chloride [$CaCl_2$ $2H_2O$], 0.5% calcium caseinate and 0.2 M calcium hydroxide [$Ca(OH)_2$]. The experiment was a randomized block design with 3 replications, each of which comprised one tree. A surfactant was added as 0.04% including the control treatment. Cherries were harvested on June 1, 2012, when they reached commercial harvest ripeness. Fruit were picked from 1.5 to 2 m above the ground and around the canopy for each replicate. Harvest was made from the outer parts of the trees to be representative of an actual harvest from mature trees. Approximately 4 kg cherries were harvested from each tree and brought to the laboratory immediately after collection.

Cracking index

Cracking indexes of cherries were previously calculated (Bilgener *et al.*, 1999). Forty cherries were randomly selected from each replicate and immersed in 2 L plastic containers filled with water (20 ± 1 °C) for 6 h. Cherries were then removed from the water; cracked fruit were counted and separated and uncracked fruit were quickly immersed again in the water. This process was repeated 3 times for each treatment. Cracking indexes were calculated according to the following formula:

Cracking index = $(5a + 3b + c) \times 100/200$

Where *a, b,* and *c* represent the number of cracked fruit after 2, 4, and 6 h, respectively. Total number of fruits immersed = 40; maximum cracking, 40 × 5 = 200.

Quality assessment

Cherries were weighted (XB-320M,Presica Instruments Ltd., Switzerland) after harvest and average fruit weight was calculated. Fruit firmness was determined on the peel of 25 cherries using a penetrometer (FT 011, Effegi, Japon) with a 4-mm diameter tip and the results were expressed in Newtons (N). Fruit removal force was measured with a penetrometer on 25 cherries selected randomly from each replicate. The total soluble solid (TSS) content of the juice was determined with a digital refractometer (PR-1, Atago, Japon) and expressed as a percentage. Titratable acidity was measured by titration with 0.1 N NaOH to pH 8.1, and the results were expressed as g malic acid/100 mL fruit juice.

Statistical analysis

All data were subjected to Analyses of Variance (ANOVA). Significant differences among groups were determined using Duncan's multiple range tests at $p<0.05$ All computation and statistical analyses were done using IBM SPSS Statistics 19 statistical software (IBM, NY, USA).

Results and discussion

Fruit Cracking

The different preharvest calcium compounds significantly ($p \leq 0.01$) decreased the cracking ratio of cherries relative to the control (Fig. 1). Calcium nitrate was the least effective application for reducing cracking; cracking index was reduced by 38% in cherries that received foliar calcium nitrate compared to those without any application. Treatment of *Napoleon* cherry with 0.5% calcium nitrate was reported to significantly decrease cracking without leaving any residue (Yamamoto *et al.*, 1992). Calcium caseinate treatment significantly decreased cracking compared to the control treatment; the cracking ratio of calcium caseinate treated cherries was half that of control cherries. A white film layer was observed on fruit and leaves after calcium caseinate applications. Although this might be considered a problem for marketing, the white film on harvested cherries can be easily washed off. It is also thought that these white films will disappear during the precooling and postharvest treatments, which use water. It was reported that preharvest spray application of calcium caseinate did not leave residue on fruit and significantly

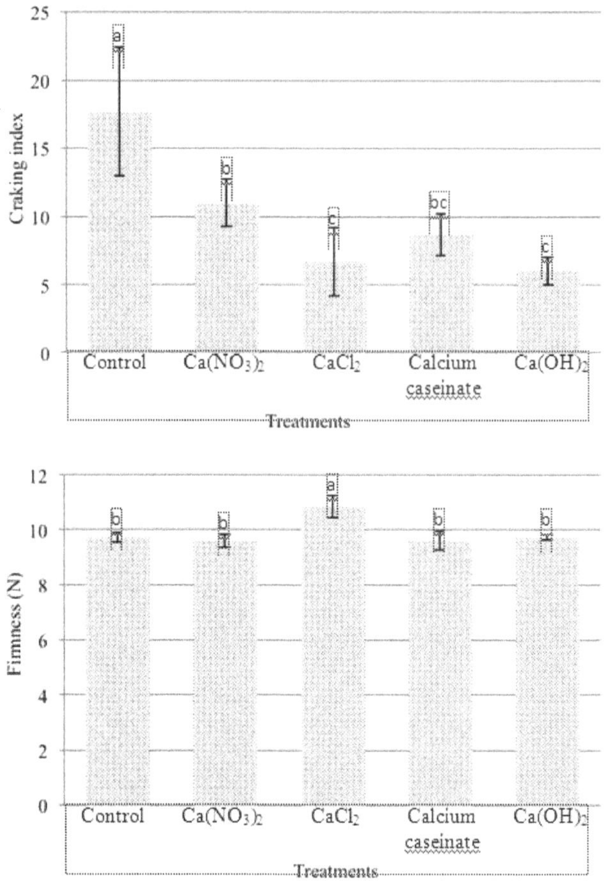

Fig. 1. Effect of pre-harvest different calcium form applications on cherry fruit cracking. Index and fruit firmness. Vertical bars indicate the standard deviation calculated from means of replicates

decreased cracking (Ono *et al.*, 1954). Calcium chloride and calcium hydroxide treatments decreased cracking of cherries (Meheriuk *et al.*, 1991). Our results revealed that the most efficient treatments for decreasing cracking were calcium chloride and calcium hydroxide. Preharvest calcium chloride spray, applied to prevent rain-induced cracking, protected cherries and significantly reduced cracking (Marshal and Weaver, 1999; Long, 2013). In this study, the cracking ratio in trees treated with calcium chloride was 6.7%, compared to 17% in the control. Cherry cracking was reported to decrease to 11% versus 33% with calcium chloride treatments (Fernandez and Flore, 1998). In locations with better climate conditions, spray application of mineral solutions (e.g., calcium chloride) at harvest time can effectively reduce the ratio of cracked fruit (Simon, 2006). It was stated that automated cyclical overtree sprinkler applications of dilute calcium chloride solutions decreased cracking ratio (Lang *et al.*, 1998). The lowest cracking ratio in this study was obtained by 6% calcium hydroxide treatment. Another study using calcium hydroxide and calcium chloride determined that calcium hydroxide treatment was more effective than calcium chloride treatment in preventing cracking (Callan, 1986).

Fruit quality traits

The various calcium treatments had no marked effects on average cherry weight, which varied from 8.62 to 9.71 g (Tab. 1). Preharvest foliar Ca treatments did not affect fruit weight in kiwi (Koutinas *et al.*, 2010) or apple (*Brown et al.*, 1996; Domagała-Świątkiewicz and Błaszczyk, 2009). Fruit firmness is the most significant quality criteria in cherries and it is used in the evaluation of the fruit (Esti *et al.*, 2002; Crisosto *et al.*, 2003).

Firmness of cherries affects shelf life and consumer acceptance (Brown *et al.*, 1988). The effects of foliar treatments on fruit firmness were significant. Except for calcium chloride, the treatments had no significant effect on fruit firmness (Fig. 1). Other researchers reported that preharvest calcium chloride application had no effect on the firmness of cherries (Meheriuk *et al.*, 1991; Rupert *et al.*, 1997) or strawberries (Chéour *et al.*, 1990), while preharvest calcium chloride application increased fruit firmness in the present study (Fig. 1). It was also reported that calcium chloride application ($p \leq 0.05$) increased the firmness of apple fruit (Malakouti *et al.*, 1999). However, preharvest calcium nitrate, calcium caseinate, and calcium hydroxide sprays did not affect fruit firmness. The firmness of cherries from the calcium hydroxide treatment did not differ from that of the control. Similarly, it was stated that preharvest calcium hydroxide treatment did not affect fruit firmness in cherries (Meheriuk *et al.*, 1991; Brown *et al.*, 1996). In contrast, it was reported that $Ca(OH)_2$ treatment increased fruit firmness (Demirsoy and Bilgener, 1998).

Fruit removal force was best protected in cherries treated with calcium hydroxide and calcium nitrate and the weakest resistance was observed in the control (Tab. 1). However, the effect of foliar treatments on the removal force of fruit was not significant.

One of the most significant parameters determining consumer preference in cherries is the ratio of TSS and TA (Crisosto *et al.*, 2003). In this study, it was determined that the effects of preharvest calcium sprays on TSS and TA in cherry were not significant (Tab. 1). The TSS value of control plants was 15.63%, while it varied between 12.90% and 14.97% in plants that received calcium treatments. Acidity was 0.75% in the control while it varied between 0.65% and 0.74% in calcium-treated cherries (Tab. 1). TSS and TA content depend on the variety; TSS ranges between 11–25%, while TA ranges between 0.4–1.5% (Esti *et al.*, 2002; Bernalte *et al.*, 2003). In parallel with the results of this study, other researchers found that preharvest calcium treatments did not change fruit quality parameters such as total soluble solids and total titratable acidity in cherry and peach (Brown *et al.*, 1996; Crisosto *et al.*, 2000; Manganaris *et al.*, 2005). In another study, preharvest calcium treatments increased soluble solids in cherries but did not affect acidity (Meheriuk *et al.*, 1991; Vangdal *et al.*, 2008).

Conclusions

Cracking, especially rain-induced, of sweet cherries is one a major problem for cherry producers in almost all producing countries due to significant amount of unmarketable yields. The results of this study have shown that preharvest calcium nitrate, calcium chloride, calcium caseinate, and calcium hydroxide applications decreased the cracked fruit ratio, significantly. Calcium applications reduced the cracking index of cherries by 38% to 66.0% compared to those that received no foliar Ca applications. The most effective preharvest foliar calcium treatments were calcium chloride and calcium hydroxide, which reduced cracking by 62% and 66%, respectively. Calcium

Tab. 1. Effect of pre-harvest different calcium form applications on cherry fruit removal force, average fruit weight, TSS and TA content

Treatments	Fruit removal force (N)	Average fruit weight (g)	TSS (%)	TA (g malic acid/100 ml)
Control	2.79	8.62	15.63	0.75
$CaNO_3$	3.27	9.71	14.70	0.74
$CaCl_2$	3.09	9.29	14.97	0.69
Calcium caseinate	2.84	8.75	12.90	0.65
$Ca(OH)_2$	3.37	9.24	14.63	0.69
Significance	NS	NS	NS	NS

NS - nonsignificant

caseinate even if reduced cracking by 50 % left a white covering over leaves and fruit that could easily be further removed during hydro-cooling and/or washing process. Preharvest calcium sprays had no negative effect on other major fruit quality parameters. Foliar treatment with calcium chloride exerted created an additional effect by increasing firmness, significantly (12 %). On the basis of these data, we can recommend use of tested preharvest calcium foliar treatments to reduce cracking without guaranteeing complete elimination.

References

Ackley WB (1956). Fruit cracking of sweet cherries. Washington State University Progress in Agriculture and Home Economics Research, Progress Report No. 9.

Bernalte MJ, Sabio E, Hernandez MT, Gervasini C (2003). Influence of storage delay on quality of 'Van' sweet cherry. Postharvest Biol Technol 28:303-312.

Bilgener Ş, Demirsoy L, Demirsoy H (1999). The effects of vapor gard, GA_3 and calcium hydroxide applications on fruit cracking in 'Türkoğlu' sweet cherry. Etkilerinin araştırılması. III. National Horticultural Congress, Turkey. p. 828-832 (in Turkish).

Brown, GS, Kitchener AE, McGlasson WB, Barnes S (1996). The effects of copper and calcium and foliar sprays on cherry and apple fruit quality. Sci Hort 67:219-227.

Brown SK, Bourne MC, (1988). Assesment of fruit firmness in selected sour cherry genotypes. HortScience 23:882-884.

Callan NC (1986). Calcium hydroxide reduces splitting of 'Lambert' sweet cherry. J Am Soc Hort Sci 111:173-175.

Chéour F, Willemot C, Arul J, Desjardins Y, Makhlouf J, Charest PM, Gosselin A (1990).Foliar application of calcium chloride delays postharvest ripening of strawberry. J Am Soc Hort Sci 115:789-792.

Cline JA, Trought M (2007). Effect of gibberellic acid on fruit cracking and quality of Bing and Sam sweet cherries. Canadian J Plant Sci 87(3):545-550.

Crisosto CH, Crisosto GM, Metheney P (2003). Consumer acceptance of 'Brooks' and 'Bing' cherries is mainly dependent on fruit SSC and visual skin color. Postharvest Biol Technol 28:159-167.

Demirsoy LK, Bilgener S (1998). The effects of preharvest calcium hydroxide applications on cracking in 0900 'Ziraat','Lambert' and 'Van' sweet cherries. Third International Cherry Symposium. Acta Hort 468(2):657-662.

Domagała-Świątkiewicz I, Błaszczyk J (2009). Effect of calcium nitrate spraying on mineral contents and storability of 'Elise' apples. Polish J Environ Stud 18(5):971-976.

Esti M, Cinquante L, Sinesio F, Moneta E, Di Matteo M (2002). Physicochemical and sensory fruit characteristic of two sweet cherry cultivars after cool storage. Food Chem 76:399-405.

FAO (Food and Agriculture Organization of the United Nations) (2014). http://faostat.fao.org.

Fernandez RT, Flore JA (1998). Intermittent application of $CaCl_2$ to control rain cracking of sweet cherry. Acta Hort 468:683-689.

Huang XM, Wang HC, Li JG, Yin JH, Yuan WQ, Lu JM, Huang HB (2005). An overview of calcium's role in lychee fruit cracking. Acta Hort 665(1):231-240.

Koutinas N, Sotiropoulos T, Petridis Almaliotis D, Deligeorgis E, Therios I, Voulgarakis N (2010). Effects of preharvest calcium foliar sprays on several fruit quality attributes and nutritional status of the kiwifruit cultivar Tsechelidis. Hort Sci 45(6):984-987.

Lang G, Guimond C, Flore JA, Southwick S, Facteau T, Kappel F, Azarenko A (1998). Performance of calcium/sprinkler-based strategies to reduce sweet cherry rain-cracking. Acta Hort 468:649-656.

Long LE (2013). A review of cherry rain cracking. http://www.cherries.co.za/wpcontent/uploads/2013/01/Publication.

Malakouti MJ, Tabatabaei SJ, Shahabil A, Fallahi E (1999). Effects of calcium chloride on apple fruit quality of trees grown in calcareous soil. J Plant Nutri 22(9):1451-1456.

Manganaris GA, Vasilakakis M, Mignani I, Diamantidis G, Tzavella-Klonari K (2005). The effect of preharvest calcium sprays on quality attributes, physicochemical aspects of cell wall components and susceptibility to brown rot of peach fruits (*Prunus persica* L. cv. Andross). Sci Hort 107:43-50.

Marshall R, Weaver E (1999). An update on the use of calcium chloride to reduce rain cracking of cherries-year 2 results. Orchardist 72:34-36.

Meheriuk M, Neilsen GH Mckenzie DL (1991). Incidence of rain splitting in sweet cherries treated with calcium or coating materials. Can J Plant Sci 71: 231-234.

Ono T, Oyazizu W, Suzuki K (1954). Studies on the reduction of cracking in sweet cherries (1). J Hort Japan 22:239-243.

Rupert M, Southwick S, Weis K, Vikupitz J, Flore J, Zhou H (1997). Calcium chloride reduces rain cracking in sweet cherries. California Agric 51:35-40.

Sekse L (1995). Fruit cracking in sweet cherries (*Prunus avium.* L.) some physiological aspects-a mini review. Scientia Hort 63: 135-141.

Shear CB (1975). Calcium-related disorders of fruits and vegetables. Hort Sci 10:361-365.

Simon G (2006). Review on rain induced fruit cracking of sweet cherries (*Prunus avium* L.), its causes and the possibilities of prevention. International J Hort Sci 12(3):27-35.

Vangdal E, Hovland KL, Borve J, Sekse L, Slimestad R (2008). Foliar application of calcium reduces postharvest decay in sweet cherry fruit by various mechanisms. Proceedings of the International Symposium on the Role of Postharvest Technology in the Globalisation of Horticulture. Acta Hort 768:143-148.

Yamamoto T, Satoh H, Watanabe S (1992). The effects of calcium and naphtalene acetic acid sprays on cracking index and natural rain cracking in sweet cherry fruits. J Japan Soc Hort Sci 61(3):507-511.

33

Xanthophyll Esters in Fruits and Vegetables

Andrea BUNEA, Carmen SOCACIU, Adela PINTEA*

*University of Agricultural Sciences and Veterinary Medicine, Department of Chemistry and Biochemistry, 3-5 Mănăştur Street, Cluj-Napoca, Romania; andrea_bunea@yahoo.com; carmen.socaciu@usamvcluj.ro; apintea@usamvcluj.ro (*corresponding author)*

Abstract

Carotenoids possessing hydroxyl groups (xanthophylls) are often found as fatty acid esters in many fruits and vegetables. The developments in high resolution chromatographic and spectroscopic techniques have led to a detailed characterization of xanthophyll esters in commonly consumed fruits and vegetables, such as apples, apricots, mandarins, mangoes, papayas, red and chili peppers, potatoes or squash. Some more rich sources have been identified, like wolfberry (goji), sea buckthorn, persimmon, whose popularity is increasing due to the high content of bioactive compounds. Esterification increases the lipophilicity of xanthophylls and contributes to the sequestration of carotenoids, to the formation of specialized structures in the chromoplasts and to an increased photoprotection. The process occurs during ripening in fruits and it is associated with a significant change in colour. Even if the specific enzymes which catalyze the esterification process were not characterized yet in fruits, detailed analytical data regarding the carotenoid composition suggested a selectivity of these enzymes for certain fatty acids and selectivity for the ring in the case of non-symmetric xantophylls. Xanthophyll esters seem to be efficiently hydrolyzed and absorbed in humans leading to a comparable bioavailability to the unesterified compounds. In addition, the xanthophyll esters preserve the antioxidant capacity of the parent compounds while having a better stability in fruits during storage and processing. All these properties are important from the perspective of the use of fruits rich in xanthophyll esters as valuable components of the human diet and as sources of bioactive compounds in the prevention of severe degenerative diseases.

Keywords: analysis, antioxidant capacity, biosynthesis and accumulation, bioavailability, stability

Introduction

Fruits and vegetables are important components of the human diet and they are associated with numerous health benefits due to their high content of micronutrients. Carotenoids are a family of widely distributed lipophilic compounds, comprising more than 750 representatives (Britton *et al.*, 2004). The yellow, orange or red colour of fruits and vegetables is due to the presence of carotenoid pigments. Carotenoids are biosynthesized by plants and some microorganisms but not by animals and human, which rely on their diet to incorporate carotenoids. It is considered that up to 90% of the carotenoid intake in humans comes from fruits and vegetables (Maiani *et al.*, 2009).

The two classes of carotenoids are carotenes and xanthophylls, which are oxygenated derivatives. Of particular importance among xanthophylls are the hydroxycarotenoids: β-cryptoxanthin, lutein and zeaxanthin. Some other well known xanthophylls are violaxanthin and neoxanthin in green tissues, capsanthin and capsorubin in pepper, astaxanthin in algae (*Haematococcus pluvialis*) or pink-flesh fish (Britton and Khachik, 2009; Dufossé, 2009). In fruits and vegetables xanthophylls are present either in a free, unesterified form, or as esters with fatty acids. In some foods like corn, spinach, broccoli or other green leafy vegetables, xanthophylls are present exclusively in unesterified form. In other fruits and vegetables such as pepper, wolfberry, sea buckthorn, apple, squash etc., xanthophylls are mostly found in esterified form (Britton and Khachik, 2009; Pérez-Gálvez and Minguez Mosquera, 2005).

Analysis of carotenoids is usually performed on saponified extracts. Saponification (alkaline hydrolysis) is used to facilitate carotenoid isolation because it is effective in removing contaminating lipids (especially in food-rich samples) and destroying chlorophylls. In the same time, the hydrolysis of carotenoid esters occurs and artefacts can be produced. Due to the fact that xantophylls present in fruits are often esterified, more and more studies are investigating the unsaponified carotenoid extracts which reflect the native composition of the plant food (Giuffrida *et al.*, 2013; Pintea, 2008; Rodriguez-Amaya, 2010).

The best documented function of carotenoids is the provitamin A activity. Carotenoids possessing at least one unsubstituted β-ring are absorbed and cleaved to form retinol in animals and humans. Food rich in carotenoids are considered to be beneficial in the prevention of severe diseases like cancer, cardiovascular diseases, degenerative diseases, age-related macular degeneration (AMD) and cataract (Krinsky and Johnson, 2005; Landrum *et al.*, 1997; Rao and Rao, 2007). Lutein and zeaxanthin accumulate in the structures of the human retina, with the highest concentration in the macula were they act as light filter pigments but also as antioxidants (Bernstein *et al.*, 2001). Observational, epidemiological and intervention trials demonstrated that a high concentration of lutein in plasma is correlated with an increase of macular pigment density and a reduced risk of AMD (Chew *et al.*, 2013; Schalch *et al.*, 2009). β-Cryptoxanthin has provitamin A activity but has also been proved to stimulate bone calcification (Yamaguchi, 2012), to have antiproliferative (Wu *et al.*, 2013) and anti-inflammatory effects (Pattison *et al.*, 2005).

Taking into account that xanthophylls esters are efficiently absorbed and cleaved into human body and that their availability is comparable with the unesterified form, food rich in esters can be considered a valuable source of biologically active xanthophylls. Furthemore, the xanthophyll esters preserve the well known antioxidant capacity of carotenoids which is associated with most of their biological properties.

Occurrence

Carotenoid esters occur in flowers, fruits and vegetables. In flower petals, xanthophylls are usually present in esterified form, the best known example being the flowers of *Tagetes* sp., where more than 90% of total carotenoids are lutein esters (Gregory *et al.*, 1986; Ohmiya, 2013; Pérez-Gálvez and Minquez-Mosquera, 2005). Lutein, β-cryptoxanthin, zeaxanthin and violaxanthin are the most frequent xanthophylls found in esterified form in fruits and vegetables and they are located in the specialized structures in chromoplasts (Breithaupt and Bamedi, 2001; Ohmiya, 2013).

The hydroxyl groups of xanthophylls are acylated with saturated fatty acids such as lauric (C12:0), myristic (C14:0), palmitic (C16:0) and stearic (C18:0) acids (Britton *et al.*, 1995a) but unsaturated fatty acids like oleic acid (C18:1), linoleic (C18:2) or linolenic (C18:3) acids were also occasionally reported in fruits, as well as the short chain fatty acids like butyric acid (C4:0).

The most common fruits and vegetables which contain remarkable amounts of xanthophyll esters are: apples (*Malus domestica*) (Delgado-Pelayo *et al.*, 2014); apricots (*Prunus armeniaca* L.) (Kurz *et al.*, 2008); mangoes (*Mangifera indica* L.) (Ornelas-Paz *et al.*, 2007), sea buckthorn berries (*Hippophae rhamnoides* L.) (Giuffrida *et al.*, 2012; Pop *et al.*, 2014; Weller and Breithaupt, 2003); mandarins (*Citrus reticulata*) (Giuffrida, 2006); gojis (*Lycium barbarum* L. and *Lycium chinense* L.) (Inbaraj *et al.*, 2008; Zhao *et al.*, 2013); papayas (*Carica papaya* L.) (Schweiggert *et al.*, 2012); oranges (*Citrus sinensis* L.) (Giuffrida *et al.*, 2010); orange, red and chilli peppers (*Capsicum* sp.) (Giuffrida *et al.*, 2013; Hornero-Méndez and Minguez-Mosquera, 2000); potatoes (*Solanum* sp.) (Breithaupt and Bamedi, 2002a; Burmeister *et al.*, 2013; Fernandez-Orozco *et al.*, 2013).

Some other less common fruits which were also reported to contain xanthophyll esters are Chinese lanterns (*Physalis alkekengi* L.) (Goodwin, 1980; Pintea *et al.*, 2005; Weller and Breithaupt, 2003); persimmon (*Diospyros kaki*) (Weller and Breithaupt, 2003); yellow raspberries (*Rubus idaeus* L.) (Carvalho *et al.*, 2013); sarsaparilla berries (*Smilax aspera* L.) (Delgado-Pelayo and Hornero-Méndez, 2012); corozo, sastra, frutita, maracuja chino, mamey roja (Murillo *et al.*, 2013); cashew apple (Pinto de Abreu *et al.*, 2013).

A very complex screening regarding β-cryptoxanthin esters was performed by Breithaupt and Bamedi (2001). β-cryptoxanthin laurate, myristate and palmitate, as well as the free β-cryptoxanthin, were quantified in several fruits: orange, blood orange, pepper and chili, clementine, papaya, persimmon, tangerine, loquat, kumkuat, nectarine. Similarly, Weller and Breithaupt (2003) determined the zeaxanthin esters (mono and diesters with laurate, myristate, palmitate and stearate) in red and orange pepper, sea buckthorn, wolfberries, Chinese lanterns and persimmon. In both studies, LC/MS/APCI was used for separation and identification, as well as standard esters obtained by semisynthesis.

Carotenoid composition and concentration in fruits is influenced by genetic and environmental factors, such as: cultivar, maturity, climate, geographic area, processing and storage (Rodriguez-Amaya, 2010). Absolute quantitative data for the carotenoid esters content are difficult to be set. However, pepper and chilli (orange, red), wolfberry, papaya, tangerines, persimmon and sea bukhthorn seem to be among the richest sources (Breithaupt and Bamedi, 2001; Weller and Breithaupt, 2003). Detailed composition of xanthophyll esters in selected fruits, including the fatty acids esterifying the hydroxyl groups are presented in Tab. 1.

Tab. 1. Distribution, extraction and analytical procedures for separation and identification of xanthophyll esters in selected fruits

Fruit	Analyte	Extraction protocol	Column (stationary phase)	Mobile phase	Detection system	References
Capsicum varieties	*Cis*-capsanthin-esters: C12:0; C14:0 Antheraxanthin esters: C12:0; C14:0) Capsanthin-5,6-epoxy esters: C14:0 Lutein esters: C14:0 Capsanthin esters: C12:0; C14:0; C16:0; C14:0-C14:0; C12:0-C16:0; C14:0-C16:0; C16:0-C16:0 Zeaxanthin esters: C12:0; C14:0; C16:0; C12:0-C12:0; C12:0-C14:0; C14:0-C14:0; C14:0-C16:0; C16:0-C16:0 β-cryptoxanthin esters: C12:0; C14:0; C16:0 Cyptocapsin esters: C14:0; C16:0 Capsorubin diesters: C14:0-C14:0; C14:0-C16:0	Acetone	YMC C30 (250x4.6 mm, 5 μm)	(A): MeOH:TBME:H_2O 82:16:2 (v/v/v) (B): MeOH:TBME:H_2O 10:88:2 (v/v/v)	HPLC-DAD-APCI-MS	Giuffrida *et al.*, 2013

Spice paprika (*Capsicum annum* L.)	Capsorubin methyl ester Cucurbitaxanthin methyl ester Capsanthin epoxide methyl ester Capsanthin methyl ester-1 Cryptoxanthin methyl ester Capsanthin ME-2ME Antheraxanthin ME Zeaxanthin ME and DE Capsanthin DE	Dichlorethane Acetone/ Methanol (2:1:1, v/v/v)	Purospher C-18 (250x4.6 mm, 3 μm)	A: H_2O B: MeOH C: Methanol/Isopropanol / Acetonitrile 10:55:35 (v/v/v)	LC-MS	Daood, *et al.*, 2014
Mango	9-*cis*-violaxanthin esters: C4:0; C4:0- C4:0 *all-trans*-violaxanthin: C4:0-C4:0; C4:0-C6:0	Methanol	YMC C30 (150 x 4.6 mm, 3 μm) 15°C	A: Water B: MeOH C: MTBE	LC-APCI$^+$-MS	Ornelas-Paz *et al.*, 2007
Sea buckthorn (*Hippophae rhamnoides* L.)	Lutein esters: C16:3; C16:2; C18:2; 18:1; C16:0-C16:0 Zeaxanthin esters: C14:0; C16:0; C14:0-14:0; C16:0-16:1; C14:0-16:0; C16:0-16:0 β-cryptoxanthin esters: C14:0; C16:0	Methanol/ Ethyl Acetate/ Petroleum Ether, (1/1/1; v/v/v)	YMC C30 (250 x 4.6 mm -5 μm)	A: Methanol/MTBE/H_2O (83:15:2, v/v/v) B: Methanol/MTBE/H_2O (8:90:2, v/v/v)	HPLC-DAD-APCI-MS	Giuffrida *et al*, 2012
	β-cryptoxanthin esters: C16:0 Zeaxanthin esters: C14:0; C16:0; C14:0-C16:0; C16:0:C16:0 Lutein esters: C16:0; C14:0-14:0; C14:0-16:0; C16:0-C18:0	Methanol/ Ethyl Acetate/ Petroleum Ether, (1/1/1; v/v/v)	LiChrosorb RP 18 (250 x 4.6 mm, 5 μm) RP 18 UPLC BEH Shield (2.1 x 150 mm, 1,7 μm)	A: AcCN:H_2O 9:1 B: AcEt A: Ac Ethyl B: Acetonitrile C: Acetonitrile: H_2O (1:1, v/v)	RP-PDA-HPLC UHPLC-PAD-ESI-MS	Pop *et al.*, 2014
Mandarin essential oil	β-cryptoxanthin esters: C12:0; C14:0; C16:0	Oil:TBME (1:1)	YMC C30 (250x4.6 mm, 5 μm)	A: MeOH B: MTBE	HPLC-PDA-APCI-MS	Giuffrida *et al.*, 2006
Sarsaparilla berries (*Smilax aspera* L.)	Zeaxanthin monoester: C14:0 β-cryptoxanthin esters: C10:0; C12:0; C14:0; C16:0; C18:0	acetone	RP C18 Mediterranea SEA 18 (200 x 4.6 mm, 3 μm); 25°C	A: Acetone B: H_2O	LC-MS (APCI+)	Delgado-Pelayo and Hornero-Méndez, 2012
Mandarin	β-cryptoxanthin esters: C10:0; C12:0; C14:0; C16:0 Mutatoxanthin isomer esters: C10:0; C12:0 Lutein esters: C14:0; C16:0; C18:0; C18:1 Mutatoxanthin esters: C16:0; C18 :0 Luteoxanthin esters: C10:0; C12:0; C14:0	Essential oil injected directly	Discovery Cyano (250x1 mm; 5μm) Chromolith Performance RP-18 (100x4.6 mm)	A: n-hexane B: n-hexane/ butylacetat/acetone (80:15:5) v/v/v A: 2-propanol B: 20% water in acetonitrile (v/v)	LC x LC, DAD/ APCI	Dugo *et al.*, 2008
Orange juices	*cis*-violaxanthin esters: C10:0; C12:0; C14:0; C16:0; C12:0-C12:0; C12:0-C14:0; C14:0- C14:0; C14:0-C16:0; C16:0- C16:0 Luteoxanthin esters: C12:0; C14:0; C16:0 Auroxanthin: C16:0 Antheraxanthin: C16:0 Mutatoxanthin (C16:0) β-cryptoxanthin (C12:0; C14:0; C16:0)	Methanol/ Ethyl Acetate/ Petroleum Ether, (1/1/1; v/v/v)	YMC C30 (250x4.6 mm, 5 μm) 23°C	A: Methanol/MTBE/Water (90:7:3; v/v/v) B: Methanol/MTBE (10:90; v/v)	LC-MS-APCI	Giuffrida *et al.*, 2010
Apricots (*Prunus armeniaca* L.)	β-criptoxanthin esters (C12:0; C18:1; C16:0; C18:1-C18:1; C16:0-C18:1) Lutein esters (C18:1- C18:1; C16:0- C18:1; C12:0- C12:0; C12:0- C14:0; C14:0- C14:0; C14:0- C16:0; C16:0- C16:0) Antheraxanthin esters (C12:0; C14:0; C16:0)	Acetone/ Hexane (1:1; v/v)	C30 YMC 150x3 mm 3μm 25°C	A:Methanol/MTBE/ H_2O (81:15:4 v/v/v) B: Methanol/MTBE/H_2O (4:92:4 v/v/v)	LC-MS	Kurz *et al.*, 2008
Yellow raspberries (*Rubus idaeus* L.)	Lutein mono and diesters: C8:0; C10:0; C12:0: C14:0; C16:0	Methanol/Tris-HCl buffer/ Chloroform (1:1:4, v/v/v)	C30 YMC (250 x 4.6 mm, 5 μm)	A: MeOH B: H_2O/MeOH (20/80 v/v) +0,2% ammonium acetate C: TMBE	LC-APCI-MS	Carvalho *et al.*, 2013
Apple (flesh and peel)	*all-trans*-Neoxanthin esters: C10:0-C10:0, C10:0-C12:0; C12:0-C12:0; C12:0-C14:0; C14:0; C14:0-C14:0; C16:0; C16:0-C16:0; C16:0-C18:0 *all-trans*-Violaxanthin esters: C10:0-C10:0; C10:0-C12:0; C12:0-C12:0; C12:0-C14:0; C14:0-C14:0; C14:0-C16:0; C16:0-C16:0; C16:0-C18:0 9-*cis*-Violaxanthin esters: C10:0-C10:0; C10:0-C12:0; C12:0-C12:0; C12:0-C14:0; C14:0-C14:0; C14:0-C16:0; C16:0-C16:0; C16:0-C18:0	N,N-dimethyl-formamide	RP-C18 Mediterranea SEA 18 (200x4.6 mm, 3 μm) 25 °C	A: Acetone B: Deionised water	LC-MS (APCI+)	Delgado-Pelayo *et al.*, 2014
Goji (*Lycium barbarum*)	Zeaxanthin esters: C16:0; C16:0-C16:0 β-cryptoxanthin esters: C16:0	Hexane/ Ethanol/ Acetone Toluene (10:6:7:7, v/v/v/v)	C30 25^{0}c	A: Dichloromethane B: Methanol/Acetonitrile / Water (84:14:5, v/v/v)	HPLC-DAD-APCI-MS	Inbaraj *et al.*, 2008

Corozo (*A. aculeate*)	Zeaxanthin esters (C10:0; C12:0; C16:0; C10:0- C10:0; C10:0- C12:0; C12:0- C12:0; C12:0-C14:0; C10:0-C16:0; C14:0-C14:0; C12:0-C16:0; C14:0-C16:0; C16:0- C16:0) β-criptoxanthin esters (C8:0; C10:0; C12:0; C14:0; C16:0)	Acetone	YMC C30 (250 x 4.6 mm, 5 μm)	A: Methanol/MTBE/H_2O (81:17:2, v/v/v) B: Methanol/MTBE/H_2O (10:88:2, v/v/v)	HPLC-DAD-APCI-MS	Murillo *et al.*, 2013
Sastra (*G. intermedia*)	β-criptoxanthin esters (C12:0; C14:0; C16:0) Lutein esters (C12:0- C12:0; C12:0-C14:0; C12:0-C16:0) Zeaxanthin esters (C14:0- C14:0; C14:0-C16:0; C16:0- C16:0)	Acetone	YMC C30 (250 x 4.6 mm, 5 μm)	A: Methanol/MTBE/H_2O (81:17:2, v/v/v) B: Methanol/MTBE/H_2O (10:88:2, v/v/v)	HPLC-DAD-APCI-MS	Murillo *et al.*, 2013
Sapote (*Q. cordata*)	n.i. (xanthophyll esters) Zeaxanthin esters (C12:0- C12:0; C12:0- C14:0; C14:0-C18:1; C14:0- C14:0; C12:0- C16:0; C16:0-C18:1; C14:0-C16:0; C16:0- C16:0; C16:0- C18:0; C18:0- C18:0)	Acetone	YMC C30 (250 x 4.6 mm, 5 μm)	A: Methanol/MTBE/H_2O (81:17:2, v/v/v) B: Methanol/MTBE/H_2O (10:88:2, v/v/v)	HPLC-DAD-APCI-MS	Murillo *et al.*, 2013
Frutita *A. psilospermus*	n.i.(cis-Apo-carotenoid- ester) (C8:0; C12:0; C14:0; C16:0) β-citraurin (C6:0; C8:0; C10:0; C12:0; C14:0; C16:0)	Acetone	YMC C30 (250 x 4.6 mm, 5 μm)	A: Methanol/MTBE/H_2O (81:17:2, v/v/v) B: Methanol/MTBE/H_2O (10:88:2, v/v/v)	HPLC-DAD-APCI-MS	Murillo *et al.*, 2013
Maracuja chino (*C. macrantis)*	n.i. (xanthophyll esters) Cryptocapsin (C12:0) β- cryptoxanthin esters (C12:0; C14:0; C16:0)	Acetone	YMC C30 (250 x 4.6 mm, 5 μm)	A: Methanol/MTBE/H_2O (81:17:2, v/v/v) B: Methanol/MTBE/H_2O (10:88:2, v/v/v)	HPLC-DAD-APCI-MS	Murillo *et al.*, 2013
Mamey rojo (*P. sapata*)	Cryptocapsin-5,6-epoxide ester: C12:0 β-cryptoxanthin-5,8-epoxide ester: C12:0; 13z/13z-Cryptocapsin ester: C14:0 Cryptocapsin esters: (C14:0; C16:0; C18:0)	Acetone	YMC C30 (250 x 4.6 mm , 5 μm)	A: Methanol/MTBE/H_2O (81:17:2, v/v/v) B: Methanol/MTBE/H_2O (10:88:2, v/v/v)	HPLC-DAD-APCI-MS	Murillo *et al.*, 2013
Papaya (*Carica papaya* L.)	β-cryptoxanthin ester: C10:0; C12:0; C14:0; C16:0	Methanol/ Ethyl Acetate/ Petroleum Ether, (1/1/1; v/v/v) 25 °C	YMC C30 (150 x 3 mm, 3μm)	A: Methanol/MTBE/H_2O (91:5:4, v/v/v) B: Methanol/MTBE/H_2O (6:90:4, v/v/v)	HPLC-DAD-APCI	Schweiggert *et al.*, 2012
Mamey (*Pouteria sapota*)	Lutein ester: C12:0-C14:0	Hexane/ Dichloromethane (1:1, v/v)	YMC C30 (150 x 4.6 mm)	A: Methanol B: MTBE	HPLC LC-MS-TOF	Yahia *et al.*, 2011
Peach	β-cryptoxanthin esters: C12:0; C14:0; C16:0	MeOH:EtAc:PE (1:1:1, v/v/v)	YMC C30 (250 x 4.6 mm -5 μm)	A: Methanol/MTBE/H_2O (83:15:2, v/v/v) B: Methanol/MTBE/H_2O (8:90:2, v/v/v)	HPLC-DAD-APCI-MS	Giuffrida *et al*, 2013
Tamarillo fruits (*Solanum betaceum* Cav.)	*All-trans*-neoxanthin esters: C14:0-C14:0; C14:0-C16:0; C16:0-C16:0 *All-trans*-5,6-epoxy-β-cryptoxanthin esters: C14:0; C16:0 *Cis*-neoxanthin esters: C14:0-C14:0; C14:0-C16:0 *All-trans*-violaxanthin esters: C14:0-C14:0; C14:0-C16:0; C16:0-C16:0 *All-trans*-β-cryptoxanthin esters: C14:0; C16:0 *All-trans*-antheraxanthin esters: C14:0-C14:0; C14:0-C16:0; C16:0-C16:0 *All-trans*-lutein esters: C14:0-C14:0; C14:0 C16:0 *All-trans*-zeaxanthin: C14:0-C14:0; C14:0-C16:0; C16:0-C16:0	Ethanol/Hexane (4:3, v/v)	YMC C30 (250 x 4.6 mm, 5 μm)	A: Water/20 mM ammonium acetate B: MeOH/20 mM ammonium acetate C: MTBE	HPLC/PDA-MS	Mertz *et al.*, 2010

Structure and properties

Carotenoids are isoprenoidic compounds which can be divided into carotenes (hydrocarbons) and oxygenated derivatives (xanthophylls). The xanthophylls of higher plants posses a C40 backbone which can bear different oxygenated functional groups: hydroxyl, aldehyde, ketone, epoxy, carboxyl. Xanthophylls containing hydroxyl groups, also named carotenol, can be esterified with carboxylic acids, mainly with long chain and saturated ones. The chemical structure of the most common xanthophyll esters is presented in Fig.1.

Most of the physical and chemical properties of carotenoids are due to the presence of the polyenic system. The long system of conjugated double bonds (chromophore) is responsible for the light absorption properties of carotenoids, in consequence also for their colouring properties. Also, the polyene chain is responsible for most of the chemical properties of carotenoids. Esterification of hydroxyl group does not modify the chromophore, in consequence the UV-VIS light absorption spectra of the esters will be identical with those of the unesterified compounds (Britton, 1995; Britton *et al.*, 2004). However, the molar absorptivity (as well as the specific absorption coefficient $A^{1\%}_{1cm}$) is different. The molar absorptivity of xanthophyll ester can be calculated as the product between the specific absorption coefficient $A^{1\%}_{1cm}$ of the unesterified xanthophyll and the ratio between the molecular weight of xanthophyll and the molecular weight of the ester (Schiedt and Liaaen-Jensen, 1995). The similar light absorption properties of esterified and free xanthophylls unable their identification based only on absorption maxima, but it requires proper identification using standards and mass spectra analysis. Mass spectra of xanthophyll diesters are relatively simple and contain six types of ions above m/z 200: $[M]^{+\cdot}$- the molecular ion (or

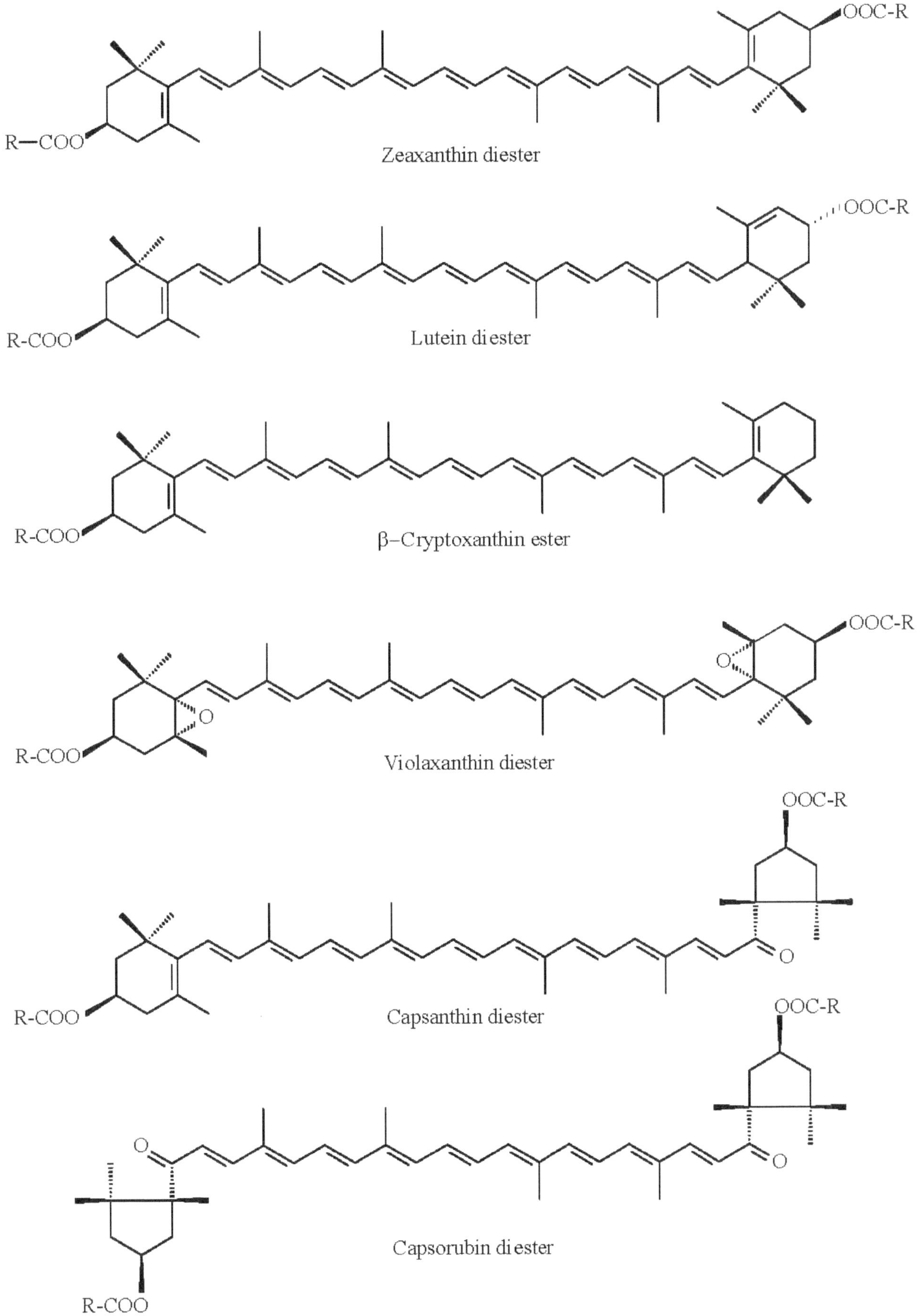

Fig. 1. Structure of major xanthophyll esters in fruits and vegetables

$[M+H]^+$ in APCI positive ionisation mode), $[M-R_1COOH]^{+\cdot}$, $[M-R_2COOH]^{+\cdot}$, $[M-R_1COOH-R_2COOH]^{+\cdot}$, $[R_1COOH]^{+\cdot}$ and $[R_2COOH]^{+\cdot}$. These fragments correspond to the ester molecule (ionized or protonated), to the consecutive loss of fatty acids moieties and to the fatty acids which esterify the hydroxyl group (Enzell and Back, 1995).

The esterification of xanthophyll decreases the polarity and increases the solubility in lipids. The high molecular weight and the very low polarity confer a special chromatographic behaviour to the xanthophyll esters, which is characterized by higher retention time than carotenes when separated on reversed phase stationary phase (C18 or C30). The chromatographic behaviour of esters depends on the polarity of the parent xanthophyll, on the chain length and unsaturation of the fatty acids. The retention times of carotenol esters on reversed-phase column increases as the number of carbon atoms in the fatty acids moiety increases and decreases with the unsaturation for the same chain length (*e.g.* linoleic acid esters of lutein elute earlier than stearic acid esters). Also, the esters of violaxanthin (which posseses epoxide group) elute in front of lutein or zeaxanthin esters (Khachik, 2009).

Xanthophyll esters can be hydrolyzed by the common alkaline hydrolysis (methanolic potassium hydroxide or sodium hydroxide) (Schiedt and Liaaen-Jensen, 1995) or by enzymatic hydrolysis with various lipases (as described in the section *Metabolism and bioavailability*) (Breithaupt *et al.*, 2002; Breithaupt *et al.*, 2007).

Most of the beneficial effects of carotenoids are attributed to their antioxidant activity which was proved by numerous *in vitro* and *in vivo* studies. Carotenoids are known as effective quenchers of singlet oxygen, of excited sensitizers and as scavengers of other reactive oxygen species. Antioxidant properties of carotenoids are due to the presence of polyenic system but they can be modified by the presence of different functional groups (Krinsky and Johnson, 2005; Yeum *et al.*, 2009).

The experimental results regarding the antioxidant capacity of xanthophyll esters are contradictory. Lutein, lutein monomyristate and lutein dimyristate were found to have similar antioxidant activities when tested by methyl linoleate hydroperoxide assay and by DPPH (1,1-diphenyl-2-picrylhydrazyl) test (Subagio and Morita, 2001). In a later work, the same authors reported that lutein and its dimyristate ester have prooxidant effect on purified triacylglycerols from corn, when heated at 40°C, in darkness. However, the prooxidant activity of free lutein was more pronounced than that of its dimyristate (Subagio and Morita, 2003). Carotenoid esters of capsanthin and capsorubin from paprika exhibited a better stability toward lipoxygenase catalyzed linoleic acid oxidation than the free xanthophylls (Biacs *et al.*, 1989). The esterified forms of capsanthin were proved to be good radical scavengers, with an antioxidant activity similar to the free form, when assessed by AMVN-induced oxidation of methyl linoleate (Matsufuji *et al.*, 1998). Radical-induced oxidation (with AIPN as initiator) occured at the same rate in the free and the esterified capsanthin and capsorubin, but faster for the zeaxanthin esters than for free zeaxanthin. The differences were explained by the differences in the fatty acids which esterified the xanthophylls. Capsanthin and capsorubin are esterified with saturated fatty acids, while zeaxanthin is esterified mainly with the unsaturated linoleic acid. This proves that beside the xanthophyll, the fatty acid moiety can influence the antioxidant properties of carotenol esters (Pérez-Gálvez and Minguez-Mosquera, 2002, 2005). The scavenging capacity of hydroxyl radical, superoxide anion and singlet oxygen (determined by chemiluminescence) proved that esterification with saturated fatty acids does not affect the antioxidant activity of β-cryptoxanthin (Fu *et al.*, 2010). Slightly lower antioxidant activity was found for the β-cryptoxanthin and zeaxanthin esters with unsaturated fatty acids (oleic and linoleic) compared to the saturated esters and to the free forms (Pintea *et al.*, 2013a, 2013b, 2014a).

Extraction and chromatographic separation

Extraction of xanthophyll esters, as well as of all the carotenoids, requires the use of organic solvents. Several factors are taken into account when choosing the solvents, such as the structure of the matrix containing carotenoid esters and the differences in the polarities of carotenoids found in the matrix. Xanthophyll esters are non-polar compounds and usually, they are extracted with non-polar solvents. However, in different matrices the presence in the matrix of the more polar xanthophylls requires the use of mixtures also containing polar solvents. Acetone, a water-miscible organic solvent dissolves both carotenes and xanthophylls in an efficient way (Rodriquez-Amaya, 2010). Most of the studies regarding carotenoid esters extraction have included acetone as extraction solvent (Delgado-Pelayo and Hornero-Mèndez, 2012; Giuffrida *et al.*, 2013; Murillo *et al.*, 2013). Usually, a mixture of polar and non-polar solvents or acetone alone is used for carotenoid esters extraction from different matrices. A mixture of acetone: hexane (1:1, v/v) was used to isolate β-cryptoxanthin, lutein and antheraxanthin esters from apricots (*Prunus armeniaca* L.) (Kurz *et al.*, 2008), and for the extraction of 9-*cis*-violaxanthin and all-*trans*-violaxanthin esters from mango (Ornelas-Paz *et al.*, 2007). Extraction of carotenoid esters from oranges juices and from sea buckthorn (*Hippophae rhamnoides* L.) was performed with a mixture of methanol:ethylacetate:petroleum ether (1:1:1, v/v/v) (Giuffrida *et al.*, 2012; Pop *et al.*, 2014; Weller and Breithaupt, 2003). Inbaraj *et al.* (2008) extracted zeaxanthin and β-cryptoxanthin esters from Goji (*Lycium barbarum*) with a mixture of hexane: ethanol:acetone:toluene (10:6:7:7, v/v/v/v). N,N-dimethyl formamide was used for xanthophyll esters extraction from potato (Fernandez-Orozco *et al.*, 2013) and a mixture of 1,2-dichloroethane:acetone:methanol (2:1:1, v/v/v) for carotenoid esters extraction from *Capsicum annum* L. species (Daood *et al.*, 2014). Almost all the procedures used the addition of $NaHCO_3$, $MgCO_3$ or $CaCO_3$ in the first steps of the extraction in order to neutralize the acidity of plant tissues and to prevent rearrangements. The extractions are repeated until the residues are colourless and the filtrates are combined, partitioned into a separatory funnel with an adequate solvent (diethyl ether). The organic phase is removed and evaporated and the residue is further dissolved in a suitable amount of solvent for HPLC analysis (Britton, 1995).

HPLC techniques are widely used for the separation and quantification of carotenoid esters. Unsaponified extracts can be injected directly into HPLC columns or after a previous fractionation by open column chromatography (Bernhard, 1995; Pintea *et al.*, 2005; Subagio *et al.*, 1996). The first attempts to separate carotenoid esters in fruits by HPLC revealed the superiority of reversed phase (octadecylsilane, C18) over the normal phase (silica) column (Philip and Chen, 1988). Khachik and Beecher (1988a, b) developed chromatographic methods for the separation of synthethic diesters with various fatty acids of violaxanthin, auroxanthin, lutein, zeaxanthin, isozeaxanthin, and beta-cryptoxanthin on C18 reversed phase column.

Most of the separation on C18 columns used binary gradients with mobile phases containing various ratio of: acetonitrile/ethyl acetate/water; acetone/water; 2-propanol/water/ acetonitrile. Good separation of esters from sarsaparilla berries (Delgado-Pelayo and Hornero-Méndez, 2012), potatoes (Fernandez-Orozco *et al.*, 2013) and apples (Delgado-Pelayo *et al.*, 2014) were obtained on C18 RP columns. Sea buckthorn unsaponified extracts containing complex mixtures of carotenes and esters were both separated on C18 column (LiChrosorb RP 18) and on a Waters Acquity UPLC BEH Shield RP18 column (2.1x150 mm, 1,7 x150 mm, 1.7 μm) (Pop *et al.*, 2014). Octadecylsilane C18 column are robust columns which allow good separation of both unesterified and esterified xanthophylls, in a relatively short time, with affordable solvents.

C30 columns of different size are however mostly used for the separation of fatty acid esters of carotenoids in native extracts from fruits or vegetables. Binary or ternary gradients are used, based on a combination in various proportion of the following solvents: methanol/tert-butyl-methyl-ether/water (with or without ammonium acetate) or dichloromethane/methanol/acetonitrile. Good separations on C30 columns of a high number of esters were obtained for apricot (Kurz *et al.*, 2008); *Capsicum* sp. (Giuffrida *et al.*, 2013); sea buckthorn (Giuffrida *et al.*, 2012; Weller and Breithaupt, 2003); exotic fruits (Murillo *et al.*, 2013). C30 columns provide better separations, especially for long-chain hydrophobic compounds (like xanthophyll esters) and for geometric isomers (Sander *et al.*, 2000). The disadvantages of C30 columns are the longer separation time and more expensive solvents. However, for very complex matrix like sea buckthorn, pepper or citrus fruits, the method of choice is the separation on C30 column with narrow particles (3 μm).

A particular case is the analysis of carotenoids in mandarin essential oil by LCxLC, where a normal phase column (Supelcosil LC-SI) was used on the first dimension and a reversed phase (Chromolith Performance RP-18) on the second dimension. Beside other carotenoids, eighteen esters of different xanthophylls were identified (Dugo *et al.*, 2008).

Carotenoids are commonly identified based on their light absorption properties using UV-VIS and PDA detectors. Carotenoid esters have the same light absorption properties like the parent compounds and can not be identified solely based on their spectra (position of absorption maxima and the fine structure %III/II).

The availability of relatively few commercial standards led to the extensive use of hyphenated chromatographic techniques: LC/MS and LC/MALDI. Mass spectrometry detectors provide the molecular peak but also the fragments which can be further used in order to respect the minimum criteria of identification for carotenoids and their esters. Several MS detection methods were reported to use continuous flow fast atom bombardment, matrix-assisted laser desorbtion/ionization (MALDI), electrospray ionization (ESI), atmospheric pressure photoionization (APPI), atmospheric solids analysis probe (ASAP) and atmospheric pressure chemical ionization (APCI). APCI method can form both positively and negatively charged molecular ions species (Amorim-Carrilho *et al.*, 2014; Cacciola *et al.*, 2012; Van Breemen *et al.*, 2012; Wingerath *et al.*, 1997). The LC/APCI-MS method, characterized by high sensitivity and superior linearity of detector response, became the standard analysis method in carotenoid esters analysis. For example, 31 esters were identified in *Capsicum* sp. (Giuffrida *et al.*, 2013) and 24 esters in apple (Delgado-Pelayo *et al.*, 2014). An overview on the extraction, chromatographic methods and detections employed for characterization of unsaponified extracts, as well as the main esters identified in different fruits and vegetables are presented in Tab. 1.

Generally, the quantification of the carotenoids in carried out using the PDA detection and standard curves obtained either with commercially available standards, standards prepared by semi-synthesis (in house) (Weller and Breithaupt, 2003) or with the parent xanthophyll. Due to the difference in absorbtivity, it is recommended that quantitative determination of esters is made by an internal standard calibration with isozeaxanthin esters (e.g. dinonanoate) prepared by acylation (Khachik, 2009). For the analysis of fatty acids profile in ester fraction or in matrices containing carotenoid esters, gas chromatography can be used (Pintea *et al.*, 2005; Pop *et al.*, 2013).

Biosynthesis and accumulation of carotenoid esters

Carotenoids are tetraterpenes which belong to the large family of isoprenoids. More than 32,000 isoprenoids and about 750 carotenoids have been identified so far. Most of the carotenoids are C40 isoprenoids and they are divided into carotenes (hydrocarbons) and oxygenated derivatives (xanthophylls). Carotenoid pigments are synthesized only by plants and microorganisms, while *de novo* biosynthesis of carotenoids could not be demonstrated in any animal organism (Britton *et al.*, 1998).

Similarly with other isoprenoids, carotenoids are biosynthesized by the condensation of the C5 precursors – the isopentenyl diphosphate (IPP) and dimethylallyl diphosphate (DMAPP). The biosynthesis of IPP and DMAPP takes place by two routes: the classical acetate mevalonate pathway (MVA) and the lately discovered 2C-methyl-D-erythritol-4-phosphate (MEP) pathway. The two biosynthetic pathways have different locations - the MVA pathway is localized into the cytosol while the MEP pathway is located in the plastids. In plants carotenoids are biosynthesized from C5 precursors generated mostly by MEP pathway (Moise *et al.*, 2013; Rohmer, 1998). The main steps of carotenoid esters biosynthesis are:

- head-to-tail condensation of C5 units (three IPPs and

one DMAPP) to form the C20 intermediate – geranylgeranyldiphosphate (GGPP);
- head-to-head condensation of two GGPP units, catalyzed by phytoene synthase (PSY), to form 15-cis-phytoene (a colourless carotenoid);
- desaturation and isomerization reactions to form lycopene;
- cyclization reactions, catalyzed by various cyclases, which leads to the formation of different cyclic carotene structures (β, ε, κ, etc.);
- chemical changes in the cycle (hydroxylation, epoxidation, etc.) to generate xanthophylls;
- esterification of hydroxyl group with corresponding acyl-coenzyme A (Britton *et al.*, 1998, Giuliano, 2014; Gomez-Garcia and Ochoa-Alejo, 2013; Moise *et al.*, 2013; Shumskaya and Wurtzel, 2013).

Significant progresses in the understanding of carotenoid biosynthesis were made during last years. Most of the enzymes involved in biosynthetic pathway were purified and characterized in several plant models like Arabidopsis, tomato, maize, pepper, daffodil, etc. Recently the gene encoding for the enzyme that catalyzes the xanthophyll esterification (PYP1 – pale yellow petal) was identified in tomato flowers. The PYP 1 protein has two domains, a hydrolase domain and an acyltransferase domain (LPAT – lysophospholipid acyltransferase). It was proved that the deletion or the alteration of the acyltransferase domain abolishes the esterification of xanthophylls and determined the disruption of normal chromoplast development (Ariizumi *et al.*, 2014)

Hydroxyl groups of xanthophylls are often esterified in fruits and flowers with mixtures of fatty acids. Carotenoids are biosynthesized by almost all the plastids, but xanthophyll esterification is associated with the formation of chromoplasts of fruits and flowers and with ripening in fruits (Minguez-Mosquera and Hornero-Méndez, 1994; Ohmiya, 2013). In chloroplasts, carotenoids are localized in the thylakoid membranes as component of the photosynthetic complexes. The major carotenoids in all green leaves are: β-carotene, lutein, violaxanthin, neoxanthin; while α-carotene and zeaxanthin are present in small amounts (Britton, 1995a). Xanthophylls of the chloroplasts are found in non-esterified form and the esterification occurs in senescent leaves, being associated with the degeneration of chloroplasts and the formation of chromoplasts (Biswal, 1995; Cardini, 1982).

Chromoplasts are differentiated from chloroplasts during the ripening of green fruits but also from leucoplasts or amyloplasts during ripening of other fruits or development of roots. Several types of chromoplasts have been described, with different morphologies and containing different types of carotenoid sequestering substructures, such as plastoglobules. It is considered that chromoplasts serve as a metabolic sink for the accumulation of carotenoids (Li and Yuan, 2013). Xanthophyll esters were found to be associated with fibrillins and polar lipids to form fibrils in the chromoplasts of red pepper (Deruère *et al.*, 1994). In the isolated chromoplast fibrils of red pepper all the xanthophylls were esterified in proportions varying from 75 to 100%. The authors performed an *in vitro* study to reconstitute chromoplast fibrils by mixing fibrillin with the lipid components and they found that xanthophyll diesters (zeaxanthin and capsanthin) were the most efficient for fibril assembly. This study demonstrated the critical role of the xanthophyll ester for the fibril formation. Later, Ytterberg *et al.* (2006) identified twenty-eight proteins in the plastoglobules of red pepper chromoplasts, among them some fibrillins (structural proteins) and an esterase which might be responsible for the esterification of xanthophylls. The esterification process increases significantly the lipofilicity of xanthophylls and has an important physiological relevance for the sequestration of carotenoids in the chromoplasts (Ariizumi *et al.*, 2014; Ohmiya, 2013). It was also proved that the carotenoids stored in plastoglobules have a higher stability to light. Also, their accumulation can contribute to the protection of plastoglobules and light sensitive constituents against blue light (Merzlyak and Solovchenko, 2002).

In red pepper (*Capsicum annuum*) the xantophylls zeaxanthin, β-cryptoxanthin, capsanthin and capsorubin are esterified with different fatty acids. Zeaxanthin, capsanthin and capsorubin, which possess two hydroxyl groups, can be present in free form, as monoesters or diesters with different fatty acids, giving rise to a complex carotenoid pattern. Zeaxanthin and β-cryptoxanthin are esterified mainly with linoleic acid (50-60 %) while capsanthin and capsorubin are esterified preferentially with middle chain saturated fatty acids. Esterification occurs not only in newly synthesized carotenoids, but also in pigment existing in green fruits, like violaxanthin (Hornero and Mosquera, 2000). Studying five red pepper cultivars, the same authors observed a 23.5 to 48 fold increase of total carotenoids during ripening process, depending on the phenotype. Also, a gradual decrease of the free (F) form of xanthophylls but an increase of partially esterified (PE, monoesters) and totally esterified (TE, diesters) fractions could be observed. Only in the last stage of ripening was a small decrease of TE noticed, as well as an increase of PE, fact that suggests the establishment of a balance between the fractions. Independent on the phenotype, the proportion of the three fractions was similar: 24.17 ± 4.06 (F), 31.48 ± 4.61 (PE) and 44.36 ± 5.05 (TE).

Similar processes were reported during ripening in apple fruits. Esters of lutein, violaxanthin and neoxanthin with various fatty acids were found in apple peel (*Malus pumila*, cv. 'Cox's Orange Pippin') (Knee, 1988). The author suggested that esters are synthesized *de novo* in ripening apples, their concentration increases during ripening independent on ethylene production so they could serve as early indicators of fruit maturity. The changes in the chlorophyll and carotenoid composition in the skin of apple fruits ('Antonovka' variety) in sun and shade ripening, both on and off the tree, were determined. Violaxanthin and neoxanthin, the major esterified xanthophylls in apples, were quantified by HPLC. In shaded apple on the three, xanthophylls esters increased from 1.4 % to 19.7 % (6 September – 11 October) while in sunlit apples there was a dramatical increase, from 5.1 % to 38.4 %. In detached fruits, xanthophylls esters reached 50 % in shaded and respectively 95 % in sunlit apples. A decrease of chlorophyll and a significant change in the carotenoid pattern was also observed, with a decrease of β-carotene and lutein and an increase of carotenoids of xanthophyll cycle (Solovchenko *et al.*, 2006). The changes in carotenoid composition of apple skin exposed to sun light could be mediated by the reactive oxygen species generated in these conditions (Merzlyak *et al.*, 2002; Solovchenko *et al.*, 2006).

Mango fruits contain as main compounds β-carotene and dibutyrates of all-*trans*-violaxanthin and 9-*cis*-violaxanthin. The changes in the carotenoid content and the changes in the color in mesocarp and epidermis were determined for two mango cultivars 'Ataulfo' and 'Manila', after harvesting during 16 days. For both cultivars there were significant increases (exponential/second order polynomial) of all carotenoids. *All trans*-violaxanthin and 9-*cis*-violaxanthin dibutyrates reached 31.97×10^{-3} g/kg and 16.81×10^{-3} g/kg in the mesocarp of 'Manila' cultivar at the end of experiment. Interestingly, at the moment of harvest/acquisition no 9-*cis*-violaxanthin esters were detected. The increases in the concentration of all carotenoids were highly correlated with a^* and $h°$ color values of the mesocarp and epidermis (Ornelas-Paz *et al.*, 2008).

Sea buckthorn berries are known as rich sources of zeaxanthin and β-cryptoxanthin esters (Pintea *et al.*, 2005; Weller and Breitphaupt, 2003). Depending on the region, the oily berries are normally harvested from August – September until late November. A study performed in Sweden on four sea buckthorn cultivars showed a significant increase of total carotenoids during ripening: from 323 µg/g (28th of July), to 1015 µg/g (15th of September). The most significant increase was recorded for the esterified fraction (from 100 to 648 µg/g) while free xanthophylls increased only from 61 to 82 µg/g (Andersson *et al.*, 2009). Even thought sea buckthorn oil is characterized by a high content of unsaturated fatty acids, mainly oleic and palmitoleic acid (Pintea *et al.*, 2005), the carotenoid ester fraction is mainly esterified with palmitic and myristic acid. Palmitoleic acid was identified together with palmitic acid in the mixed ester of zeaxanthin, while oleic acid and other unsaturated fatty were found to esterify only lutein and in very small amount sustaining the idea of selective esterification of xanthophylls in plants (Giuffrida *et al.*, 2012). The accumulation of esters during ripening was proved by HPLC-DAD analysis in *Physalis alkekengi* L. sepals (Bunea et al., 2007).

Interesting data regarding xanthophyll esters accumulation were obtained in the case of red fleshed papaya fruits (*Carica papaya* L.). Unless in the other fruits, β-cryptoxanthin esters (caprate and laurate) were already present in early ripening stages of papaya fruits and representing 42% of total carotenoids. The amount of total carotenoids increased significantly, from 130 µg/100g in ripening stage 1 preharvest to 6214 µg/100g in ripening stage 7 postharvest, as well as the amount of β-cryptoxanthin (free and esterified). However, the proportion of esters decreased to 19-20%, as well as the ratio β-cryptoxanthin esters/β-cryptoxanthin. Also, the accumulation of esters seems to be dependent on the fatty acids since β-cryptoxanthin laurate accumulated with much higher rate than other esters (Schweiggert *et al.*, 2011).

Recently Fernandez-Orozco *et al.* (2014) analyzed sixty potato cultivars and found an average of 23.1% carotenoids esters. The highest xanthophyll ester content was found at 867.6 lg/100 g dry wt, which represents 58.6% of total carotenoids. Moreover, the authors found a direct correlation between the total carotenoid content and the ester fraction, suggesting once again the importance of esterification for the sequestration and accumulation of xanthophylls within the plastids. An interesting finding is the high proportion of linoleic acid as esterifying fatty acid (39.0%), similar with the ester fraction of tritordeum grains were lutein was extensively esterified with linoleic acid (Mellado-Ortega and Hornero-Mendez, 2012). Other fatty acids, like oleic acid (14% of total fatty acids in grains) could not be detected in carotenoid ester fraction. Based on these studies the authors suggest a high specificity of xanthophyll acyl transferase responsible for the formation of xanthophyll esters in plants. On the other hand, it was proved that the fatty acids preferentially esterify the hydroxyl group of the β-ring in the monoesters fraction of lutein (Khachik *et al.*, 1988b). A preference for esterification of lutein with lauric acid was observed in yellow varieties of raspberry (Carvalho *et al.*, 2013).

Metabolism and bioavailability/bioaccessibility

Fruits and vegetables are the major sources of dietary carotenoids for human. Due to their lipophilic character, carotenoids are metabolized following the same pathways as dietary lipids. The main steps of carotenoid metabolism are: the release from the food matrix; solubilization and incorporation into micelles; carotenoid uptake by intestinal mucosal cells and absorption (passive and active diffusion); transport in the blood by incorporation into chylomicrons and other lipoproteins; distribution and accumulation in tissues; structural modification and excretion. Bioavailability of carotenoids is affected by several factors, some of the related to the food: the food matrix and the location of carotenoids, the chemical nature of carotenoids (including isomerism), the food processing, the amount of lipid in the food, the interaction between carotenoids and with other components of the food (proteins, fibers). Additionally, the bioavailability/bioaccessibility of carotenoids is affected by human factors, like age, gender, health status (parasitic infections), habits (smokers), etc. All these factors, and probably some others (the existence of "low" or "non-responders"), contribute to a large variability in the carotenoid bioavailability in humans an in the response of human subjects to the supplementation with carotenoids (Canene-Adams and Erdman, 2008; Férnandez-Garcia *et al.*, 2012; Schweigert, 1998).

Xanthophyll esters have to be hydrolyzed before intestinal absorption. *In vitro* studies have proved that xanthophyll esters (lutein, zeaxanthin and β-cryptoxanthin) can be hydrolyzed by pancreatic lipase, microbial lipase and by cholesterol esterase (Breithaupt *et al.*, 2002, 2007; Jacobs *et al.* 1982). In a model of simulated digestion, zeaxanthin esters purified from *Lycium chinense* were partially hydrolyzed by carboxyl ester lipase (porcine, bovine, colipase or combined) and the hydrolysis increased the efficiency of micellarization. The hydrolysis of esters also increased significantly the uptake of zeaxanthin from micelles by Caco-2 human intestinal cells (Chitchumroonchokchai and Failla, 2006).

In humans, carotenoids esters seem to be efficiently hydrolyzed before absorption. β-cryptoxanthin concentrations in chylomicrons increased after the ingestion of a tangerine concentrate rich in β-cryptoxanthin esters, but no esters were found in blood (Wingerath *et al.*, 1995). Similarly, only small amounts of lutein esters were found in blood (Granado *et al.*, 1998) and in skin (Wingerath *et al.*, 1998) after a long term dietary supplementation with a mixture of lutein esters.

Most of the studies regarding the bioavailability of esters proved that xanthophylls esters have comparable bioavailability with the corresponding free xanthophylls and a comparable increase of plasma free xanthophylls can be observed after supplementation with food or products rich in esters. Lutein diesters were found to be with 61.6% more bioavailable than unesterified after administration of a single dose (Bowen *et al.*, 2002) but was influenced by the type of formulation (amount of oil), as reported also in a previous study (Roodenberg *et al.*, 2000). Administration of paprika oleoresin (1 g oleoresin with 5 g oil) containing esterified xanthophyll in healthy volunteers lead to a significant increase of zeaxanthin, β-cryptoxanthin and β-carotene in the chylomicrons fraction of plasma. However none of them could be identified in the esterified form, even if in the oleoresin the xanthophylls were mainly present as esters with fatty acids (Pérez-Gálvez *et al.*, 2003). Plasma β-cryptoxanthin increased significantly after administration of a single dose of both, free or esterified xanthophyll from papaya, supporting their comparable bioavailability and the effective enzymatic cleavage of esters (Breithaupt *et al.*, 2003).

A randomised, single-blind cross over study was performed on volunteers that received equal doses (5 mg in yoghurt) of zeaxanthin dipalmitate (purified from wolfberry) or unesterified zeaxantin. The plasma samples analyzed by chiral HPLC showed a better bioavailability of esterifed form, even if both forms determined a significant increase of zeaxanthin in plasma (Breithaupt *et al.*, 2004). The bioavailability of zeaxanthin dipalmitate from wolfberry was tested in a human supplementation trial in which the subjects receive 15 g/day whole fruits (containing about 3 mg zeaxanthin) for 28 days. In the supplementation group the plasma zeaxanthin concentrations increased from 0.038 to 0.096 μmols/l, demonstrating that wolfberry is a bioavailable source of carotenoids (Cheng *et al.*, 2005). The significant increase of zeaxanthin in plasma concentration after supplementation with fruits has a particular importance since zeaxanthin bioavailability from spinach or corn is low (Bone *et al.*, 2000; Mozaffarieh *et al.*, 2003). An interesting study on carotenoid bioaccessibility was performed using an *in vitro* digestion model on carrot, mango, papaya and tomato. The mentioned sources of carotenoids have a different chromoplast morphology which was supposed to influence the release of carotenoids from the food matrix. Xanthophylls and their esters accumulate in liquid-crystalline and lipid dissolved state in globular-tubular substructures of the chromoplasts in the fruits of mango and papaya. Carotenes (β-carotene and lycopene) are localized in crystalloid structures in chromoplasts of carrots, tomato and papaya. Generally, the bioaccessibility of all carotenoid was higher for mango and papaya for but xanthophylls (β-cryptoxanthin and lutein) and their esters had a superior bioaccessibility then carotenes. The addition of oil improved the bioaccessibility of carotenoids without changing the ranking. It seems that the accumulation of carotenoids (including xanthophylls and their esters) in liquid-crystalline and lipid dissolved state of globular-tubular substructures have a positive effect on their bioaccessibility from fruits and vegetables (Schweiggert *et al.*, 2012).

Stability in model systems and food

Fruits and vegetables which provide high carotenoid intake are considered very important to maintain health and to reduce the risks of several degenerative diseases or cancer (Krinsky and Johnson, 2005; Rao and Rao, 2007). The carotenoid content in plant derived food is influenced by several factors, such as genetic origin, environmental conditions, cultivation practice and post-harvest treatment. Post-harvest treatment, including transport, storage, processing and cooking, is a key step in preserving the nutritional value of plant food. Carotenoids are more stable in vegetables and fruit than in isolated (purified) form. In plant tissues, carotenoids are protected by molecular interaction with other compounds (proteins, lipids) and by the presence of other antioxidants. Disruption of the tissues exposes carotenoids to light, oxygen and oxidizing enzymes like lipoxygenase (Britton and Khachik, 2009; Caris-Veyrat, 2008; Mercadante, 2008). Xanthophyll esters, as all the carotenoids, are susceptible to degradation under high temperature, light, acidic pH and reactive oxygen species. The mechanisms involved in carotenoids degradation in food are similar with the degradation of pure compounds and occur mainly through oxidation, geometrical isomerisation, but also by rearrangement of 5,6-epoxides and dehydration. Enzymic oxidation is due to the contact of carotenoids with lipoxygenase which leads to the bleaching of pigments. Non-enzymic oxidation occurs as a consequence of direct interactions of carotenoids with dioxygen molecule or with reactive oxygen species (superoxide anion radical, hydroxyl radical, singlet oxygen and hydrogen peroxide). The non-enzymic oxidation of carotenoids results in the formation of different epoxides and apocarotenals. High temperature, light and low pH determines the geometrical isomerisation of carotenoids. The position of the double bond of the polyene chain which is isomerized and also the proportion between the *Z* isomers which are formed depends on the applied treatment (Britton and Khachik, 2009; Caris-Veyrat, 2008; Mercadante, 2008).

The stability of free carotenoids in model systems or in different food systems were reviewed in very comprehensive publications (Maiani *et al.*, 2009; Mercadante, 2008). Compared to free carotenoids, the studies on the stability of xanthophylls esters are relatively scarce and the approaches used are very different. Thermal stability of xanthophyll esters was evaluated using purified esters or on food containing esters. Using purified lutein mono- and dimyristate Subagio *et al.* (1999) found that they were more stable toward heat treatment and UV light than free lutein, suggesting that esterification with saturated fatty acid stabilize the highly unsaturated lutein. More recently, the thermal stability of β-cryptoxanthin esters with saturated fatty acids compared to free β-cryptoxanthin was determined in tetrahydrofuran solutions during incubation at 50 °C, in dark, for up to 72 hours. The stability of pigments, determined by HPLC-PDA, showed that β-cryptoxanthin esters were more stable than free β-cryptoxanthin (Fu *et al.*, 2010). Thermal stability tests were performed at different temperatures on β-cryptoxanthin and zeaxanthin esterified with saturated (myristic and palmitic acids) and unsaturated fatty acids (oleic and linoleic acids) compared with the free xanthophylls. At all temperatures, the esters with saturated fatty acids showed a significantly better stability than the free xanthophylls, while the esters with unsaturated fatty acids exhibited lower stability than the saturated ones, but still higher than the free β-cryptoxanthin and zeaxanthin (Bunea *et al.*, 2013; Pintea *et al.*, 2013). Lutein esters (mono- and dimyristate were found to be more stable toward UV light than free lutein (Subagio *et al.*, (1999). Contrarily, the photostability toward UVA irradiation was found to be lower for β-cryptoxanthin palmitate then for free cryptoxanthin, both in hexane solution and in liposomes containing carotenoids. The degradation rate was higher for carotenoids in liposomes and the highest degradation rate was recorded for β-cryptoxanthin palmitate (Arita *et al.*, 2004).

Relatively few papers were published on the stability of xanthophyll esters in fruits or vegetables. Khachik and Beecher (1988a, b) found that xanthophylls esters in squash were more stable than free carotenoids. Diesters of violaxanthin were more stable than monoesters during cooking and no significant izomerization nor epoxide rearrangement were observed. Stability studies were performed on red peppers (*Capsicum annuum* L.) and hot chilli peppers (*Capsicum frutescens* L.) pods minced, heated for 5 min at 80, 90 or 100 °C and then lyophilised. Caspsanthin, capsorubin, zeaxanthin and β-cryptoxanthin mono- and diesters from both types of peppers showed comparable photo- and thermal stability, in all the cases higher than for non-esterified forms (Schweiggert *et al*, 2007).

A different approach was applied for investigation of thermal lability of carotenoids of tamarillo fruits (*Solanum betaceum* Cav). In this study, the samples were heated (at 80, 90, 95 °C) in hermetically sealed amber vials, in both degassed and not degassed nectar, in which the final level of oxygen was determined. HPLC-PDA/MS analysis of samples before and after treatment showed that zeaxanthin esters appeared to be the less thermo-labile carotenoids, in both experimental conditions (Mertz *et al.*, 2010).

Most of the stability studies performed on pure esters (isolated or synthesized), in esters fractions or in plant materials rich in xanthophyll esters demonstrated the protective effect of esterification, especially with saturated fatty acids.

Conclusions

Xanthophyll esters represent an important fraction of total carotenoids in many fruits and vegetables. The major xanthophylls present in esterified form are lutein, zeaxanthin, β-cryptoxanthin, violaxanthin, capsanthin and capsorubin. Commonly consumed fruits and vegetables, like apples, apricots, mandarins, mangoes, papayas, red and chilli peppers, potatoes or squash, contain xanthophylls mainly in esterified form. Some other rich sources have been identified, like: wolfberry (goji), sea buckthorn, persimmon and exotic fruits (sapote, maracuja, frutita, mamey, sastra, corozo, tamarillo, etc).

Identification of carotenoid esters in very complex unsaponified extracts requires powerful chromatographic techniques and mass spectrometry detectors. Significant progresses were reported during last years due to the technical development of separation and identification tools. Also, there is an increasing interest in the characterization of unsaponified extracts which, in contrast to the saponified ones, reflects the native composition of food. Lutein esters have been identified as major compounds in squash; zeaxanthin esters are present in wolfberry, sea buckthorn and red pepper; β-cryptoxanthin esters in apricots, sea buckthorn, mandarin; violaxanthin and neoxanthin esters in apple and capsanthin and capsorubin esters in pepper.

Esterification of xanthophylls has physiological significance for the chromoplast formation. The increased lipophilicity of esters is important for the sequestration of carotenoids and for the formation of specialized structures (plastoglobules, fibrils) in the chromoplasts. The esters also provide protection against photooxidative damages in plants. Esterification process and accumulation of esters occurs during ripening in fruits and is associated with significant change in the colour of fruits. Even if the specific enzymes which catalyze the esterification process were not characterized yet, detailed analytical data regarding the carotenoid composition suggests a selectivity of these enzymes for certain fatty acids and selectivity for the ring in the case of non-symmetric carotenoids like lutein.

Xanthophyll esters seem to be efficiently hydrolyzed and absorbed in humans leading to a comparable bioavailability to the unesterified compounds. Also, the xanthophyll esters preserve the antioxidant capacity of the parent compounds while having a better stability in fruits during storage and processing. All these properties are important from the perspective of the use of fruits rich in xanthophyll esters as valuable components of the human diet and as sources of bioactive compounds in the prevention of severe degenerative diseases.

Acknowledgements

This work was supported by the grants of the Romanian National Authority for Scientific Research CNCS-UEFISCDI, project number PN-II-ID-PCE-2011-3-0721.

References

Amorim-Carrilho KT, Cepeda A, Fente C, Regal P (2014). Review of methods for analysis of carotenoids. TRAC-Trend Anal Chem 56:49-73.

Andersson S, Olsson ME, Johansson E, Rumpunen K (2009). Carotenoids in sea buckthorn (*Hippophae rhamnoides* L.) berries during ripening and use of Pheophytin a as a maturity marker. J Agric Food Chem 57:250-258.

Ariizumi T, Kishimoto S, Kakami R, Maoka T, Hirakawa H, Suzuki Y, Ozeki Y, Shirasawa K, Bernillon S, Okabe Y, Moing A, Asamizu E, Rothan C, Ohmiya A, Ezura H (2014. Identification of the carotenoid modifying gene *Pale Yellow Petal 1* as an essential factor in xanthophyll esterification and yellow flower pigmentation in tomato (*Solanum lycopersicum*). Plant J 79:453-465.

Arita S, Otsuki K, Osaki K, Murata Y, Shimoishi Y, Tada M (2004). Reduction in photostability by esterification of β-cryptoxanthin. Biosci Biotechnol Biochem 68(2):451-453.

Bernhard K (1995). Chromatography: Part II Column Chromatography, p. 81-108. In: Carotenoid. Vol. 1A: Isolation and Analysis, Britton G, Liaaen-Jensen S, Pfander H (Eds.). Birkhauser Verag, Basel.

Bernstein PS, Khachik F, Carvalho LS, Muir GJ, Zhao DY, Katz NB (2001). Identification and quantitation of carotenoids and their metabolites in the tissues of the human eye. Exp Eye Res 72(3):215-223.

Biacs PA, Daood HG, Pavisa A, Hajdu F (1989). Studies on the carotenoid pigments of paprika (*Capsicum annuum* L. var Sz-20). J Agric Food Chem 37:350-353.

Biswal B (1995). Carotenoid catabolism during leaf senescence and its control by light. J Photoch Photobio B 30(1):3-13.

Bone RA, Landrum JT, Dixon Z, Chen Y, Llerena CM (2000). Lutein and zeaxanthin in the eyes, serum and diet of human subjects. Exp Eye Res 71:239-245.

Bowen PE, Herbst-Espinosa SM, Hussain EA, Stacewicz-Sapuntzakis M (2002). Esterification does not impair lutein bioavailability in humans. J Nutr, 132(12):3668-3673.

Breithaupt DE, Bamedi A (2001). Carotenoid Esters in Vegetables and Fruits: A Screening with Emphasis on β-Cryptoxanthin Esters. J Agr Food Chem 49: 2064-2070.

Breithaupt DE, Bamedi A (2002a). Carotenoids and carotenoid esters in potatoes (*Solanum tuberosum* L.): new insights into an ancient vegetable. J Agric Food Chem 50:7175-7181.

Breithaupt DE, Bamedi A, Wirt U (2002). Carotenol fatty acid esters: easy substrates for digestive enzymes. Comp Biochem Physiol B 132:721-28.

Breithaupt DE, Weller P, Wolters M, Hahn A (2003). Plasma response to a single dose of dietary β-cryptoxanthin esters from papaya (*Carica papaya* L.) or non-esterified β-cryptoxanthin in adult human subjects: a comparative study. Br J Nutr 90:795-801.

Breithaupt DE, Weller P, Wolters M, Hahn A (2004). Comparison of plasma responses in human subjects after the ingestion of 3R, 3R9-zeaxanthin dipalmitate from wolfberry (*Lycium barbarum*) and non-esterified 3R, 3R9-zeaxanthin using chiral high-performance liquid chromatography. Br J Nutr 91:707-713.

Breithaupt DE, Alpman A, Carrière F (2007). Xanthophyll esters are hydrolyzed in the presence of recombinant human pancreatic lipase. Food Chem 103:651-656.

Britton G (1995a). Worked examples of isolation and analysis. Example 1: Higher plants. In Carotenoid, Vol. 1A: Isolation and Analysis, Britton G, Liaaen-Jensen S, Pfander H (Eds.). Birkhauser Verlag, Basel, p. 201-214.

Britton G (1995b). UV-Visible Spectroscopy. In: Carotenoid, Vol. 1B: Spectroscopy, Britton G, Liaaen-Jensen S, Pfander H (Eds.). Birkhauser Verlag, Basel, p. 13-62.

Britton G (1998). Overview of carotenoids biosynthesis, p. 13-140. In: Carotenoids. Vol. 3: Biosynthesis and Metabolism, Britton G, Liaaen-Jensen S, Pfander H (Eds.). Birkhauser Verlag, Basel.

Britton G, Liaaen-Jensen S, Pfander H (2004). Carotenoids – Handbook. Birkhauser Verlag: Basel, Switzerland.

Britton G, Khachik F (2009). Carotenoids in food. In Carotenoids. Vol. 5 Nutrition and Health, Britton G, Liaaen-Jensen S, Pfander H (Eds.). Birkhauser Verlag, Basel, p. 45-66.

Bunea A, Andrei S, Rugina D, Pintea A (2013). Thermal and light stability of β-cryptoxanthin esters. Proceedings of the 7th International Congress on Pigments in Food. Novara, Italy, 170-173.

Bunea A, Pintea A, Socaciu C (2007). HPLC separation of carotenoids from *Physalis alkekengi* L. in different stages of maturation. Annals of West University of Timişoara, Series of Chemistry 16(2):129-134.

Burmeister A, Bondiek S, Jerz G, Fleischmann P (2013). Carotenoid ester profiles in *Solanum tuberosum* and *Solanum phureja* cultivars. Proceedings of the 7th International Congress on Pigments in Food. Novara, Italy, 26-29.

Cacciola F, Donato P, Beccaria M, Dugo P, Mondello L (2012). Advanced in LC-MS for Food Analysis. LC GC Eur. 15-24 (special issue).

Canene-Adams K, Erdman JW (2009). Absorption, Transport, Distribution in Tissues and Bioavailability, p. 115-148. In Carotenoids, Vol. 5 Nutrition and Health, Britton G, Liaaen-Jensen S, Pfander H (Eds.). Birkhauser Verlag, Basel.

Cardini F (1982). Carotenoids in ripe green and in autumn senescing leaves of apple tree: I - Qualitative composition of free carotenoids, xanthophyll esters and fatty acids of esters. Giornale Botanico Italiano 116:97-115.

Caris-Veyrat C (2008). Antioxidant and prooxidant actions and stabilities of carotenoids in vitro and in vivo and carotenoids oxidation products, p. 177-192. In Food Colorants: Chemical and Functional properties. Socaciu C (Ed.). CRC Press, Taylor & Francis Group, Boca Raton, London, New York.

Carvalho E, Fraser PD, Martens S (2013). Carotenoids and tocopherols in yellow and red raspberries. Food Chem 139:744-752.

Cheng CY, Chung WY, Szeto YT, Benzie IFF (2005). Fasting plasma zeaxanthin response to *Fructus barbarum* L. (wolfberry; Kei Tze) in a food-based human supplementation trial. Br J Nutr 93:123-130.

Chew EY, Bressler S, Fish G, Hubbard B, Klein M, Chandra S, Blodi B, Domalpally A, Friberg T, Wong W, Rosenfeld P, Clemons TE, Agron E, Toth C, Bernstein P, Sperduto R, SanGiovanni JP, Danis R, Ferris FL, Elman M, Antoszyk A, Ruby A, Orth D (2013). Lutein plus zeaxanthin and omega-3 fatty acids for age-related macular degeneration the age-related eye disease study 2 (AREDS2) Randomized Clinical Trial. JAMA-J Am Med Assoc 309:2005-2015.

Chitchumroonchokchai C, Failla ML (2006). Hydrolysis of Zeaxanthin Esters by Carboxyl Ester Lipase during Digestion Facilitates Micellarization and Uptake of the Xanthophyll by Caco-2 Human Intestinal Cells. J Nutr 136:588-594.

Daood HG, Palotás G, Palotás G, Somogyi G, Pék Z, Helyes L (2014). Carotenoid and antioxidant content of ground paprika from indoor-cultivated traditional varieties and new hybrids of spice red peppers. Food Res Int DOI: 10.1016/j.foodres.2014.04.048

Delgado-Pelayo R, Hornero-Méndez D (2012). Identification and quantitative analysis of carotenoids and their esters from sarsaparilla (*Smilax aspera* L.) Berries. J Agr Food Chem 60:8225-8232.

Delgado-Pelayo R, Gallardo-Guerrero L, Hornero-Méndez D (2014). Carotenoid pigments in peel and flesh of commercial apple fruit varieties. Food Res Int http://dx.doi.org/10.1016/j.foodres.2014.03.025

Deruère J, Römer S, d'Harlingue A, Backhaus RA, Kuntz M, Camara B (1994). Fibril assembly and carotenoid overaccumulation in chromoplasts: a model for supramolecular lipoprotein structures. Plant Cell 6:119-133.

Dufossé L (2009). Microbial and Microalgal Carotenoids as Colourants and Supplements. In Carotenoids, p. 83-98. Vol. 5 Nutrition and Health, Britton G, Liaaen-Jensen S, Pfander H (Eds.). Birkhauser Verlag, Basel.

Dugo P, Herrero M, Kumm T, Giuffrida D, Dugo G, Mondello L (2008). Comprehensive normal-phase x reversed-phase liquid chromatography coupled to photodiode array and mass spectrometry detection for the analysis of free carotenoids and carotenoid esters from mandarin. J Chromatogr A 1189:196–206.

Enzell CR, Back S G (1995). Mass Spectrometry, p. 261-320. In: Carotenoid, Vol. 1B: Spectroscopy, Britton G, Liaaen-Jensen S, Pfander H (Eds.), Birkhauser. Verlag, Basel.

Férnandez-Garcia E, Carvajal-Lérida I, Jarén-Galan M, Garrido-Fernandez J, Pérez-Galvez A, Hornero-Méndez D (2012). Carotenoids bioavailability from foods: From plant pigments to efficient biological activities. Food Res Int 46:438-450.

Fernandez-Orozco R, Gallardo-Guerrero L, Hornero-Méndez D (2013). Carotenoid profiling in tubers of different potato (*Solanum* sp) cultivars: Accumulation of carotenoids mediated by xanthophyll esterification, Food Chem 141:2864-2872.

Fu HF, Xie B, Gan F, Ma S, Zhu XR, Pan SY (2010). Effect of esterification with fatty acid of β-cryptoxanthin on its thermal stability and antioxidant activity by chemiluminescence method. Food Chem 122:602-609.

Giuffrida D, La Torre L, Stelitano M, Pellicanò TM, Dugo G (2006). Application of HPLC-APCI-MS a C-30 reversed phase column for the characterization of carotenoid esters in mandarin essential oil. Flavour Fragr J 21:319-323.

Giuffrida D, Dugo P, Salvo A, Saitta M, Dugo G (2010), Free carotenoid and carotenoid ester composition in native orange juices of different varieties. Fruits 65:277-284.

Giuffrida D, Pintea A, Dugo P, Torre G, Pop RM, Mondello L (2012). Determination of carotenoids and their esters in fruits of seabuckthorn (*Hippopae rhamnoides* L.) by HPLC-DAD-APCI-MS. Phytochem Anal 23(3):267-273.

Giuffrida D, Dugo P, Torre G, Bignardi C, Cavazza A, Corradini C, Dugo G (2013). Characterization of 12 *Capsicum* varieties by evaluation of their carotenoid profile and pungency determination. Food Chem 140:794-802.

Giuffrida D, Torre G, Dugo P, Dugo G (2013). Determination of the carotenoid profile in peach fruits, juice and jam. Fruits 68:39-44.

Giuliano G (2014). Plant carotenoids: genomics meets multi-gene engineering. Curr Opin Plant Biol 19:111-117.

Gomez-Garcia MR, Ochoa-Alejo N (2013). Biochemistry and molecular biology of carotenoid biosynthesis in chili peppers (*Capsicum* spp.). Int J Mol Sci 14:19025-19053.

Goodwin TW. 1980. Carotenoids in higher plants. In: The Biochemistry of the Carotenoids, Vol. I. Chapman and Hall: New York.

Granado F, Olmedilla B, Gil-Martinez E, Blanco I (1988). Lutein ester in serum after lutein supplementation in human subjects,. Br J Nutr 80:445-449.

Gregory GK, Chen TS, Philip T (1986). Quantitative analysis of lutein esters in marigold flowers (*Tagetes erecta*) by high performance liquid chromatography. J Food Sci 51:1093-1094.

Hornero-Mendez D, Gomez-Landron De Guerva R, Mínguez-Mosquera MI (2000). Carotenoid biosynthesis changes in five red pepper (*Capsicum annuum* L.) cultivars during ripening. Cultivar selection for breeding. J Agric Food Chem 48:3857-3864.

Inbaraj BS, Lu H, Hung CF, Wu WB, Lin CL, Chen BH (2008). Determination of carotenoids and their esters in fruits of *Lycium barbarum* Linnaeus by HPLC-DAD-APCI-MS. Journal Pharmaceut Biomed 47:812-818.

Jacobs PB, LeDoeuft RD, McCommast SA, Tauber JD (1982). The cleavage of carotenoid esters by cholesterol esterase. Comp Biochem Physiol 72B:157-160.

Khachik F, Beecher GR (1988a). Separation and identification of carotenoids and carotenol fatty acids esters in some squash products by liquid chromatography. 1. Quantification of carotenoids and related esters by HPLC. J Agric Food Chem 36:929-937.

Khachik F, Beecher GR, Lusby WR (1988b). Separation and identification of carotenoids and carotenol fatty acid esters in some squash products by liquid chromatography. 2. Isolation and characterization of carotenoids and related esters. J Agric Food Chem, 36:938-946.

Khachik F. (2009). Analysis of carotenoids in nutritional studies, p. 7-44. In: Carotenoids, Vol. 5 Nutrition and Health, Britton G, Liaaen-Jensen S, Pfander H (Eds.). Birkhauser Verlag, Basel.

Knee M (1988). Carotenol esters in developing apple fruits. Phytochem 27(4):1005-1009.

Krinsky NI, Johnson EJ (2005). Carotenoid actions and their relation to health and disease. Molec Aspect Med 26:459-516.

Kurz C, Carle R, Schieber A (2008). HPLC-DAD-MS

characterization of carotenoids from apricots and pumpkins for the evaluation of fruit product authenticity. Food Chem 110: 522-530.

Landrum JT, Bone RA, Joa H, Kilburn MD, Moore LL, Sprague KE (1997). A one year study of the macular pigment: the effect of 140 days of a lutein supplement. Exp Eye Res 63:57-62.

Li L, Yuan H (2013). Chromoplast biogenesis and carotenoid accumulation. Arch Biochem Biophys 539:102-109.

Maiani G, Caston MJP, Catasta G, Toti E, Cambrodon IG, Bysted A, Granado-Lorencio F, Olmedilla-Alonso B, Knuthsen P, Valoti M, Böhm V, Miebach EM, Behsnillan, Schlemmer U (2009). Carotenoids: Actual knowledge on food sources, intakes, stability and bioavailability and their protective role in humans, Mol Nutr Food Res 53:S194-S218.

Matsufuji H, Nakamura H, Chino M, Tkeda M (1998). Antioxidant activity of capsanthin and the fatty acid esters in paprika (*Capsicum annuum*). J Agric Food Chem 46:3468-3472.

Mellado-Ortega E, Hornero-Méndez D (2012). Isolation and identification of lutein esters, including their regioisomers, in tritordeum (Tritordeum Ascherson et Graebner) grains: Evidence for a preferential xanthophyll acyltransferase activity. Food Chem 135:1344-1352.

Mercadante AZ (2008). Carotenoids in Foods: Sources and stability during processing, p. 241–276. In: Food Colorants: Chemical and Functional properties, Socaciu C (Ed.). CRC Press, Taylor & Francis Group, Boca Raton, London, New York.

Mertz C, Brat P, Caris-Veyrat C, Gunata Z (2010). Characterization and thermal lability of carotenoids and vitamin C of tamarillo fruit (*Solanum betaceum* Cav.). Food Chem 119:653-659.

Merzlyak MN, Solovchenko AE (2002). Photostability of pigments in ripening apple fruit: a possible photoprotective role of carotenoids during plant senescence. Plant Sci 163:881-888.

Mínguez-Mosquera MI, Hornero-Méndez D (1994). Changes in carotenoid esterification during the fruit ripening of *Capsicum annuum* cv. Bola. J Agric Food Chem 42:640-644.

Moise AR, Al-Babili S, Wurtzel ET (2014). Mechanistic aspects of carotenoid biosynthesis. Chemical Reviews 114:164-193.

Mozaffarieh M, Sacu S, Wedrich A (2003). The role of the carotenoids, lutein and zeaxanthin, in protecting against age-related macular degeneration: a review based on controversial evidence. Nutr J 2:20-31.

Murillo E, Giuffrida D, Menchaca D, Dugo P, Torre G, Meléndez-Martinez AJ, Mondello L (2013). Native carotenoids composition of some tropical fruits. Food Chem 140:825-836.

Ohmiya A (2013). Qualitative and quantitative control of carotenoid accumulation in flower petals. Sci Hortic-Amsterdam 163:10-19.

Ornelas-Paz JI, Yahia ME, Gardea-Bejar A (2007). Identification and quantification of xanthophyll esters, carotenes, and tocopherols in the fruit of seven mexican mango cultivars by Liquid Chromatography-Atmospheric Pressure Chemical Ionization-Time-of-Flight Mass Spectrometry (LC-(APcI+)-MS). J Agric Food Chem 55:6628-6635.

Ornelas-Paz JJ, Yahia EA, Gardea AA (2008). Changes in external and internal color during postharvest ripening of 'Manila' and 'Ataulfo' mango fruit and relationship with carotenoid content determined by liquid chromatography–APcI+-time-of-flight mass spectrometry. Postharvest Biol Tec 50:145-152.

Pattison DJ, Symmons DPM, Lunt M, Welch A, Bingham SA, Day NE, Silman AJ (2005). Dietary β-cryptoxanthin and inflammatory polyarthritis: results from a population-based prospective study. Am J Clin Nutr 82:451-455.

Perez-Gálvez A, Minguez-Mosquera MI (2002). Degradation of non-esterified and esterified xanthopylls by free radicals. Biochim Biophys Acta 1569:31-34.

Pérez-Gálvez A, Martin HD, Sies H, Stahl W (2003). Incorporation of carotenoids from paprika oleoresin into human chylomicrons. Br J Nutr 89:787-793.

Pérez-Gálvez A, Minguez-Mosquerra MI (2005). Esterification of xanthophylls and its effect on chemical behavior and bioavailability of carotenoids in the human. Nutr Res 25:631-640.

Philip T, Chen TS (1988). Separation and quantitative analysis of some carotenoid fatty esters of fruits by liquid chromatography. J Chromatogr 435(1):113-136.

Pintea A, Varga A, Stepnowski P, Socaciu C, Culea M, Diehl HA (2005). Chromatographic analysis (HPLC, GC) of carotenol fatty acids esters in *Physalis alkekengi* and *Hippophae rhamnoides*, Phytochem Anal 16:188-195.

Pintea A, Bunea A, Socaciu C. (2013a). Effect of esterification on thermal stability and antioxidant activity of zeaxanthin. Proceedings of the 7th International Congress on Pigments in Food. Novara, Italy, 174-177.

Pintea A, Rugină D, Bunea A, Andrei S (2013b). Impact of esterification on the antioxidant Capacity of β-cryptoxanthin. Bulletin USAMV Series Animal Science and Biotechnology 70(1):79-85.

Pintea A, Bunea A, Socaci S, Socaciu C (2014). Effect of esterification on the antioxidant activity of zeaxanthin. Carotenoid Science Proceedings Abstract 17th international symposium on carotenoids 10:142.

Pintea A (2008). Food Colorants derived from Natural Sources by Processing. In: Food Colorants - chemical and functional properties (ed. Socaciu C). CRC Press, Taylor and Francis, p. 329-347.

Pinto de Abreu F, Dornier M, Dionisio AP, Carail M, Caris-Veyrat C, Dhuique-Mayer C (2013). Cashew apple (*Anacardium occidentale* L.) extract from by-product of juice processing: A focus on carotenoids. Food Chem 138:25-31.

Pop RM, Weesepoel Y, Socaciu C, Pintea A. Vincken JP,

Gruppen H (2014). Carotenoid composition of berries and leaves from six Romanian seabuckthorn (*Hippophae rhamnoides* L.) varieties. Food Chem 147:1-9.

Rao AV, Rao RG (2007). Carotenoids and human health. Pharm Res 55:207-216.

Rodriguez-Amaya DB (2010). Quantitative analysis, *in vitro* assessement of bioavailability and antioxidant activity of food carotenoids – A review. J Food Comp Anal 23:726-740.

Rohmer M (1999). The discovery of a mevalonate-independent pathway for isoprenoid biosynthesis in bacteria, algae and higher plants. Nat Prod Rep 16:565-574.

Roodenberg AJC, Leenen R. van het Hof KH, Weststrate JA, Tijburg LBM (2000). Amount of fat in the diet affects the bioavailability of lutein esters but not of α-carotene, β-carotene, and vitamin E in humans. Am J Clin Nutr 71:1187-1193.

Sander LC, Sharpless KE, Pursch M. (2000). C30 stationary phases for the analysis of food by liquid chromatography. J Chromatog A 880:189-202.

Schalch W, Landrum JT, Bone RA (2009). The Eye, p. 301-334. In: Carotenoids, Vol. 5 Nutrition and Health, Britton G, Liaaen-Jensen S, Pfander H (Eds.). Birkhauser Verlag, Basel.

Schiedt K, Liaaen-Jensen S (1995). Isolation and Analysis, p. 81-108. In Carotenoid, Vol. 1A: Isolation and Analysis, Britton G, Liaaen-Jensen S, Pfander H (Eds.). Birkhauser Verlag, Basel.

Schumskaya M, Wurtzel ET (2013). The carotenoid biosynthetic pathway: Thinking in all dimensions. Plant Science 208:58-63.

Schweigert FJ (1998). Metabolism of Carotenoids in Mammals, p. 249-284. In: Carotenoids, Vol. 3 Biosynthesis and Metabolism, Britton G, Liaaen-Jensen S, Pfander H (Eds.). Birkhauser Verlag, Basel.

Schweiggert RM, Steingass CB, Mora E, Esquivel P, Carle R (2011). Carotenogenesis and physico-chemical characteristics during maturation of red fleshed papaya fruit (*Carica papaya* L.). Food Res Int 44:1373-1380.

Schweiggert RM, Mezger D, Schimpf F, Steingass CB, Carle R (2012). Influence of chromoplast morphology on carotenoid bioaccessibility of carrot, mango, papaya, and tomato. Food Chem 135:2736-2742.

Schweiggert RM, Steingass CB, Esquivel P, Reinhold C (2012). Chemical and morphological characterization of Costa Rican papaya (*Carica papaya* L.) Hybrids and lines with particular focus on their genuine carotenoid profiles. J Agric Food Chem 60:2577-2585.

Schweiggert U, Kurz C, Schieber A, Carle R (2007). Effects of processing and storage on the stability of free and esterified carotenoids of red peppers (*Capsicum anuum* L.) and hot chilli peppers (*Capsicum frutescens* L.). Eur Food Res Technol 225:261-270.

Solovchenko AE, Avertcheva OV, Merzlyak MN (2006). Elevated sunlight promotes ripening-associated pigment changes in apple fruit. Postharvest Biol Tec 40:183-189.

Subagio A, Morita N (2001). No effect of esterification with fatty acid on antioxidant activity of lutein. Food Res Int 34:315-320.

Subagio A, Morita N (2003). Prooxidant activity of lutein and its dimyristate esters in corn triacylglyceride. Food Chem 81:97-102.

Subagio A, Wakaki H, Morita N (1999). Stability of lutein and its myristate esters, Biosci Biotechnol Biochem 63(10):1784-1786.

Van Breemen RB, Dong L, Pajkovic ND (2012). Atmospheric pressure chemical ionization tandem mass spectrometry of carotenoids. Int J Mass Spectrom 312:163-172.

Weller P, Breithaupt DE (2003). Identification and quantification of zeaxanthin esters in plants using liquid chromatography-mass spectrometry. J Agric Food Chem 51:7044-7049.

Wingerath T, Sies H, Stahl W (1998). Xanthophyll esters in human skin. Arch Biochem Biophys 355:271-274.

Wingerath T, Stahl W, Kirsch D, Kaufmann R, Sies H (1997). Fruit Juice Carotenol Fatty Acid Esters and Carotenoids As Identified by Matrix-Assisted Laser Desorption Ionization (MALDI) Mass Spectrometry. J Agric Food Chem 44:2006-2013.

Wingerath T, Stahl W, Sies H (1995). β-Cryptoxanthin selectively increased in human chylomicrons upon ingestion of tangerine concentrate rich β-cryptoxanthin esters. Arch Biochem Biophys. 324:385-390.

Wu C, Han L, Riaz H, Wang S, Cai K, Yang L (2013). The chemopreventive effect of β-cryptoxanthin from mandarin on human stomach cells (BGC-823). Food Chem 136:1122-1129.

Yahia EM, Gutierrez-Orozco F, Arvizu-de Leon C (2011). Pytochemical and antioxidant characterization of mamey (*Pouteria sapota* Jacq. H.E. Moore & Stearn) fruit. Food Res Int 44:2175-2181.

Yamaguchi M. (2012). Role of carotenoid β-cryptoxanthin in bone homeostasis. J Biomed Sci 19:36.

Yeum KJ, Aldini G, Russel RM, Krinsky NI (2009). Antioxidant/Pro-oxidant actions of carotenoids, p. 235-268. In: Carotenoids, Vol. 5 Nutrition and Health, Britton G, Liaaen-Jensen S, Pfander H (Eds.). Birkhauser Verlag, Basel.

Ytterberg AJ, Peltier J, Van Wijk KJ (2006). Protein profiling of plastoglobules in chloroplasts and chromoplasts: a surprising site for differential accumulation of metabolic enzymes. Plant Physiol 140:984-997.

Zhao L-Q, Qiu Z-Q, Narasimhamoorthy B, Greaves JA (2013). Development of a rapid, high-throughput method for quantification of zeaxanthin in Chinese wolfberry using HPLC–DAD. Ind Crop Prod 47:51-57.

Permissions

All chapters in this book were first published in NBHAC, by University of Agricultural Sciences and Veterinary Medicine, Cluj-Napoca; hereby published with permission under the Creative Commons Attribution License or equivalent. Every chapter published in this book has been scrutinized by our experts. Their significance has been extensively debated. The topics covered herein carry significant findings which will fuel the growth of the discipline. They may even be implemented as practical applications or may be referred to as a beginning point for another development.

The contributors of this book come from diverse backgrounds, making this book a truly international effort. This book will bring forth new frontiers with its revolutionizing research information and detailed analysis of the nascent developments around the world.

We would like to thank all the contributing authors for lending their expertise to make the book truly unique. They have played a crucial role in the development of this book. Without their invaluable contributions this book wouldn't have been possible. They have made vital efforts to compile up to date information on the varied aspects of this subject to make this book a valuable addition to the collection of many professionals and students.

This book was conceptualized with the vision of imparting up-to-date information and advanced data in this field. To ensure the same, a matchless editorial board was set up. Every individual on the board went through rigorous rounds of assessment to prove their worth. After which they invested a large part of their time researching and compiling the most relevant data for our readers.

The editorial board has been involved in producing this book since its inception. They have spent rigorous hours researching and exploring the diverse topics which have resulted in the successful publishing of this book. They have passed on their knowledge of decades through this book. To expedite this challenging task, the publisher supported the team at every step. A small team of assistant editors was also appointed to further simplify the editing procedure and attain best results for the readers.

Apart from the editorial board, the designing team has also invested a significant amount of their time in understanding the subject and creating the most relevant covers. They scrutinized every image to scout for the most suitable representation of the subject and create an appropriate cover for the book.

The publishing team has been an ardent support to the editorial, designing and production team. Their endless efforts to recruit the best for this project, has resulted in the accomplishment of this book. They are a veteran in the field of academics and their pool of knowledge is as vast as their experience in printing. Their expertise and guidance has proved useful at every step. Their uncompromising quality standards have made this book an exceptional effort. Their encouragement from time to time has been an inspiration for everyone.

The publisher and the editorial board hope that this book will prove to be a valuable piece of knowledge for researchers, students, practitioners and scholars across the globe.

List of Contributors

Sercan PAZARLAR
University of Ege, Faculty of Agriculture, Department of Plant Protection, 35100 Bornova, İzmir, Turkey

Mustafa GÜMÜŞ
University of Ege, Faculty of Agriculture, Department of Plant Protection, 35100 Bornova, İzmir, Turkey

Gölgen Bahar ÖZTEKİN
University of Ege, Faculty of Agriculture, Department of Horticulture, 35100 Bornova İzmir, Turkey

Roman HOŁUBOWICZ
Poznan University of Life Sciences, Department of Plant Pathology Seed Science and Technology, Division of Seed Science and Technology, Szamotulska 28, Baranowo 62-081 Przezmierowo, Poland

Leszek KUBISZ
Poznan University of Medical Sciences, Department of Biophysics, Poland, Fredry 10, 61-710 Poznan, Poland

Marlena GAUZA
Poznan University of Medical Sciences, Department of Biophysics, Poland, Fredry 10, 61-710 Poznan, Poland

Yilin TONG
Poznan University of Life Sciences, Department of Plant Pathology Seed Science and Technology, Division of Seed Science and Technology, Szamotulska 28, Baranowo 62-081 Przezmierowo, Poland

Dorota HOJAN-JEZIERSKA
Poznan University of Medical Sciences, Department of Biophysics, Poland, Fredry 10, 61-710 Poznan, Poland

Özlem AYTÜRK
Marmara University, Department of Biology, Göztepe Campus, 34730 Istanbul, Turkey

Meral ÜNAL
Marmara University, Department of Biology, Göztepe Campus, 34730 Istanbul, Turkey

Gülden YILMAZ
Trakya University, Faculty of Science, Department of Biology, Balkan Campus, 22030, Edirne/Turkey

Mehmet TEKIN
Cumhuriyet University, Faculty of Science, Department of Biology, 58140, Sivas, Turkey

Sevgi CALISKAN
Department of Plant Production and Technologies, Nigde University, 51240, Nigde, Turkey

Halit YETISIR
Erciyes University, Faculty of Agriculture, Department of Horticulture, Kayseri, Turkey

Sema KARANLIK
Mustafa Kemal University, Faculty of Agriculture, Department of Soil Science, 31000 Hatay, Turkey

Duygu BAYRAM
Sinop University, Department of Biology, Faculty of Art and Science, 57000, Sinop, Turkey

Burcu SECKIN DINLER
Sinop University, Department of Biology, Faculty of Art and Science, 57000, Sinop, Turkey

Eda TASCI
Sinop University, Department of Biology, Faculty of Art and Science, 57000, Sinop, Turkey

Ionela Catalina GUȚA
National Research and Development Institute for Biotechnology in Horticulture, 37 Calea Bucuresti, Stefanesti-Arges, România

Elena-Cocuța BUCIUMEANU
National Research and Development Institute for Biotechnology in Horticulture, 37 Calea Bucuresti, Stefanesti-Arges, România

Emilia VISOIU
National Research and Development Institute for Biotechnology in Horticulture, 37 Calea Bucuresti, Stefanesti-Arges, România

Adrian I. TIMOFTE
University of Oradea, Faculty of Environmental Protection, 26 Gen. Magheru St., 410058 Oradea, Romania

Simona VICAŞ
University of Oradea, Faculty of Environmental Protection, 26 Gen. Magheru St., 410058 Oradea, Romania

Jolanta PATAMSYTĖ
Vilnius University, Department of Botany and Genetics, 21 M.K. Čiurlionio Str., LT-03101 Vilnius, Lithuania

Tatjana ČĖSNIENĖ
Vilnius University, Department of Botany and Genetics, 21 M.K. Čiurlionio Str., LT-03101 Vilnius, Lithuania

Donatas NAUGŽEMYS
Vilnius University, Botanical Garden, 43 Kairėnų Str., LT-10239 Vilnius, Lithuania

Violeta KLEIZAITĖ
Vilnius University, Department of Botany and Genetics, 21 M.K. Čiurlionio Str., LT-03101 Vilnius, Lithuania

Virginija TUNAITIENĖ
Vilnius University, Department of Botany and Genetics, 21 M.K. Čiurlionio Str., LT-03101 Vilnius, Lithuania

Virginija VAITKŪNIENĖ
Vilnius University, Department of Botany and Genetics, 21 M.K. Čiurlionio Str., LT-03101 Vilnius, Lithuania

Vytautas RANČELIS
Vilnius University, Department of Botany and Genetics, 21 M.K. Čiurlionio Str., LT-03101 Vilnius, Lithuania

Rita MIKALIŪNAITĖ
Šiauliai University, Department of Environmental Research, 19 P. Višinskio Str., LT-77156, Šiauliai, Lithuania

Donatas ŽVINGILA
Vilnius University, Department of Botany and Genetics, 21 M.K. Čiurlionio Str., LT-03101 Vilnius, Lithuania

Marouf KHALILI
University of Payame Noor, Department of Agriculture, PO BOX 19395-3697, Tehran, Iran

Alireza POUR-ABOUGHADAREH
Imam Khomeini International University, Faculty of Engineering and Technology, Department of Plant Breeding and Production, Qazvin, Iran

Mohammad Reza NAGHAVI
University of Payame Noor, Department of Agriculture, PO BOX 19395-3697, Tehran, Iran

Esmail MOHAMMAD-AMINI
University of Tabriz, Faculty of Agriculture, Department of Biotechnology and Plant Breeding, Tabriz, Iran

Beata JANOWSKA
Department of Ornamental Plants, Faculty of Horticulture and Landscape Architecture, Poznan University of Life Sciences, Dabrowskiego 159, Poznan, Poland

Natalia MANSFELD
Department of Ornamental Plants, Faculty of Horticulture and Landscape Architecture, Poznan University of Life Sciences, Dabrowskiego 159, Poznan, Poland

Roman ANDRZEJAK
Department of Phytopathology, Seed Science and Technology, Faculty of Horticulture and Landscape Architecture, Poznan University of Life Sciences, Dabrowskiego 159, Poznan, Poland

Ozlem ALAN
Ege University, Odemis Vocational Training School, Izmir, Turkey

Gulcan KINACI
Eskisehir Osmangazi University, Faculty of Agriculture, Department of Field Crops, Eskisehir, Turkey

Engin KINACI
Eskisehir Osmangazi University, Faculty of Agriculture, Department of Field Crops, Eskisehir, Turkey

Imren KUTLU
Eskisehir Osmangazi University, Faculty of Agriculture, Eskisehir, Turkey

Zekiye BUDAK BASCİFTCİ
Eskisehir Osmangazi University, Faculty of Agriculture, Department of Field Crops, Eskisehir, Turkey

Kenan SONMEZ
Eskisehir Osmangazi University, Faculty of Agriculture, Department of Horticulture, Eskisehir, Turkey

Yasemin EVRENOSOGLU
Eskisehir Osmangazi University, Faculty of Agriculture, Department of Horticulture, Eskisehir, Turkey

Mohamad Al HASSAN
Instituto de Biología Molecular y Celular de Plantas (UPV-CSIC), Universitat Politècnica de València, CPI, edificio 8E, Camino de Vera s/n, 46022 Valencia, Spain

Andrea PACURAR
Instituto de Biología Molecular y Celular de Plantas (UPV-CSIC), Universitat Politècnica de València, CPI, edificio 8E, Camino de Vera s/n, 46022 Valencia, Spain
Permanent address: University of Agricultural Sciences and Veterinary Medicine, Faculty of Horticulture, 3-5 Manastur Str., 400372 Cluj-Napoca, Romania

Alexandra GASPAR
Instituto de Biología Molecular y Celular de Plantas (UPV-CSIC), Universitat Politècnica de València, CPI, edificio 8E, Camino de Vera s/n, 46022 Valencia, Spain
Permanent address: University of Agricultural Sciences and Veterinary Medicine, Faculty of Horticulture, 3-5 Manastur Str., 400372 Cluj-Napoca, Romania

Oscar VICENTE
Instituto de Biología Molecular y Celular de Plantas (UPV-CSIC), Universitat Politècnica de València, CPI, edificio 8E, Camino de Vera s/n, 46022 Valencia, Spain

Monica BOSCAIU
Instituto de Biología Molecular y Celular de Plantas (UPV-CSIC), Universitat Politècnica de València, CPI, edificio 8E, Camino de Vera s/n, 46022 Valencia, Spain

Instituto Agroforestal Mediterráneo (UPV), Universitat Politècnica de València, CPI, edificio 8E, Camino de Vera s/n, 46022 Valencia, Spain

Erzsébet BUTA
University of Agricultural Sciences and Veterinary Medicine, 3-5 Manastur Str., 400372, Cluj-Napoca, Romania

Anamária TÖRÖK
University of Babes Bolyai, Faculty of Chemistry and Chemical Engineering, Arany János 11, Cluj-Napoca, Romania

Árpád CSOG
University of Babes Bolyai, Faculty of Chemistry and Chemical Engineering, Arany János 11, Cluj-Napoca, Romania

Bilassé ZONGO
University of Ouagadougou, Laboratory of Biology and Ecology of Plants, 09 BP 848 Ouagadougou 09, Burkina Faso, Western Africa

Maria CANTOR
University of Agricultural Sciences and Veterinary Medicine, 3-5 Manastur Str., 400372, Cluj-Napoca, Romania

Mihai BUTA
University of Agricultural Sciences and Veterinary Medicine, 3-5 Manastur Str., 400372, Cluj-Napoca, Romania

Cornelia MAJDIK
University of Agricultural Sciences and Veterinary Medicine, 3-5 Manastur Str., 400372, Cluj-Napoca, Romania

Maria Antonia Machado BARBOSA
Núcleo de Pesquisa Vegetal Básica e Aplicada, Universidade Federal Rural da Amazônia, Paragominas, Brazil

Allan Klynger da Silva LOBATO
Núcleo de Pesquisa Vegetal Básica e Aplicada, Universidade Federal Rural da Amazônia, Paragominas, Brazil

Thaís Soares PEREIRA
Núcleo de Pesquisa Vegetal Básica e Aplicada, Universidade Federal Rural da Amazônia, Paragominas, Brazil

Gélia Dinah Monteiro VIANA
Núcleo de Pesquisa Vegetal Básica e Aplicada, Universidade Federal Rural da Amazônia, Paragominas, Brazil

José Ricardo Santos BARBOSA
Núcleo de Pesquisa Vegetal Básica e Aplicada, Universidade Federal Rural da Amazônia, Paragominas, Brazil

Kelly Nayara Nascimento COELHO
Núcleo de Pesquisa Vegetal Básica e Aplicada, Universidade Federal Rural da Amazônia, Paragominas, Brazil

Leila Sobral SAMPAIO
Núcleo de Pesquisa Vegetal Básica e Aplicada, Universidade Federal Rural da Amazônia, Paragominas, Brazil

Benedito Gomes dos SANTOS FILHO
Núcleo de Pesquisa Vegetal Básica e Aplicada, Universidade Federal Rural da Amazônia, Paragominas, Brazil

Joaquim Albenísio Gomes SILVEIRA
Laboratório de Metabolismo de Plantas, Departamento de Bioquímica e Biologia Molecular, Universidade Federal do Ceara, Fortaleza, Brazil

Zeliha KÜÇÜKYUMUK
Department of Soil Science and Plant Nutrition, Agricultural Faculty, Suleyman Demirel University, 32260, Isparta, Turkey

Hülya ÖZGÖNEN
Department of Plant Protection, Agricultural Faculty, Suleyman Demirel University, 32260, Isparta, Turkey

Ibrahim ERDAL
Department of Soil Science and Plant Nutrition, Agricultural Faculty, Suleyman Demirel University, 32260, Isparta, Turkey

Figen ERASLAN
Department of Soil Science and Plant Nutrition, Agricultural Faculty, Suleyman Demirel University, 32260, Isparta, Turkey

Marcel PARVU
Babes-Bolyai University, Faculty of Biology and Geology, Department of Biology, 42 Republicii St, 400015 Cluj-Napoca, Romania

Laurian VLASE
Iuliu Hatieganu University of Medicine and Pharmacy, Faculty of Pharmacy, Department of Pharmaceutical Technology and Biopharmaceutics, 12 Ion Creanga St, 400010 Cluj-Napoca, Romania

Laszlo FODORPATAKI
Babes-Bolyai University, Hungarian Department of Biology and Ecology, 1 Mihail Kogalniceanu St, 400084 Cluj-Napoca, Romania

Ovidiu PARVU
Babes-Bolyai University, Faculty of Mathematics and Computer Science, 1 Mihail Kogalniceanu St, 400084 Cluj-Napoca, Romania

Oana ROSCA-CASIAN
Babes-Bolyai University, A. Borza Botanical Garden, 42 Republicii St, 400015 Cluj-Napoca, Romania

Csaba BARTHA
Babes-Bolyai University, Hungarian Department of Biology and Ecology, 1 Mihail Kogalniceanu St, 400084 Cluj-Napoca, Romania

Lucian BARBU-TUDORAN
Babes-Bolyai University, Electron Microscopy Center, 5-7 Clinicilor St, 400006 Cluj-Napoca, Romania

Alina Elena PARVU
Iuliu Hatieganu University of Medicine and Pharmacy, Faculty of Medicine, Department of Pathophysiology, 3 Victor Babes St, 400012 Cluj-Napoca, Romania

Judita CERNIAUSKIENE
Institute of Agriculture and Food Sciences, Aleksandras Stulginskis University, Studentu str. 11, LT-53361 Akademija, Kaunas distr., Lithuania

Jurgita KULAITIENE
Institute of Agriculture and Food Sciences, Aleksandras Stulginskis University, Studentu str. 11, LT-53361 Akademija, Kaunas distr., Lithuania

Honorata DANILCENKO
Institute of Agriculture and Food Sciences, Aleksandras Stulginskis University, Studentu str. 11, LT-53361 Akademija, Kaunas distr., Lithuania

Elvyra JARIENE,
Institute of Agriculture and Food Sciences, Aleksandras Stulginskis University, Studentu str. 11, LT-53361 Akademija, Kaunas distr., Lithuania

Edita JUKNEVICIENE
Institute of Agriculture and Food Sciences, Aleksandras Stulginskis University, Studentu str. 11, LT-53361 Akademija, Kaunas distr., Lithuania

Jixiang LIN
Alkali Soil Natural Environmental Science Center, Northeast Forestry University/Key Laboratory of Saline-alkali Vegetation Ecology Restoration in Oil Field, Ministry of Education, Harbin, 150040, China

Zhuolin LI
Key Laboratory of Vegetation Ecology of Ministry of Education, Institute of Grassland Science, Northeast Normal University, Changchun, 130024, China

Shuai SHAO
Alkali Soil Natural Environmental Science Center, Northeast Forestry University/Key Laboratory of Saline-alkali Vegetation Ecology Restoration in Oil Field, Ministry of Education, Harbin, 150040, China

Yingnan WANG
Alkali Soil Natural Environmental Science Center, Northeast Forestry University/Key Laboratory of Saline-alkali Vegetation Ecology Restoration in Oil Field, Ministry of Education, Harbin, 150040, China

Chunsheng MU
Alkali Soil Natural Environmental Science Center, Northeast Forestry University/Key Laboratory of Saline-alkali Vegetation Ecology Restoration in Oil Field, Ministry of Education, Harbin, 150040, China

Esmaeil GHOLINEZHAD
Department of Agronomy, Payame Noor University, Tehran, Iran

Reza DARVISHZADEH
Department of Plant Breeding and Biotechnology, Urmia University, Urmia, Iran

Iraj BERNOUSI
Department of Plant Breeding and Biotechnology, Urmia University, Urmia, Iran

Xiaojun YUAN
School of Life Science, Shanghai University, Shanghai 200444; Shanghai Key Laboratory of Bio-Energy Crops, Shanghai University, China

Zhenjing BAO
School of Agriculture and Biology, Shanghai Jiaotong University, Shanghai 200240, China

Yali HE
School of Agriculture and Biology, Shanghai Jiaotong University, Shanghai 200240, China

Qun CHEN
School of Agriculture and Biology, Shanghai Jiaotong University, Shanghai 200240, China

Gang WANG
School of Life Science, Shanghai University, Shanghai 200444; Shanghai Key Laboratory of Bio-Energy Crops, Shanghai University, China

Culia SÎRBU
University of Agricultural Sciences and Veterinary Medicine Iasi, Faculty of Agriculture, 3, Mihail Sadoveanu Alley, 700490, Iasi, Romania

Adrian OPREA
University "Alexandru Ioan Cuza", Botanical Garden, 7-9, Dumbrava Rosie Street, 700487, Iasi, Romania

Cristian Valeriu PATRICHE
Romanian Academy, Department of Iasi, Geography Group, B-dul Carol I, 8, 700505, Iasi, Romania

Costel SAMUIL
University of Agricultural Sciences and Veterinary Medicine Iasi, Faculty of Agriculture, 3, Mihail Sadoveanu Alley, 700490, Iasi, Romania

Vasile VÎNTU
University of Agricultural Sciences and Veterinary Medicine Iasi, Faculty of Agriculture, 3, Mihail Sadoveanu Alley, 700490, Iasi, Romania

Apostolos P. KYRIAZOPOULOS
Democritus University of Thrace, Department of Forestry and Management of the Environment and Natural Resources, 193 Pantazidou str.,68200 Orestiada, Greece

Michail ORFANOUDAKIS
Democritus University of Thrace, Department of Forestry and Management of the Environment and Natural Resources, 193 Pantazidou str.,68200 Orestiada, Greece

Eleni M. ABRAHAM
Aristotle University of Thessaloniki, Department of Forestry and Natural Environment, Laboratory of Range Science (236), 54124 Thessaloniki, Greece

Zoi M. PARISSI
Aristotle University of Thessaloniki, Department of Forestry and Natural Environment, Laboratory of Range Science (236), 54124 Thessaloniki, Greece

Nikoleta SERAFIDOU
Democritus University of Thrace, Department of Forestry and Management of the Environment and Natural Resources, 193 Pantazidou str.,68200 Orestiada, Greece

Zehra AYTAC
Eskisehir Osmangazi University, Faculty of Agriculture, Department of Field Crops, Eskisehir, Turkey

Nurdilek GULMEZOGL
Eskisehir Osmangazi University, Faculty of Agriculture, Eskisehir, Turkey

Zeynep SIREL
Eskisehir Osmangazi University, Faculty of Agriculture, Department of Field Crops, Eskisehir, Turkey

Inci TOLAY
Akdeniz University, Faculty of Agriculture, Department of Soil Science and Plant Nutrition, Antalya, 07058, Turkey

Ayfer ALKAN TORUN
Cukurova University, Faculty of Agriculture, Department of Soil Science and Plant Nutrition, Adana, 01030, Turkey

Lajos HELYES
Szent István University, Horticultural Institute, Páter K.út 1., H-2103 Gödöllo, Hungary

Andrea LUGASI
Budapest Business School, College of Commerce, Catering and Tourism, Department of Catering; Alkotmány u. 9-11., Budapest, H-1054, Hungary

Hussein G. DAOOD
Szent István University, Regional Knowledge Center, Páter K. u. 1., Gödöllo, H-2100, Hungary

Zoltán PÉK
Szent István University, Horticultural Institute, Páter K.út 1., H-2103 Gödöllo, Hungary

Meizhi ZHAI
Northwest Agriculture & Forestry University, Walnut Research Center, College of Forestry, Yangling, Shaanxi, China

Zhenyuan WANG
Northwest Agriculture & Forestry University, Walnut Research Center, College of Forestry, Yangling, Shaanxi, China

Dan WANG
Northwest Agriculture & Forestry University, Walnut Research Center, College of Forestry, Yangling, Shaanxi, China

Jing XU
Northwest Agriculture & Forestry University, Walnut Research Center, College of Forestry, Yangling, Shaanxi, China

Guanzhao SHI
Northwest Agriculture & Forestry University, Walnut Research Center, College of Forestry, Yangling, Shaanxi, China

Marek GAJEWSKI
Warsaw University of Life Sciences - SGGW, Department of Vegetable and Medicinal Plants, Nowoursynowska 166, 02-787 Warszawa, Poland

Katarzyna MAZUR
Warsaw University of Life Sciences - SGGW, Department of Vegetable and Medicinal Plants, Nowoursynowska 166, 02-787 Warszawa, Poland

Jadwiga RADZANOWSKA
Warsaw University of Life Sciences - SGGW, Department of Vegetable and Medicinal Plants, Nowoursynowska 166, 02-787 Warszawa, Poland

Katarzyna KOWALCZYK
Warsaw University of Life Sciences - SGGW, Department of Vegetable and Medicinal Plants, Nowoursynowska 166, 02-787 Warszawa, Poland

Monika MARCINKOWSKA
Warsaw University of Life Sciences - SGGW, Department of Vegetable and Medicinal Plants, Nowoursynowska 166, 02-787 Warszawa, Poland

Klaudyna RYL
Warsaw University of Life Sciences - SGGW, Department of Vegetable and Medicinal Plants, Nowoursynowska 166, 02-787 Warszawa, Poland

Karolina KALOTA
Warsaw University of Life Sciences - SGGW, Department of Vegetable and Medicinal Plants, Nowoursynowska 166, 02-787 Warszawa, Poland

Memet INAN
Adiyaman University, Kahta Vocational School, Medicinal and Aromatic Plants, 02400, Adiyaman, Turkey

Ahmet Zafer TEL
Adiyaman University, The Faculty of Science and Letters, Department of Biology, 02400 Adiyaman, Turkey

Andras NEMENYI
Szent István University, Faculty of Agricultural and Environmental Sciences, Institute of Horticulture, Páter K. u. 1., Gödöllő, 2100 Hungary

Eva STEFANOVITSNE-BANYAI
Corvinus University of Budapest, Faculty of Food Science, Department of Applied Chemistry, Villányi út 64-70., Budapest, 1118 Hungary

Szonja Szimona BURJAN
Szent István University, Faculty of Agricultural and Environmental Sciences, Institute of Horticulture, Páter K. u. 1., Gödöllő, 2100 Hungary

Zoltan PEK
Szent István University, Faculty of Agricultural and Environmental Sciences, Institute of Horticulture, Páter K. u. 1., Gödöllő, 2100 Hungary

Attila HEGEDUS
Corvinus University of Budapest, Faculty of Horticultural Science, Department of Genetics and Plant Breeding, Villányi út 64-70., Budapest, 1118 Hungary

Csaba GYURICZA
Szent István University, Faculty of Agricultural and Environmental Sciences, Institute of Plant Production, Páter K. u. 1., Gödöllő, 2100 Hungary

Lajos HELYE
Szent István University, Faculty of Agricultural and Environmental Sciences, Institute of Horticulture, Páter K. u. 1., Gödöllő, 2100 Hungary

Milica G. ACIMOVIC
University of Belgrade, Faculty Agriculture, Nemanjina 6, 11080 Zemun, Serbia

Jasna KORAC
University of Novi Sad, Faculty of Agriculture, Trg Dositeja Obradovica 8, 21000 Novi Sad, Serbia

Goran JACIMOVIC
University of Novi Sad, Faculty of Agriculture, Trg Dositeja Obradovica 8, 21000 Novi Sad, Serbia

Snezana OLJACA
University of Belgrade, Faculty Agriculture, Nemanjina 6, 11080 Zemun, Serbia

Lana DJUKANOVIC
Institute for Plant Protection and Environment, Teodora Drajzera 9, 11000 Belgrade, Serbia

Vesna VUGA-JANJATOV
Agricultural High School Sabac, Vojvode Putnika 56, 15000 Sabac, Serbia

Mehmet SINCIK
Uludag University, Faculty of Agriculture, Department of Field Crops, 16059, Bursa, Turkey

A. Tanju GOKSOY
Uludag University, Faculty of Agriculture, Department of Field Crops, 16059, Bursa, Turkey

Andreea Nicoleta BOCA
"Iuliu Hatieganu" University of Medicine and Pharmacy, Department of Pharmacology, Toxicology and Clinical Pharmacology, 6 Pasteur Street, Cluj-Napoca, Romania

Alexandru TATARU
"Iuliu Hatieganu" University of Medicine and Pharmacy, Department of Dermatology, 3-5 Clinicilor Street, Cluj-Napoca, Romania

Anca Dana BUZOIANU
"Iuliu Hatieganu" University of Medicine and Pharmacy, Department of Pharmacology, Toxicology and Clinical Pharmacology, 6 Pasteur Street, Cluj-Napoca, Romania

Carlo PINCELLI
University of Modena and Reggio Emilia, Department of Medicine, Institute of Dermatology, Laboratory of Cutaneous Biology, 71 Via Del Pozzo, Modena, Italy

Carmen SOCACIU
University of Agricultural Sciences and Veterinary Medicine, Faculty of Food Science and Technology, 3-5 Manastur St., Cluj-Napoca, Romania
Center for Applied Biotechnology CCD-BIODIATECH, Proplanta Cluj-Napoca, Romania

Deniz EROGUL
Ege University, Faculty of Agriculture, Department of Horticulture, 35100, Izmir, Turkey

Andrea BUNEA
University of Agricultural Sciences and Veterinary Medicine, Department of Chemistry and Biochemistry, 3-5 Manastur Street, Cluj-Napoca, Romania

Carmen SOCACIU
University of Agricultural Sciences and Veterinary Medicine, Department of Chemistry and Biochemistry, 3-5 Manastur Street, Cluj-Napoca, Romania

Adela PINTEA
University of Agricultural Sciences and Veterinary Medicine, Department of Chemistry and Biochemistry, 3-5 Manastur Street, Cluj-Napoca, Romania